International Directory of Building Research, Information and Development Organizations

JOINT EDITORS

Gy. Sebestyén
Secretary General

AND

C.E. Pollington
Deputy Secretary General

FIFTH EDITION

International Council for
Building Research, Studies
and Documentation (CIB)

CRC Press
Taylor & Francis Group
Boca Raton London New York

CRC Press is an imprint of the
Taylor & Francis Group, an **informa** business
A TAYLOR & FRANCIS BOOK

First published in 1959 by the International Council for Building Research, Studies and Documentation
2nd Edition 1963
3rd Edition 1971
4th Edition 1979

Fifth Edition published in 1986 by Talyor & Francis

Published 2019 by CRC Press
Taylor & Francis Group
6000 Broken Sound Parkway NW, Suite 300
Boca Raton, FL 33487-2742

First issued in paperback 2020

ISBN 13: 978-0-367-58033-9 (pbk)
ISBN 13: 978-0-419-12990-5 (hbk)

British Library Cataloguing in Publication Data

International directory of building research, information and
development organizations.—5th ed.
1. Building research—Directories
I. International Council for Building Research,
Studies and Documentation
690′.072 TH1

ISBN 0–419–12990–1

Library of Congress Cataloging in Publication Data

International directory of building research, information and
development organizations.
Includes index.
1. Building research—Directories
I. International Council for Building Research,
Studies and Documentation
TH12.I58 1986 624′.072 86–12

ISBN 0–419–12990–1

Contents

Contents

Foreword

CIB exists for international cooperation in building research, studies and documentation. This Directory, the 5th edition of what is by now a well-established series, is an important resource for international cooperation. It gives concise information on the programs and resources of 640 building research organizations in 54 countries. This new edition provides a subject index to help users find quickly the organizations active in one or another area.

The global span of the entries reflects the worldwide influence of CIB and it is this that makes it such an effective means for bringing together researchers and practitioners from so many countries to address problems and research opportunities of common interest. In fact it is no exaggeration to say that the contents of this Directory link the reader to most of the world's building research.

On behalf of CIB, I thank all those who have provided the entries for the Directory and the staff of the General Secretariat for its efforts in compiling this useful and attractive document. Readers are encouraged to bring changes and new entries for the next Directory to the attention of the Secretariat.

R.N. Wright
President of CIB

Codes for the Representation of Names of Countries

The codes for the representation of names of countries that have been used in this Directory are according to International Standard ISO 3166–1974 plus Amendment 1 published 1977–08–01.

The codes and the corresponding countries are given below.

(ARG)	Argentina
(AUS)	Australia
(AUT)	Austria
(BEL)	Belgium
(BRA)	Brazil
(CAN)	Canada
(CHN)	China
(COL)	Colombia
(CUB)	Cuba
(CSK)	Czechoslovakia
(DNK)	Denmark
(EGY)	Egypt
(ETH)	Ethiopia
(FIN)	Finland
(FRA)	France
(DDR)	German Democratic Republic
(DEU)	Germany, Federal Republic of
(GHA)	Ghana
(GRC)	Greece
(GTM)	Guatemala
(HUN)	Hungary
(ISL)	Iceland
(IND)	India
(IDN)	Indonesia
(IRN)	Iran
(IRQ)	Iraq
(IRL)	Ireland
(ISR)	Israel
(ITA)	Italy
(JAM)	Jamaica
(JPN)	Japan
(KEN)	Kenya
(KOR)	Korea, Republic of
(NDL)	Netherlands
(NZL)	New Zealand
(NGA)	Nigeria
(NOR)	Norway
(PAK)	Pakistan
(PHP)	Philippines
(POL)	Poland
(PRT)	Portugal
(ROM)	Romania
(SGP)	Singapore
(ZAF)	South Africa
(ESP)	Spain
(SWE)	Sweden
(CHE)	Switzerland
(TZA)	Tanzania, United Republic of
(TUR)	Turkey
(SUN)	Union of Soviet Socialist Republics
(GBR)	United Kingdom
(USA)	United States
(VEN)	Venezuela
(YUG)	Yugoslavia

(INT) International Organizations – this section starts on page 1.

CIB Membership

Membership of CIB can be in one of three categories appropriate to the scale of that member's direct participation in CIB activities. The categories are: Full Member (FM), Associate Member (AM) and Unattached Member (UM). In this Directory, the abbreviations in brackets are used to indicate those entries which are from a CIB member.

INTERNATIONAL ORGANIZATIONS

INT.1

International Council for Building Research, Studies and Documentation (CIB)
Conseil International du Bâtiment pour la Recherche, l'Etude et la Documentation (CIB)
PO Box 20704, Weena 704, 3001 JA Rotterdam, Netherlands
Telephone: 010–4110240
Telex: 22530 bouwcentrum rt
Date of establishment
1953

An international, non-governmental, non-profit-making organization whose purpose is to encourage, facilitate and develop international co-operation in building, housing and planning research, studies and documentation, covering not only the technical but also the economic and social aspects of building and the related environment.

Main fields of work
CIB is the primary mechanism through which the international building community exchanges findings on building research and practice.

In carrying out its objectives, CIB:

- Promotes international co-operation between all types of building research organizations
- Encourages the international exchange of research personnel
- Fosters exchanges of building and construction information and research documentation
- Sponsors international symposia on topical themes and holds a triennial congress.

Working Commissions: The main thrust of CIB's work is carried out through a network of some 40 Working Commissions which are composed of specialist representatives of CIB member institutions and of other co-opted experts. These deal with a variety of topics selected from CIB's field of interest. The figure of 40 is the actual number operating in 1984 but it should be emphasized that the overall programme of the Working Group's activities is continually reviewed by the Programme Committee. A working group that has fulfilled its main purpose is terminated; new ones are continually being established.

The titles of the currently existing Working Groups are: *Structural Safety*: Fire; Heating and Climatization. *Timber Structures*: Industrialized Building Techniques in Concrete. *Wall Structures*: Modular Co-ordination; Concrete Surface Finishings; Heat and Moisture Transfer; Multi-Storey Buildings; Human Requirements; Tolerances; Acoustics; Building Economics; Lightweight Constructions; Building Documentation; Performance Concept; Water Supply and Drainage; Low Cost Housing; Organization and Management of Construction; Industrial Buildings. *Energy Conservation*: Energy Conservation in Developing Countries; Housing Sociology; Maintenance and Modernization; Building Climatology. *Urban Planning*: Earthquake Hazard Mitigation; Information Coordination; Mechanization in Building; Utilization of Local Materials and Industrial Wastes; Utilization of Local Materials and Industrial Wastes in Developed Countries; Physical (Indoor) Environment; Computer-Aided Design; Control of Building Services, Service Life Prediction; Actions on Structures; Long Term Forecasting; Single Layer Roofing; Building Concept for the Handicapped, Structural Deformations.

A selection of activities currently (1984) going on, without a Working Commission having been formally established is: Building Defects (Pathology); Windows; Quality Assurance and Liability; Electrical Equipment in Buildings; Drafting Standards for Computer Storage; Industrialization of Construction; Register of Unique Laboratory Equipments; Air Quality.

Congresses, colloquia and symposia: CIB holds its congress every three years and these meetings are recognized as being events of major importance in the international world of building research. They attract several hundred practitioners including research workers, architects, engineers, surveyors, builders and bring to public notice the results achieved by the Working Commissions. Seven congresses have been held to date. The years and venues are listed below together with the central theme of each:

● 1959	Netherlands	Building Research and Information
● 1962	United Kingdom	Innovation in Building
● 1965	Denmark	Towards Industrialized Building
● 1968	Canada and USA	Building, Cost and Quality
● 1971	France	Research into Practice – the Challenge of Application
● 1974	Hungary	The Impact of Research on the Built Environment
● 1977	United Kingdom	Construction Research International
● 1980	Norway	Building Research World Wide
● 1983	Sweden	To Build and Take Care of What We Have Built with Limited Resources

In addition to these congresses, a series of colloquia, symposia, etc. are continually being organized. These are staged by individual Working Commissions and provide the means to maximize the benefit that can be drawn from the collective expertise of the members of that Working Commission and to disseminate it among a much wider circle. It is also CIB's practice to join forces with other international scientific and technical organizations in co-sponsoring or co-organizing events of this nature.

Activities in developing countries: CIB is increasingly taking steps to extend collaboration by developing countries in its work. A (limited) fund has been set up to promote this involvement. CIB works in concert with the appropriate bodies in the United Nations and other organizations specializing in the field.

In general terms it can be said that the area of interest of CIB comprises building, the built environment and the building industry and therefore a wide range of considerations enter the scope of CIB. An analysis of these would include:

- The planning, utilization and adaptation of the environment in so far as it directly influences the building activity
- The entire domain of building science and technology, including techniques and engineering, economics and problems of management in the industry, the sociological and social questions related to the analysis of building needs, etc.
- The bridging between scientific innovation and full-scale practical application; preparation and provision of information for different categories of user; development of information techniques and systems; and development work involving jointly research and industry.

There are three categories of membership in CIB, namely full membership, associated membership and unattached membership.

Full members are organizations engaged on a substantial scale in the fields of interest of CIB and having a large, qualified graduate staff of researchers.

Associate members are organizations engaged in work on particular aspects of the fields of interest of CIB or within its general field but on a scale that does not qualify them for full membership.

Unattached members are individuals whose experience in one or more aspects of the fields of interest of CIB qualifies them to make individual contributions to international co-operation. This category is reserved for those experts who cannot easily establish a liaison via a member institution.

At the beginning of 1984 the total membership of CIB stood at about 300 (mainly institutional) members in over 70 countries around the world.

As indicated, membership of CIB is essentially institutional and virtually all the major national building research establishments in the world are members. Countries with several eligible research organizations may have more than one full member; others may have one or more full members as well as associate members. Countries and their governments, as such, are therefore not represented in CIB.

Total staff: General Secretariat: 5
Director or Chief Executive
Secretary General: Professor Dr Gy. Sebestyén
Principal sources of finance
Financed mainly by annual membership fees.

Principal related bodies

The General Assembly is the governing body in which the authority of CIB is vested and is composed of representatives of full members. This body normally meets at least every three years.

The Board is responsible for the management of the affairs and property of CIB. It consists of four Officers, the immediate Past President and not more than nineteen other members.

The Officers for the period 1983–1986 are:

President: Dr R.N. Wright, USA
Vice Presidents: Dr Gy. Kunszt, Hungary
G.C. Mathur, India
Honorary Treasurer: P. Chemillier, France

Committees of the Board assist in the work of managing the activities of CIB; the Board appoints an Administrative Committee and a Programme Committee.

The Administrative Committee assists the Board with regard to general, administrative and financial matters including applications for membership.

The Programme Committee assists the Board in planning and execution of the work programme of CIB, from the viewpoint both of examining proposals for new activities and of itself stimulating particular projects to be undertaken on the international level in relation to developing trends and needs.

General Secretariat is a permanent office charged with the execution of CIB's decided policy of action. It also assists, serves and advises the General Assembly, the Board and the Committees of CIB.

Publications

Bâtiment International Building Research and Practice is published six times per year, in English or French, depending on the original language of the article.

CIB Information – Newsletter for Members is published six times per year.

The titles listed below are only a selection from the comprehensive series of Reports, etc. that have been published.

- Proceedings of CIB Congresses
- A Master List of Properties for Building Materials and Products
- Survey of Meteorological Information for Architecture and Building
- CIB Master Lists for structuring documents relating to buildings, building elements, components, materials and services
- Methodology of Cost Comparisons
- The SfB System
- Industrial Buildings
- Tall Buildings
- The Analysis of Climatological Data for the Building Industry
- Climatology and Building
- Indexing Guidelines
- Guidelines for the Elaboration and Development of Thesauri
- Energy Conservation in the Built Environment
- Draft Recommendations for Trade Literature
- Computer Languages in Building
- Principles of Fire Safety in Buildings
- Requirements for Statistical Data on Fires in Buildings
- Evaluation of Information Systems for the Construction Industry
- Directory of Building Research Information and Development Organizations
- Heating Systems in Practice
- Smoke Control in Compartmented Buildings
- Structural Timber Design Code
- Systems of Maintenance Planning
- International Recommendations for Masonry Structures

INT.2

Comité Euro-International du Béton (CEB)

Euro-International Committee for Concrete
EPFL – Case Postale 88 – CH1015 Lausanne
Telephone: (021) 472747
Telex: 24478EPFVDCH
Date of establishment
1953

The Euro-International Committee for Concrete is an International technico-scientific association under Swiss law (Head office in Geneva). Its aim is to develop, at an international level, the study of scientific and technical problems to encourage progress in the concrete construction industry.

The Committee is concerned with coordinating research findings with a view to their practical implementation by constructors, in the hope of improving the technology and economics of concrete construction, and in particular with establishing international standardized sets of guidelines, practical recommendations and implementation manuals on the calculations involved in concrete construction and on the execution of structures, based on performance concepts and taking into account the maximum wear and tear capacities and breaking points of concrete components. This aim is achieved in cooperation with other existing international technico-scientific associations. International activity of the Committee extends worldwide and is not limited to the participation of European nations.

Main fields of work

The aims of the Euro-International Committee for Concrete, which seeks to promote the development and progress of scientific and technical questions in the field of the concrete construction industry are, more specifically, as follows:

Research Synthesis of scientific knowledge and coordinating of research programmes and findings in the various member countries.

Theory Establishment of new principles and data for structural safety in constructions, through implementation of the latest developments in theoretical knowledge in the spheres of experiment, statistics and probability, taking into particular account the basic demands of the quality of human life, requisite guarantees regarding the serviceability and durability of structures, and the new urgent problem of economy of energy and raw material consumption.

Calculation Practical methods of structural calculation and optimalization of structures.

Regulations Standardized recommendations and standardized international sets of guidelines for reference in the design, calculation and building of concrete structures, with practical applications in the different techniques involved in the structural uses of concrete. More specifically, the Euro-International Committee for Concrete plays a technical role in all efforts to standardize and bring into line the civil engineering regulations of different nations.

Practical application Publication of international calculation and technology manuals, developing practical methods of application for new technico-scientific concepts with a view to future progress in techniques and in the quality of construction.

Director or Chief Executive

President: Professor T. Tassios
Président Délégué: Yves Saillard
Directeurs Techniques: Professor Jean Perchat
Rüdiger Tewes

Principal sources of finance

Statutory subscriptions from participating national groups and, in some countries, extra contributions and benefits (from either professional or government sources).

Principal related bodies

Board of Directors: technical and administrative management of the Association's activities with a view to carrying out its statutory aims.

'Research' and 'Development' Advisory Committee: definition and orientation of technico-scientific priorities for Commissions and working groups.

Plenary sittings: debate and ratification of technical and scientific resolutions and documents; endorsing the achievements of the Association's work.

Publications

- *International sets of guidelines and recommendations* in particular, Code Modèle International Unifié CEB-FIP-1978, Annexe 'Calcul sismique', Annexe 'Calcul au feu', etc.
- *Application manuals*
  Titles in print: *Lightweight Aggregate Concrete, Autoclaved Aerated Concrete, Buckling and Instability, Concrete Reinforcement Technology, Bending and Compression*
  Titles to be published: *Structural effects of time-dependent behaviour of Concrete, Cracking and Deformation*
- *News bulletins* (about 6 per year): State-of-the-art reports from commissions and working groups; international proposals and recommendations (for inquiry and debate among participating national groups).

INT.3

Commission Internationale de l'Eclairage (CIE)
International Commission on Illumination
52 Bd Malesherbes, 75008 Paris
Telephone: 387.21.21
Date of establishment
1913

CIE is an organization devoted to international co-operation and exchange of information among its member countries on all matters relating to the art and science of lighting. Its membership consists of 30 countries, each having a national committee of individual members who devote their time and talent to the objectives of the organization. In addition, individuals from eight other countries have associate member status.

The objectives of the CIE are:

- To provide an international forum for the discussion of all matters relating to science, technology and art in the fields of light and lighting
- To develop basic standards and procedures of metrology in the fields of light and lighting
- To provide guidance in the application of the basic principles and procedures to the development of international and national standards in the fields of light and lighting
- To prepare and publish reports, recommendations and standards concerned with all matters relating to science, technology and art in the fields of light and lighting
- To maintain liaison and technical interaction with other international organizations concerned with matters related to science, technology and art in the fields of light and lighting

The work of the CIE is carried out by 26 technical committees, each of which is assigned to a member country.

Main fields of work

The CIE covers subjects ranging from those which involve basic and fundamental matters to all types of lighting applications.

Director or Chief Executive

President: Professor J.B. de Boer
Executive Secretary: P. Lemaigre-Voreaux

Principal sources of finance

Dues from national committees

Publications

- CIE Journal, two issues per year
- Technical committee reports
- Recommendations (Standards)
- Proceedings of the quadrennial sessions

INT.4

Council on Tall Buildings and Urban Habitat
Institute for the Study of the High-Rise Habitat, Building 13, Lehigh University, Bethlehem, PA 18015, USA
Telephone: 215/861-3515
Telex: 5106514740LEHIU
Date of establishment
1969 as Joint Committee on Tall Buildings; 1976 as Council on Tall Buildings and Urban Habitat

The Council on Tall Buildings and Urban Habitat, an international activity sponsored by engineering, architectural, and planning professionals, was established to study and report on all aspects of the planning, design, construction and operation of tall buildings. Such structures find application primarily as commercial and residential facilities in both industrial and developing countries.

Prior to 1976, the Council on Tall Buildings and Urban Habitat was known as the Joint Committee on Tall Buildings, a group originally formed by ASCE and IABSE in 1969. In 1973, as a result of the increased emphasis on planning and environmental criteria, AIA, APA, IFHP and UIA were invited to join the forming organizations as equal participants.

In 1974 the Council was admitted as a consulting non-governmental organization to United Nations' UNESCO under Category C. In 1979 it was admitted to Category B.

The Steering Group is the Board of the Council. An executive committee carries out the policy of the Steering Group. The headquarters are located at Lehigh University.

Main fields of work

The major thrusts of the Council are:

- *International information source:* comprehensive monograph high-rise building data base on building heights and other statistics; bibliography; conference proceedings; planning and design data base.
- *Identification and stimulation of needed research:* survey and report on current research and on future research needs; stimulation of research on the most important topics.
- *Utilization of findings:* implementation of the results of Council work; application to design practice; incorporation into codes, specifications, and standards.
- *Planning and environmental criteria:* improved methodology considering social effects of the environment; role of the high-rise, decision-making parameters.
- *Exchange of information:* international, regional, and local conferences with proceedings; periodic technical forums; dissemination of reports and proceedings; newsletter.
- *Approaches for developing countries:* identification of unique solutions appropriate to developing nations; case studies.

Total staff: 12
Professionally qualified staff: 6
Director or Chief Executive
Director: Dr Lynn S. Beedle

Publications

A comprehensive Monograph on tall buildings is published, for use by those responsible for their planning and design. Other publications include conference proceedings, reports and a newsletter.

INT.5

European Organisation for the promotion of New Techniques and Methods in Building (EUROBUILD)
26 rue La Pérouse, 75116 Paris
Telephone: 1-720-10-20
Telex: FEDEBAT 611975F
Date of Establishment
1965

The Organisation is devoted to improving the functional effectiveness and economic efficiency of building methods and techniques throughout the European Construction Industry.

Main fields of work

- To promote the use of new building techniques and methods in its member countries and other countries in the world.
- To exchange relevant commercial and technical information as requested by members.
- To disseminate international information and contacts, in order to obtain a co-ordination of national works and studies and a promotion of technical and commercial contacts between associations, industrial groups, firms and individuals.
- To acquaint governments, clients, industrialists, architects, engineers and other professional bodies and also the general public with the essential advantages of new building techniques and methods.
- To organize colloquia, conferences, technical study visits and working groups (permanent or otherwise) in order to facilitate and promote development of the building industry. Most of those events will be done with the close collaboration of member associations and any national or international organisations dealing with the same subjects.

President: W.S. Forgham
Vice Presidents: P.M. van Onna
K. Jingrap
General Secretary: M. Vandenkerckhove
Principal source of finance
Annual subscriptions from members
Publications

- Internal Information Bulletins
- Conference proceedings are published annually

INT.6

European Cement Association (CEMBUREAU)
Association Européenne du Ciment
2, rue Saint-Charles, 75740 Paris Cedex 15, France
Telephone: 579.28.66
Telex: 250630 CEMBURO
Cables: CEMBUREAU
Date of establishment
1947

A non-profit making association. Supported by membership fees. The General Assembly comprises of the 19 member countries.

Main fields of work
The exchange of information and experience within the European Cement Industry on common problems, with particular reference to the uses of cement and concrete, the organization of special studies and other joint activities as required. Permanent and ad hoc committees dealing with market development, technical and economic questions. In addition, the Association compiles statistics and provides library and information services in support of its members' work.

Director or Chief Executive
Directors: H.C. Collis and P. Dutron
Principal sources of finance
Supported by membership fees. A non-profit making association.
Publications

- Technical publications in connection with meetings and special studies, as well as compilations of standards, research, etc.
- Directories giving details about cement companies throughout the world, production, trade, statistics, etc.

INT.7

European Committee for Standardization (CEN)
Comité Europeéen de Normalisation
Europäisches Komitee für Normung
Rue Brederode 2, Bte 5, 1000 Bruxelles, Belgium
Telephone: 02/513.55.64
Telex: 26257 cenlec b
Date of establishment
1961 as European Standards co-ordinating Committee; 1971, present title; 1975, reconstituted in Brussels and registered according to the Belgian law of 25 October 1919 (modified 6 December 1954).

Non-profit-making association comprising the national standards organizations of 16 countries of the European Economic Communities (EEC), the European Free Trade Association (EFTA) and Spain: Austria, Belgium, Denmark, Finland, France, Federal Republic of Germany, Greece, Ireland, Italy, Netherlands, Norway, Portugal, Spain, Sweden, Switzerland, United Kingdom.

Aims: implementation of harmonized standards on a European basis in order to facilitate the exchange of goods and services by elimination of obstacles caused by technical requirements.

Means: harmonization of national standards, promoting uniform implementation of international standards (ISO); preparation of European Standards and harmonization documents when justified. Providing the CEN conformity certification marking system and the CEN system of mutual recognition of test and inspection results.

Main fields of work
Standardization in the fields of: sanitary appliances, fittings, valves, taps, drainage equipment (dimensions, quality); baths, shower trays from acrylic materials; doors, windows, shutters, building hardware (tests); valves and fittings to equip radiators; prefabricated district heating pipe systems; wood preservatives (tests); particle boards, adhesives for wood and derived products; glass fibre reinforced plastics (tests); wallcoverings; ceramic tiles; cement, ready-mixed concrete, concrete; classification of building materials according to their thermal properties; hoists, passenger goods and service lifts; mobile elevating work platforms; scaffolding; ladders; portable fire extinguishers, automatic fire detection systems; central heating boilers, space heaters, water heaters (gas fired); oil stoves, oil burners; gas cooking appliances, large kitchen appliances using gaseous fuels; homogeneous elastomers for static seals in domestic appliances using combustible gases; domestic refrigerators (function, safety, testing); oil and gas pipelines (design, construction); hoses and hose couplings for petroleum products; steel drums; petroleum products (test methods); injection moulding machines (safety); portable grinding machines (safety); dental products; respiratory protective devices, eye protective equipment; sterilizers for medical purposes; safety of toys; packages for washing and cleaning powders; capacities of glass jars, metal cans.

Total staff: 10
Professionally qualified staff: 10
Director or Chief Executive
Secretary General: H.J. Worch
Principal related bodies
General Assembly; Administrative Board; Central Secretariat; Technical Committees and their secretariats; Advisory Committees as the Technical Board on technical matters and the CENCER Steering Committee on matters of CEN certification.
Principal sources of finance
Membership contributions of the 16 national standards organizations which are members of CEN.
Publications

- European Standards: a full list is available.
- Annual Report on the activities of the Technical Committees
- National publication of European Standards

INT.8

European Committee of Consulting Firms
Comité Européen des Bureaux d'Ingénierie (CEBI)
Boulevard de Waterloo 103, 1000 Brussels, Belgium
Telephone: 02-538.6030
Telex: 22810 CDI
Date of establishment
Founded in 1971

CEBI was founded with the object of bringing together the consultancy associations of member countries of the European Community. Membership now covers 1034 firms with a total staff of 158 110. The firms cover a range of professional activities, including technical, architectural and industrial studies; feasibility studies; programming, management, economic and financial studies; agronomy. Multidisciplinarity is an essential characteristic of CEBI.

In the French version of CEBI's title the word 'Ingénierie' was deliberately chosen. It is taken to mean 'the whole range of intellectual activities whose object is to obtain the best results from an investment'. The achievement of members make a substantial contribution to the part played by Europe in world-wide development.

Main fields of work

CEBI maintains close contact with the European Communities, in particular with the European Development Fund, to give professional advice. It also represents European interests to the United Nations organizations and to international and regional development funds, and collaborates on a permanent basis with other consulting associations, maintaining day-to-day contact with related professions.

Members:

Belgium: Comité des Bureaux d'Etudes de Belgique (BUROBEL asbl), Boulevard de Waterloo 103, 1000 Bruxelles. Tel. 02.538.60.30; Telex 22810 CDI (35 members)

Denmark: Danish International Consultants Organization (DICO), 45 Teknikerbyen, DK 2830 Virum. Tel. 45.2857311; Telex 37280 Cowi (30 members)

France: Chambre Syndicales des Sociétés d'Etudes et de Conseils (SYNTEC), Rue Léon Bonnat 3, 75016 Paris. Tel. 524.43.53. Association Française des Sociétés d'Etudes et de Conseils (ABETEX), Rue Léon Bonnat 3, 75016 Paris. Tel. 524.43.53 (276 members).

Germany: Verband unabhängig beratender Ingenieufirmen (VUBI), Winston Churchillstrasse 1, 5300 Bonn 12 (Postfach 1204064). Tel. (0228) 217064-65; Telex 0886551; Telegram: vubi bonn (147 members)

Greece: Hellenic Association of Consulting Firms (HACF), 24 Stratiotikou Syndesmou St, P.O. Box 471, Athens 136. Tel. (301) 362.3091; Telex 21.5227 (37 members).

Italy: Associazione delle Organizzazioni di Ingegneria e di Consulenza Tecnico-Economica (OICE), Largo Messico 7 (Parco di Villa Grazioli), 00198 Rome. Tel. 851685-860547; Telex 614004 Agroti; Telegram ASSOCIE Roma (81 members)

Netherlands: Orde van Nederlandse Raadgevende Ingenierus-Afdeling Buitenland (ONRIF.S.), Javastraat 44, 2585 Den Haag. Tel. 70-637565; Telex 31580 (25 members)

United Kingdom: British Consultants Bureau (BCB), Westminster Palace Gardens, 1–7 Artillery Row, London SW1P 1RG. Tel. 44-1.222.3651/4; Telex 24552 for britburo (288 members)

Spain: Asociacion Espanola de Empresas de Ingenieria y Consultores (ASEINCO), Claudio Coello 86, Madrid 6. Tel. 34.1.2263754; Telegram ASEINCI (65 members)

Switzerland: Community of Swiss Exporting Consultants (SWEXCO), PO Box 720 CH 1001 Lausanne. Tel. (021) 203231; Telex 25425 (50 members)

Total staff

Permanent Secretariat 3

Director or Chief Executive

President: Baron E. Coppee

Vice-President: H. Talbot-Pedersen

Secretary General: P. Carlier

Executive: F. Fonseca

Heads of department

Standing Joint Committee (CEBI–CEDIC–FED relations): Chairman: A. Sharman

Rural Development Group: Chairman: L. Gould

Urban Planning Group: Chairman: M. Welbank

Principal sources of finance

Annual membership fees: a turnover in 1980 of more than $6000 million.

Publications

- *Journal:* CEBI News, quarterly
- CEBI Directory of Consultants
- Directory of African Caribbean and Pacific Consultants
- Proceedings of the Conference held in Rome in May 1982 on the Development of the ACP Countries and the Future of Europe

INT.9

European Communities

Communautés Européennes

European Coal and Steel Community (ECSC)

European Economic Community (EEC)

European Atomic Energy Community (EURATOM)

Organizations common to the three communities:

Commission of the European Communities, 200 rue de la Loi, 1049 Bruxelles, Belgium

Telephone: 235 11 11

Telex: COMEU B 21877

European Parliament, Bâtiment R. Schuman, Kirchberg, Luxembourg (GD)

Telephone: 43021

Telex: EUPA 3697

Court of Justice, Kirchberg-Luxembourg, PO Box 96, Luxembourg (GD)

Telephone: 43031

Telex: CURIA-LU 2510

Council of Ministers, 170 rue de la Loi, 1048 Bruxelles, Belgium

Telephone: 2 34 61 11

INT.10

Construction Economics European Committee

Comité Européen des Economistes de la Construction

8 avenue Percier, 75008 Paris, France

Secretariat: G.F. Wheatley, 266 rue de Faubourg St Honore, 75008 Paris, France

Telephone: Paris (1) 764 17 27

Telex: 250302

Date of Establishment

29 June 1979

The Committee is an association of the leading professional body in each EEC member state which is nationally recognized within each such state as representing the profession of construction economics.

The aim of the Committee is to facilitate the exchange of experience and information between professionally qualified persons who are responsible for construction economics in the EEC member states and to initiate studies with a view to promoting the training and qualification of persons who are responsible for construction economics and drawing up proposals for the harmonization and acceptance of standards of training and qualification; establishing guidelines for the definition, content, control and practice of construction economics; ensuring adequate representation of qualified persons who are responsible for construction economics in the EEC Commission and in other European institutions; studying existing and proposed legislation and regulations relating to construction economics, with a view to their harmonization; co-ordinating working methods; establishing European statistics relating to costs and types of construction, procedures and materials.

Main fields of work

- Representation of the profession of construction economics in the Commission of the EEC and other European bodies
- Preparation of a concordance document for the exchange of building cost information between member states
- Preparation and proposed publication of a European building cost data bank
- Initiation of studies into global costs, building development time studies and computer applications related to the profession of construction economics.

Director or Chief Executive

Chairman: C.J. Willis

Administrative Secretary: G.F. Wheatley

Heads of department

Belgian delegation: Professor F. Peters
German delegation: BDB, Head of Delegation: Herr H. Kieselack
Danish delegation: Bjorn Bindslev
French delegation: UNTEC, Head of Delegation: Louis Plagnol
British delegation: RICS, Head of Delegation: B.E. Drake
Italian delegation: Signor Guido Dandri
Irish delegation: RICS, Head of Delegation: N. McDonagh
Dutch delegation: NVBK, Head of Delegation: W. Meijer

Principal sources of finance

Annual subscription of member societies.

INT.11

Economic and Social Committee of the European Communities

2 rue Ravenstein, 1000 Brussels, Belgium
Telephone:512 39 20 or 513 95 95
Telex: 25 983 CESEUR
Cables: ECOSEUR

Date of establishment

Treaty establishing European Economic Community, 25 March 1957; Treaty establishing European Atomic Energy Community (Euratom), 25 March 1957.

The Economic and Social Committee, a consultative assembly of representatives of economic and social groupings in the member states, forms an integral part of the European Community framework. It can also issue opinions on its own initiative. The Committee was set up to involve the various economic and social interest groups in the establishment of the Common Market and give them an institutional vehicle for putting across their point of view on the various issues to the Commission and the Council.

The Committee consists of representatives of *employers* (Group 1), *workers* (Group II), *various interest groups* (Group III), such as agriculture, transport, trade, small enterprises, the professions and consumers. Its 156 members come from the ten member states. The members are proposed by the member states' governments and appointed by the Council of the European Communities for a term of four years. They may be reappointed. The Committee members live and work in their home countries and only come to Brussels for Committee meetings.

The Committee's Chairman and Bureau are elected by the Committee from among its own members for a term of two years. The Bureau, consisting of 24 members, has the job of organizing the Committee's work.

Main fields of work

Under the EEC Treaty, the Economic and Social Committee has to be consulted on questions relating to agriculture, freedom of movement for workers, right of establishment, transport, approximation of laws, social policy, European Social Fund and vocational training.

Under the Euratom Treaty, it has to be consulted on certain areas of the nuclear field: research and training programmes, establishment of schools for training specialists, health and safety, investment, access to skilled jobs and insurance cover against nuclear risks.

The Council and the Commission may also consult the Committee on other matters when they deem this appropriate. In addition, the Committee has obtained the right to advise on its own initiative on all questions affecting the work of the Community. On average the Committee now delivers 120 opinions per year.

The Committee is divided into sections responsible for the principal fields covered by the Rome Treaties (there are at present nine such sections):

- Economic and financial questions
- External relations
- Social questions
- The protection of the environment, public health and consumer affairs
- Agriculture
- Regional development
- Industry, commerce, crafts and services
- Transport and communications
- Energy and nuclear questions.

The Committee also organizes conferences of specific interest: one of these, held on 11 and 12 September 1980, dealt with the problems of the building industry in Europe.

Total staff: 391
Professionally qualified staff: 44 administrators and 82 linguists

Director or Chief Executive

Secretary General: Roger Louet

Principal sources of finance

European Community Budget

Principal related bodies

European Community institutions

Publications

- *Journal:* Bulletin, monthly
- *Recommendations:* Opinions delivered by the Economic and Social Committee
- Studies and reports

INT.12

European Convention for Constructional Steelwork (ECCS)

Convention Européenne de la Construction Metallique (CECM)
Europaische Konvention for Stahlbau (EKS)
avenue Louise 326, bte 52, B-1050 Brussels, Belgium
Telephone: 2-6499081
Telex: 24346

Date of establishment

1955

An association of national bodies active in the field of steel construction. European organizations can become full members, non-European organizations can become associate members. The general aims are to:

- Prepare recommendations and guidelines for the design and execution of steel structures in view of an international harmonization of standards and codes (Eurocodes, ISO. . .)
- Promote co-ordination and co-operation with other European and international organizations concerned with construction, standardization and testing of materials
- Undertake research and investigations of international interest, prepare and publish practical results and solutions which are directly applicable by structural steel designers and fabricators and which lead to the development of the steel construction industry.

Main fields of work

The ECCS achieves its aims mainly through the activities of committees whose members are appointed by the Convention member countries, are experts in the professions, the research and academic world, government departments and the steelwork industry.

The ECCS has 14 committees, 1 economic, 3 advisory and 10 technical committees; for specific tasks committee chairmen may create Working Groups and appoint their members.

For its technical activities, the ECCS has the following advisory (AC) and technical (TC) Committees:

- Multi-storey Buildings (AC1)
- Industrial-type Buildings (AC2)
- Bridges (AC3)
- Structural Safety (TC1)
- Aluminium Alloy Structures (TC2)
- Fire Safety (TC3)
- Corrosion (TC4)
- Fatigue (TC6)
- Cold-formed Thin-walled Sheet Steel (TC7)
- Structural Stability (TC8)
- Bolted and Welded Connections (TC10)
- Wind (TC12)
- Seismic Behaviour (TC13)

Director or Chief Executive
General Secretary
Publications
- Manual on Stability of Steel Structures, 1976
- European Recommendations for Steel Construction, 1977
- Model code for Composite Structures (steel and concrete), Construction Press, 1981
- European Recommendations for the Fire Safety of Steel Structures: Calculation of the Fire Resistance of Load Bearing Elements and Structural Assemblies exposed to the Standard Fire, 1983
- European Recommendations for Steel Construction: Buckling of Shells, 2nd edition, 1983
- European Contract Conditions, 1976
- A dictionary of about 3000 words (in Dutch, English, French, German, Italian and Spanish) on steel construction and steelwork fabrication and erection, is planned.

INT.13

European Industrial Research Management Association (EIRMA)
Association Europeenne pour l'Administration de la Recherche Industrielle
38 Cours Albert 1er, 75008 Paris, France

Telephone: (1) 225.60.44
Telex: 643 980
Date of establishment
1966

The Association aims to:
- increase the productivity and effectiveness of research development and innovation in industry.
- provide a unique forum for contacts and professional enrichment to those responsible for industrial innovation in European industry.
- exchange and make available experience on the techniques and practice of Research and Development management.
- add to the stock of knowledge on R & D management practice.
- improve industrial innovation through closer association of R & D with the other company functions and the corporate planning process.
- explore technological trends in an interdisciplinary fashion.
- foster cooperation between industrial R & D and the academic and institutional research communities.
- explore the socio–economic environment of industrial innovation in Europe.
- make industrial R & D better understood as a force for the benefit of society in Europe.

Members are industrial firms carrying out a substantial amount of scientific research and technological development in Europe. Present membership is about 160. Member companies are officially represented by their chief executive responsible for R & D and by one or two alternate representatives. Participation in the work programme is open to the representatives and alternates and all other qualified staff nominated by these.

Work is carried out through conferences, working groups and meetings of different type and length, usually reserved to members.

Main fields of work
- techniques and practice of Research and Development management,
- relations between R & D and other company functions,
- relations between industrial R & D and academia/scientific institutes, governments/international organisation, in the fields of science and technology, the public.
- general (horizontal) technical and technological developments.

Total staff: 7
Professionally qualified staff: 3
Principal sources of finance
Membership fees
Publications
- Information Bulletin (2 per year)
- A comprehensive list of publications is available

INT.14 AM

European Insulation Manufacturers Association (EURIMA)
Registered Office: 8 rue Nicolas Welter, Luxembourg
Secretariat: Algade 5–7, DK 4000 Roskilde, Denmark

Telephone: 45 2 46 67 36
Telex: 16600 Fotex DK
Cables: Att. EURIMASEKR, Roskilde
Date of establishment
1961 first Constitution agreed

Association sans but lucratif, registered according to Luxembourg law.

EURIMA maintains contact with governmental as well as non-governmental organizations which have an interest in the conservation of energy, the protection of the environment, and standardization within the fields of thermal, sound and fire insulation.

In 1983 EURIMA had members in sixteen countries in Western Europe. It is estimated that the production of the 34 Member Companies covers more than 95% of the total production of mineral fibre insulation materials in this area.

Exchange of information between Members of EURIMA and contacts with other parties involved in the work of the Association are channelled through the permanent Secretariat.

Staff and structure
Secretariat (3 persons) and Management Board and Committees
Total staff: 3
Director or Chief Executive
Secretary General: Werner Linander
Principal sources of finance
Member Companies
Principal related bodies
Management Board and Committees: Technical Committee and Working Groups; Market Development Committee and Working Groups; Health and Safety Committee.
Publications
Journal: EURIMA News

INT.15

European Economic Community (EEC)
Communauté Economique Européenne (CEE)
200 rue de la Loi, 1049 Bruxelles, Belgium

Telephone: 235 11 11
Telex: Comeu B 21877

Decisions in accordance with the Treaty are taken by the Commission and the Council of Ministers, which, following the institutional merger of July 1967, are common to the three Communities (EEC, Euratom and ECSC). The Commission is the permanent body responsible for securing implementation of the Treaty and for representing the Community. The Commission is composed of 14 members appointed by the member states for 4 years. The Commission makes an annual report to the European Parliament, which is now directly elected. The Council of Ministers is composed of 10 members, representatives of the governments of the 10 member states (Belgium, Denmark, Federal Republic of Germany, Greece, France, Ireland, Italy, Luxembourg, the Netherlands and the United Kingdom). The Commission and the Council are aided by an Economic and Social Committee composed of representatives of the various categories of economic and social life, in particular representatives of producers, agriculturalists, transport operators, workers, manufacturers, artisans, the liberal professions and of other generally interested bodies.

The aims are to promote through the Community a harmonious development of economic activities, a continuous and balanced expansion, an increase in stability, an acceleration in the rise of the standard of living and closer relations between its member states, by establishing a customs union over a transitional period of 12 to 15 years (in fact, it was established on 1 July 1968) and progressively approximating the economic policies of member states.

Main fields of work

The Community has become one of the world's largest economic units. Its unified market of over 270 million people accounts for about a fifth of world trade, leaving aside the trade among the ten member states themselves.

The Community is currently reviewing its agricultural policy, strengthening its policies in fields such as research, industry and new technologies, regional development, employment creation. Spain and Portugal are currently negotiating membership with the Community.

Publications

- Bulletin of the European Communities, monthly
- Journal Officiel des Communautés Européennes
- General Report
- Studies and pamphlets on social, technical and economic problems of the Common Market
- Reports by the Commission and European Parliament committee and on parliamentary debates
- Information Service and Statistical Office publications

INT.16

European Group for Construction Prospects (EURO-CONSTRUCT)

Groupe Européen de Prospective pour la Construction (EURO-CONSTRUCT)

Secretariat: c/o BIPE, 122 avenue Charles de Gaulle, 92522 Neuilly-sur-Seine Cedex, France

Telephone: 747.11.66

Main fields of work

Euro-Construct associates economic and technical research institutes, to elaborate short or medium-term prospects for the construction activity in Europe. It systematically embraces the European framework for its analyses and forecastings; these are put into concrete form in the five-year forecasts for the European building and civil engineering market, the latest edition of which is now available (1975–1980).

Meetings take place every six months and are devoted to short-term prospects.

Each member institute may call on its associates' competence in order to realize, upon request from public or private organizations, investigations ranging over technical and economical problems in the building and civil engineering field.

Member institutes are:

- BIPE, Bureau d'Informations et de Prévisions Economiques, France
- CRESME, Centro Richerche Economiche, Sociologiche e di Mercato nell'Edilizia, Italy
- CSTC, Centre Scientifique et Technique de la Construction, Belgium
- DULBEA, Département d'Economie Appliquée de l'Université Libre de Bruxelles, Belgium
- EIB, Economisch Instituut voor de Bouwnijverheid, Netherlands
- IFO, Ifo-Institut für Wirtschaftsforschung, Federal Republic of Germany
- IFS, Institute for Future Studies, Denmark
- NEDO, National Economic Development Office, United Kingdom
- James Capel Research, United Kingdom

INT.17

European Organization for the Promotion of Prefabrication and other Industrialized Building Methods (Europefab)

26 rue la Pérouse, 75116 Paris

Telephone: 1-720-10-20

Telex: FEDEBAT 611 975F

Date of establishment ·

1965

A non-profit association, with staff provided by member organizations. Management is by an elected council which works through a general secretariat.

Main fields of work

The field of work covers the aims of Europrefab which are to:

- Exchange technical information between member associations; to exchange information on regulations and on research into user experience; to circulate articles of particular interest; to encourage the exchange of statistical information in order to give members a broader view of the amount of building under construction and also an estimate of future levels of demand for building industrialization
- Advance international information and contacts, in order to obtain a co-ordination of national works and studies and a promotion of technical and commercial contacts between associations, industrial groups, firms and individuals
- Make known to industrialists, architects, engineers and other professional bodies and also the general public, the essential advantages of building industrialization
- Organize conferences, exhibitions, technical study visits and, more generally, to take any action which may be useful for the promotion of building industrialization

Director or Chief Executive

President: W.S. Forgham, United Kingdom
General Secretary: M. Vandenkerckhove, France
Other members of the Board of Management are:
F. Aguirre, Spain, Immediate Past-President
D. Casarico, Italy, Vice-President
P.M. van Onna, The Netherlands, Vice-President
I. Briones, Spain
J. Bräuer, Czechoslovakia
F. Brink Laursen, Denmark
R.v. Halasz, Federal Republic of Germany
L. Kolaric, Yugoslavia
L. Pozsgai, Hungary
H.F. Ritter, Switzerland
A. Skjelle, Norway

Principal sources of finance

Annual subscriptions

INT.18

Fédération Européenne des Fabricants de Tuiles et de Briques (TBE)

European Association of Brick and Tile Manufacturers (TBE)

Obstgartenstrasse 28, PO Box 217, Ch 8035 Zurich, Switzerland

Telephone: 01 361 96 50

Date of establishment

1951

Private association of organizations in the heavy clay industry of European countries. The purpose of the Association is the promotion and thorough study of technical knowledge, and the spreading of information for the products of the brick and tile industry with the aim of utilizing and promoting the production of the industry as extensively and effectively as possible.

The Association seeks to achieve this aim:

- By exchanging relevant data (research results and experience in the field of heavy clay products as well as competitive products and – as

far as this is possible – by co-ordinating these activities),
- By the development of technical and vocational training
- By further measures which are designed to promote the production and utilization of heavy clay products.

Main fields of work
- Research in the field of production, quality and application of heavy clay products (bricks, tiles, floor blocks, pipes) by the laboratories of the member associations; exchange of results and co-ordination by a special committee of the Heads of Laboratories
- Studies in common of some problems of application; and publications
- Studies on marketing systems of the heavy clay industry
- Exchange of statistics concerning production and sales

Total staff: Secretary's office 2. Staff of laboratories and offices included in national member associations.

Director or Chief Executive
Director: Dr Karl Heusi

Heads of department
Management Committee: Chairman: E.A. Bäumer, Osnabrück, Federal Republic of Germany
Committee of Laboratories Heads: Chairman: H.W.H. West, Stoke-on-Trent, United Kingdom
Committee of Directors of member associations: Chairman: Dr K. Heusi, Zurich, Switzerland

Principal sources of finance
Annual contributions by the member associations, fixed by the Management Committee.

Principal related bodies
Association Européenne de l'Industrie Céramique (European Association of the Ceramic Industry).

Publications
- Recommendations for the design and execution of masonry, 1977/78
- Recommendations for clay block floors, including roofs and similar structures, 1982
- Research reports by the national member associations

INT.19

Federación Interamericana de la Industria de la Construcción
Inter-American Federation of the Construction Industry
Fédération Interaméricaine de l'Industrie de la Construction
Colima 254, Colonia Roma, CP 06700 Mexico, DF

Telephone: 533-00-60; 533-17-10
Telex: 1772742 CNIC MEX

Date of establishment
November 5, 1960

International non-profit confederation of 18 national and private associations representative of the construction industry in 18 countries on the American Continent

Main fields of work
The Federation's objective is to promote the optimum development of the construction industry on the American Continent. To this end, it will:
- Develop links between the construction industries
- Promote the creation of national associations representative of the private construction organizations
- Watch over the technical improvement of the construction industry as well as over the reduction of costs
- Stimulate harmonious relations between contractors and workers in the realization of private or public construction works
- Promote the use of private construction enterprises for the realization of private or public construction works
- Organize Inter-American congresses, regional conventions and local meetings
- Maintain relations with international bodies, especially with the Organization of American States
- Contribute to the solution of problems relative to studies, projects and works that are of common interest to American countries
- Manage the collection and exchange of data, information, opinions and studies in all matters relevant to technical, legal, economic and social problems, as concerns the American construction industry.

Director or Chief Executive
Secretary General: José Luis Guerra
Assistant to the Secretary: Luis Raúl Diaz González

Total staff: 2
Professionally qualified staff: 2

Principal sources of finance
The Federation is financed by members' subscriptions

Principal related bodies
Congress (every 2 years), the ruling body; Board of Directors, composed of representatives of all member bodies; General Secretariat, the executive body, the address of which is as above.

Publications
Journal: Boletin de la Federación Interamericana de la Industria de la Construcción, monthly

INT.20

Fédération Internationale de Documentation (FID)
International Federation for Documentation
Prins Willem-Alexanderhof 5, 2595 BE The Hague, PO Box 90402, 2509 LK The Hague, Netherlands

Telephone: 140671

Date of establishment
Founded 1895 in Brussels as the Institut International de Bibliographie; in 1931 as the Institut International de Documentation; in 1938 as the Fédération Internationale de Documentation.

The purpose of the Federation is to promote, through international co-operation, research in and development of documentation, which includes inter alia the organization, storage, retrieval, dissemination and evaluation of information, in whichever way it is recorded, in the fields of science, technology, social science, arts and humanities. International non-governmental organization.

Main fields of work
Central Classification Committee with 62 international revision committees for the Universal Decimal Classification (UDC).

Study Committees: Classification Research, Terminology of Information and Documentation, Education and Training, Information for Industry, Informetrics, Linguistics in Documentation, Patent Information and Documentation, Research on the Theoretical Basis of Information, Social sciences Documentation, Broad Systems of Ordering (BSO); Task Forces on Information Systems and Network Design and Management, Study of User Needs.

Regional Commission for Asia and Oceania, and Latin America.

Total staff: 7
Professionally qualified staff: 2

Director or Chief Executive
Secretary General: Ms Stella Keenan

Principal sources of finance
Financed by members' dues, sales of publications, grants.

Principal related bodies
General Assembly
Council
Executive Committee

Publications
- FID News Bulletin, monthly
- International Forum on Information and Documentation, quarterly
- Newsletter on Education and Training for Information Personnel, quarterly
- R & D Projects in Documentation and Library, bi-monthly
- FID Directory, biennial
- FID Publications List, irregular

- Proceedings of congresses
- Studies, reports, bibliographics, etc.

INT.21

Federation of Representatives of European Heating and Ventilating Associations (REHVA)

Clearing Centre: Prinses Julianaplein Ia, 3811NM Amersfoort, Netherlands

Telephone: 033-634388

Date of establishment

1963

An international institution at the disposal of all associations and private individuals interested in the science and practice of heating, ventilating and airconditioning.

Members are *technical* associations in 17 European countries each represented in the General Assembly by 2 or more delegates.

Standing committees are the Technical Programme Committee and the Documentation Committee.

Main fields of work

Promotion of the science and practice of heating, ventilating and airconditioning internationally, including all associated engineering service in the built environment and industrial processes.

The federation seeks to achieve its objective by organizing and co-ordinating meetings, by publishing the proceedings and promoting education.

In charge of the Federation is a Board of four, assisted by a Clearing Centre (2)

Total staff: 6

Director or Chief Executive:

In charge of the Clearing Centre and Secretary General: W.J.F. Oudijn

Principal sources of finance

Subscriptions of members

Revenues from organizing congresses and seminars

Publications

- REHVA Newsletter
- REHVA Yearbook
- REHVA Dictionary of Heating, Ventilating and Airconditioning (10 languages)
- International Building Services Abstracts (bimonthly)

INT.22

Fédération Internationale de la Précontrainte (FIP)

International Federation of Prestressed Concrete

Wexham Springs, Slough SL3 6PL, United Kingdom

Telephone: 02816 2727

Telex: 848352 Advicret

Date of establishment

August 1952, Cambridge, England

The aims of FIP are to promote groups for the advancement and development of concrete, of prestressing and related materials and techniques in all countries; to act as a link between associations, committees and federation groups on all aspects of prestressing; to exchange opinions on scientific and technical problems on prestressing, in particular by the organization of international congresses and symposia; to promote research, development work and investigations; and to keep members informed of the latest developments in prestressing and related developments by the publication of FIP congress and symposia proceedings, technical papers and other publications.

Main fields of work

FIP works through its technical Commissions. At the present time these are as follows:

- Concrete Sea Structures
- Seismic
- Nuclear Pressure Vessels
- Concrete Storage Vessels
- Concrete – including Lightweight and Very High Strength, and Special Concretes
- Prefabrication
- Practical Construction
- Practical Design
- Steels for Prestressing

Total staff: 6

Director or Chief Executive

Secretary-General: B.W. Shacklock

Principal sources of finance

Subscriptions from members

Publications

- FIP notes, quarterly
- Comprehensive lists of all publications, including recommendations and reports, are issued.

INT.23

Fédération Internationale Européenne de la Construction (FIEC)

International European Construction Federation (IECF)

9 rue La Pérouse, 75016 Paris, France

Telephone: (1) 720.10.20

Telex: FIEC 613456 F

Date of establishment

1905 as an international association

In 1949 the Federation was granted advisory status as a non-governmental body within the UNO. The members are national federations of building and public works employers from 18 Western European countries.

Main fields of work

The objects of the Federation are to serve as a permanent link between all the federated groups; to collect economic, commercial, industrial, technical and social information concerning the building and public works industries, and make it available to members; and to be of assistance generally to the industry which it represents.

Total staff: 7 officers

Steering Committee

Director or Chief Executive

President 1982–1985: Fernando Piccinini

Secretary General: Eric LePage

Heads of department

Secretariat: Eric LePage

Permanent Committee for the European Common Market: Paul Willemen

International European Contractors Section: Jean de Greef

Vocational Training Commission: Cees Slingerland

Working Party: Image of the Construction Industry: Paul Naudo

Principal sources of finance

Annual subscription per country

Publications

- *Journal:* L'Enterprise Européenne, quarterly
- International Conditions of Contract for Works of Civil Engineering Construction, 3rd Edn
- Temporary European Grouping – European Contract of Sub-Contracting, 1981
- Main Contracting in Europe, 1982
- Image of the Construction Industry
- Investment or Consumption, 1981
- Situation of the Construction Industry in EEC countries, half yearly

INT.24

Génie Climatique International (GCI)
International Union of the Associations of Heating, Ventilating and Air Conditioning Contractors
9 rue La Pérouse, 75784 Paris, Cedex 16, France
Telephone: 720.10.20
Telex: Fédébat 611975F
Date of establishment
1935

The objects of GCI are to provide permanent contact between national member associations; to study professional problems of common interest which present themselves to national member associations and to submit to them the results of its work, in the form of recommendations where appropriate; to arrange international congresses that serve the objects of GCI; and to defend and promote, whether in relation to the bodies or institutions concerned or in relation to other associations, the interests of GCI.

The General Assembly consists of the President of GCI, and the national delegations, consisting of a president and up to three delegates. Fourteen national associations are GCI members.

Main fields of work
The study of political, commercial and educational problems of the profession.

GCI is a member of the Comité Européen des Equipements Techniques du Bâtiment (CEETB), the European Committee of Technical Services for Buildings.

CEETB is an association formed by GCI; the International Union de la Couverture et Plomberie (UICP); and the Association Internationale des Entreprises d'Equipement Electrique (AIE). The CEETB President is M. Newman. FEETB defends the interests of the contractors dealing with technical equipment in buildings. It has close contact with CEE, especially for contracts problems in subcontracting. The CEETB Secretary General is M. Fabre (BP 128 – 59703 Marcq-en-Baroeul Cedex, France).

An agreement between CEETB and FIEC (European International Federation of Construction) regarding the letting of sub-contracts in the building industry was signed in 1982.

Director or Chief Executive
President 1982–84: Ady Kieffer
Secretary General: François Clain
Principal sources of finance
Fees from the national associations
Principal related bodies
CEETB

INT.25

International Association for Bridge and Structural Engineering (IABSE)
Association Internationale des Ponts et Charpentes
Internationale Vereinigung für Brückenbau und Hochbau
ETH-Hönggerberg, CH - 8093 Zurich, Switzerland
Telephone: 01 377 26 47
Telex: 822 186 IABS CH
Cables: IABSE, CH-8093 Zurich
Date of establishment
1929

The aims of the Association are the promotion of international collaboration between engineers and researchers and particularly representatives of science, industry and public authorities; the encouragement of the awareness and the responsibility of its members towards the needs of society; the promotion of the exchange of technical and scientific knowledge; and the publication of experiments and results of research.

Main fields of work
At present, there are seven Working Commissions, with the objective of promoting the international exchange of technical knowledge in their fields of concern, as below:

- *Structural Performance, Safety and Analysis:* structural performance (serviceability, durability); actions on structures (climatic actions, earthquake force, traffic loading, collision loading, fire and explosion, soil-structure interaction); structural safety (failure or collapse under static or dynamic loading, quality assurance and inspection, risk assessment, reliability-based approach); structural analysis (modelling, analytical theory, methods of analysis)
- *Steel, Metal and Timber Structures:* planning, design, analysis, construction, maintenance, repair and dismantling of civil engineering structures in steel, metal and timber (buildings, bridges, industrial structures, offshore structures, etc.)
- *Concrete Structures:* planning, design, analysis, construction, maintenance, repair and dismantling of reinforced and prestressed concrete structures (buildings, bridges, industrial structures, offshore structures, tunnels, retaining walls, hydraulic and utility structures, reservoirs, reactor-vessels, etc.)
- *Construction Management:* management of resources and planning, construction and implementation of important buildings and heavy construction projects and programmes; health and safety in construction; design and construction of temporary works
- *Design Methods and Processes:* design methods (selection and prescription of the elements of a structure involving creative work in integrating ideas in structural concepts leading to a synthesis of objectives, resources and environment); design process (rational and inter-personal processes within the tasks of design); organization of the design process, decision making, communication and the feedback from successes or failures
- *Informatics in Structural Engineering:* application of electronic data processing (informatics) to design, analysis, construction and maintenance of structures (computer aided management, computer aided design, quality assurance of structural computations, technical and legal responsibility of computer calculations)
- *Building Physics:* phenomena of heat and energy, moisture, air, quality acoustics, fire and daylight occurring in the interior of rooms, in buildings and structural components and in the environment of structures (physics in urban environment)

Total staff: 6
Director or Chief Executive
Executive Director: Alain Golay
President of IABSE: B. Thürlimann
Heads of department
President of Executive Committee: B. Thürlimann
Chairman, Technical Committee: H. Wittfoht
Chairmen of Working Commissions: E. Giangreco, J. Brozzetti, P. Richard, W. Oberndorfer, R. Weidle, M. Fanelli, K. Gertis
Principal sources of finance
Membership fees, conference fees, sale of publications
Principal related bodies
CEB, ECCS, FIP, IASS, RILEM, within the Liaison Committee
Publications

- IABSE Periodica, quarterly, containing five series: IABSE Surveys, IABSE Journal, IABSE Proceedings, IABSE Structures, IABSE Bulletin
- IABSE reports: published in relation to a conference
- IABSE Congress reports: published in relation to a Congress every four years
- Structural Engineering Documents: irregular series in structural engineering with comprehensive presentation by well-known specialists

INT.26

International Association of Electrical Contractors (AIE)
Association Internationale des Entreprises d'Equipement Electrique (AIE)
5 rue Hamelin, 75116 Paris
Telephone: (1) 727 97 49
Telex: FELEC 620 993F
Date of establishment
1954

The object of the AIE is to serve as a permanent link between all national associations. In all countries of the world the heads of enterprises devote the whole of their lives to their profession. The establishment of bonds of mutual esteem and friendship between them is one of the aims of AIE.

Main fields of work
The object of the Association is to promote exchanges of opinions on technical and economic problems, and problems concerning human relations between employers and salaried workers, in particular by the organization of periodic congresses.
Total staff: 1
Director or Chief Executive
General Secretary: B. Clermont

INT.27

International Association for Housing Science (IAHS)
PO Box 340254, Coral Gables, Florida 33114, USA
Telephone: (305) 554-2764
Date of establishment
Founded in 1972 in St Louis, Missouri

A non-profit association registered in the state of Missouri as a General Not for Profit Corporation; belongs to all its active individual members

An Advisory Board formed by a member for each country acts as the liaison group between IAHS and the countries.

Main fields of work
To generate new knowledge through research and assemble all other available information on housing science and develop means to disseminate it to all concerned.

To sponsor and organize international conferences and workshops on housing problems at proper places and times to help generate interest and present innovative ideas and methods.

Director or Chief Executive
President: Dr Oktay Ural
Heads of department:
Treasurer: LeRoy Thompson
Directors of regional offices are:
Cairo, Egypt: Dr Adel Fareed
Monrovia, Liberia, West Coast of Africa: C. Pryde Davis
Levant-Istanbul, Turkey: Dr Hasan Karatas
Bangkok, Thailand: Dr Ricardo Pama
Caracas, Venezuela: Dr Fernando Tinoco
Vienna, Austria: Dr Robert Krapfenbauer
Principal sources of finance
Membership sponsored
Principal related bodies
Board of Directors
Publications
- *Journal:* International Journal for Housing and Building, quarterly
- Proceedings of international symposia
- Studies

INT.28

International Association for Shell and Spatial Structures (IASS)
Secretariat: Alfonso XII, No 3, Madrid 7, Spain
Telephone: 4 67 20 12
Date of establishment
1960

A non-profit organization, founded by Eduardo Torroja.

Main fields of work
The aim of the Association is the advancement of the technology of shell and spatial structures. The term technology here takes its widest possible meaning, including architecture, conceptual design, analysis, research and all aspects of realization of projects involving shell and spatial structures of any material, including methods of execution. The aims of the IASS are advanced through the exchange of ideas and information, using the appropriate means such as:

- Organization of conferences, symposia and colloquia
- Publication of relevant literature
- Creation of working groups to investigate and report on special topics of the relevant technology
- Collaboration with other similar organizations
- Other suitable means as determined from time to time

Working groups are devoted to: pipes and tanks; industrial chimneys; cooling towers; masts and towers; recommendations for shell structures (with a subcommittee on stability); tension structures; pneumatic structures; spatial steel structures; industrial fabricable spatial structures; panel structures; ferrocement; timber structures; numerical methods on shell structures.

The results of the investigations of each Working Group are submitted to the Executive Council who have the right to approve, disapprove or even reject the corresponding publication under the auspices of the Association.

Director or Chief Executive
President: Professor Rafael Lopez Palanco
Secretariat: Luis M. Ortega (Secretary)
Principal sources of finance
Membership subscriptions, donations, subsidies, grants, interest on accumulated capital and from any other lawful source
- *Journal:* IASS bulletin, 3 per year (in English)
- Recommendations of the Working Groups
- Proceedings of the IASS Congresses

INT.29

International Association for the Study of People and their Physical Surroundings (IAPS)
Association Internationale pour l'etude de l'homme et de son environnement
c/o Dr Martin Symes, The Bartlett School of Architecture and Planning, University College London, Gower Street, London WCIE 6BT, United Kingdom
Telephone: 01 387 7050
Telex: 28722.UC.PHYS.
Date of establishment
First conference held 1970; English Company Incorporated: 5th April 1984.

An international association of academics, practitioners and others. It holds conferences, supports study groups and publishes a newsletter.

Main fields of work
Over the past few years a number of conferences have been held in various countries bringing together people from all over the world who share an interest in the significance of the physical surroundings for the people who experience those surroundings. International conferences have been held in a number of places (Kingston 1970, Surrey 1972, Lund 1983, Strasbourg 1976, Louvain-la-Neuve 1979, Surrey 1979,

Barcelona 1982, and West Berlin 1984). Because of the current and developing significance of research and practice in this area, IAPS was formed as an international organization which would bring together the great variety of work being done by those actively involved in studies of people and their physical surroundings, such as psychologists, sociologists, architects and planners.

The objectives of the Association are:

To facilitate communication among those concerned with the study of the relationship between people and their physical surroundings and by so doing to maintain and improve the quality of research being conducted in the cause of improving the physical environment and human well-being.

In order to achieve its objectives, the Association:

- Facilitates contact and exchange of ideas between members all over the world.
- Develops relationships with similar organisations.
- Holds regular conferences, and also smaller specialized seminars as required.
- Publishes a newsletter, conference and seminar proceedings, and a biannual directory.

Membership is open to anyone who shares the objectives of the Association.

Director or Chief Executive
Chairman: Gilles Barbey

Heads of departments
Secretary: Martin Symes
Membership Secretary: Sue Ann Lee
Treasurer: Peter Ellis
Newsletter Editor: Sue Ann Lee
Executive Committee: 5 members
Board of Directors: 15 members

Principal sources of finance
Membership subscriptions

Publications
IAPS Newsletter three times per annum.

INT.30

International Energy Agency (IEA)
Agence Internationale de l'Energie (AIE)

Château de la Muette, 2 rue André-Pascal, F-75775 Paris, Cedex 16, France
Telephone: 524 82 00
Telex: 630 190 Energ A
Date of establishment
15 November 1974

The IEA is an autonomous agency within the Organization for Economic Co-operation and Development (OECD). The growing need for international co-operation in energy led to its establishment as a forum for the 21 participating countries to co-ordinate their energy planning. The IEA provides a framework within which the co-operative efforts of its participating countries will reinforce one another to improve the overall energy situation.

The Agency's principal decision-making body is the Governing Board, composed of senior level representatives from each participating country. The Governing Board also meets at ministerial level from time to time. A Secretariat, with headquarters in Paris, has a permanent staff of energy experts drawn from member countries, and supports the work of the Governing Board and subordinate bodies. The Secretariat is headed by the Executive Director, who is appointed by the Governing Board.

The IEA Secretariat collects and analyses energy data, reviews member countries' domestic energy policies and programmes, makes forecasts, undertakes and publishes special studies.

The countries co-operating in the IEA are: Australia, Austria, Belgium, Canada, Denmark, Germany, Greece, Ireland, Italy, Japan, Luxembourg, Netherlands, New Zealand, Norway, Portugal, Spain, Sweden, Switzerland, Turkey, the United Kingdom and the United States. The Commission of the European Communities also participates in the work of the IEA.

Main fields of work
The Agency carries out the International Energy Program and a Long-Term Co-operation Agreement and works:

- To mould a better world energy supply and demand structure, now and for the future
- To prepare participating countries against the risk of oil supply disruptions and to share remaining oil supplies in a period of severe supply disruption
- To develop alternative energy sources and to increase the efficiency of energy use through co-operative research and development programmes
- To promote co-operative relations with oil-producing nations and other oil-consuming countries.

The Governing Board regularly reviews the world energy situation as well as domestic energy policies in order to assess future energy supply and demand patterns, and to formulate appropriate policies to meet changing conditions.

Total staff: 125
Professionally qualified staff: 60

Director or Chief Executive
Executive Director: Dr Ulf Lantzke

Heads of department
Information and Emergency Systems Operations: T. Sunami
Long-Term Co-operation and Policy Analysis: D. Jones
Oil Market Developments: B. Barth
Energy Research and Development and Technology Application: E. Willis

Principal sources of finance
Contributions by member countries.

Publications
- World Energy Outlook, 1982
- Energy Policies and Programmes of IEA Countries, 1982 Review
- Energy Research Development and Demonstration in the IEA Countries, 1981 Review, 1982
- Natural Gas, Prospects to 2000
- Heat Pumps Systems, A Technology Review, 1982

INT.31

International Federation of Asian and Western Pacific Contractors Associations (IFAWPCA)

3rd Floor, Padilla Building, Emerald Avenue, Ortigas Commercial Complex, Pasig, Metro Manila, Philippines
Telephone: 682-1657; 60-22-33/34 Locals 29 and 31
Cables: IFAWPCA MANILA
Date of establishment
March 1956

An international federation of 13 national contractors' associations organized by private individuals for the purpose of regional complementation, co-operation and friendship.

Main fields of work
IFAWPCA is active in the following areas:

- *Contract conditions (domestic and international):* Difficulties arising from contract conditions are gathered and discussed, and revisions are recommended to appropriate bodies such as the Confederation of International Contractors' Associations (CICA).
- *Labour:* labour laws and regulations applicable to the construction industry are brought up for the member nations to reach a compromise beneficial to their individual or collective interests. Training programmes are compared and contrasted with a view to eventual profit from the experiences and applications of the group.
- *Improvement of the industry's image:* IFAWPCA works closely with its

members in the pursuit of its foremost objective – to foster international co-operation.

- *Liaison:* the Premanent Secretariat serves as a constant link between the members and the Executive Board. It disseminates information and notices; prepares working papers, agenda for meetings and convention proceedings. The Secretariat also undertakes various researches for the members, as well as providing proper co-ordination with sister federations such as the International European Construction Federation (FIEC), the Inter-American Federation of the Construction Industry (FIIC) and the Associated General Contractors of America (AGC).
- *Construction technology:* concerned about the industry's image in the public mind, IFAWPCA, in its own small way, supports the central idea of improving construction technology.

Total staff: 8
Professionaly qualified staff: 6
Director or Chief Executive
Executive Director: Perfecto Mañalac
Heads of department
Executive Administrator: Miss Constancia J. Sales
Research Director: Rodante D. Marcoleta
Principal sources of finance
Membership dues, contributions and donations
Principal related bodies
The Executive Board is the policy-making body, composed of the incumbent officers of the Federation, Past-Presidents, Founder Members and the appointed representative. It meets four or five times during its term.
Publications

- *Journal:* Asia Pacific Contractor, monthly
- Construction Industry in Asia and the Western Pacific
- Comparative presentation of:
- Selected labour laws and regulations in the IFAWPCA region
- Training programmes adopted by member nations
- Basic conditions of contract/tendering procedures
- Registration system practised in member nations
- Difficulties arising from domestic/international conditions of contract
- Fluctuation/Escalation clauses

INT.32

International Federation of Building and Woodworkers
Fédération Internationale des Travailleurs du Bâtiment et du Bois
27–29 rue de la Coulouvrenière, CH - 1204 Genève, Switzerland
Telephone: 21 16 11
Telex: 428 577
Cables: interbuild
Date of establishment
Founded 1891, Brussels, present title adopted 1934

Director or Chief Executive
General Secretary: John Löfblad
Principal source of finance
Affiliation fees
Members are National Trade Unions and others.
There are 119 affiliates in 60 countries
Publications

- *Journal:* Building and Wood Bulletin, quarterly
- Occasional papers
- Congress reports

INT.33

International Federation of Consulting Engineers
Fédération Internationale des Ingénieurs-Conseils (FIDIC)
Internationale Vereinigung Beratender Ingenieure
PO Box 86, CH-1000, Lausanne 12, Chailly, Switzerland
Telephone: 021-335003
Telex: 24698 FIDI CH
Cables: FIDICHOUSE – Lausanne
Date of establishment
1913, Gent

Federation of national associations of Consulting Engineers in 37 countries. Registered according to Swiss law. The aims are to study, promote and protect in common the professional interests of the national associations; to encourage the setting up of standards and rules of professional conduct in conformity with the responsibility and dignity of the profession and to watch over their application and observance; to facilitate co-operation with other organizations collaborating in technical work. Membership is restricted to one professional association per country.
Main fields of work
Co-signatory with International European Construction Federation of International Conditions of Contract for Civil Engineering Construction Works. The International Conditions have received the approval and ratification of the International Federation of Asian and Western Pacific Contractors Associations, the Associated General Contractors of America, the International Road Federation and the Inter-American Federation of the Construction Industry. Languages used are English, French, German.
Total staff: 10 paid
Director or Chief Executive
Managing Director: Burt Campbell
Principal sources of finance
Members' dues
Publications

- Conditions of Contract for Works (civil engineering, electrical and mechanical works)
- International Forms of Agreement between Client and Consulting Engineer
- Recommended Tendering Procedures
- Various guidelines and seminar reports
- Consulting Engineer's Guide to International Agencies

INT.34

International Federation for Housing and Planning (IFHP)
Fédération Internationale pour l'Habitation, l'Urbanisme et l'Aménagement des Territoires
Internationaler Verband für Wohnungswesen, Städtebau und Raumordnung
43 Wassenaarseweg, 2596 CG, The Hague, The Netherlands
Telephone: 070-244557
Telex: 31578 inter nl ref. I–999
Cables: IFHP THE HAGUE
Date of establishment
1913

Individual and corporate members in 65 countries. Official languages: English, French, German. Consultative status with: UN-ECOSOC, UN-WHO, UNESCO. Close ties with EAROPH (Eastern Regional Organization for Planning and Housing), SIAP (Sociedad Interamericana de Planificación, Latin America). The overall objective of IFHP is to improve general knowledge in housing, planning and related fields – and thereby to improve housing and planning practice throughout the world. IFHP gives its members easy access to internationally available sources in these fields and provides opportunities for meeting colleagues on an international level and for establishing contacts by correspondence.

Main fields of work
Meetings: annually a world-wide conference with 800–1000 participants (occasionally two conferences in one year, in different parts of the world); specialized international seminars; traning courses; study tours.

Standing Committees and Working Parties: Housing, Traffic Problems, Urban Renewal, Urban and Building Climatology, New Towns, Urban Land Policy, Managing the Metropolis, Women's Perspectives in Planning, Energy and Environment, Research on Housing Costs and Financing. These bodies and their different 'task forces' meet 1–3 times yearly in order to prepare specialist studies. Membership by invitation, for IFHP members only.

Library: IFHP and its neighbour maintain a library of current books and reports of some 20 000 titles on housing, planning and public administration – plus some 650 journals, newsletters and reviews from all over the world. Publications are available on loan for members only.

Specialized services: addresses for professional contacts/visits; identification of projects for specialized study visits; identification of suitable speakers for various national and international meetings; intermediary for exchange of (summer) holiday housing; intermediary for international exchange of jobs.

Total staff: 7
Professionally qualified staff: 2
Director or Chief Executive
President: Professor Hch. Kunz
Secretary General: Jon H. Léons
Principal sources of finance
Membership contributions, governmental subsidies, income from services and activities
Publications
- IFHP News Sheet, 7 issues yearly
- Full congress and seminar reports
- Directory of members, biennial
- Special publications for Standing Committees and Working Parties
- Journal: Planning & Administration, twice a year with the International Union of Local Authorities (IULA)
- Congress and seminar reports
- Joint publications IFHP-CRRU (Centre de Recherche et de Rencontres d'Urbanisme)

INT.35
International Federation of Surveyors (FIG)
Fédération Internationale des Géomètres (FIG)
PO Box 5378, Postal Station E, Edmonton, Alberta, Canada T5P 4C9
Telephone: (403) 451-6465
Telex: 037-41399 FIG BUREAU EDM
Date of establishment
Founded in Paris, France, 18 July 1878.

The International Federation of Surveyors (FIG) is an international non-governmental organization whose purpose is to support international collaboration for the progress of surveying in all fields and applications.

It was set up by delegates from the national professional associations of surveyors of Belgium, Germany, France, United Kingdom, Italy, Switzerland and Spain.

Main fields of work
The technical and scientific work of the Federation is carried out by nine Commissions and one permanent institution as follows:
Commission 1: Professional Practice, Organization and Legal Systems
Commission 2: Professional Education and Literature
Commission 3: Land Information Systems
Commission 4: Hydrographic Surveying
Commission 5: Survey Instruments and Methods
Commission 6: Engineering Surveys
Commission 7: Cadastre and Rural Management
Commission 8: Urban Land Systems Planning and Development
Commission 9: Valuation and Management
OICRF (Office International du Cadastre et du Régime Foncier)
(International Office of the Cadastre and Land Registry)
President: mr. ir. J.L.G. Henssen
Total staff: 2
Professionally qualified staff: 1
The administrative arm of the Federation is the Bureau comprised of the following for the years 1985–1987: President C.H. Weir, Secretary General: C.W. Youngs, Treasurer: M. Gaudreault, Congress Director: S.E. Daykin, Vice Presidents: T.D.W. McCulloch, I.S. Katzarsky, J. Talvitie.
Commission Chairmen
Commission 1: M. Neil Franklin
Commission 2: Dr Seppo Härmälä
Commission 3: Professor G. Eichhorn
Commission 4: Jean Riemersma
Commission 5: R. Oluwole Coker
Commission 6: Professor Detreköi
Commission 7: Professor Hopfer
Commission 8: J. Hippenmeyer
Commission 9: C.W. Jonas
Principal sources of finance
Annual fees paid by the member associations, supplemented normally by a grant from the government of the country in which the Bureau is located for a three year term.
Principal related bodies
FIG is a non-governmental organization in consultative status with the United Nations.
Publications
- FIG Bulletins (two per year)
- Technical standards
- Technical papers presented at a Congress (every three years) are published in a separate volume for each Technical and Scientific Commission.
- Research reports: Contained in the technical papers published during a Congress.

INT.36
International Institute
8715 Via de Commercio, Scottsdale, Arizona 85258
Telephone: (602) 998-8267
Telex: 165-092
Date of establishment
1975

An independant, non-profit association of real-estate professionals, established for the purposes of disseminating information in the specialized areas of real estate finance, development, investment, and valuation. The basic task of the Institute is to obtain, process, and put into practice scientific and technical information on the specialized areas of real-estate investment, development, finance, and valuation, and to provide professional recognition for our members, both within their nation as well as internationally, and to provide a medium to communicate on an international level.
Professional designations:
- RIM – Registered International Member
- SCV – Senior Certified Valuer
- RPM – Registered Property Manager

Main fields of work
- To initiate, encourage, promote, and conduct research and educational programmes relating to the specialized areas of real estate.
- To offer services in the areas relevant to the fields of activity of the Institute.
- To provide training programmes at different levels and its field of activity in collaboration with other organizations to co-operate with similar national, regional, and international organizations and bodies in its fields of activity.

The Institute performs the following specific functions:

- Conducts research in real-estate finance, development, investment, and valuation.
- Provides teaching, research, and training for students and research workers in the areas relevant to the fields of activity.
- Helps in the publication of research reports, and the dissemination of information to all interested bodies on a regional and international level.
- Provides a medium to communicate on an international level.

Total staff: 98
Advisory Board
Professionally qualified staff: One representative from every nation
Director or Chief Executive
Executive Director: Robert G. Johnson
Managing Director: Stephen L. Schneck
Three Organizational Divisions: International Real-Estate Journal; International Real-Estate Institute; International Institute of Valuers
Principal sources of finance
Annual Subscription
Publications

- 'International Income Property Report'
- International Real-Estate Newsletter
- 'International Real-Estate Journal', 3 times per year
- International Academy of Real-Estate Studies, annually in Stockholm, Sweden
- World Real-Estate Congress
- Guide to International Real-Estate Leasing
- Registry of Members
- World Directory of Real-Estate Professionals.

INT.37

International New Towns Association (INTA)
Associations Internationale des Villes Nouvelles
Wassenaarseweg 45, 2596 CG The Hague, The Netherlands
Telephone: (070) 246521
Telex: 32504 INTA NL
Date of establishment
1975

The Association is incorporated in Paris (France) as a non-profit international organization which, to this effect, is governed by the French decree of 12 April 1939.

It is a non-political association. It provides a forum for the exchange of information among all individuals, groups and corporations concerned with the planning and development of new and renewing communities.

The phrase 'new and renewing communities' is used to describe projects and developments throughout the world that have been conceived for political or economic reasons or are intended to relieve the pressure on a large urban centre or to exploit an undeveloped area or a national resource. The phrase includes projects that are built outside or within urban areas which are based on an existing conurbation or are entirely new developments or are designed to accommodate a few thousand or hundreds of thousands of inhabitants.

Members of INTA are primarily the administrative bodies and agencies (ministries and regional or local authorities), planning and development organizations (whether public or private), and the people who are responsible for planning, building and management, as well as specialists, consultants and university professors.

To ensure the implementation of its aims and role, the Association acts in different ways and on different levels: it acts as a structure which collects, adapts and dispatches information, according to the needs of its members. It seeks to encourage the exchange of information and facilitate contact between members in four main ways:

- It prepares and circulates a bulletin to all members describing new town activities, projects and developments, and giving details of studies, seminars, etc. that are taking place in order to ensure a continuous flow of up-to-date international information on new community development.
- Its Secretariat in the Hague, the Netherlands, helps members to obtain answers to individual questions by, for example, putting them in touch with the appropriate person or organization, thus acting as an international relations department for each of its members.
- It organizes annual conferences for planners and developers throughout the world. INTA can also help to set up seminars, study tours, exhibitions and meetings.
- It encourages international studies on new town development and circulates the findings of such studies to its members.

The Secretariat is in the Hague, the Netherlands; the registered office is in Paris, France.
Total staff: 2
Professionally qualified staff: 2
Director or Chief Executive
Secretary General: J.P. Zapasnik
Principal sources of finance
Membership fees and national government subscriptions
Principal related bodies
General Assembly composed of all members: determines policy; Board of Governors composed of all national members: reports annually to the General Assembly; Executive Committee composed of five members: reports to the Board of Governors.
Publications

- Newsletter, 5 issues a year
- Reports on New Town policies and developments in France, the Netherlands and the United Kingdom. Proceedings of conferences on specialized subjects related to urban development
- Bibliographies on urban problems.

INT.38

International Organization for Standardization (ISO)
Organisation internationale de normalisation
1 rue de Varembé, case postale 56, CH 1211 Genèva 20, Switzerland
Telephone: (022) 34 12 40
Telex: 23 887 iso ch
Cables: ISORGANIZ
Date of establishment
1947

Legally constituted as a non-governmental organization in accordance with Articles 60 et seq. of the Swiss Civil Code. Aims are to promote the development of standardization and related activities in the world with a view to facilitating international exchange of goods and services and to developing co-operation in the sphere of intellectual, scientific, technological and economic activity.

Main fields of work
ISO is the specialized international agency for standardization, comprising the national standards bodies of nearly 90 countries.

The scope of ISO covers standardization in all fields except electrical and electronic engineering standards, which are the responsibility of the International Electrotechnical Commission (IEC).

The following ISO technical committees are responsible for standardization in areas of particular interest in connection with building:

ISO/TC 5 Ferrous metal pipes and metallic fittings
ISO/TC 10 Technical drawings
ISO/TC 12 Quantities, units, symbols, conversion factors and conversion tables
ISO/TC 17 Steel
ISO/TC 21 Equipment for fire protection and fire fighting
ISO/TC 35 Paints and varnishes
ISO/TC 43 Acoustics
ISO/TC 55 Sawn timber and sawlogs
ISO/TC 59 Building construction

ISO/TC 71 Concrete, reinforced concrete and pre-stressed concrete
ISO/TC 74 Cement and lime
ISO/TC 77 Products in fibre reinforced cement
ISO/TC 86 Refrigeration
ISO/TC 87 Cork
ISO/TC 89 Fibre building boards
ISO/TC 92 Fire tests on building materials, components and structures
ISO/TC 96 Cranes, lifting appliances and related equipment
ISO/TC 98 Bases for design of structures
ISO/TC 99 Semi-manufactures of timber
ISO/TC 105 Steel wire ropes
ISO/TC 108 Mechanical vibration and shock
ISO/TC 109 Oil burners and associated equipment
ISO/TC 115 Pumps
ISO/TC 116 Space heating appliances
ISO/TC 117 Industrial fans
ISO/TC 118 Compressors, pheumatic tools and pneumatic machines
ISO/TC 127 Earth-moving machinery
ISO/TC 136 Furniture
ISO/TC 138 Plastic pipes, fittings and valves for the transport of fluids
ISO/TC 139 Plywood
ISO/TC 142 Cleaning equipment for air and other gases
ISO/TC 144 Air distribution and air diffusion
ISO/TC 151 Particle boards
ISO/TC 152 Gypsum, gypsum plasters and gypsum products
ISO/TC 160 Glass in building
ISO/TC 161 Control and safety devices for heat generating systems
ISO/TC 162 Doors and windows
ISO/TC 163 Thermal insulation
ISO/TC 165 Timber structures
ISO/TC 167 Steel and aluminium structures
ISO/TC 178 Lifts, escalators, passenger conveyors
ISO/TC 179 Masonry
ISO/TC 180 Solar energy
ISO/TC 182 Geotechnics

The principal bodies within ISO are the General Assembly, Council and the Executive Committee. The principal officers are the President, the Vice-President, the Treasurer and the Secretary-General. The first three are honorary appointments; the Secretary-General, appointed by Council, is the head of the Central Secretariat.

The preparation of ISO International Standards is undertaken in ISO technical committees, sub-committees and working groups. The Central Secretariat in Geneva is responsible for the general administration and organization, for the day-to-day planning and co-ordination of the technical work, for the publication International Standards and Technical Reports, for the secretariats of the Council committees, for relations with other international organizations and for public relations and information services on international standardization.

In 1974, a special unit for developing countries was set up at the ISO Central Secretariat; it deals more particularly with those countries' standardization needs. ISO Council has set up a number of advisory committees to deal with particular aspects of standardization, comprising experts nominated by member bodies and appointed by Council. Membership is limited to one body per country.

Total staff: 500 full-time staff, including the 130 permanent staff at the Central Secretariat, in Geneva.

Director or Chief Executive
Secretary;General: Olle Sturen

Principal sources of finance
Financed by the contributions of member bodies, plus income from sales of publications; since a large percentage of ISO's member bodies are financed by their governments, it may be said that ISO is indirectly supported by governments.

Principal related bodies
ISO has a member body or a correspondent member in each of 89 countries.

Publications
- ISO Bulletin, monthly
- 5000 International Standards
- ISO Catalogue, annual
- Specialized bibliographies
- ISO Report of Activities, annual
- Memento, annual; general information on the administration of ISO; list of member bodies; technical bodies and scope of their work; trilingual, including a Russian section
- Liaisons, international organizations in liaison with ISO
- Standards Marks, Marques de Conformité aux Normes, compilation and use of various national and international standards marks
- Member bodies, Comités Membres, basic data
- Constitution, trilingual, including a Russian section
- Directives for the Technical Work of ISO, Directives pour les Travaux Techniques de l'ISO
- Guide for the Presentation of International Standards and Technical Reports, Guide pour la Présentation des Normes Internationales et des Rapports Techniques
- Brochures, leaflets, bibliographies, listings, films, etc.

INT.39

International Road Federation (IRF)

525 School Street, SW, Washington, DC 20024 and 63 rue de Lausanne, CH 1202, Geneva, Switzerland
Telephone: (202) 554-2106 (Washington); (022) 31 71 50 (Geneva)
Telex: 440336 INRFUI (Washington); 27 590 IRF-CH (Geneva)
Cables: Introadfed (both)
Date of establishment
1948

IRF is a non-profit, non-political service organization whose purpose and continuing objective is to encourage better road and transportation systems world wide, and to help in the application of technology and management practices which will give the maximum economic and social return from national road investments. IRF is the international point of affiliation for national road associations in 66 countries.

IRF is a consultant to the United Nations and the Council of Europe, is an accredited co-operating agency of the Organization of American States, and works closely with the European Conference of Ministers of Transport and with the Organization for Economic Co-operation and Development.

Main fields of work
IRF sponsors and conducts world meetings at four-year intervals. In addition, IRF sponsors and conducts regional meetings, seminars and workshops in many countries.

The Federation supports advanced training programs for graduate transportation engineers from overseas, and almost 800 foreign students have graduated from its IRF Fellowship Program over the last 33 years.

IRF collects, publishes and distributes technical, economic, statistical and educational material. For 29 years IRF has made an inventory of current research under way in the areas of roads and road transport. This annual publication lists an average of almost 5000 research and development programs in over 50 countries.

In addition to the World Survey of Current Research and Development on Roads and Road Transport, IRF has carried out 30 in-depth studies as part of its research and development program.

Total staff: 12 (Washington), 12 (Geneva)

Director or Chief Executive
President: W. Gerald Wilson
Director-General: Count F. Arco

Heads of department
IRF Fellowship Program: Vice-President: Merron L. Latta
Communications: Director: Hugh M. Gillespie
Engineering: Director: I. Robert Nehmad

Principal sources of finance
Dues from sustaining members
Principal related bodies
Executive Committee: Board of Directors; 66 Affiliated national road associations.
Publications
- World Highways, 9 times a year (Washington)
- Rue de Monde, 8 times a year (Geneva)
- World Survey of Current Research and Development on Roads and Road Transport (Annual Report, Washington)
- Annual Report (Geneva)
- World Road Statistics (Geneva)
- Limits of Motor Vehicle Sizes and Weights (Washington)

INT.40
International Society of City and Regional Planners (ISoCaRP)
Association Internationale des Urbanistes
Internationale Gesellschaft der Stadt- und Regionalplaner
Van Speijkstraat 29, 2518 EV Den Haag, The Netherlands
Telephone: (70) 46.26.54
Date of establishment
23 January 1965

A non-governmental organization granted official relationship with UNESCO and the Council of Europe. The Society's activities are international and are aimed at: the promotion of the profession of planning, of planning research, of planning education, and the protection of professional interests over the international fields.
Main fields of work
- To carry out a co-ordinated action programme on work themes and research subjects, the conclusion of which, drawn up on the occasion of seminars and congresses, are laid down in reports published and subsequently distributed
- Exhibitions, publications, study tours, etc.
- Meetings with planners and other experts specializing in planning matters in the host countries or during special conferences
- Comparative studies on policies, methods, planning legislation, etc.
- To collect up-to-date documentation material on every country, whether represented in ISoCaRP or not.
- To take part in major national or international events dealing with planning
- Evaluation of evolution in professional planning practice.

Total staff: 2
Director or Chief Executive
Executive Secretary: Mrs C.W.E. Steer-Diederen
Principal sources of finance
Membership fees/subsidies from Governments and private organizations
Principal related bodies
Executive Committee, with a President, Secretary General and five Vice Presidents; Bureau, composed of the Executive Committee and one member of each national delegation
Publications
- News Bulletin, twice a year
- Congress Proceedings

INT.41
International Society for Soil Mechanics and Foundation Engineering
University Engineering Department, Trumpington Street, Cambridge, CB2 1PZ United Kingdom
Telephone: 0223-355050
Telex: 81239 DEPENG G
Date of establishment
1936

Promotion of international co-operation in the field of soil mechanics and geotechnical engineering
Main fields of work
Soil mechanics and geotechnical engineering
Director or Chief Executive:
Secretary General: Dr R.H.G. Parry
Secretarial staff: 2
Principal sources of finance
Member society subscriptions
Principal related bodies and their roles
Member societies representing countries; Regional vice presidents; Technical Committees working on special problems.
Publications
- ISSMFE News (quarterly)

INT.42
International Union of Architects
Union Internationale des Architectes (UIA)
51 rue Raynouard, 75016 Paris, France
Telephone: 524.36.88
Telex: 614855F
Cables: UNIARCH
Date of establishment
1948

A foreign association (Law of 1901) whose bodies are the Assembly, supreme legislative body of the Union; the Council, superior executive body; the Bureau, select executive body; the College of Delegates; the Secretariat General.

Members: 78 National Sections, representing approximately 300 000 architects, grouped into five geographical regions (Western Europe, Eastern Europe and the Middle East, Americas, Asia and Australia, Africa). Each architect who is a member of a national organization affiliated to the Union is, by virtue of this, a member of the Union and may benefit from its services.
Main fields of work
In accordance with the aims set down in its statutes, the UIA must principally participate in the improvement of men's living conditions so as to fulfil their aspirations for material and spiritual well-being, particularly in the developing countries. It must promote the evolution of the methods of teaching of architecture and town planning and encourage exchanges between professionals of various disciplines, users, teachers and students.

Activities consist of:
- Triennial Programme: deals in priority with special subjects, through Directing Bodies, Working Groups and Regional Commissions.
- Working Groups: town planning; teaching establishments; housing; work places; construction and industrialization; public health; sports and leisure. They are composed of members delegated by the National Sections for their competence in the subject and individual members, architects and specialists in other disciplines.
- The Regional Commissions focus more generally on the evolution of the profession.
- Congresses and regional meetings; organized under the auspices of the UIA by a National Section and the specialist organizations of the country responsible for their preparation, they deal with a general theme, grouped into sub-topics, relating if necessary to the activities of the Working Groups
- Competitions, prizes, distinctions: organized at national and international level
- International relations: UIA collaborates officially with ECOSOC (with whom it has category II Consultative Status), UNESCO (it is classified in Category A of UNESCO, UNIDO, WHO, etc.

Total staff: 6 for Secretariat
Director or Chief Executive
President: Rafael de la Hoz
General Secretary: Michel Lanthonie

Publications
- UIA Bulletin, monthly, gives detailed information on the UIA activities
- Reports, published after congresses, seminars and studies of the Working Groups
- Publications: statutes and regulations, annuals, information leaflets, etc.

INT.43
Organization for Economic Co-operation and Development (OECD)
Organisation de Cooperation et de Developpement Economiques (OCDE)
2 rue André Pascal, 75775 Paris, Cedex 16, France
Telephone: 524.82.00
Telex: 620160 OCDE PARIS
Cables: Developeconomie
Date of establishment
30 September 1961

Established in Paris by the Convention on the Organization for Economic Co-operation and Development (signed 14 December 1960); successor body to Organization for European Economic Co-operation (OEEC), established 16 April 1948, Paris. Council composed of one representative for each member country. Decisions and Recommendations are adopted by mutual agreement of all members of the Council. Executive Committee composed of 13 members of the Council designated annually by the latter. Specialized committees and working parties (more than one hundred). Languages used are English, French. Aims to promote economic and social welfare throughout the OECD area by assisting its member Governments in the formulation of policies designed to this end and by co-ordinating these policies; to stimulate and harmonize its members' aid efforts in favour of developing countries.

Main fields of work
The Commission of the European Communities usually takes part in the work of OECD under a Protocol signed at the same time as the OECD Convention. Official relations with inter-governmental organizations. Special arrangements establishing close links with Council of Europe concluded in 1962. Under a decision of the Council of 13 March 1962, international non governmental organizations which are deemed to be widely representative in general economic matters or in a specific economy sector can be granted a consultative status enabling them to discuss subjects of common interest with a Liaison Committee chaired by the Secretary-General, and be consulted in a particular field by the relevant OECD Committee or its Officers.

Director or Chief Executive
Secretary-General: Jean Claude Paye

Publications
- Annual Reports, including the Secretary-General's Reports on the Activities of OECD and Economic Surveys (by country)
- OECD Observer, bi-monthly
- News from OECD, monthly
- Economic Outlook, twice yearly
- Information, technical and statistical bulletins
- Numerous specialized books and reports on all aspects of the economy.

INT.44
Road Transport Research Programme of the Organization for Economic Co-operation and Development
Programme de Recherche en matière de Routes et de Transports Routiers de l'Organisation de Coopération et de Développement Economiques
2 rue André-Pascal, 75775 Paris, Cedex 16, France
Telephone: 524 92 44
Telex: 620.160 OCDE PARIS
Date of establishment
1967 decision by OECD Council

A special programme of the Directorate for Science, Technology and Industry of the Organization for Economic Co-operation and Development in which 24 member governments participate.

The Programme has two main fields of activity:
- the international co-operation in road and road transport research to provide scientific support for decisions by Member governments and international governmental organizations, and to assess future strategies concerning roads and road transport problems and the priority policy concerns of Member countries;
- the information and documentation programme (IRRD – International Road Research Documentation), a co-operative scheme that provides a mechanism for the systematic world-wide exchange of information on scientific literature and current research programmes.

Main fields of work
- assessment of urban and inter-urban road transport strategies;
- development and management of road traffic control and driver communication systems to enhance network efficiency and quality of service;
- formulation and evaluation of integrated road and traffic safety programmes;
- construction, preservation and rehabilitation of road infrastructure.

The aim of the IRRD system, established in 1965 and in which institutions from 26 countries participate, is to assemble and disseminate information of interest to the road research community. This includes:
- informative abstracts of published and unpublished documents (serial and non serial publications, reports, books, conference proceedings, theses, standards, patents, etc.) on road planning, design, construction and maintenance, traffic and transport, traffic safety and vehicles;
- summaries of current research projects.

The IRRD information is prepared by IRRD Members according to uniform rules, using coded keywords. The IRRD database contains about 120 000 records with an annual addition of 10 000 to 12 000. The information is regularly updated.

The IRRD system is a de-centralized scheme with three co-ordinating centres in the Federal Republic of Germany, France and the United Kingdom. It is tri-lingual. Since October 1982 the database can be accessed on line through the Information retrieval service of the European Space Agency in Frascati, Italy.

Total staff: 9
Professionally qualified staff: 4
Director or Chief Executive
Acting Head of Division: B. Horn
Principal sources of finance
Contributions by OECD member governments
Principal related bodies and their roles
The governing body is the Steering Committee for Road Transport Research. Chairman: Mr A. Plaza.
Publications
- Catalogues of Research Reports etc are available.

INT.45 AM
Regional Secretariat of the Regional Network in Asia for Low Cost Building Materials Technologies and Construction Systems Projects
Ministry of Human Settlements,
5th Floor, Agustin Bldg,
Emerald Avenue, Pasig, Metro Manila, Philippines
Telephone: 673-23-40; 673-19-36 up to 50 Loc. 355
Date of establishment
November 1982

A special operating unit within the Ministry of Human Settlements which provides secretariat support services to the Regional Network in Asia for Low Cost Building Materials Technologies and Construction Systems projects of the United Nations Industrial Development Organization.

Main fields of work

Generally serves as Secretariat of the Technical Board. It is primarily responsible for report preparation and submission; collation, integration and dissemination of requisite information; review, co-ordination and monitoring of activities and accomplishments. It also performs liaison work between the Technical Board and all National Focal Points and other institutions within and outside the network.

The Secretariat is headed by a Regional Co-ordinator. It has four sections: Information and Documentation, Research and Development, Technical Services and Administrative Services

Total staff: 10

Director or Chief Executive

Regional Co-ordinator: Lilia Gc. Casanova

Heads of department

Assistant Co-ordinator: Lorna B. Balina

Research and Documentation Officer: Julieta M. Dimayuga

Publications Officer: Nelia Paculan

Information System Officer: Felicitas A. Tan

Programmer: Fe Vida N. Dy-Liaco

Principal sources of finance

The Ministry of Human Settlements provides the personnel complement and the maintenance and operating costs of the Secretariat. The United Nations Development Programme serves as the principal source of finance.

Principal related bodies

Technical board, the policy making body of the Regional Network; United Nations Development Programme, funding agency; United Nations Industrial Development Organization, project implementing agency; Ministry of Human Settlements, Philippines, Host of the Regional Secretariat.

Publications

- Network Monitor, a quarterly newsletter

INT.46

Réunion Internationale des Laboratoires d'Essais et de Recherches sur les Materiaux et les Constructions (RILEM)

International Union of Testing and Research Laboratories for Materials and Structures

12 rue Brancion, 75737 Paris, Cedex 15, France

Telephone: 531 18 07

Telex: CEBTP 250 071 F

Date of establishment

1947

RILEM is a non-profit-making international association bringing together scientists, engineers and industrialists for the service of the building community.

Main fields of work

RILEM is concerned with:

- Research on the properties and performance of materials and structures, in laboratory and service conditions
- The unification of test methods on an international scale and the development of new experimental procedures
- The application of research results to the structural use of materials in building and civil engineering fields.

It aims at:

- Promoting transfer of knowledge and experience through analytical study and experimentation to industrial practice
- Developing prestandardization as a means of communication and exchange so that materials, components and structures are improved

Symposia: 76 up to 1982. Languages used are French and English. Members consist of 200 laboratories and 500 individuals in 75 countries.

Total staff: 4

Director or Chief Executive

Secretary General: M. Fickelson

Deputy Secretary General: J. Bresson

Heads of department

Secretariat General: Mrs G. Calliger

Editorial Secretariat: Mrs A. Mairesse

Principal sources of finance

Membership fees

Principal related bodies

Bureau: exercises control of day-to-day affairs of RILEM and ensures that the statutes and rules are observed. General Council: principal organ, composed of one delegate and one national representative from each member country (both being full members of RILEM).

Publications

- *Journal:* Materials and Structures, every 2 months
- *Technical standards/recommendations:* Recommendations for unified test methods directed towards establishing the scientific and technical bases for the international standardization of building products
- *Research reports:* State-of-the art reports (also issued by RILEM technical committees) consisting of up-to-date surveys of a given field of study so that gaps are identified and the needs for future research pinpointed. They include: properties of materials at high temperatures, aggregates, resin adherence to concrete, soiling and cleaning of building façades.
- *Bibliographies/catalogues:* Lists of: recommendations, proceedings of symposia, final reports of committees; various indexes; published in the last issue of the year.
- Directory of member laboratories and institutions
- Directory of individual members
- Compendium of technical committees
- Newsletter, quarterly
- Technical Dictionary in 3 languages

INT.47

Union Européenne pour l'Agrément Technique dans la Construction (UEAtc)

4 avenue du Recteur Poincaré, 75782 Paris, Cedex 16, France

Telephone: 524.43.02

Telex: PAR 610 710 F

Cables: CESITEBA- PARIS

Date of establishment

1960

Federation of the institutes that are entitled to deliver technical agréments in their country and to approve agréments delivered by other member institutes.

Main fields of work

For products to be used in construction, the UEAtc:

- Determines common directives according to which Agréments are delivered or confirmed in the different member countries
- Synthesizes the results of studies and research by its Members
- Promotes the reciprocal confirmation of Agréments for information exchange and consultations

The principal subjects which have been covered are: houses, façades, windows, partitions, coating, insulations, sealants, ducts, structural fittings

Total staff

No permanent staff

Director or Chief Executive

President: Ruy J. Gomes (LNEC)

Secretary General: Emile Farhi (CSTD)

Principal related bodies

The operation is ensured by the Co-ordination Commission (which is the ruling body), the President and the General Secretariat. Specialized Commissions establish common directives for Agrément. Confirmation Commissions are in charge of examining, on the initiative of the member institutes, the requests for confirmation brought forward by a member institute to another member institute with a view to facilitate their adoption.

- The member institutes are:
- Bundesanstalt für Materialprüfung (BAM), Federal Republic of Germany
- The British Board of Agrément (BBA), United Kingdom
- Centre Scientifique et Technique du Bâtiment (CSTB), France
- Forschungsgesellschaft für Wohnen, Bauen und Planen (FGW), Austria
- Istituto Centrale per l'Industrializzazione e la Tecnologia Edilizia (ICITE), Italy
- Instituto Eduardo Torroja de la Construcción y del cemento (IETcc), Spain
- Institute for Industrial Research and Standards (IIRS), Ireland
- Union Belge pour l'Agrément technique dans la Construction (INL-NIH) (UBAtc) (BUtgb), Belgium
- Stichting voor Onderzoek, Beoordeling en Keuring van Materialen en Constructies (KOMO/O), Netherlands
- Laboratorio Nacional de Engenharia Civil (LNEC), Portugal
- Statens Byggeforskningsinstitut (SBI), Denmark

Publications

- UEAtc Information, half-yearly report
- Directives UEAtc pour l'Agrément (translated into all the languages of the member institutes (at present there are more than 25 Directives)

Principal sources of finance

Financed from each member institute

INT.48

Union Internationale des Centres du Bâtiment
International Union of Building Centres
26 Store Street, London, WC1E 7BT, UK
Telephone: 01 637 1022
Telex: 261507 Ref.No. 3324
Date of establishment
1958

The Union provides a connecting link between national and regional Building Centres and kindred organisations.

Main fields of work

Development of building information activities and services on an international level. Through its Building Information Network the Union maintains an international chain of building information organisations by strengthening ties between member Centres and by encouraging the exchange of information about all developments within the international building industry.

Total staff: 2
Professionally qualified staff: 1
President, Executive Committee (7), Secretary General, Secretary
Secretary General: Douglas Clayton

Principal sources of finance

Annual subscriptions from members

Publications

- Bulletin, 3 times/year

INT.49

Union of International Technical Associations (UITA)
Union des Associations Techniques Internationales (UATI)
Union de Asociaciones Técnicas Internacionales (UATI)
UNESCO annex, 1 rue Moillis 75015, Paris, France
Telephone: (1) 577 16 10
Telex: 204461 Paris
Cables: Unesco Paris
Date of establishment
2 March 1981

Established in Paris as Union of International Engineering Organizations (UIEO), following the conference international engineering organizations convened by UNESCO 12 October 1950, which adopted constitution, amended 18 February 1981, registered in accordance with the French law. Now called UITA, the Union is a non-profit making non-governmental organization.

It co-ordinates, by common agreement, the activities of the member organizations, such as information, documentation and terminology; and supports their actions for helping developing countries in connection with the specialized agencies of the United Nations.

The Executive Council of the Union meets at least once a year.

Director or Chief Executive

Secretary General: Maurice Remillieux

INT.50

United Nations (UN)
Nations Unies (ONU)
Secretary-General: Javier Perez de Cuellar, United Nations, New York, NY 10017, USA
Telephone: 212 754 1234
Cables: Unations, New York
Other address:
Director-General United Nations Office at Geneva: Eric Suy, Palais des Nations, 1211 Genève 10, Switzerland
Date of establishment
Declaration by United Nations of 1 January 1942; UN Charter at the Conference on International Organization, San Francisco, 25 April to 26 June 1945.

The UN officially came into existence on 24 October 1945. The UN is an organization composed of States which have accepted the obligations contained in the Charter of the UN, drawn up and signed at the San Francisco Conference in 1945.

Membership is open to all peace-loving nations which accept and, in the judgement of the Organization, are able and willing to carry out the obligations of the Charter. The official languages of the UN are Arabic, Chinese, English, French, Russian and Spanish.

The four principal purposes of the UN are to:

- Maintain international peace and security
- Develop friendly relations among nations
- Co-operate internationally in solving international economic, social, cultural or humanitarian problems and in promoting respect for human rights and fundamental freedoms
- Be a centre for harmonizing the actions of nations in attaining these common ends

The UN has six principal organs: General Assembly, Security Council, Economic and Social Council, Trusteeship Council, International Court of Justice, Secretariat.

Main fields of work

The United Nations Development Programme is the world's largest channel for multilateral technical assistance. It is active in some 150 countries and territories, and in thousands of assistance projects concerned with housing and building, farming, fishing, forestry, mining, power, transport, communications, health and environmental sanitation.

In 1978 the General Assembly established the United Nations Centre for Human Settlements (UNCHS, Habitat), with headquarters in Nairobi, to co-ordinate human settlements activity within the UN. Habitat assists developing countries in such areas as urban and regional planning, rural and urban housing and infrastructure development, slum upgrading, low-cost building technology, and water supply and sanitation systems.

Publications

The United Nations publishes hundreds of reports and studies each year. Catalogues listing sales publications may be obtained from United Nations Publications, United Nations, NY 10017, or Palais des Nations, CH-1211, Geneva 10, Switzerland.

INT.51

United Nations Centre for Human Settlements (UNCHS, Habitat)

PO Box 30030, Nairobi, Kenya
Telephone: 333930
Telex: 22996
Cables: UNHABITAT
Date of establishment
12 October 1978

The United Nations Centre for Human Settlements (Habitat) was established by the General Assembly as the secretariat to the intergovernmental policy-making body, the United Nations Commission on Human Settlements.

The Centre:

- Assists the United Nations Commission on Human Settlements in co-ordinating human settlements activities in the United Nations system and provides substantive support to the Commission;
- Executes human settlements projects;
- Provides the focal point for a global exchange of information about human settlements;
- Deals with interregional human settlements matters;
- Promotes collaboration with, and the involvement of, the world scientific community concerned with human settlements;
- Maintains a global directory of consultants and advisers in the field of human settlements and assists in the recruitment of experts;
- Initiates public information activities on human settlements;
- Promotes the further and continual use of audio-visual material relating to human settlements.

Main fields of work
UNCHS (Habitat) works in the field of human settlements. Its main areas of concern are: settlements policies and strategies; settlements planning; shelter and community services; land; development of the indigenous construction sector; low-cost infrastructure for human settlement development; and human settlements institutions and management. The Centre acts through programmes of research, training, technical co-operation and information dissemination. Research and Development Division carries out research, analysis and appraisal on substantive issues in the field of human settlements and disseminates the results through technical publications, seminars, workshops and expert group meetings; has main responsibility for the Centre's training functions. Technical Co-operation Division carries out the Centre's programme of technical assistance to developing countries, including execution of a large number of projects in those countries funded by the United Nations Development Programme and other funding agencies of the United Nations systems. Information, Audio-Visual and Documentation Division disseminates globally through publications, radio programmes and audio-visual material information concerning the activities of the Commission and the Centre, including results of projects, seminars and workshops; maintains a global information network through regional information officers in all continents.
Total staff: 250 at headquarters and 300 in field assignments
Professionally qualified staff: 76 at headquarters and 300 in field posts.
Director or Chief Executive
Executive Director: Dr Arcot Ramachandran
Heads of department
Office of the Executive Director: Chief: Darshan Johal
Technical Co-operation Division: Deputy Director: Sergei Stepanov
Research and Development Division: Chief: Bruce Hyland
Information, Audio-visual and Documentation Division: Chief: Jim Carney
Division of Administration: Officer-in-charge: Jean-Louis Lawson
Habitat and Human Settlement Foundation: Deputy Administrator: Sumihiro Kuyama
Principal sources of finance
There are three main sources of funding for the Centre's activities: allocations from the regular budget of the United Nations which supports the regular staff of the organization and its core activities; extra-budgetary resources, mostly from UNDP allocations to countries; and voluntary contributions to the Habitat and Human Settlements Foundation.
Principal related bodies
United Nations Commission on Human Settlements, United Nations Economic and Social Council
Publications

- *Journal:* Habitat News, three times a year (free)
- Development of Indigenous Construction Sector, 1982 (free)
- Human Settlements Management with Special Reference to Rehabilitation of Existing Housing Stock, 1982 (free)
- Building Operations in Low-Cost Housing Construction, 1981
- Survey of Slums and Squatter Settlements, 1982
- The Construction Industry in Developing Countries, 1984
- Bibliographic Notes (supplement to Habitat News) 3, times a year
- UNCHS (Habitat) Bibliography, every two years 1980–1982–1984
- Bibliography on Local Building Materials, Plants and Equipment, 1983
- Community Participation Bibliography, 1983
- Technical Notes – irregular (free)
- Guide to Information Sources: Building and Construction, 1980
- Guide to Information Sources: Housing, 1980
- Directory of National Research Organizations in Construction Materials and Technologies in Developing Countries, 1980
- Directory of Information and Documentation Centres in the Field of Human Settlements, 1980
- Building Codes and Regulations in Developing Countries:
- How Building Codes and Regulations Can be Adapted to Meet the Basic Needs of the Poor, 1981
- Thesaurus in the field of human settlements – English, French, Spanish 1984.

INT.52

United Nations Economic and Social Commission for Asia and the Pacific

United Nations Building, Rajadamnern Avenue, Bangkok 10200, Thailand
Telephone: 2829161 - 200
Telex: 82392 ESCAP TH, 82315 ESCAP TH
Cables: ESCAP BANGKOK
Date of establishment
28 March 1947 by resolution of the Economic and Social Council

ESCAP is the regional office of the United Nations in Asia and the Pacific. It reports through ECOSOC to the General Assembly of the United Nations.

Main fields of work
ESCAP has a variety of activities in the fields of research, training and advisory services. What follows describes its work concerning human settlements.

The Division of Industry, Human Settlements and Technology in the ESCAP secretariat at Bangkok is responsible for the implementation of human settlements activities and work programmes for the ESCAP region. Its work includes integrated settlement policies and planning, and development of shelter, infrastructure and land, and stimulating institutional capabilities and public participation. It conducts surveys, research, seminars and conferences on the significant human settlements issues of Asia and the Pacific, such as upgrading slums and squatter settlements, physical planning in rural areas, indigenous building materials and construction technologies, strengthening planning capacities of local authorities, transportation in urban areas, energy conservation and generation in human settlements.

During the 1980–1981 biennium, the Human Settlements Section completed 13 projects, with financial assistance from extrabudgetary resources and UNCHS in Nairobi. Among these projects was a study-cum-review of the human settlements situation in the ESCAP region, for regional input to the UNCHS global review of human settlements. The review consists of an analysis of current programmes and policies in regional countries and has a statistical annex. There is also a human settlements atlas, graphically depicting this situation in a number of major cities in the region. The review, published in 1982, is designed as a policy guide for various activities in this field.

The Regional Congress of Local Authorities for Development of Human Settlements, held at Yokohama, Japan, in June 1982, reviewed the role of local authorities and made recommendations on the nature and degree of their responsibilities and the types of suitable resources to be made available to them.

The principal committees associated with the work on human settlements are: Agricultural Development;Development Planning; Industry, Technology, Human Settlements and the Environment; Natural Resources; Population; Social Development; Statistics; Trade; Shipping, and Transport and Communications. There is also an ad-hoc conference, and a special body on Land-Locked Countries.

Director or Chief Executive

Executive Secretary: S.A.M.S. Kibria

Heads of department

Division of Administration: Kylin Chhor
Agriculture Division: S.Z. Khan
Development Planning Division: Edward Van Roy
ESCAP/UNIDO Division of Industry, Housing and Technology: V.J. Ram
International Trade Division: B.B. Arzadon
Natural Resources Division: Le Nian Fan
Population Division: Boonlert Leoprapai
Division of Shipping, Ports and Inland Waterways: D. Koludrovic
Social Development Division: Boonlert Leoprapai
Statistics Division: M.A. Sahib
Technical Co-operation Division: H.S. Wanasinghe
Transport, Communication and Tourism Division: Ebni Ali
ESCAP Liaison Office for the South Pacific, Nauru: Paula Sotutu

Publications

- In the field of human settlements:
- Ad hoc Technical studies on human settlements issues
- A human settlements atlas for Asia and the Pacific is being prepared
- A human settlements bibliography is being prepared

INT.53

United Nations Economic Commission for Africa (ECA)

Commission Economique des Nations Unies pour l'Afrique (CEA)

PO Box 3001, Addis Ababa, Ethiopia

Telephone: 447200
Telex: 21029 UNECA ET
Cables: ECA, Addis Ababa

Date of establishment

29 April 1958 in pursuance of a resolution of the Economic and Social Council (ECOSOC) of the United Nations.

One of the five regional commissions under the Economic and Social Council.

Main fields of work

ECA, acting within the framework of the policies of the United Nations and subject to the general supervision of ECOSOC, provided that the Commission takes no action with respect to any country without the agreement of the government of that country, is authorized to assist member States to deal, as appropriate, with matters relating to economic and social development taking account of the interrelationship of economic and social factors.

The Housing, Construction and Physical Planning Section within the Joint ECA/UNIDO Industry Division is in charge of ECA activities relating to human settlements policies and programmes. The activities cover spatial planning, including regional planning, urban planning and planning for integrated rural development; housing in urban and rural areas with special emphasis on slum and squatter settlements, low-cost housing, co-operative and self-help housing; building materials and construction industries; financing of human settlements development including housing credit facilities.

The assistance is provided to member States through consultancy, field projects, collection and dissemination of information, training activities.

Assistance in the field of building materials and construction is provided through a special development programme which aims at assisting member States to progress towards self-sufficiency. An important element of this programme is the promotion of research and development activities. Directories of building and building materials research organizations and of research scientists and technologists in Africa have been published. The establishment of joint subregional research centres is being given active promotional support.

The Section also serves as the secretariat for the Joint Intergovernmental Regional Committee on Human Settlements and Environment which has responsibility for the formulation of regional and subregional policies and programmes in the African region and for their implementation.

The Housing, Construction and Physical Planning Section has a regular staff of four professionals and two secretaries; in addition, experts under short-term contracts are recruited as needed.

Total staff: 1000 approximately
Professionally qualified staff: 300 approximately

Director or Chief Executive

Executive Secretary: Professor Adebayo Adedeji
Housing, Construction and Physical Planning Section: Max Faladé

Principal sources of finance

The Commission is financed by the regular budget of the United Nations approved by the General Assembly. Funds from bilateral and multilateral sources are also secured for specific development projects in the African region.

INT.54

United Nations Economic Commission for Europe

Palais des Nations, CH-1211, Geneva 10, Switzerland

Telephone: 34 60 11 and 31 02 11
Telex: 28 96 96
Cables: UNATIONS GENEVE

Date of establishment

20 March 1947

The ECE is a regional subsidiary body of the United Nations Economic and Social Council. It is an inter-governmental body composed of representatives of 34 member countries, including all European countries, Canada and the United States.

Main fields of work

The ECE and its 15 principal subsidiary bodies undertake studies and promote co-operation and exchange of information among governments in all the main sectors of the national economy. Of most direct concern to CIB is the work of the ECE Committee on Housing, Building and Planning. The Committee was set up in 1948 and was, inter alia, instrumental in establishing the CIB. At present the Committee's activities comprise projects in the following four main areas: 1. Integrated settlements, policies and strategies; 2. Urban and regional planning; 3. Housing policy; 4. Building

Projects in work area 1 are being dealt with by the Committee itself. The Working Party on Urban and Regional Planning deals with projects in work area 2, the Working Party on Housing with projects in work area 3 and the Working Party on Building with projects in work area 4. Work area 2 includes a major project aimed at the promotion of international co-operation in the field of urban and regional research. This project is monitored by the Group of Experts on Urban and Regional Research which prepares quadrennial conferences on urban and regional research, promotes the circulation of annotated lists of research projects as well as research digests and organizes research colloquia on specific topics. Work area 4 contains a major project aimed at the international harmonization of building regulations and standards which is carried out in co-operation with a large number of technical non-governmental international organizations, including CIB.

There are nine Divisions, of which the Environment and Human Settlements Division is responsible for work on environment, water and human settlements problems.

The Division consists of four units: The Human Settlements Section, the Environment Section, the Water Unit and the Air Pollution Unit.

Total staff: 250

Professionally qualified staff: 125

Director or Chief Executive

Executive Secretary: Klaus A. Sahlgren

Heads of department

Environment and Human Settlements Division: C. Lopez-Polo
Human Settlements Section: Y. Palm

Principal sources of finance

United Nations regular budget.

Principal related bodies

The ECE secretariat co-operates with the secretariats of the other regional commissions (Economic Commission for Africa, the Economic Commission for Western Asia, the Economic Commission for Latin America and the Economic and Social Commission for Asia and the Pacific), as well as with the United Nations Centre for Human Settlements (UNCHS). Close contacts and co-operation are also maintained with a large number of other UN bodies, specialized agencies, intergovernmental organizations and non-governmental organizations.

Publications

The Journal of the ECE is published quarterly in co-operation with Pergamon Press. The Annual Bulletin of Housing and Building Statistics for Europe is published in September every year. The ECE Compendium of Model Provisions for Building Regulations is being issued in loose-leaf form with insertions being circulated as and when approved by the Working Party on Building. A list of ECE/HBP publications is issued. An up-dated list of bibliographies is issued every year.

INT.55

United Nations Economic Commission for Latin America (ECLA)

Comisión Económica para América Latina (CEPAL)

Headquarters: Edificio Naciones Unidas, Av. Dag Hammarskjold s/n

Mailing address: Casilla 179-D, Santiago

Telephone: 485051 to 61

Telex: 340295 (Transradio); 441054 (ITT)

Cables: UNATIONS SANTIAGO

Other addresses:

Latin American Institute for Economic and Social Planning: Director: Alfredo Costa-Filho

Latin American Demographic Centre: Director: Oscar J. Bardeci

Mexico Office: Av. Presidénte Masaryk 29, 11570 Mexico, DF
Cables: CEPAL MEXICO
Mailing address: Apartado Postal 6-718, 06600 Mexico, DF
Telephone: 250 12 31 (Administration); 250 12 56 (Director); 250 15 55 (General)
Director of Office: Gert Rosenthal

ECLA Subregional Headquarters for the Caribbean: Salvatori Building, Room 300, 2 Frederick Street, Port of Spain, Trinidad and Tobago
Cables: ECLA PORT OF SPAIN
Mailing address: PO Box 1113, Port of Spain, Trinidad and Tobago
Telephone: 62-35595; 62-35482
Telex: 394
Director of Office: Clyde Applewhite

Washington Office: 1801 K Street, NW Suite 1261, Washington, DC 20006, USA
Cables: ECLAWASH WASHINGTON
Telephone: (202) 296-0822
Telex: 440034
Chief of Office: Joaquín Izcúe

Montevideo Office: Juncal 1305, Piso 10, Escritorio 1002/03, Montevideo, Uruguay
Cables: CEPAL MONTEVIDEO
Mailing address: Casilla de Correo 1207, Montevideo
Telephone: 907473; 903876
Telex: 808
Chief of Office: José María Punno

Bogotá Office: Carrera 10 (A) No 70–48, Bogotá, Colombia
Cables: CEPAL BOGOTA
Mailing address: Apartado Aéreo 75471, Bogotá, Colombia
Telephone: 2492257; 2557989
Telex: 43316
Chief of Office: Alejandro Power

Buenos Aires Office: Callao 67, 4 Piso, 1022 Buenos Aires, Argentina
Cables: UNATIONS BAIRES
Mailing address: Casilla de Correo 4191, 1000 Buenos Aries, Argentina
Telephone: 456181; 456238; 400431
Telex: 9170
Chief of Office: Alfredo Eric Calcagno

Brasilia Office: Setor Bancario Sul, Edificio Seguradoras, 14 Andar, 70 000 Brasilia, DF, Brazil
Cables: CEPAL BRASILIA
Telephone: 612-2269568; 612-2269502; 612-2269523
Telex: 06111181
Chief of Office: Alejandro Vera-Vassallo

Date of establishment

The Commission was established by Resolution 106(VI) of the Economic and Social Council of the United Nations, adopted on 25 February 1948.

A subsidiary body of the Economic and Social Council of the United Nations. Member States of the Commission are Antigua and Barbuda, Argentina, Bahamas, Barbados, Belize, Bolivia, Brazil, Canada, Colombia, Costa Rica, Cuba, Chile, Dominica, Dominican Republic, Ecuador, El Salvador, France, Grenada, Guatemala, Guyana, Haiti, Honduras, Jamaica, Mexico, Netherlands, Nicaragua, Panama, Paraguay, Peru, Portugal, Saint Lucia, Saint Christopher and Nevis, Saint Vincent and the Grenadines, Spain, Suriname, Trinidad and Tobago, Uruguay, United Kingdom, United States of America, Venezuela. Associate members are Netherlands Antilles, British Virgin Islands, United States Virgin Islands, Montserrat.

Main fields of work

To promote the economic development of the Latin American and Caribbean countries and improve the standard of living of their peoples. Also to strive to strengthen economic relations, both among countries in the region and with other nations of the world.

The Commission of the Whole meets every two years in one of the Latin American capitals. The Committee of the Whole meets for special purposes and usually in years in which no session of the Commission has been scheduled.

The Central American Economic Co-operation Committee (Committee for Economic Co-operation of the Central American Isthmus), established in August 1952 as a permanent organ of the ECLA, is charged with advising the Central American Governments and providing technical assistance in connection with the Central American Economic Integration Program. The Mexico City Office serves as secretariat for the Committee.

The Trade Committee was established in 1955 to facilitate the intensification of inter-Latin American trade.

The Committee of High-Level Government Experts, established in 1971, meets every two years to appraise the International Development Strategy for the UN Second Development Decade as it concerns Latin America.

The Caribbean Development and Co-operation Committee, established in 1975, is in charge of advising Governments in the Caribbean region with regard to integration and their relationship with the rest of Latin America. It acts as a co-ordinating body for activities relating to development and co-operation in the Caribbean, and carries out research and studies in connection with social and economic development needs.

Director or Chief Executive

Executive Secretary: Enrique V. Iglesias
Deputy Executive Secretary for Economic and Social Development: Norberto González
Deputy Executive Secretary for Co-operation and Support Services: Robert T. Brown
Chief, Programme Planning and Co-ordination Office: Ricardo Cibotti
Secretary of the Commission: Daniel S. Blanchard
Chief, Information Office: Jorge Reiner

Heads of department

ECLA/UNIDO Industry and Technology Division: Oscar Altimir
Economic Projections Centre: Eligio Alves
Operations Division: Nessim Arditi
Economic Development Division: Andrés Bianchi
Documents and Publications Service: Mrs Marta Boeninger
Transport and Communications Division: Robert T. Brown
Latin American Documentation Centre: Claudionor Evangelista
Division of Administration: Giorgio Gamberini
Computer Centre: J. Michael Hewitt
Joint ECLA/FAO Agricultural Development Division: Luis López Cordovez (FAO)
International Trade and Development Division: Guillermo Maldonado
Acting Chief, CTNC/ECLA Unit of Transnational Corporations: Luiz Claudio Marinho
Natural Resources and Energy Division: Michael Nelson
Social Development Division: Germán Rama
Statistics and Quantitative Analyses Division: Pedro Sáinz
Co-ordinator, Joint CEPAL/UNEP Environment Unit: Osvaldo Sunkel

Principal sources of finance

United Nations regular budget.

Publications

General publications are:

- Economic Survey of Latin America
- Economic Bulletin for Latin America, semi-annual
- Statistical Yearbook for Latin America
- Notas sobre la economia y el desarrollo de América Latina, semi-monthly
- CEPAL Review

A wide variety of other documents is also published.

INT.56

United Nations Economic Commission for Western Asia (ECWA)

Commission Economique des Nations Unies pour L'Asie Occidental (CEAO)

PO Box 27, Baghdad, Iraq

Telephone: 5569400–449
Telex: 213303 UNECWA IK
Cables: UNECWA, Baghdad

Date of establishment

9 August 1973 by resolution of the Economic and Social Council of the United Nations (ECOSOC)

To initiate and participate in measures for facilitating concerted action for the economic reconstruction and development of Western Asia, for raising the level of economic activity in Western Asia and for maintaining and strengthening the economic relations of the countries of that area, both among themselves and with other countries of the world.

The Commission is empowered to make recommendations directly to member governments and to specialized agencies of the United Nations, but no action can be taken with respect to any state without the agreement of that state.

The Commission is an intergovernmental regional organization whose membership is composed of: Bahrain, Iraq, Jordan, Kuwait, Lebanon, Oman, Qatar, Saudi Arabia, Syria, United Arab Emirates, Yemen Arab Republic, Democratic Yemen and the Palestine Liberation Organization.

ECWA is a subsidiary organ of the Economic and Social Council of the United Nations.

Main fields of work

Draft work programmes are prepared by the secretariat and submitted to member states for consideration and adoption at the Commission's annual sessions which are held at ministerial or senior official level. The Commission directs its attention to those fields which have priority in the region, such as agriculture and the development and improvement of water resources; human settlements and social development; industrial development; the transfer and development of technology; development and planning; statistics; housing; communications transport and tourism; international trade; natural resources and technical co-operation.

Total staff: 264
Professionally qualified staff: 104

Director or Chief Executive

Executive Secretary: Mohamed Said Al-Attar

Heads of department

Development Planning Division: Baseem Hannush
Joint ECWA/UNIDO Industry Division: R. Abu El-Haj
Joint ECWA/FAO Agriculture Division: Abdel S. Alwan
Division of Administration: T.N. Francois
Natural Resources, Science and Technology Division: Officer-in-Charge, Ertuna Cengiz
Human Settlements Division: Marwan Moshen
Transport, Communications and Tourism Division: Mohammed Fayyad
Social Development and Population Division: Hamid Ammar
Statistics Unit: Officer-in-Charge, Akil Al-Akel
Technical Co-operation Unit: L. Abdunnur
Programme Planning and Co-ordination: Acting Chief, George Haddad
Programme Evaluation Unit: Richard van der Graaf
Environmental Co-ordination Unit: Officer-in-Charge, Raja Gedeon

Principal sources of finance

United Nations

Principal related bodies

The Commission is composed of representatives of member countries which meet annually to review the Secretariat's activities and provide policy guidance.

Publications

Listings of the Journal and Research Reports, etc. are regularly issued

INT.57

United Nations Educational, Scientific and Cultural Organization (UNESCO)

Organisation des Nations Unies pour l'Education, la Science et la Culture

Place de Fontenoy, 75700 Paris, France

Telephone: (1) 577 16 10
Telex: 204461 Paris
Cables: Unesco Paris

Date of establishment

4 November 1946, when Greece, the twentieth State to do so, deposited the instrument of ratification in London.

An international, intergovernmental organization, Unesco is one of the specialized agencies of the United Nations. It is at present made up of 158 Member States. The official liaison between the Secretariat of Unesco and the Member State is the National Commission, located in the Member State itself. The National Commission has a Secretary General, and frequently several other members, some of whom are specialists in the main fields of interest covered by the Organization.

Unesco has three main components: the General Conference, the Executive Board and the Secretariat. Part of the Secretariat is in the Paris headquarters, and part in the field (that is, in many of the Member

States). In an effort to decentralize much of the work, regional offices have been established in a number of regions of the world. Unesco has highly qualified experts working on specific projects in the field.

Main fields of work

The moral aims of Unesco, as set out in its Constitution, are the promotion of peace, human rights and international understanding. These are to be achieved by collaboration among nations in the fields of education, science and culture, as well as those of social sciences and communication.

The medium-term objectives contributing to the solution of these world problems are laid down in Unesco's six-year plan of action, and are the guidelines for the concrete projects to be carried out in each of the above-mentioned spheres. These projects range from the eradication of illiteracy campaigns, technical co-operation with Member States in educational planning and educational building programmes, the international scientific programmes such as Man and the Biosphere, ocean and coastal marine systems, the preservation of the cultural heritage, the promotion of cultural identy, the development of information, institutions and research relating to peace, a round table on the problems of youth in the 80s, to the establishment of a new world information and communication order, to name but a few in each sphere of activity.

Unesco collaborates with a large number of non-governmental organizations, and following the status that has been accorded to these NGOs, enjoys either 'consultative and associate relations', or 'information and consultative relations', or 'mutual information relations'.

Total staff: 3658

Director or Chief Executive

Director General: Amadou-Mahtar M'Bow

Heads of department

Education Sector: Assistant Director-General: S. Tanguiane
Science Sector: Assistant Director-General: A. Kaddoura
Sector of Culture: Assistant Director-General: M. Makagiansar
Social Science Sector: Acting Asst. Director-General: Mrs Z. N'Kanza
Communication Sector: Assistant Director-General: G. Bolla

Principal sources of finance

Contributions of Member States make up the regular budget.

Principal related bodies

The General Conference made up of representatives of each Member State meets every two years and votes the programme and budget for the following biennium. It also votes its agreement for the objectives set out in the medium-term plan covering a period of 6 years. Between sessions the Executive Board (made up of representatives from 45 member States) meets regularly to supervise the execution of the programme by the Secretariat.

Publications

There are 10 paying periodicals, including the Unesco Courier. The Office of Unesco Presse publishes about 200 titles a year. Unesco also produces over 15 000 different documents a year, including conference reports, studies, series, etc. Unesco publications and documents are indexed in a quarterly computerized catalogue (with title, personal name and subject indexes), and the paying publications are listed in the Unesco Publications Catalogue.

INT.58

United Nations: Food and Agriculture Organization (FAO)

Organisation des Nations Unies pour l'Alimentation et l'Agriculture (OAA)

Viale delle Terme di Caracalla, 00100, Rome, Italy
Telephone: 57971
Telex: 61181 Foodagri
Cables: Foodagri
Regional Offices:
Africa: PO Box 1628, Accra, Ghana
Cables: Foodagri
Liaison UN ECA: PO Box 3001, Addis Ababa, Ethiopia
Cables: Eca
Asia and the Far East: Maliwan Mansion, Phra Atit Road, Bangkok, Thailand
Telephone: 817844
Cables: Foodagri
Latin America: Casilla 10095, Santiago, Chile
Telephone: 462061
Cables: Foodagri
Near East: FAO Headquarters, Rome, Italy
Telephone: 57971
Cables: Foodagri
North America Liaison Office: 1776 F Street, NW, Washington, 20437, DC, USA
Telephone: 3762309
Cables: Foodagri
Liaison Office with UN: Room 1125, DC Building, New York, NY 10017, USA
Telephone: (212) 754-1234
Cables: Foodagri

Date of establishment

16 October 1945, Quebec.

The main components of FAO are:
Conference; Council, Statutory Subsidiary Bodies. Conference appoints Director-General who is assisted by an international staff. Statutory bodies are the council commissions, committees, working parties and panels.

As a specialized agency of the UN, FAO reports annually to ECOSOC and co-operates closely with the former and sister agencies. FAO has entered into formal relations with non-governmental organizations that can and do help to promote the purposes of the Organization.

Main fields of work

- Raising the levels of nutrition and standards of living of the peoples under the respective jurisdiction of the Member Governments
- Securing improvements in the efficiency of production and distribution of all food and agricultural products
- Bettering the conditions of rural populations; thus contributing towards an expanding world economy and ensuring humanity's freedom from hunger.

Conferences: Regular sessions held biennially since 1949.

Library containing 700 000 volumes. 5000 current periodicals, reference service.

Director or Chief Executive

E. Saouma

Publications

In English, French, Spanish, Arabic.

INT.59

United Nations: Office of the United Nations Disaster Relief Co-ordinator (UNDRO)

Bureau du Coordonnateur des Nations Unies pour les secours en cas de catastrophe (UNDRO)

Palais des Nations, 1211 Geneva 10, Switzerland
Telephone: (022) 34 60 11 or 33 20 10 (emergency only)
Telex: 28148 undr ch
Cables: UNDRO, Geneva

Date of establishment

14 December 1971 by United Nations General Assembly Resolution 2816 adopted at its twenty-sixth session in New York.

UNDRO is a separate entity within the United Nations Secretariat. Its main function is to ensure prompt, effective and efficient response to a Government's need for assistance, at the time of a natural disaster or other disaster situation. Its second function is to assist governments in pre-disaster planning including prevention and preparedness.

Main fields of work

Under UNDRO's Mandate, the main functions of the Office are as follows:

- To mobilize, direct and co-ordinate relief activities of the various organizations of the United Nations system in response to a request for disaster assistance from a stricken state
- To co-ordinate United Nations assistance with assistance given by inter-governmental and non-governmental organizations, in particular by the International Red Cross
- To assist the Government of the stricken country to assess its relief and other needs and to evaluate the priority of those needs, to disseminate that information to prospective donors and others concerned, and to serve as a clearing house for assistance extended or planned by all sources of external aid
- To promote the study, prevention, control and prediction of natural disasters, including the collection and dissemination of information concerning technological developments
- To assist in providing advice to Governments on pre-disaster planning in association with relevant voluntary organizations, particularly with the League of Red Cross Societies, and to draw upon United Nations resources available for such purposes.

The Office is headed by a Disaster Relief Co-ordinator with the rank of an Under-Secretary-General who acts on behalf of the UN Secretary-General and reports directly to him. It is composed of the Relief Co-ordination and Preparedness Branch and the Prevention and Support Services Branch.

Total staff: 51

Professionally qualified staff: 23

Director or Chief Executive

United Nations Disaster Relief Co-ordinator: M'Hamad Essaafi
Director and Deputy to the Co-ordinator: Hans W. Einhaus

Heads of department

Relief Co-ordination and Preparedness Branch: Gerald A. Dunn
Prevention and Support Services Branch: Vit Karnik
Data Processing and Communications Unit: Charles Kerpelman
Administrative Support Unit: Jean-Pierre Goulut
New York Liaison Office: Ridha Zoubeidi

Principal sources of finance

The Office is financed mainly from the Regular United Nations Budget, supplemented by a Trust Fund, established by General Assembly Resolution 3243 (XXIX) of 1974, and financed entirely by voluntary contributions of Member States of the United Nations.

Principal related bodies

The Resident Co-ordinators/Representatives of the United Nations Development Programme (UNDP) represent UNDRO in developing countries and serve as regional focal points. UNDRO has also entered into agreements (Memoranda of Understanding) with other UN agencies, defining areas and means of co-operation, in order to strengthen the collective response of the UN system to disasters.

Publications

- *Journal:* UNDRO NEWS, published six times a year
- Case Reports
- Disaster Assessment Mission Reports
- Proceedings of Seminars and Expert Meetings
- Disaster Prevention and Mitigation: a Compendium of Current Knowledge (12 volumes)
- Guidelines for Disaster Prevention (3 volumes)
- Shelter after Disaster – Guidelines for Assistance
- Disasters and the Disabled
- Handbook on Volcanic Emergencies

INT.60

United Nations: World Health Organization (WHO)
Nations Unies: Organization Mondiale de la Santé (OMS)
1211 Geneva 27, Switzerland
Telephone: (022) 91 21 11
Telex: 27821
Cables: UNISANTE GENEVE
Regional Offices:
Africa: PO Box 6, Brazzaville, Congo
Telephone: 81 38 60–65
Telex: 5217; 5278
Cables: UNISANTE BRAZZAVILLE (E or F)
Americas (Pan American Sanitary-Bureau): 525, 23rd Street, NW, Washington, DC, 20037, USA
Telephone: 861-3200
Telex: 248338
Cables: OFSANPAN WASHINGTON (E)
Eastern Mediterranean: PO Box 1517, Alexandria, Egypt
Telephone: 802318/807843
Telex: 54028
Cables: UNISANTE ALEXANDRIA (E or F)
Europe: 8 Scherfigsvej, DK 2100 Copenhagen 0, Denmark
Telephone: 29-01-11
Telex: 15348
Cables: UNISANTE COPENHAGEN (E or F)
South-East Asia: World Health House, Indraprastha Estate, Mahatma Gandhi Road, New Delhi 110002, India
Telephone: 27 01 81-88
Telex: 312241; 312195
Cables: WHO NEW DELHI (E)
Western Pacific: PO Box 2932, 12115 Manila, Philippines
Telephone: 59 20 41–59 37 41
Telex: 27652; 63260; 40365
Cables: UNISANTE MANILA (E)

Date of establishment

On 22 July 1946, the Constitution was signed by 61 States, 51 of which were also Members of the UN. The WHO Interim Commission was simultaneously established. On 7 April 1948, the Constitution entered into force, having been ratified by 26 Member States of the United Nations.

The World Health Organization is a specialized agency within the United Nations system. Its status is inter-governmental.

The Constitution of WHO defines health as 'a state of complete physical, mental and social well-being and not merely the absence of disease or infirmity'. Article 1 of the Constitution states: 'The objective of WHO shall be the attainment by all peoples of the highest possible level of health'.

In 1977, the World Health Assembly decided that the main health target of governments and of WHO should be the attainment by all the people of the world by the year 2000 of a level of health that will permit them to lead a socially and economically productive life, popularly known as 'Health for all by the year 2000'.

Main fields of work

The Seventh General Programme of Work, covering the period 1984–9, was adopted in 1982 and is the first of the three general programmes of work of WHO needed to cover the period until the target date of the year 2000. The Programme constitutes WHO's support to the national and regional strategies for attaining health for all by the year 2000 and to the global strategy. It emphasizes 'health' as defined in the WHO's Constitution, rather than the mere control of specific diseases and has the following broad functions.

Direction, co-ordination and management will concern itself with the formulation of the policy of WHO, and the promotion of this policy among Member States and in international political, social and economic forums, as well as the development, co-ordination and management of the Organization's general programme.

Health system infrastructure will aim at establishing comprehensive health systems based on primary health care and the related political, administrative and social reforms, including a high degree of community involvement. It will deal with:

- The establishment, progressive strengthening, organization and operational management of health system infrastructures, including the related manpower, through the systematic application of a well-defined managerial process and related health systems research, and on the basis of the most valid available information.
- The delivery of well-defined countrywide health programmes
- The absorption and application of appropriate technologies that form part of these programmes
- The social control of the health system and the technology used in it.

Health science and technology, as an association of methods, techniques, equipment and supplies, together with the research required to develop them, constitutes the content of a health system. Health science and technology programmes will deal with:

- The identification of technologies that are already appropriate for delivery by the health system infrastructure
- The research required to adapt or develop technologies that are not yet appropriate for delivery
- The transfer of appropriate technologies
- The search for social and behavioural alternatives to technical measures
- The related aspects of social control of health science and technology

Programme support will deal with informational, organizational, financial, administrative and material support.

Total staff: 3616

Director or Chief Executive

Director-General: Dr H. Mahler

Principal sources of finance

Contributions from Member States

Principal related bodies

The work of the Organization is carried out by: The World Health Assembly (the supreme directing organ of WHO); the Executive Board; the Secretariat (headed by the Director-General).

Publications

Basic Documents (Constitution, Regulations, Agreements, etc: 33rd edition) in Arabic, Chinese, English, French, Russian, Spanish; Handbook of Resolutions and Decisions of the World Health Assembly and of the Executive Board (Vol. 1, 1948–1972; Vol. 2, 5th edition, 1973–1982) in Arabic, Chinese, English, French, Russian, Spanish; Official Records of the World Health Organization, Nos 1–250, series discontinued in 1978 (replaced by separate documents containing the proceedings of the World Health Assembly, and the Executive Board, the proposed Programme Budget, financial reports, etc.); Biennial report of the Director-General; Reports on the World Health Situation in Arabic, Chinese, English, French, Russian, Spanish; The First Ten Years of the World Health Organization (1958); The Second Ten Years of the World Health Organization (1968) in English, French, Russian, Spanish; International Health Regulations, 3rd annotated edition 1983, in Arabic, English, French, Russian, Spanish.

Periodicals: World Health (illustrated monthly) in Arabic, English, French, German, Portuguese, Russian, Spanish; International Digest of the Health Legislation, quarterly, in English, French; WHO Chronicle (six numbers per year) in Arabic, Chinese, English, French, Russian, Spanish; World Health Statistics Report, quarterly, in English, French; World Health Statistics Annual in English, French, Russian; Weekly Epidemiological Record in English, French; Bulletin of the World Health Organization, 6 numbers per year, articles in English or French, Russian; World Health Forum, quarterly, in Arabic, Chinese, English, French, Russian, Spanish.

Serial Publications: Health For All series in Arabic, Chinese, English, French, Russian, Spanish (8 titles up to end 1982); Monograph Series in English and French; some also in Arabic, Russian, Spanish (63 titles published up to mid-1976); Public Health Papers in English, French, Spanish, some also in Arabic and Russian (75 titles published in 1981); Technical Report Series in English, French, Russian Spanish. Reports of Expert Committees, Scientific Groups, etc. in English, French, Russian and Spanish; some also in Arabic, 683 published in 1982.

Offset publications in English and French; some also in Arabic and Spanish, 70 published up to 1982.

INT.61

United Nations: World Meteorological Organization (WMO)

Nations Unies: Organisation Météorologique Mondiale (OMM)

Case postale No. 5, avenue Giuseppe-Motta 41, 1211 Genève 20, Switzerland

Telephone: 346400

Telex: 23260

Cables: METEOMOND GENEVA

Date of establishment

11 October 1947 (the former International Meteorological Organization, a non-governmental organization, was established in 1873).

A specialized agency of the United Nations organization.

Main fields of work

WMO takes care of various relevant aspects with a view to coordinating, standardizing and improving world meteorological and related activities and to encouraging an efficient exchange of meteorological and related information between countries in the aid of human activities. More specifically, there is a Working Group on Urban and Building Climatology, established by the Technical Commission for Climatology at its eighth session in 1982. A meeting of experts on Urban and Building Climatology (December 1982) prepared a programme with seven projects on: Promoting co-operation with users; Effects of climate on buildings and human settlements; Education and training of meteorologists and users; Requirements for meteorological information; Guidance for data books; Dissemination of information on pertinent activities; and International co-operation and activities.

Director or Chief Executive

Secretary General: G.O.P. Obasi

Principal related bodies

The World Meteorological Congress; the Executive Council; regional meteorological associations; Technical Commissions; the Secretariat.

Publications

General: Basic documents; Final Reports of meetings; Technical Notes; Manuals, Guides, Nomenclatures and Atlases. Annual Reports (to members of the UN); WMO Bulletin, quarterly illustrated information bulletin; World Climate Programme reports and others.

Urban and building climatology: Proceedings of WHO/WMO Symposium on Urban Climates (Vol. 1) and Building Climatology (Vol. 2), Brussels 1968; Selected Bibliography on Urban Climate, 1970; Review of Urban Climatology, 1968–73; Proceedings of WMO Symposium on Meteorology as Related to Urban and Regional Land-Use Planning, Asheville, NC 1975; Weather, Climate and Human Settlements, 1976; Urban Climatology and its Relevance to Urban Design, 1976; Applications of Building Climatology to the Problems of Housing and Building for Human Settlements, 1976; Review of Urban Climatology, 1973–76; Bibliography of Urban Climate, 1977–80; Climate/Energy Graphics, Climate Data Applications in Architecture, 1982; Report of the Meeting of Experts on Urban and Building Climatology, 1982.

NATIONAL ORGANIZATIONS

ARG.62 AM

Industrialized Building in the Habitat Research Centre
Centro de Investigacion de la Construccion Industrializada en el Habitat (CICIHA)
Casilla de Correo 157, Suc. San Martin, CP 1650, Provincia de Buenos Aires
Telephone: 752-5701; 755-6161
Telex: 021859 INTIAR
Cables: INTI BAIRES
Date of establishment
Constitution agreement of 13 July 1978

The Centre is a member of the Centre System of the National Institute for Industrial Technology (Instituto Nacional de Technologia Industrial (INTI)). It was set up in mid-1978 in response to a request made by a group of institutes: the Argentine Chamber for Economical Housing (Camara de la Vivienda Economica de la Republica Argentina (Cavera)), the Argentine Portland Cement Institute (Instituto del Cemento Portland Argentino (ICPA)), the Red Ceramic Argentine Institute (Instituto Argentino de la Ceramica Roja (INCER)) and the Research and Development Institute for Materials and Techniques applied to Dry Building (Instituto para la Investigacion y Desarrollo de Materiales y Tecnicas Aplicables a la Construccion en Seco (INCOSE)). They were aware of the necessity for an institution which could provide studies, research, development and dissemination of information. CICIHA intends to develop the 'Casa-Partes' industry or standardized components industry. The components come from different manufacturers and can be combined or interchanged without difficulty.

Main fields of work
Development of the Casa-Partes project: this is orientated towards the use of an up-to-date building industry technology. The development of the 'modular components open approach' by means of standardized components, called Casa-Partes, to obtain the best and cheapest building production, is the result of participation by professional, industrial and contractor sectors. They join in periodic meetings of specific work groups belonging to sub-commissions on building systems, panels, carpentry, roofs and sanitary systems.

Organization of technical services: courses, working days, lectures anywhere in the country; tests and physical mechanical appraisals of materials, components and systems; technical assistance and consultancy for private and government organizations; systematic technical commercial information on products; subscriptions to the Mail Information Service (Servicio de Informacion Postal (SIP)); bibliographical search and Xerox service; specialized library and periodical library; production and publishing of documents: technical sheets, orientation summaries, reports, books and catalogues; distribution and selling of publications; collaboration and co-operation with similar organizations in Argentina and abroad.

Total staff: 9
Professionally qualified staff: 4
Director or Chief Executive
Technical Director: Lidia Edith Barrio
Technical Sub-Director: Juan Pereyra
Principal sources of finance
INTI, promoter associations, members, and service charges and fees
Principal related bodies
Faculty of Architecture and Urbanization (Facultad de Arquitectura y Urbanismo (FAU)) (architectural design, construction techniques); Sub Secretariat of Urban Development and Building (SUBSECRETARIA de Desarrollo Urbano y Vivienda (SSDUV)) (official building plans, national building Fund) (Fondo Nacional de la Vivienda (FONAVI)); Provincial Building Institutes (Institutos Provinciales de Vivienda (IPV)) (provincial building plans, national building fund (Fondo Nacional de la Vivienda (FONAVI)); Argentine Institute for Rationalization of Materials (Instituto Argentino de Racionalizacion de Materiales (IRAM)) (rules and regulations), National Mortgage Bank (Banco Hipotecario Nacional (BHN)) (official credits); International Work Office (Oficina Internacional del Trabajo (OIT))/Research and Documentation Interamerican Centre for Professional Development (Centro Interamericano de Investigacion y Documentacion sobre Formacion Profesional (CINTERFOR)) (senior executive development and distribution); Argentine Chamber for Economical Housing (Camara de la Vivienda Economica de la Republica Argentina (CAVERA)) (promoter association); Argentine Portland Cement Institute (Instituto del Cemento Portland Argentino (ICPA)) (promoter association); Red Ceramic Argentine Institute (Instituto Argentino de la Ceramica Roja (INCER)) (promoter association); Research and Development Institute for Materials and Techniques applied to Dry Building (Instituto para la Investigacion y Desarrollo de Materiales y Tecnicas Aplicables a la Construccion en Seco (INCOSE)) (promoter association).
Publications
- Mail Information System (Servicio de Informacion Postal SIP), once four months
- Orientation Summaries (Resumenes de Orientacion, RO), once every four months
- Technical sheets (Hojas Tecnicas, HT), once every 4 months
- Informative documents, once every 4 months
- Casa-Partes catalogues: Roofs, 1982; Sanitary Systems, 1983; Panels, 1983; Carpentry in wood and different kinds of metal, 1983
- Lists of Publications are issued.

ARG.63 AM

Instituto del Cemento Portland Argentino
Argentine Portland Cement Institute
San Martín 1137, 1004 Buenos Aires
Telephone: 312-3048
Other addresses:
Laboratory: Capitán Bermudez 3958, Olivos (Buenos Aires)
Branches:
Seccional Bahia Blanca, Luis Ma Drago 23-3er Piso Oficina 4/5/6, Bahia Blanca
Seccional Córdoba, Avenida General Paz 70, 3er Piso, Córdoba
Seccional Corrientes, Catamarca 1515, Corrientes
Seccional Mendoza, San Lorenzo 170, PB Local L, Mendoza
Seccional La Plata, Calle 48 No 632, 50 Pisa Oficina 65, La Plata (Buenos Aires)
Seccional Rosario, San Lorenzo 1047, 10 A, Rosario (Santa Fé)
Seccional San Juan, Ignacio De la Rosa 194 Oeste, San Juan
Seccional Tucumán, 25 de Mayo 30, 1er Piso Oficina 5/6, San Miguel de Tucumán
Seccional Neuquén, Avenida Argentinas 251, 1er Piso Local 116/117, Neuquén
Date of establishment
1939

A scientific and technical non-profit-making institute, for the promotion and development of Portland cement.

Main fields of work
- Studies and research for the development of concrete and Portland cement applications
- Dissemination of technological developments of cement and concrete.
- Courses, seminars

The principal departments are:
- Technical Promoting Division. Roads, Structures, House building; Prefabrication; Prestressed concrete; Technological development
- Research Division: Laboratory for technology; Study and testing of structures.
- Publications, Advertisement, Statistics
- Library
- Card-index file, Dispatching
- Accountancy

Total staff: 100
Professionally qualified staff: 64
Director or Chief Executive
Technical Director: Carlos Ernesto Duvoy

Heads of department
Roads: Eng. Colombo
Laboratory: Eng. Aubert
Principal sources of finance
Financed by the cement industry
Principal related bodies
Executive Committee formed by representatives of the cement industry.
Publications
- Publicaciones Técnicas (Technical Publications), pamphlets or books, Numbers 1–66
- Cemento Portland, Revista del Instituto del Cemento Portland Argentino (Portland Cement, Review of the Argentine Portland Cement Institute), Numbers 1–76, yearly
- Boletin del Cemento Portland, Actualidades e Informaciones. Publicación del Instituto del Cemento Portland Argentino (Bulletin of Portland Cement. Events and information. Publication of the Institute), Numbers 1–52, bi-monthly.
- Bibliographies, yearly
- Library Catalogue, 1958

ARG.64
Instituto de Mecánica Aplicada (IMA)
Institute of Applied Mechanics
Base Naval Puerto Belgrano 8111
Telephone: 80-1324
Date of establishment
5 February 1975

Centre for research and development. Established as the result of an agreement between the Argentine National Council for Scientific and Technical Research and the Naval Service for Research and Development.

Approximately 60% of the research and development effort is devoted to in-house projects. The remainder consists of research and development work co-ordinated with the National (Argentine) Atomic Energy Commission and other agencies, consultancy work for private industry, organizing seminars for graduate engineers, etc.
Main fields of work
- Vibrations of mechanical systems
- Dynamics of structures
- Applied mathematics (heat conduction, waveguide theory)
- Bio-engineering

Researchers of IMA hold two patents (assignors to the National Atomic Energy Commission) and have developed a concept for the protection of bridge piers when crossing waterways (featured in Lloyd's List, 19 February 1982 and Safety at Sea, June 1982).
Total staff: 11
Professionally qualified staff: 9
Director or Chief Executive
Professor Dr Patricio A.A. Laura
Heads of department
Experimental Vibrations Division: Research Engineer J.L. Pombo
Bio-engineering: Research Engineer Laureano C. Nava
Principal sources of finance
CONICET, private foundations, private industry, etc. CONICET is the equivalent of the US National Science Foundation
Publications
Over 240 publications, approximately half in Journal of Sound and Vibration, Applied Acoustics, Journal of the Acoustical Society of America, etc.

ARG.65
Instituto De Mecánica Aplicada y Estructuras (IMAE)
Institute of Applied Mechanics and Structures
Berutti y Riobamba CUR, Rosario 2000, Provincia de Santa Fe
Telephone: (041) 81-3194; (041) 81-2121
Date of establishment
Resolution of National University of Rosario: 1962

A research institute in the Faculty of Engineering within the National University of Rosario (Universidad Nacional de Rosario, UNR)

Supervisory Boards of Directors of the Laboratories indirectly influence the policy of the Institute
Main fields of work
Areas in which the IMAE is active are: materials testing and soil mechanics, structures, experimental stress analysis, road research and metallurgy
Total staff: 40
Professionally qualified staff: 20
Director or Chief Executive
Director: Ing. Carlos Luis Torregiani
Heads of department
Materials Technology and Testing Laboratory: Ing. Emilio Gavilán
Structural Laboratory: Ing. Héctor Cervera
Experimental Stress Analysis Laboratory: Ing. Rodolfa Rasia
Road Research Laboratory: Ing. Hugo Poncino
Soil Mechanics Laboratory: Ing. Carlos Bessone
Metallurgy Laboratory: Ing. Hugo Villagra
Principal sources of finance
Research funds are received from the University for approved programmes. About 70% of the total funds come from tests and consultancy work performed for third parties.
Publications
Most of the research results from laboratory investigations are presented at the annual South-American Congress of Structural Engineering (Jornadas Sudamericanas de Ingenie ría Estructural) and some are also published in international journals.

AUS.66

Australian Building Systems Appraisal Council Ltd (ABSAC)
332 Albert Street, East Melbourne, Victoria 3002
Telephone: (03) 419.4555
Telex: AA 32449
Cables: MBA-VIC
Date of establishment
7 February 1978. Incorporated under the Companies Act 1961.

The Company was established jointly by the Master Builders Federation of Australia, the CSIRO Division of Building Research, and the Australian Institute of Building Surveyors, to carry out appraisal of innovations in the building industry. The Housing Industry Association and the Insurance Council of Australia are now part of the Company.

It is a national organization, incorporated in Victoria as a public company, limited by guarantee, non-profit-making, registered in all States and Territories of Australia. It is non-governmental in that through the Board of Directors, it determines its own destiny. There are two secretariats: an administrative and a technical.

Main fields of work
Hitherto it had been difficult for an innovation in the building industry to achieve approval for its use. By its very nature, a product, material, system, etc. which was the subject of neither an Australian Standard nor a Building Regulation in Australia was subject to the judgement of individual building surveyors in each municipality where its use was proposed. There was no uniform basis on which a building surveyor could exercise his discretionary approval powers. So the need existed for an organization such as ABSAC.

In 6 years, it has received over 100 applications for appraisal, and produced some 40 Technical Opinions. There is general acceptance of its work on the basis of a 'one stop shop'.

The actual appraisal is carried out by a qualified Appraisal Officer with expertise in the specific or related relevant discipline, under the direction of the Technical Advisory Committee (TAC), which is reprseentative of a wide cross-section of the industry and various authorities.

Total staff: 4
Professionally qualified staff: 2
Director or Chief Executive
Chairman: K.G. Deacon
Secretary: D.G. Forbes
Heads of department
Chairman, Technical Advisory Committee: Dr F.A. Blakey
TAC Secretaries: B.L. Schafer; Mrs B.L. Budgen
Principal sources of finance
Self-funding by application fees and appraisal fees.
Principal related bodies
CSIRO DBR (Technical Secretariat and provision of Appraisal Officers); EBS (provision of Appraisal Officers)
Publications

- *Journal:* ABSAC news, quarterly (free publication)
- Technical Opinions (available only from the applicant for the appraisal resulting in the Technical Opinion)
- Consolidated list (published annually)
- Annual Report.

AUS.67

Australian Institute of Steel Construction Ltd (AISC)
1st floor, BP House, 110 Alfred Street, Milsons Point, New South Wales
Postal address: PO Box 434, Milsons Point 2061
Telephone: (02) 929 6666
Date of establishment
1962

A non-profit-making public company incorporated in the State of New South Wales with capital limited by guarantee of the Members.

Main fields of work
The Institute engages in a wide range of promotional and educational activities, including technical meetings, conferences and seminars, student lectures and so on. AISC acts for the industry in the preparation of Australian codes and standards. Special attention is given to areas such as tolerances, inspection standards, etc. to ensure a balance between theoretical ideals and what is practically achievable, with emphasis on fitness for purpose.

Total staff: 8
Professionally qualified staff: 3
Director or Chief Executive
Director: G.A. Day
Principal sources of finance
Membership fees, sales of publications.
Publications

- *Journals:* Steel Construction, quarterly; Steel Fabrication Journal, quarterly

Others:

- Bolting of Steel Structures
- Composite Construction Design Guide
- Design of Structural Connections
- Economical Structural Steelwork
- How to use Steel in Homes
- Safe Load Tables for Structural Steel
- Source Book for Australian Steel Structures Code
- Standardized Structural Connections.

AUS.68

Australian National University, Urban Research Unit
Box 4, PO, Canberra, Act, 2000
Telephone: 492297
Date of establishment
1966

A research unit within the Research School of Social Sciences at the Australian National University.

Main fields of work
Social science research into urban development in Australia. Much of the work has been related to policy in areas such as housing, urban land, transport, city size, urban planning and urban government. The research is closely concerned with the relationships between government and the private sector in urban development, and with equity issues. Most of the research is carried out by individuals or by small teams of two or three people.

Total staff: 10
Professionally qualified staff: On average 9
Director or Chief Executive
Head of the Unit: P.N. Troy
Principal sources of finance
Largely through the University's normal budget.
Publications

- Monographs containing detailed research results
- A social atlas of Syndey based on computer mapping
- Journal articles
- Special research reports.

AUS.69

Australian Road Research Board (ARRB)
500 Burwood Highway, Vermont South, Victoria 3133
Postal address: PO Box 156, Bag 4, Nunawading 3131, Victoria
Telephone: (03) 233 1211
Telex: (AA) 33113
Cables: ROADRESEARCHMELB
Date of establishment
1960; registered in 1965 as a non-profit company in the State of Victoria under the appropriate Companies Act.

A road/transport research establishment established as the national centre for road research and information, carrying out research on behalf of the States and Commonwealth of Australia.

The current research programme includes aspects of road research ranging from pavement construction and materials studies, through driver behaviour and traffic control, to transport and travel demand. Many studies are of interest to local government, including residential area design, drainage and low cost road development. ARRB disseminates information through its conferences, symposia and publications. It maintains an extensive library of road literature, operates computer-based indexes listing all Australian published material (Australian Road Index, ARI) and research projects (Australian Road Research in Progress, ARRIP) in the road and road transport subject area. ARRB is the Australian member of the International Road Research Documentation (IRRD) system. ARRB maintains close contacts with a number of road research organizations overseas, including the Transportation Research Board in the USA, the Permanent International Association of Road Congresses (PIARC) and the Transport and Road Research Laboratory (TRRL) in the UK.

Main fields of work
ARRB undertakes, arranges and disseminates road and road transport research, as defined by its constitution. The four main areas of research are: road engineering, traffic engineering, transport planning, and instrumentation. ARRB also operates as a national documentation centre for road-related literature and research in progress, and is the national member of the OECD's Road Research Programme, and International Road Research Documentation system.

The ARRB Three Year Research and Development Strategy Plan outlines the intended research programme for the next three years. Version 13, the 1982–84 Strategy, continues to reflect an interdisciplinary approach to roads and road transport. Within this overall approach a number of areas of concentrated effort have developed, principal amongst which are investigations of community travel patterns by assembling individual person/household behaviour data from a disaggregate base and the exploration of pavement design and performance in relation to dynamic loads.

The strategy groups project into coherent groups of topics, which may be grouped in a number of ways, of which the most application-orientated is related to the planning, providing and operating of the road and road transport infrastructure.

The planning group of topics includes studies of vehicle fleet composition and usage, individual response to the social, economic and transport environment, and appraisal of transport systems, management policies and practices. Development of studies of freight movements should yield better predictions of future truck traffic operations but will require a shift of resources. Traffic loads provide a positive link to the other groups by looking to the definition of loads imposed on the road and bridge infrastructure by trucks of large mass and complex axle configurations. An energy audit of traffic and transport management proposals has been developed.

Studies of heavy vehicles, with particular emphasis on the roll-over behaviour of trucks, is the first topic of the operation group where on-road behaviour is related to performance and safety in the traffic stream. Rural road operations topic looks at the characteristics of the traffic stream more broadly, examining the interaction of traffic flow and road design elements, including improving techniques directed at specific operational problems. Road user information and guidance collects together projects covering the link between operation and behaviour, particularly the signs, road markings, delineation and lighting provided to the driver for his guidance. Road user behaviour attempts to relate driver characteristics in behavioural terms to their use of the road system and examines, for example, how a free speed is selected in the urban situation. A more direct control of behaviour results from transport and traffic management where the initial emphasis of current projects is directed towards examining the effects of techniques on parameters other than simple traffic delay.

The topics on infrastructure provision assemble together the materials, construction and maintenance studies.

Total staff: 110
Professionally qualified staff: 50%
Director or Chief Executive
Dr M.G. Lay
Principal sources of finance
Commonwealth government (Department of Transport and Construction); State governments (six State Road Authorities).
Publications
- *Journal:* Australian Road Research, quarterly
- Research reports ARR, series
- Special reports, series
- Technical manual ATM, series
- Internal reports AIR, series
- Australian Road Index, quarterly; annual cumulation
- Australian Road Research in Progress, annual

Other publications include Proceedings of the ARRB biennial conference, and of other symposia; and certain specialist monographs such as the Source Book for Australian Roads, 1981.

AUS.70

Brick Development Research Institute (BDRI)
Architecture Building, University of Melbourne, Parkville, Victoria 3052
Telephone: (03) 347-6796
Date of establishment
2 November 1965

A research organization established by, and supported by, almost every clay brick maker in Australia. The Institute conducts research and testing in its own laboratories and where appropriate sponsors work in other places.

The Institute is also available to give advice to brick specifiers and users.

Main fields of work
- The use of clay bricks in construction and landscaping
- The measurement of the properties of clay bricks: size and size variability; water absorption properties; efflorescence characteristics; liability to lime spalling; compressive strength; transverse strength; long-term dimensional changes (brick expansion)
- The properties of mortar, particularly as they affect bond strength
- The properties of wall-ties
- The properties of brickwork
- The preparation or improvement of material standards for bricks, brickwork and associated materials
- The improvement of building regulations or ordinances as they affect brickwork
- Advising brick specifiers and users.

Total staff: 9
Professionally qualified staff: 6
Director or Chief Executive
Director: Tom McNeilly
Heads of department
Publications Officer, Publications Department: Digby Hughes
Laboratory Manager, Laboratory: Stephen Zsembery
Regional Architect, New South Wales Branch: John Macintosh
Regional Officer, Queensland Branch: Graham Slater
Regional Engineer, South Australian Branch: Roger Taggart

Principal sources of finance
Members' subscriptions; an approximately matching grant from the Australian government; fees for laboratory testing; publication sales.
Principal related bodies
Close liaison is maintained with the CSIRO Division of Building Research and the Department of Transport and Construction, Experimental Building Station. The Chiefs of both organizations are members of the Brick Development Research Institute governing Council.
Publications
- Technical Notes on Clay Bricks, about 3 per year
- Brickwork Facts about 2 per year
- Techniques, less than 1 per year
- BDRI Specification and Methods of Test for Clay bricks
- Research reports are occasionally prepared: most are reprints of papers presented at conferences.

AUS.71
Building Science Laboratory
Department of Architecture, The University of Adelaide, PO Box 498, Adelaide, South Australia
Telephone: (08) 228 5742
Telex: UNIVAD AA 89141
Cables: UNIVAD
Date of establishment
1959

The Building Science Laboratory is a part of the Department of Architecture in the University of Adelaide. Teaching and research in all aspects of environmental science applied to architecture and building. Where facilities are available, these are open to application for usage by the architectural profession and others; the laboratory may perform tests on request.
Main fields of work
Emphasis is placed on activities related to daylight/daylighting and electric lighting; solar radiation/thermal/climate and energy problems; architectural and building acoustics and noise control; and aspects of building science applied to urban design and planning.

Four academic staff are regular users of the facilities and are responsible to the Chairman of the Department. There are two full-time Laboratory Technicians in the Building Science Laboratory.
Total staff: 6
Professionally qualified staff: 4
Director or Chief Executive
J.D. Kendrick
Principal sources of finance
From Government sources through the University and from the University Research Grant and occasional 'outside' research grants e.g. Australian Research Grants Scheme, State Energy Research Council, and South Australian Housing Trust.
Publications
Research papers, conference papers, or special lectures are published from time to time in appropriate journals e.g. *Architectural Science Review*, Proceedings of Conference of Australian and New Zealand Architectural Science Association; or individual special reports.

Also: *Daylighting Newsletter* (twice annually).

AUS.72
Cement and Concrete Association of Australia
147 Walker Street, North Sydney, New South Wales, 2060
Telephone: 923 1244
Date of establishment
1928

The Cement and Concrete Association of Australia is a non-profit organization formed to promote and develop the uses of cement and concrete.
Main fields of work
- Technical advisory service to users of cement and concrete
- Funding of research on cement and concrete
- Education of those involved in the building industry
- Participation in the preparation of national standards
- Promotion of new uses of cement and concrete.

Total staff: 45
Professionally qualified staff: 19
Director or Chief Executive
Director: K.J. Cavanagh
Heads of department
Federal Office: K.J. Cavanagh
NSW Regional Office: J.W. Cruickshank
Queensland Regional Office: I.H. Orchard
Victorian Regional Office: K.D. Campbell
SA Regional Office: D.J. Amey
WA Regional Office: D.R. Buchanan
Tasmania Regional Office: S.E. Gill
Principal sources of finance
Financed solely by the Portland cement industry in Australia.
Publications
Journal: Constructional Review, quarterly

The Association publishes a series of Technical Manuals, Technical Reports, Constructive Notes and Information Sheets on an intermittent but continuing basis; also promotional literature.

AUS.73 FM
Commonwealth Scientific and Industrial Research Organization, Division of Building Research (CSIRO DBR)
PO Box 56, Highett, Victoria 3190
Telephone: 61 3 555 0333
Telex: AA33766
Date of establishment
1945 by decision of the then Council for Scientific and Industrial Research

A statutory agency of the Commonwealth Government of Australia available to assist all federal, state and municipal services, industry and private individuals. The Division concentrates mainly on mission-orientated strategic work with a time horizon of about 5 years.
Main fields of work
The Division endeavours to cover in its research all the activities of the construction sector of the economy which is understood to include housing, building, civil engineering works, and the repair and maintenance of these. More specifically, the activities range from basic studies in building materials through such design disciplines as structural engineering, acoustics, and building services, to planning activities, operations research, and social and economic studies related to the built environment.

There are two principal committees: the Building Research Committee, which co-ordinates building research activities financed by the Commonwealth Government; and the Building Research and Development Advisory Committee (with its four standing committees), which is the principal liaison with private industry.
Total staff: 195
Professionally qualified staff: 88
Director or Chief Executive
Chief: F.A. Blakey
Heads of department
Design for Durability: K.G. Martin
Life Cycle Performance: Dr F.J. Bromilow
Safety and Risk: Dr R.H. Leicester
Shelter and Infrastructure: Dr J.F. Brotchie
Principal sources of finance
By appropriation from consolidated revenue of the Commonwealth Government of Australia
Publications
- *Journal:* Rebuild, twice monthly
- Research papers, annually
- *Research reports:* about 30 a year, plus journal articles

AUS.74

Concrete Institute of Australia
147 Walker Street, North Sydney, New South Wales 2060
Telephone: (02) 923 1244
Date of establishment
17 April 1970

An association of individuals and firms or companies with a common interest in the development of concrete technology and practice. Control is maintained by the elected Federal Council and State Branch Committees.

Main fields of work
One of the objectives of the Institute is to promote, encourage, foster and develop concrete technology and practice. The Institute does this by holding lectures, conventions and publishing technical papers and manuals.
Total staff: 7
Professionally qualified staff: 7
Director or Chief Executive
Federal Council, Federal Secretary/Treasurer: K.J. Cavanagh
Principal sources of finance
Membership fees.
Principal related bodies
Cement and Concrete Association of Australia providing Administration and Technical Secretariat.
Publications
- *Journal:* CIA News, quarterly
- Reference Specification for Concrete Work
- Recommended Practice Grouting of Prestressing Ducts and Ground Anchors, 1982
- Technical Committee Reports
- Technical Papers.

AUS.75 AM

Jennings Industries Ltd
690 Springvale Road, Mulgrave, Victoria 3170
Telephone: 566 8060
Telex: AA 31532
Cables: JENNINGSCO
Date of establishment
1932

Public company engaged in building and manufacturing research connected to support such work.

Main fields of work
Housing (detached; retirement); non-housing (all types of general building as project manager, builder, financial); civil engineering; manufacturing (plastic, timber and clay products).
Professionally qualified staff: 15
Director or Chief Executive
Director of research: Victor Edward Jennings
Principal sources of finance
Internal

AUS.76 AM

New South Wales Institute of Technology (NSWIT), Building Studies Research Unit
New South Wales Institute of Technology, PO Box 123, Broadway, New South Wales 2007
Telephone: Australia 02-20930
Telex: 75004
Date of establishment
1984

To act as a vehicle for research for staff of the School of Building Studies in order to: attract research funds; supervise research projects; formulate research policy for the School. The research unit is a body within the School of Building Studies responsible to the Head of School.

Main fields of work
Research projects currently being undertaken are:
- Egress behaviour in buildings in the event of fire
- Career structures in the building industry
- The role of the quantity surveyor in Australia
- The management of large projects.

Total staff: 15
Professionally qualified staff: 15
Director or Chief Executive
Dr V. Ireland
Principal sources of finance
$80 000 initial grant. Ad hoc grants from various sources.
Publications
Ad hoc research reports

AUS.77 UM

University of New South Wales, School of Building
PO Box 1, Kensington, New South Wales 2033
Telephone: 662 2189
Telex: AA26054
Cables: Unitech, Sydney
Date of establishment
1971

The School of Building provides degree courses that prepare students for professional and executive employment within one of Australia's largest industries, the building industry. Careers in a wide variety of areas, in both private enterprise and the public sector are available to building graduates. These areas include the management and supervision of building developments, the production of building materials, building technology, finance and administration.

The School has excellent teaching and research facilities including a building science laboratory, a computer service and graphics facility, and a structures laboratory.

Main fields of work
The main field of work is the preparation of students for the undergraduate degree of Bachelor of Building (BBuild) degree, and the graduate degrees of Master of Science (Building) (MSc (Building)), Master of Building (MBuild) and Doctor of Philosophy (PhD).

In 1982, a Building Research Centre was set up with the School, funded initially by the University. Funds have also been received from outside sources. Since its inception, three research workers have been engaged to work on research projects. The School's academic staff also participate in the Centre's work. To achieve its aims, the Centre maintains close links with industry groups, and at the individual project level, wherever appropriate, suitable people from industry, government and other institutions are invited to participate in project teams.

In general, the Centre is concerned with:
- the development of close working relationships with design, construction, manufacturing and entrepreneurial groups within the building industry, as well as with public and statutory authorities, and
- the development of a research programme dealing with matters such as the evaluation of building failures; computer programs for project management; domestic construction; construction estimating; the nature, incidence and allocation of risk in construction contracts including methods of management and the financial consequences; integration of the design and construction processes; project management; problems of contract preparation and administration.

Total staff: 15
Professionally qualified staff: 12
Director or Chief Executive
Head of School, School of Building: Professor A.R. Toakley
Heads of department
Undergraduate Course Co-ordinator: D.N.H. Hassall
MSc (Building) Course Co-ordinator: Dr J.M. Hutcheson

Director, Building Research Centre: Professor A.R. Toakley
Deputy Director, Building Research Centre: M. Marosszeky
Principal sources of finance
The University of New South Wales.
Publications
Some research work is written up initially in the School of Building Working Papers prior to publication in local or overseas journals. It is intended that the Building Research Centre will commence its own series of publications in the near future.

AUS.78 AM

New South Wales University School of Civil Engineering
Kensington 2033
Telephone: 663-0351
Date of establishment
January 1973 Foundation Chair of Civil Engineering Construction and Management

A teaching and research establishment in civil engineering construction and management. It also provides a consultative service for private and public industry.
Main fields of work
Field studies of construction operations. Impact of field conditions and operator skill level on equipment performance and fleet productivities in haulage operations; the filming and analysis of construction operations; study of factors which affect the speed of rock drilling.

Concrete placement systems. Factors influencing cost and selection of methods. Formwork/falsework systems. Load determination and the design, construction and management of formwork systems.

Developing of planning and resources management methodologies. Problem definition, development of criteria and systematic evaluation, decision-making theory and practice, economic analysis and the effect of time pressure on quality of decisions.

Equipment management. Development of policies and procedures for selection and replacement. Investigation into the interaction between the life and performance of equipment and its working environment.

Studies relating to the construction industry as a whole. Analysis of trends in litigation with the civil engineering construction industry, causes and remedies of contractural disputes. Industrial relations and the work environment. A study of the history of civil engineering construction in Australia.
Total staff: 13
Professionally qualified staff: 11
Director or Chief Executive
Professor R.W. Woodhead
Principal sources of finance
University Research Support Funds supplemented by sponsored research.
Principal related bodies
Australian Federation of Construction Contractors.
Publications
Research papers are presented at civil engineering seminars and symposia and to meetings of professional bodies.

AUS.79

University of Sydney, Department of Architectural Science
New South Wales 2006
Telephone: 692 2686
Telex: FISHLIB SYDNEY 20056
Date of establishment
1953

The Department of Architectural Science is part of the Faculty of Architecture within the University. The principal body is the Departmental Committee.
Main fields of work
The Department carries out undergraduate and postgraduate teaching and research in the field of the application of scientific knowledge to the requirements of buildings. This includes work on structural design, acoustics, lighting, thermal environment, building services and building materials. The Department specializes in model analysis for architectural structures for wind effects, natural lighting and sun control.

The Department also specializes in computer application to building and architecture through the use of simulation and optimization techniques; in particular, algorithmic design procedures applied to various building subsystems are being investigated.
Total staff: 20
Professionally qualified staff: 15
Director or Chief Executive
Professor Henry J. Cowan
Principal sources of finance
Principally the Australian Government through the Universities Commission.
Publications

- Departmental Reports in the subject areas of computers, structures, models, physical environment, education, general. These supersede the Bulletins which were reprints issued before 1970
- Newsletter, published twice a year
- Architectural Science Review (quarterly periodical) edited by the staff of the Department.

AUS.80

Sydney University, School of Civil and Mining Engineering
University of Sydney, Sydney, New South Wales 2006
Telephone: (02) 692 2136
Telex: AA 20056
Date of establishment
University incorporated by Act of the Legislature of New South Wales 1 October 1850. Engineering Department established in 1883.

A School in the Faculty of Engineering of the University providing teaching for BE degree in Civil Engineering, and in Mining Engineering MEngSc and PhD degrees, and carrying out research and professional investigation services.
Main fields of work
Structural engineering; geotechnical engineering; piling; concrete technology; stability of metal structures; wind loading on structures; mine economics; dust explosions.
Total staff: 66
Professionally qualified staff: 30
Director or Chief Executive
Head of School: Professor D. Campbell-Allen
Heads of department
Mining Engineering: Professor A.E. Jenkins
Geotechnical Engineering: Professor H.G. Poulos
Structural Engineering: Professor N.S. Trahair
Principal sources of finance
Government funding to University; external research grants
Principal related bodies
Civil and Mining Engineering Foundation providing industry support for School activities.
Publications
Lists of research reports are available on enquiry.

Research results are published in technical journals in Australia, the USA and Europe.

AUS.81

Western Australia University, Department of Civil Engineering

Nedlands, Western Australia 6009
Telephone: (092) 380 3072
Telex: AA92992

Date of establishment
1913

University department.

Main fields of work

Concrete and prestressed concrete structures; metal structures; structural systems; earthquake design; bases of design codes; theory of the mathematical models of structural design. The Department has always recognized a responsibility within the State of Western Australia as a source of specialist expertise in an isolated community. For many years it ran the only testing service for physical tests on construction materials and it is still the only calibrating authority for testing machines in the State.

Total staff: 30
Professionally qualified staff: 18

Director or Chief Executive
Head of Department: G.C. Reynolds

Principal sources of finance
Operating funds are part of University budget. Larger research projects are funded by research grants from other bodies.

Publications
The members of the Department publish in professional and scientific journals and their contributions to conferences and symposia appear in the conference publications; there is no in-house publication.

AUT.82 AM

Forschungsgesellschaft für Wohnen, Bauen und Planen (FGW)

Research Society for Housing, Building and Planning

Löwengasse 47, PO Box 164, A 1031 Wien

Telephone: 0222/72-62-51, 72-67-41

Date of establishment

Established in 1956 on the initiative of the Federal Ministry for Trade and Reconstruction and the Federal Ministry of Social Administration. The establishment of the society as well as its status were approved by the appropriate authority under sections 4 and 7 of the Federal Law on associations, BGB1 NR 233/1951

Non-profit-making association.

Main fields of work

Inter-disciplinary research in the field of housing.

Total staff: 296 collaborators, 70 supporters

Director or Chief Executive

Professor Ewald Liepolt

Heads of department

Division of Basic Research and Documentation: Ewald Liepolt, Professor

Technological Division: Ludwig Helmut Komoll; Alfred Pauser

Planning Division: Otto Ventruba; Hugo Potyka; Walter Blaha

Financial Division: Herbert Loidolt

Legal Division: Josef Hrebec; Hans Neuhofer; Elisabeth Langer

Principal sources of finance

Membership fees, proceeds from research work and the sale of publications.

Principal related bodies

Public authorities of the Republic, of the federal provinces (states) and of industry, represented on the Steering Committee.

Publications

- *Journal:* Wohnbauforschung in Österreich (WBFÖ), 6 editions a year; Österreichische Dokumentation für Wohnen, Bauen und Planen (ODW), 6 editions a year
- Monographs, serial publications, working papers

A list of publications is issued

AUT.83

Forschungsinstitut des Vereins der Österreichischen Zementfabrikanten

Austrian Cement Research Institute

Reisnerstrasse 53, A 1030 Wien

Telephone: 75 66 81–0

Date of establishment

1951

Research institute of the Austrian cement industry.

Main fields of work

- Official quality control of Austrian cement and other hydraulic binding agents according to the standards
- Research on cement and concrete and development of new fields of application for cement and concrete
- Dissemination of knowledge and research results gathered by the Institute, in Austria and elsewhere, through articles, lectures, consultancies, courses and seminars.

Total staff: 36

Professionally qualified staff: 14

Director or Chief Executive

Dr Hermann Sommer

Principal sources of finance

Cement industry

Principal related bodies

Association of Austrian Cement manufacturers

Publications

- *Journal:* Zement und Beton, quarterly
- Bulletins of the Austrian Cement Research Institute
- Annual list of reports and publications

AUT.84

Laboratorium für Kunststofftechnik am Technologischen Gewerbemuseum (LKT-TGM)

Laboratory for Plastics Technology at Technological Industrial Museum

Wexstrasse 19–23, 1200 Wien

Telephone: (0222) 35 2106-3

Telex: 13 18 24

Date of establishment

1957

A research and testing laboratory in the field of plastics technology.

Main fields of work

- *Training:* courses in plastics technology; special industrial development programmes in plastics technology; in-plant group training programmes (with UNIDO)
- *Research:* general research in plastics technology including building materials
- *Development:* applied R and D programmes.

Total staff: 60

Director or Chief Executive

Director: Professor Dr H. Hubeny

Principal sources of finance

Membership fees, research projects, industrial development projects, subventions.

Publications

Numerous papers are published in scientific journals.

AUT.85

Österreichisches Holzforschungsinstitut (ÖHFI)

Austrian Wood Research Institute

Arsenal, Franz-Grill-Strasse 7, A 1030 Wien

Telephone: (0222) 78 26 23

Date of establishment

1953

The Austrian Wood Research Institute is operated by the Austrian Wood Research Society.

Main fields of work

The whole field of chemical and mechanical technology of wood and wood products, including pulp and paper.

The principal divisions are: Division of Mechanical Technology of Wood (wood in the building industry, composite wood, adhesives, finishing, etc.); Division of Chemical Technology of Wood (chemistry of wood, glues and varnishes; pulp and paper; wood protection); Administration, Documentation and Information Division.

Total staff: 38

Professionally qualified staff: 25

Director or Chief Executive

Scientific Director: Professor Herbert Neusser

Commercial Director: Herbert Posch

Principal sources of finance

Receipts from research work, investigations and tests for industry, by research promoting fund, by governmental subventions and by membership fees of the Austrian Wood Research Society.

Publications

- *Journal:* Holzforschung und Holzverwertung, bi-monthly
- Schrifttumskarteidienst (Card Index Service), quarterly.

AUT.86

Österreichisches Institut für Bauforschung (ÖIB)
Austrian Institute for Building Research
An den langen Lüssen A 1190 Wien
Telephone: (0222) 32 57 88
Date of establishment
1959

A non-profit corporation.
Main fields of work
Research by the Institute ranges over all aspects of building and housing, with the emphasis on the reduction of building and maintenance costs, the rationalization of the building process, and the improvement of the quality of buildings and the built environment.

Recent and current research projects are the saving of energy in housing, building defects, the use of solar energy for room heating and warm water production, architectural competitions and organizational forms in the building industry.
Total staff: 14, of whom one third are support staff. The Institute also employs about 30 freelance scientists, architects, etc. for particular projects.
Director or Chief Executive
Scientific Director: Dr Michael Wachberger
Administrative Director: G. Spielmann
Principal sources of finance
Financed partly by membership fees of the firms and institutions which are members of the Institute, partly by research grants from the Austrian Ministry for Building and Technology and from the Ministry for Science and Research.
Principal related bodies
Board; Board of Trustees.
Publications
- ib-Nachrichten, the news bulletin of the Austrian Institute for Building Research, irregular.
- Jahresberichte, an annual report on the activities of the Institute
- Forschungsberichte, an irregular report on research projects of the Institute.

AUT.87

Österreichisches Kunststoffinstitut im Chemischen Forschungsinstitut der Wirtschaft Österreichs (OKI CFI)
Austrian Plastics Institute
Arsenal, Objekt 213, Franz-Grill-Strasse 5, A 1030 Wien
Telephone: (0222) 78 16 01
Date of establishment
1953

A research and testing institute authorized by the Austrian Government.
Main fields of work
Research and testing work in the whole field of plastics and composites, e.g. plastics in constructions (fibre reinforced plastics), plastics in buildings (plastic concrete a.s.o.), plastics in medicine (biocompatible materials), plastics in sports (skies, rackets). Elaboration of basic and practical principles for the application of plastics.
Total staff: 25
Professionally qualified staff: 20
Director or Chief Executive
Assistant Professor Otto Hinterhofer
Heads of department
Physics and Mechanical Technology: K. Binder
Chemical Technology and new products: Th. Fucik
Plastics in Medicine: O. Hinterhofer
Plastics in Buildings: M. Schmidt
Chemistry and Analytics: D. Loidl
Principal sources of finance
Private and governmental funds; directly paid services to industry.
Principal related bodies
Co-operation agreement with the Technical University of Vienna and Innsbruck.
Publications
- *Journal:* Österreichische Chemie-Zeitschrift, monthly
- Chemie – Kunststoffe – Aktuell Brochures

AUT.88

Physical-Technical Institute for Research on Heat and Sound Technology
Physikalisch-technische Versuchsanstalt für Wärme- und Schalltechnik
Wexstrasse 19–23, A 1200 Wien
Telephone: 33 92 36 or 35 35 11–411
Telex: 13 – 1824
Cables: Gewerbemuseum Wien
Date of establishment
1925

A public institution performing tests and research work upon request of all state, provincial and municipal services, civil services, industry and private individuals.
Main fields of work
Measurements and research in sound and heat insulation in building, technical and economical aspects, noise and noise abatement (traffic, machinery, industry), room acoustics.
Total staff: 10
Professionally qualified staff: 6
Director or Chief Executive
Dr Judith Lang
Principal sources of finance
Federal Government
Publications
Research work is published in different serials, journals, etc.

AUT.89

Technische Universität Graz. Institut für Bodenmechanik, Felsmechanik und Grundbau
Technical University of Graz, Institute for Soil Mechanics, Rock Mechanics and Foundation Engineering
Rechbauerstrasse 12, A 8010 Graz
Telephone: (0316) 77511 6230, 6231
Telex: 31221 tugraz a
Date of establishment
1964 (the Institute was previously connected with several others).

The Institute is a department within the Faculty of Civil Engineering.
Main fields of work
- Teaching: basic and advanced to provide the latest knowledge of new processes
- Soil testing: routine or special experiments, as necessary, are carried out
- Large-scale testing in the field
- Research: several theoretical and experimental investigations.

Total staff: 15
Professionally qualified staff: 8
Director or Chief Executive
University Professor, M. Fuchsberger
Principal sources of finance
Funded partly by governmental contributions, partly through soil testing or large-scale testing for the State or local authorities or for private persons.

Publications
- *Journal:* Mitteilungen, irregular
- *Research reports*, irregular

Many papers are published in various journals and international conferences. Some thesis papers are published by the Institute.

AUT.90

Technische Universität Wien, Institut für Hochbau und Industriebau, Abteilung Hochbau

Technical University of Vienna, Institute for Building Construction

Karlsplatz 13, A 1040 Wien

Telephone: (0222) 5601–3451

Date of establishment

1819

The Institute is part of the Technical University of Vienna.

Main fields of work
- Construction in civil engineering
- Fixing and jointing
- Building physics and acoustics
- Building information

Total staff: 11

Professionally qualified staff: 7

Director or Chief Executive

Director: University Professor Alfred Pauser

Principal sources of finance

The Institute is financed by the Ministry of Science and Research; additional 10% from research funds.

Publications
- *Research reports:* Veröffentlichungen der Abteilung Hochbau (Publications of the Institute), twice a year
- Publication of the Director, quarterly.

AUT.91

Technische Universität Wien, Institut für Strassenbau und Verkehrswesen

Technical University of Vienna, Institute for Road Construction and Transport Research

Gusshausstrasse 28–30, A 1040 Wien

Telephone: (0222) 56 01/4014

Date of establishment

1969, combined with Institute for Traffic Engineering in 1979.

Institute within the Faculty of Civil Engineering.

Main fields of work

Studies: lectures and exercises for students in civil engineering, geodesy and mechanical engineering

Science and research: work at the Institute is concentrated on the following: projects; alignment of roads depending on the dynamics of cars, driving properties and driving comfort; vehicle and road surface, especially evenness and skid resistance; traffic noise measurements, traffic analysis, traffic survey, traffic network planning, traffic prognosis, traffic assignment; some collaborators are working in several departments of the Austrian Highway Research Board.

Total staff: 14

Professionally qualified staff: 10

Director or Chief Executive

University Professor Paul Petrović

University Professor Josef R. Dorfwirth

Heads of department

Department for Road Construction: University Professor P. Petrović

Department for Traffic Engineering: University Professor J.R. Dorfwirth

Principal sources of finance

Regular endowments in the budget assigned for the University by the Ministry of Science and Research

Publications

Papers in the Oesterreichische Ingenieurzeitschrift

Schriftenreihe Strassenforschung of the Federal Ministry of Construction and Technology.

AUT.92

Versuchsanstalt für Baustoffe am Technologischen Gewerbemuseum (TGM VA Baustoffe)

Institute for Building Materials at the Technological Industrial Museum

Wexstrasse 19–23, A 1200 Wien

Telephone: (0222) 35 35 11

Telex: 13 – 1824

Date of establishment

1889

The Institution is governmental. The main purposes are:
- Testing all kinds of materials used for construction and equipment of buildings
- Testing during the development of new materials and new constructions
- Research projects of public interest
- Taking part in research for private industry
- Working out new standards in material and construction development
- Analysing deficiencies in building constructions
- Organizing seminars and the education of working staff.

Main fields of work
- *Testing:* building materials, building construction parts, special products such as pipes, building equipment, ceramics, floors, etc.
- *Building research:* construction problems and other special problems
- *Analysing deficiencies, e.g.:* water resistance of constructions, and all kinds of other properties.

Total staff: 8

Professionally qualified staff: 5

Director or Chief Executive

Chief: Professor Gerhard Kraml

Vice-chief: Professor Helmut Filz

Principal sources of finance

Governmental (Ministry of Education)

Principal related bodies

Ministry of Education (head of administration), Ministry of Building Construction (head in problem-related questions).

Publications
- Individual members' publications and results of research projects by the Ministry.
- Research for the Ministry of Building Construction published by the Ministry.
- Annual report of TGM.

BEL.93

Association Belge pour l'Etude, l'Essai et l'Emploi des Matériaux (ABEM)
Belgian Association for the Investigation, Testing and Use of Materials
Secretariat:
CSTC: rue du Lombard 41, 1000 Bruxelles
Telephone: (02) 511 06 83
Other address:
KUL, University of Leuven, Instituut voor Bouwkunde, De Croylaan 2, 3030 Heverlee
Telephone: 016/23.19.08
Date of establishment
1927

A non-profit association
Main fields of work
Aim: to initiate, encourage, manage and transmit research work leading to a better knowledge of building materials and all other materials notably by investigation and testing and in view of the use of these materials, to promote the study of all questions relative to determining their essential qualities.

Activities: ABEM organizes courses and conferences on subjects related to its social object, either directly, or under the auspices of groups and commissions which are attached to it.
Director or Chief Executive
Secretary (CSTC): H. Motteu
Secretary (Univ. Leuven): Professor D. Van Gemert
Publications
Communications addressed to the ABEM tribune.

BEL.94 AM

Association Royale des Gaziers Belges (ARGB)
Royal Belgian Gas Association
Avenue Palmerston 4, B 1040 Bruxelles
Telephone: (02) 230.43.85
Date of establishment
1877

Non-profit technical association of the piped gas industry in Belgium.
Director or Chief Executive
President: R. Bawin
General Secretary: R. de Tollenaere
Heads of department
Marketing: R. Vekeman
Laboratories: M. Desprets
Principal sources of finance
Gas industry subsidies.
Principal related bodies
Technical committees: Revue Générale du Gaz, Marketing and Information, Standards.
Publications

- *Journal:* Revue Générale du Gaz, quarterly
- Approval requirements for natural gas appliances in the domestic, commercial and industrial fields
- Manual USG (Utilisation Specifiques du Gaz)
- Chaudières AGB – Haut Rendement.

BEL.95 AM

Bureau de Contrôle pour la Sécurité de la Construction (SECO)
Office of Control for the Security of Construction
Rue d'Arlon 53, 1040 Bruxelles
Telephone: (02) 230.30.10
Telex: 61 723
Date of establishment
1934

A co-operative society, the members of which belong to the three branches of the building industry: architects, consulting engineers and contractors.

The aim of the society is to offer technical control of design, construction and materials of any type of construction and civil engineering works. Control is intended to reduce the risk of accidents and allows insurance coverage of builder's and owner's liabilities during the construction and during the 10 years' liability period.
Main fields of work

- Technical control of projects, calculations and execution of building construction and civil engineering works, such as buildings, industrial buildings, bridges, embankment walls, land locks. Controlled works include: foundations, concrete, reinforced concrete, masonry works, roofing, miscellaneous finishing works
- Control of building services (electricity, elevators, central heating and air-conditioning, sanitary installations)
- Quality control of manufactured materials for the building industry: this control allows the delivery of quality certificates (in particular BENOR and Agrément)
- Research on problems relative to the technical control of constructions
- Within the framework of l'Union Belge pour l'Agrément technique dans la construction, UBAtc, SECO controls operations linked to the technical agrément in Belgium.

Departments are organized on the following lines:

Three divisions in charge of the control of buildings and civil engineering works; one division in charge of the control of metal construction and non-corrosive paint applications; one division in charge of the control of material quality and quality certification; one specialized service for works equipment; general services.
Total staff: 140
Professionally qualified staff: 100
Director or Chief Executive
Chairman of the Board: R. Maes
Chairman of the Technical Board: Professor Dr ir D. Vandepitte
General Manager: J.R. Richelle
Principal sources of finance
Fees from control of works and materials.
Publications
Articles in technical reviews, conferences, congress communications, seminars.

BEL.96

Centre Belge d'Etude de la Corrosion (CEBELCOR)
Belgian Centre for Corrosion Study (CEBELCOR)
Avenue Paul Héger, grille 2, B 1050 Bruxelles
Telephone: 32 2 6496396 and 6493516
Telex: 23069 b UNILIB
Date of establishment
11 July 1951

A private centre specializing in corrosion research, consulting, testing, education and international co-operation.

Research is carried out under contract from government or international bodies, industrial groups or industries.
Main fields of work
Metallic corrosion, electrochemistry, electrochemical thermodynamics. Mechanisms of different types of corrosion. Behaviour of material–environmental systems; atmospheric corrosion, steel in concrete, underground corrosion, roofs, siding, heating and water supply systems. Studies, research, advice, testing. Principal divisions: Research, Consulting.
Total staff: 6
Professionally qualified staff: 3

Director or Chief Executive
Director: Antoine Pourbaix
Principal sources of finance
Research contracts (national or international bodies, industries or groups of industries); consulting; membership fees.
Principal related bodies
Board of directors, general assembly.
Publications
Rapports Techniques CEBELCOR
Research reports, conferences, discussions etc., issued irregularly (free to members).

BEL.97

Centre Belge d'Etude et de Documentation de l'Eau (CEBEDEAU)
Rue A. Stévart 2, B 4000 Liège
Telephone: (041) 52.12.33
Date of establishment
1947

CEBEDEAU is a non-profit association and a centre of basic and applied research, housed at the University of Liège, which studies the environment and its problems from different aspects.
Main fields of work
Activities include water control, health and sanitation studies, diagnosis of water treatment and sewage plants, waste treatment, rivers and waterways pollution, interaction of water with materials, and teaching and training.
Total staff: 15
Professionally qualified staff: 6
Director or Chief Executive
Director: F. Edeline
Heads of department
Water and Waste Water Department: L. van de Venne
River Department: F. Edeline
Corrosion and Waste Department: J. Hissel
Information Department: D. van den Ackerveken
Principal sources of finance
Self-supporting. CEBEDEAU works with numerous organizations, some private (50%) and some public (50%).
Publications
- *Journal:* La Tribune du CEBEDEAU, international monthly; Environnement & Pollution, bi-monthly bulletin.
- Monographs.

BEL.98

Centre National de Recherches Scientifiques et Techniques pour l'Industrie Cimentière (CRIC)
National Centrum voor Wetenschappelijk en Technisch Onderzoek der Cementnijverheid
National Cement Industry Research Centre
rue César Franck 46, 1050 Bruxelles
Telephone: 02/649.98.50
Date of establishment
1961: application of the Law of 30 January 1947

Private industrial technical centre founded by the cement manufacturers at the disposal of the cement industry, public services and private individuals.
Main fields of work
Control, tests and studies in the field of cement and concrete.
Total staff: 34
Professionally qualified staff: 15
Director or Chief Executive
President: J. Urlings
Director: Y. Dechamps
Heads of department
Documentation: A. Mommens
Laboratory: J. Plisnier
Research: Y. Dechamps
Principal sources of finance
Levy on cement manufacturers; research grants from the public authority; invoicing clients.
Principal related bodies
Technical Committees: supervision and policy of the research activities.
Publications
- Research reports (51 up to 1982)
- Copy of the CRIC documentary card system and index of authors and key words, monthly
- Annual report
- Specially published works (sometimes in co-operation with other Belgian organizations)
- Reprints from journal papers or symposium proceedings.

BEL.99 FM

Centre Scientifique et Technique de la Construction (CSTC)
Wetenschappelijk en Technisch Centrum voor het Bouwbedrijf (WTCB)
Belgian Building Research Centre
Headquarters:
Rue du Lombard 41, B 1000 Bruxelles
Telephone: 02/511.06.83
Telex: 25.416 CETEX B
Experimental station:
Avenue P. Holoffe, B 1342 Limelette
Telephone: 02/653.88.01
Telex: 25.682 CETEX B
Date of establishment
Founded in 1959 by the National Building Confederation. Recognized by application of the Law of 30 January 1947.

An industrial technical centre, created to promote the technical progress of contracting firms by all means including research, documentation, assistance, education, etc.

The Centre is organized in three executive departments: Research and Development (at the Experimental station, Limelette); Management of building companies and sites (Operational Research); Information (advisory service).
Main fields of work
- Research projects in the following areas: technology of the construction of buildings and civil engineering works; user needs and building performance; environmental and natural resource protection; energy savings
- Development assistance to contractors and manufacturers; technical agrément (through UBAtc: Union Belge pour l'Agrément technique dans la construction, created jointly with INL and Bureau SECO)
- Building site organization, working time study, cost establishment and control.
- Technical assistance to contractors, technical–commercial information bank on construction materials, scientific and technical documentation
- Courses and seminars
- International co-operation, notably with CIB, CEB, UEAtc, RILEM, FIEC, EEC.

Total staff: 100
Professionally qualified staff: 50
Director or Chief Executive
President: P. Willemen
Director: A. Legros
Principal sources of finance
A levy on contractors, state grant (through IRSIA) and sale of services.
Principal related bodies
General Council; Permanent Committee; Technical Committees (build-

ing structures and civil engineering; heating and air-conditioning; insulation; sanitary and industrial plumbing; painting; woodwork, shutters; glass; wall and floor covering; watertightness; stone and marble; operational division); Working Groups.

Publications

- CSTC Magazine, quarterly. A review of general information for disseminating the results of Belgian and foreign research and for publishing the annual report
- Technical information notes (NIT), 145 published
- Study and research reports, 27 published
- Articles in the weekly paper La Construction
- Miscellaneous publications, notably: General specifications for private buildings, technical clauses, Building site planning, Preparation and control of building sites, Practical guide for contractors, Fire resistance of building elements, Recommendations for the calculation and execution of industrial buildings in prefabricated concrete elements, R-D Construction 1973, Guide of Performance, Deep foundations.

BEL.100

Coatings Research Institute (CoRI)
Avenue P. Holoffe, B 1342 Limelette
Telephone: (2) 653 09 86
Date of establishment
March 1957

A private institution whose aim is to promote scientific development and technical progress in the field of surface coatings, paints and inks.

CoRI is a non-profit organization, supported by both the public and the private sector.

Main fields of work

- Research: Collective research for members of the Institute: anticorrosion, industrial paints (high solids, solventless), new technologies; Confidential research contracts; Improving existing products
- Technical assistance: Technological guidance; Documentation service; Bibliographic studies
- Staff training: Scientific and technological training seminars, in Belgium and foreign countries; Paint courses for industrial engineers
- Analysis, testing, expertise: Analysis and testing in accordance with interantional standards; Formulation and control of technical specifications; Consultancy fo insurance purposes.

Total staff: 20
Professionally qualified staff: 12
Director or Chief Executive
Managing Director: Dr P. Janssen Bennynck
Heads of department
Research Divisions: Dr M. Piens, Dr D.Y. Perera, Dr A. Toussaint, Dr P. Cuypers
Staff Training: Dr A. Toussaint
Technical Assistance: Ing R. Haegeman
Principal sources of finance
Private sector 50%, public sector 50%.
Publications

- CORI-INFO, free of charge
- Research report, annually, for members only.

BEL.101* FM

National Housing Institute (NHI)
Institut National du Logement (INL)
National Instituut voor de Huisvesting (NIH)
Boulevard Saint-Lazare 10, 1030 Bruxelles
Telephone: (02)218 45 70
Date of establishment
27 June 1956

The Institute is a semi-public organization, established by the Law of 27 June 1956, incorporated in the Housing Code of 10 December 1970.

Main fields of work

Field of work based on the Law of 27 July 1956:

- To keep a permanent inventory of the population's housing needs and study the means of satisfying these needs
- To propose housing standards with the help of competent public services
- To study public health, security, sanitation, insulation–conservation of energy and equipment problems relating to housing
- To follow, promote and eventually charge to third parties, studies and technical research in order to reduce the cost of construction and improve quality of design
- To create a documentation and information centre
- To provide at his request or with his agreement whoever in Belgium is concerned with housing construction the necessary advice in order to realize the best possible housing conditions
- To advisę on all questions relating to housing, either at its own initiative or by Government request
- To subsidize public building societies in order to experiment with building methods or new architectural conceptions and urban planning

Field of work based on the Decree of 18 July 1970:

Technical agrément (the instruction of which is organized within the limits of L'Union Belge pour l'Agrément technique dans la Construction, UBAtc).

Secretaryship of the Belgian Committee for CIB

Director or Chief Executive
Acting Secretary-General: I. Jacques Bouillon
Acting Deputy Secretary-General: I. Vanderstraeten
Heads of department
Management: I. Vanderstraeten
General Services: Mrs Baecke, Director
Research Department (legal, sociological, etc.): Speltinckx, Director
Information Department: Despontin, Director
Technical Department: Franssens, Eng Chief Director
Local Services
Statistical Department: Van Impe
Principal sources of finance
Annual deduction of the sums invested by social housing companies; public subsidies and income allowances.
Principal related bodies
The Supreme Council, which manages the Institute and advises on all questions relating to housing, either on its own initiative or by Government request, on the basis of reports from special sections; the Managing Committee, which executes those decisions of the Supreme Council delegated to it.
Publications

- Magazines Habiter and Wonen, quarterly
- Information Bulletin, monthly
- Documentation booklets, including a technical series covering notably reports on studies, directives for the technical agrément of the UBAtc and UEAtc
- STS: Unified technical specifications (for conditions of contract)
- Technical Agréments and information sheets
- Catalogues of the library of the National Housing Institute.

* At the time of going to press, notification was received of the official disbandment of this institute and its reconstitution in another body.

BEL.102

Institut National du Verre (INV)

National Glass Institute

Boulevard Defontaine 10, B 600 Charleroi

Telephone: (071) 31 00 41

Telex: 51430 INAVER B

Date of establishment

1947

Non-profit association which aims to promote the development of basic and applied research for the benefit of the Belgian glass industry.

Main fields of work

Research and test activities requested by the principal Belgian glass-producing and manufacturing companies.

Principal subjects of research: thermo-physics of glassmaking processes; physical and chemical properties of glass and its surface; mathematical modelling of industrial processes; statistical analysis of production data; fusion and refining of white and coloured glass; use of the computer to optimize compositions, properties; study of redox properties associated with dissolved gases; study of all poblems associated with the physics and chemistry of the surface of glass; strengthening surface treatment; techniques for lubricating the surface of hollow glass; manufacture of mirrors; homologation of automotive glass (hardened and laminated) and double glazing.

There is also a documentation department which participates in a pool of glass abstracts and produces information profiles for the Belgian glass industry.

Total staff: 60

Director or Chief Executive

President: M. t'Serstevens

Director: M. Verburgh

Principal sources of finance

Financed by the Belgian glass industry. Receives the support of the Institut Belge pour l'Encouragement de la Recherche Scientifique dans l'Industrie et l'Agriculture, and own receipts.

Principal related bodies

Management Committee; Board of Directors; General Assembly.

Publications

Research reports

Co-operates in a pool of glass abstracts.

BEL.103

Institut pour l'Encouragement de la Recherche Scientifique dans l'Industrie et l'Agriculture (IRSIA)

Instituut tot Aanmoediging van het Wetenschappelijk Onderzoek in Nijverheid en Landbouw (IWONL)

Institute for the Promotion of Scientific Research in Industry and Agriculture

Rue de Crayer 6, 1050 Bruxelles

Telephone: (02) 648 81 45

Date of establishment

1945

A public organization whose main mission consists in encouraging, promoting and generating by financial aid applied research activities leading to an improvement in the technical level and hence the economic situation of Belgian industry and agriculture. It also provides research, specialization and travel scholarships. These grants contribute exclusively to the training of research workers specializing in industrial and agricultural technology.

Main fields of work

The efforts of the Institute are orientated to all branches and areas of industrial and agricultural activity in Belgium.

Total staff: 64

Professionally qualified staff: 22

Director or Chief Executive

Director: J. van Keymeulen

Secretary-General: A. Marchal

Heads of department

Industry Department: M. Barthels

Agriculture Department: J. Ferange

Principal sources of finance

Funds are made available by the Ministry of Economic Affairs for the industrial orientated projects and by the Ministry of Agriculture for the agricultural orientated projects.

Principal related bodies

Board of Directors, made up of 18 members nominated by the King, 5 of whom are proposed by industrial companies, 3 by agricultural organizations, 5 by the Trade Unions, and 5 are elected among the Deans of Faculties for Applied Science and of Agriculture Faculties or Institutes of the Universities or similar schools.

Publications

Reports on programmes subsidized by the Institute and not published in scientific reviews

Annual report containing a table of all the subsidies accorded during the year, as well as notes assigned to each of the subsidized programmes.

BEL.104 AM

Katholieke Universiteit te Leuven, Departement Architektuur, Stedebouw en Ruimtelijke Ordening

Catholic University of Leuven, Department of Architecture, Urban and Regional Planning

Kasteel Arenberg, Kardinaal Mercierlaan 94, B 3030 Heverlee

Telephone: (016) 22 09 31 ext. 1361

Telex: 25941 elekul b

Date of establishment

Since 1929 (by law)

The Department of Architecture is part of the University of Leuven, and also carries out testing and research upon request.

Main fields of work

- Education
- Research: building physics; rationalization of the design and building process; conservation of historic towns and buildings; human settlements in developing countries; theory and history of architecture
- Urban and regional planning.

Total staff: 34

Professionally qualified staff: 34

Director or Chief Executive

President of the Department: Professor Jan Delrue, Dr Engineer Architect

Head of School of Architecture: Professor Herman Neuckermans, Dr Engineer Architect

Head of School of Urban and Regional Planning: Professor Louis Albrechts, Dr Urban Planning

Heads of department

Post graduate centre: Human Settlements: Dr H. Verschure

Centre for the Conservation of Historic Towns and Buildings: Professor R. Lemaire

Laboratory for Building Physics: Professor H. Hens

Institute of Town and Regional Planning: L. Albrechts

Principal sources of finance

University; Government.

Publications

- Research work results in specialized reviews.
- Scientific Report (1979–1983) in English.

BEL.105

Katholieke Universiteit Leuven, Departement Bouwkunde
Catholic University of Leuven, Civil Engineering Department
Park van Arenberg, de Croylaan 2, B 3030 Leuven, Heverlee
Telephone: (016) 23.19.09–23.19.08
Date of establishment
1932; 1975 (hydraulics)

The Materials Testing and Hydraulics Laboratories are part of the Civil Engineering Department of the University. Tests requested by industries are carried out in the Laboratories through the administrative and financial intermediary of Leuven Research and Development, a non-profit association. In addition, a Research Unit for Traffic Engineering and Planning and an integrated Research Unit for Numerical Techniques and Graphic Systems are active. The activities of the Department and its laboratories can be summarized as: Education; Fundamental and applied fundamental research; Research for industries and administrations; Testing of materials, structures and constructions.

Main fields of work

- Experimental and theoretical research into the static, dynamic and stability behaviour of structures; shell and space structures; numerical techniques and application of finite elements and boundary elements; foundation behaviour in earthquake computations.
- Physical and mathematical modelling of: river quality (e.g. thermal pollution); the storm effect on the North Sea and the river Scheldt; flow in open channels and canal networks (e.g. sewer networks); hydraulic transients in pipeline systems
- Theoretical and experimental research on reinforced and prestressed concrete; shrinkage and creep of concrete, concrete and reinforcement in cryogenic circumstances, endblocks, limit state analysis, deformation of concrete beams; design of concrete and steel structures
- Research into amelioration of the road infrastructure quality; development of survey and data aquisition methods using microcomputer and video technology; research into the operational and environmental aspects of main infrastructure projects
- Experimental and theoretical studies related to the structural restoration or strengthening of masonry and reinforced concrete structures; injection of soils; experimental and numerical analysis of composite materials
- Numerical methods in structural analysis; dynamic non-linear behaviour of structures; plastic design of steel frames; heat transfer

Total staff: 65
Professionally qualified staff: 38
Director or Chief Executive
President: Professor J. Berlamont
Heads of department
Reyntiens Laboratory for Materials Testing: Professor F. Mortelmans; Professor D. Van Gemert
Hydraulics Laboratory: Professor J. Berlamont
Numerical Techniques and Computer Aided Design: Professor M. Van Leathem; Professor G. De Roeck
Traffic Engineering and Infrastructure Planning Unit: Professor J. Mortelmans
Principal sources of finance
University of Leuven; Governmental research contracts; private firms
Publications

- Internal Reports
- Contributions to national and international Journals
- Communications at international congresses
- PhD Theses, Masters' Theses.

BEL.106

Katholieke Universiteit Leuven, Laboratorium voor Akoestiek en Warmtegeleiding
Catholic University of Leuven, Acoustics and Heat Conduction Laboratory
Celestijnenlaan 200 D, B 3030 Heverlee
Telephone: (016) 20 10 15
Date of establishment
1946

University Laboratory consisting of academic, scientific (permanent and temporary), technical and administrative staff and research personnel and students.

Main fields of work

- All aspects of acoustics and noise abatement
- Heat conductivity of insulating and building materials
- Water vapour transmission
- Co-operation with other national and international research institutions
- Participation in the national and international standardization of measuring methods.

Total staff: 15
Professionally qualified staff: 5
Director or Chief Executive
Professor Dr H. Myncke; Dr A. Cops
Principal sources of finance
Financed by the Ministry of Education and from sponsored research and measurements for industry and official departments.
Publications
In various scientific and technical reviews.

BEL.107

Rijksuniversiteit te Gent, Laboratorium De Meulemeester voor Technologie der Textielstoffen RUG
State University of Ghent, De Meulemeester Laboratory for Textile Technology RUG
St Pietersnieuwstraat 41, 9000 Gent
Telephone: (091) 23 38 21
Telex: RUGENT 12754
Date of establishment
1929

The laboratory is a part of the Faculty of Applied Science of the State University of Ghent. It is at the disposal of, and is recognized by, the state services, the textile industry and private individuals, and it performs testing upon request.

Main fields of work
Textiles: apart from fibre, yarn and fabric testing, the laboratory is also equipped to test and classify carpets according to properties such as wear, durability, propensity for static electricity, thermal isolation, etc. The laboratory is also equipped to perform tests in the field of flammability with small sources of ignition. The laboratory makes arrangements to perform tests on geotextile products.
Total staff: 13
Professionally qualified staff: 4
Director or Chief Executive
Head of the Laboratory: Dr ir T. Fransen
Heads of department
Industrial testing: chemical and physical: E. van Cleemput
Carpet testing and flammability: Not available at time of going to press
Research: L. Verschraege
Principal sources of finance
The State; tests carried out on a fee basis; funds for research projects.
Principal related bodies
Administration council of the State University.
Publications
The laboratory works actively on establishing standards which are

published by the Belgian Normalisation Institution, Brabançonnelaan 29, 1040 Brussels.

BEL.108 **AM**

Rijksuniversiteit te Gent, Laboratorium voor Aanwending der Brandstoffen en Warmte-overdracht

State University of Ghent, Laboratory for Fuel Technology and Heat Transfer

St. Pietersnieuwstraat 41, 9000 GENT

Telephone: 091 233821 2334

Date of establishment

1957

University laboratory for thermal research in the Department of Engineering of the State University of Ghent. Besides research, routine testing is performed on request.

Main fields of work

- *Heat transfer:* heat conduction in solids: measuring techniques for heat resistance and thermal diffusivity
- *Fuel technology:* determination of the combustion parameters of the different kinds of fuel that are available at this moment: calorific value, viscosity, hardness, etc.
- *Fire research:* routine tests on building materials and structures; research on fire safety in new and existing buildings; research on the fire behaviour of electrical wiring of all kinds.

Total staff: 26

Director or Chief Executive

Director: Professor ir R. Minne

Heads of department

Fuel Technology: Dr ir N. Willems

Heat Transfer: Ir E. Tavernier

Fire Safety: Dr ir P. Vandevelde

BEL.109

Rijksstation voor Landbouwtechniek

National Institute of Agricultural Engineering

Burg. van Gansberghelaan 115, 9220 Merelbeke

Telephone: 091 52 18 21

Date of establishment

1951

Governmental research institute.

Main fields of work

- Layout and equipment of animal houses
- Environmental research in livestock breeding
- Water management and field drainage in agriculture
- Climate control in glasshouses

Professionally qualified staff: 6

Director or Chief Executive

Director: Dr ir A. Maton

Principal sources of finance

Ministerial budget (agriculture).

Publications

Scientific communications on animal housing, manure processing, field drainage and climate control in glasshouses

Book on animal housing (600 pp).

BEL.110

Rijksuniversiteit te Gent, Faculteit van de Toegepaste Wetenschappen, Laboratorium voor Vervoertechniek

State University of Ghent, Faculty of Engineering, Laboratory of Transportation Technics

St. Pietersnieuwstraat 41, B 9000 Gent

Telephone: (091) 23 38 21

Telex: 12.754 RUGENT

The Laboratory is part of the Faculty of Engineering of the State University of Ghent and performs testing upon request.

Main fields of work

- Measurement of airtightness, watertightness and mechanical resistance of windows and curtain walls, in accordance with all national and international standards, with delivery of a test certificate
- Measurement of airtightness, thermal and mechanical ageing of air-registers and smoke valves
- Research on the behaviour of windows at extreme temperature differences between the inside and the outside
- Measurement of the coefficient of heat transmission of windows and other building elements
- Endurance tests on the fastening of windows.

Total staff: 5

Director or Chief Executive

Director: Professor Dr Ir Ch. van Aken

Principal sources of finance

Tests commissioned by industry.

Principal related bodies

Board of Directors of the State University of Ghent.

BEL.111

Rijksuniversiteit te Gent, Laboratorium Magnel voor Gewapend Beton

Ghent State University, Magnel Laboratory for Reinforced Concrete

Grotesteenweg Noord s, B 9710 Gent (Zwijnaarde)

Telephone: (091) 22 57 55

Date of establishment

1926

The Laboratory is attached to the Chair for reinforced and prestressed concrete of the Faculty of Applied Sciences at the Ghent State University and its role is threefold: education, research and work as a public institution for testing building materials.

Main fields of work

- Teaching of concrete technology and technology of other building materials, calculation of reinforced and prestressed concrete structures
- Research work in the field of concrete, reinforced concrete and prestressed concrete
- Experimental studies, measurements, controls and acceptance tests for the industry and public services, in the Laboratory and on the building site
- Participation in national and international standardization of test methods and design codes.

Total staff: 34

Professionally qualified staff: 9

Director or Chief Executive

Director: Professor Ir I I. Lambotte

Principal sources of finance

Allowances from the Ministry of Education; subsidies from research organizations; fees for tests on behalf of the industry and public services.

Publications

Articles in specialized Belgian and foreign reviews.

BEL.112

Rijksuniversiteit te Gent, Laboratorium voor Modelonderzoek

Ghent State University, Laboratory of Model Research
Grotesteenweg Noord 2, 9710 Zwijnaarde
Telephone: (091) 22 57 55
Date of establishment
1958 – Ghent University

The research laboratory is part of the Faculty of Applied Sciences of Ghent State University and performs testing upon request.

Main fields of work

- Problems of instability of structures (mainly of shells)
- Tests on models of structures and parts of structures
- Tests on full-size steel structures
- Tests on structures in situ.

Total staff: 15
Professionally qualified staff: 7
Director or Chief Executive
Director: Professor Ir D. Vandepitte
Principal sources of finance
Ministry of Education; National Fund for Scientific Research; Institute for Scientific Research in Industry and Agriculture.
Publications
Numerous papers in scientific journals.

BEL.113 FM

Société Nationale du Logement (SNL)

Nationale Maatschappij voor de Huivesting (NMH)
National Housing Society (NHS)
Breydelstraat 12, 1040 Bruxelles
Telephone: 02/230 51 25
Date of establishment
11 October 1919

A public organization but not directly founded by law. It has the form of a joint stock organization.

The statues were approved by royal decree of 22 April 1920 in application of the law of 11 October 1919 at present abrogated and included with the Housing Code, stipulating that the Government was authorized to approve the statutes of the National Housing Society.

Main fields of work

The Society is a public service, aiming to build groups of social dwellings for rent or for sale, for people of modest means. This is carried out with the help of local or regional building societies. Normally the homes are not built by the Society but by its approved building societies. However, law authorizes the Society itself to build dwellings, these being reserved for members of the personnel of new industries or industries that are expanding.

The Society is essentially a financing organization; it collects and distributes among its approved building societies the necessary capital to build these social dwellings. It should be noted that neither the Society nor its associated societies provide loans to individuals. The Society can effect all operations connected with its goal, as well as all other aims relating to the construction of social dwellings, for which it will be authorized by law.

Total staff: 221
Professionally qualified staff: 38
Director or Chief Executive
President: W. D'Havé
Director-General: H. De Smet
Assistant-Director General: H. Vicaise
Heads of department
Administrative management: P. Du Bois
Financial management: M. Corbisier
Technical management: E. Vanderelde
Principal sources of finance
Generally financed by issuing state-guaranteed private or public loans. The Society also has a policy of self-financing by the reinvestment of earnings from the sale of its houses.
Principal related bodies
Board of Directors: execution of government housing policy; Government commissioners: government control; Supervisory committee: financial control; Management Council (director general, assistant director general and three department directors): daily management.
Publications

- Annual Report
- Information Bulletin, quarterly
- General and special specifications
- Technical instructions
- Memorial volumes.

BEL.114

Université Catholique de Louvain, Bureau d'Etudes (CREAT)

Louvain Catholic University, Centre for Applied Studies
Bâtiment Vinci, Place du Levant, 1, B 1348 Louvain-la-Neuve
Telephone: 010 41 81 81, ext. 2134
Date of establishment
1965

CREAT is a research body of the teaching unit for Architecture at the Faculty of Applied Science.

CREAT specializes in local planning and also conducts theoretical research on planning detail. Its practical experience constitutes an indispensable source of raw material for teaching in the subject.

Main fields of work

Urbanistic consulting, local planning detail, urban–rural and industrial renovation. Studies on energy conservation, transportation systems, urbanistic practice, public facilities, etc.

Total staff: 8
Professionally qualified staff: 6
Director or Chief Executive
Director: P. Govaerts
Principal sources of finance
University and government services.
Publications
Préoccupation Energétique en Urbanisme-Service de la Programmation de la Politique Scientifique, Bruxelles, 1981.

BEL.115

Université Catholique de Louvain, Centre de Recherche en Architecture (CRA)

Catholic University of Louvain, Centre for Architectural Research (CRA)
Bâtiment Vinci, Place du Levant 1, B 1348 Louvain-la-Neuve
Telephone: (010) 41 81 81 ext.2341
Telex: 59037 UCL B
Date of establishment
1967

Research body of the teaching unit for Architecture at the Faculty for Applied Science.

Main fields of work

CRA tries to reply to pedagogic and scientific questions in architectural education and to resolve the problems encountered in the practice of architecture.

Research projects in architecture, town planning, building engineering, energy saving and alternation technologies.

Field of application: local and third world.

Total staff: approx. 30
Professionally qualified staff: 23

Director or Chief Executive
J.F. Mabardi, Professeur ordinaire UCL
Principal sources of finance
University, public and industrial research funds, Government services
Publications
Articles in several reviews
Research reports:

- Concevoir avec l'industrialisation de la construction (Proposal for a building materials policy)
- Element vitré dans le cadre de la climatisation naturelle
- Experiences conflictuelles de l'espace (IAPC)
- Architecture moderne en Belgique
- Analyse et proposition pour un village algérien (Ait Ali ou Mhend) (7 tomes)
- Architecture hospitalière (évolution du plan à travers l'histoire) (3 volumes).

BEL.116

Université de l'Etat à Liège, Centre de Recherches d'Architecture et d'Urbanisme (CRAU)
State University of Liege, Research Centre for Architecture and Town Planning
Quai Banning 6, 4000 Liège
Telephone: (041) 52 01 80, ext.292
Telex: 41397 UNIVLG
Date of establishment
1967

Research Centre in the framework of the architectural branch.
Main fields of work
Research either in perfecting new methods applicable to architecture and town planning, both past (renovations) and future, or by building up original construction methods or techniques with the help of traditional or new materials
Services in favour of third parties.
Total staff: 15 permanent plus personnel from the Architecture Section.
Director or Chief Executive
Directors: Professor J. Englebert and Professor F. Peters
Principal sources of finance
The University, the Ministry of French Culture and industry
Publications
Non-periodical.

BEL.117

Université de l'Etat à Liège, Faculté des Sciences Appliquées, Institut du Génie Civil, Services de Mécanique des Matériaux et Statique des Constructions et Mécanique des Structures
Liège State University, Faculty of Applied Sciences, Institute of Civil Engineering, Department of Strength of Materials and Structural Mechanics
Quai Banning 6, B 4000 Liège
Telephone: (041) 52 01 80
Telex: 41397 UNIVLG
Date of establishment
Services established in 1947

University Department belonging to the Faculty of Applied Sciences in charge of teaching: Applied statics and kinematics; Strength of materials; Theory of elasticity; Solid mechanics; Theory of structures; Theory of elastic stability; Computer analysis of structures; Paraseismic design. Research activities and collaboration with industry in the same areas.
Main fields of work

- *Education:* courses, practical work and laboratory sessions
- *Laboratory:* study of structures and their constituent parts: full scale or model size. The experimental facilities allow a combination of static or dynamic forces in any direction
- *Computing:* development and exploitation of programs on behalf of research organizations, consulting firms and administrations; implementation and maintenance on users' computers; training of personnel in computer analysis of structures; computer-aided design.

Total staff: 29
Professionally qualified staff: 29
Director or Chief Executive
Professor G. Fonder and Professor R. Maquoi
Principal sources of finance
Long-duration research contracts (1–4 years); laboratory experiments and/or computer analyses for specific objects; minor state subvention.
Principal related bodies
National Centre for Research in Metallic Construction (CRIF)
Publications
Lecture notes, students' final projects, technical notes, internal research reports, communications to various colloquia, publications in national or international journals.

BEL.118

Université de l'Etat à Liège, Laboratoire de Physique du Bâtiment
Liège State University, Building Physics Laboratory
15 avenue des Tilleuls, Bât. D1, B 4000 Liège
Telephone: (041) 52 01 80, ext. 367
Telex: 41746 enviro b
Date of establishment
September 1977

The Laboratory is part of the Faculty of Applied Sciences of the State University of Liège. It carries out research works requested by research institutions.
Main fields of work

- Any research work in thermodynamics and architecture related to the knowledge of physical and basic human phenomena occurring inside and outside actual buildings
- Research on classical and non-conventional climatization equipment
- Building stock description; audit; typology; data banks
- CAD applied to architecture and town planning
- Experimental and theoretical studies (simulation) of thermal systems

Total staff: 30
Professionally qualified staff: 20
Director or Chief Executive
Professor A. Dupagne
Heads of department
Climatization Equipment Division: J. Hannay
Thermophysical Division: J. Lebrun
Architectural Division: A. Dupagne
Principal sources of finance
Services de Programmation de la Politique Scientifique (SPPS); CEE: DG XII, DG III; IRSIA.
Principal related bodies
Applied Sciences Faculty Council; University Council.
Publications
A complete list of reports and publications of the Laboratory is issued every 6 months in the brochure LPB News (obtainable from the Laboratory on request).

BEL.119

Université de l'Etat à Liège, Laboratoires d'Essais du Génie Civil

State University of Liège, Civil Engineering Test Laboratories
Quai Banning 6, 4000 Liège
Telephone: (041) 52 01 80; (041) 52 21 60
Telex: 41488 B
Other address:
Route de Saint-Hubert, 6603 Recogne-Libramont
Telephone: (061) 22 32 91
Date of establishment
1926

The Laboratories have the following departments: Hydraulic construction and applied hydraulics laboratory; Geotechnics and rock mechanics laboratory; Chemistry, paint and hydrocarbon materials laboratory; Bridge and structural engineering laboratory; Building materials (including plastics) laboratory; and the following centres: Documentation; Publications.

Main fields of work

- International relations
- Teaching activities
- Research activities (theoretical and experimental), studies related to civil engineering problems: infrastructure, soil mechanics, rock mechanics, bridges, structural engineering, reinforced concrete, prestressed concrete, welding, metal constructions, buidling material, waterproofing, hydraulic, river and marine constructions, consultant mission, expertise, etc.

Total staff: 122
Professionally qualified staff: 37
Director or Chief Executive
President: Professor A. Lejeune
Vice-President: Professor R. Baus
Members: Professor F. Bonnechere, Professor W. Chapeau, Professor N.M. Dehousse, Professor V. de Kosinsky and Professor A. Fagnoul
Heads of department
Hydraulics: N.M. Dehousse, A. Lejeune, V. de Kosinsky
Geotechnics: A. Fagnoul, F. Bonnechere
Chemistry, and Building materials: J.M. Rigo, R. Degeimbre, J.C. Dotreppe
Bridges: R. Baus, W. Chapeau
Fire Resistance: J. Bertrand
Principal sources of finance
Financed by fees paid by Authorities, industries, public and private services, companies and private individuals who use the Laboratory services, under short or long-term contracts; also by research subsidies through IRSIA, the National Fund for Scientific Research (FNRS) and the Fund for Collective Fundamental Research (FRFC).
Publications

- Memoirs of the Centre for Scientific Studies, Research and Testing of the University of Liège (CERES)
- Collection of the Faculty for Applied Science of the University of Liège
- All reviews concerned with construction and its connected sciences
- Staff members' publications
- Seminar.

BEL.120

Université de l'Etat à Liège, Service de Résistance des Matériaux et Statique des Constructions

State University of Liège, Department of Strength of Materials and Theory of Structures
Institut du Génie Civil, Quai Banning 6, B 4000 Liège
Telephone: (041) 52 01 80, ext. 238
Telex: 41397
Date of establishment
1950

The laboratory is a public institution in which the state, provincial and municipal services, as well as civil services, industry and private individuals can find some guidance for the art of construction or perform tests in fields related to the strength and the stability of structures.

Main fields of work

Theoretical and numerical: static and dynamic analysis of structures (small strains and displacements, elastic behaviour); rigid-plastic analysis of structures (small strains); non-linear analysis of structures, including effects of plasticity and of large displacements; analysis of rubber-type materials and structures (large displacements, large strains); numerical analysis of hot rolling process; finite strips method; optimum design; shells and space structures; ultimate strength of plate and box girders; connections in steel structures; interactive buckling in thin-walled tubes; design rules for stability of steel structures; cable structures; computer aided design; design for earthquake actions.

Experimental: ultimate strength of plate girders with diagonal stiffeners and with closed section stiffeners; determination of residual stresses in hollow sections and in thick-walled H sections; biaxial buckling of H columns and of thin-walled tubes; punched holes; new technology for connections in steel structures; buckling of unsymmetrical sections.

Total staff: 27
Professionally qualified staff: 17
Heads of department
Strength of Materials: Professor Ch. Massonnet
Structural Mechanics: Professor G. Fonder
Principal sources of finance
Financed by State subsidies for research in Belgian Universities and by contracts with international organizations or private firms.
Principal related bodies
Faculty of Applied Sciences presided over by the Dean; Administrative Council of the University presided over by the Chancellor.
Publications
Numerous papers are published in professional or scientific journals. Internal reports are also published. They give more details about experimental projects or represent progress reports of theoretical studies.

BEL.121

Université Libre de Bruxelles

Brussels Free University
Service Génie Civil CP 194/4, avenue Adolphe Buyl 87, B 1050 Bruxelles
Telephone: 2 – 6472033
Telex: 23069 unilib Brux.
Date of establishment
1920

Education in civil engineering and testing of materials; research in civil engineering and testing in materials.

Main fields of work

The study of structures (concrete and metallic) and their constituent parts. The civil engineering laboratory is mainly orientated towards load tests on constructive assemblies or elements, until the breaking point. The tests are performed on full-scale or model size structures. The installation allows testing with combined, static or dynamic forces, in any direction.

The laboratory for materials testing is equipped with several testing machines, to accommodate all kinds of application. It is equipped in particular for fracture mechanics testing.

Total staff: 10 permanent
Professionally qualified staff: 6
Director or Chief Executive
Director: Professor P. Halleux
Heads of department
Materials Testing Laboratory: J.P. Elinck
Civil Engineering Laboratory: B. Espion
Principal sources of finance
The University and research subsidies (governmental and private).

Publications
- Research reports
- Final study reports
- PhD Theses.

BEL.122
Université Libre de Bruxelles, Laboratoire Jacques Verdeyen
Brussels Free University, Jacques Verdeyen Laboratory
Avenue Adolphe Buyl 87, B 1050 Bruxelles
Telephone: 02 647 94 82
Date of establishment
1958

The laboratory is part of the ULB Civil Constructions Institute. In addition to educational tasks it performs tests for research, public and industrial needs.
Main fields of work
Soils and rocks mechanics, applied geology, geotechnical cartography: education and experimental training; research; applied research and laboratory tests.
Total staff: 14
Professionally qualified staff: 8
Director or Chief Executive
Director: Professor Jean Nuyens
Principal sources of finance
University funds, research scientific subsidies and industrial applied research income.
Publications
Technical notes (non-periodical)
Research reports.

BRA.123 **AM**

Comissao Nacional da Indústria da Construcão Civil (CNICC)

National Commission for the Civil Construction Industry

Rua Mariz e Barros, 13–3°andar, Rio de Janeiro, 20.270 RJ

Telephone: (021) 273-0841

Telex: (021) 21298 MNIC

Date of establishment

Decree No. 75.204 of 9 January 1975

The Commission was set up under the Ministry of Industry and Commerce, with its composition and attributions being duly specified.

The make-up of the agency, as a collective body on which the private sector has two representatives, recognizes the importance of uniting efforts in the attainment of mutual objectives.

The Commission has a collective decision-making unit – the Board and an Executive Secretariat responsible for its operations, especially as regards the effecting or promoting of studies within its range of operation.

Main fields of work

The main duties of the Commission are:

- To propose policies for developing the sector, especially as regards problems of modernization, sizing, costs and financing for companies in the sector
- To monitor the civil construction market as regards jobs and basic inputs, co-ordinating with the competent agencies measures required for proper functioning of the sector
- To advise the government on the effects of government job programming in the sector, so that by a systematic process of following up civil works contracted for by the Direct and Indirect Federal Administration it may be possible to co-ordinate goals with availability of production factors
- To promote studies for up-dating and consequent adoption of processes, standards and technical rules covering civil works, compatible with the stage of economic and social development of Brazil and intended for adequate coverage of the requirements of the population
- To foster the establishment of conditions for setting up a national information system on the construction materials market
- To effect or promote other studies required for the development of the sector.
- The Executive Secretariat is made up of: a Board for the Co-odination of Economic and Financial Studies, a Board for the Co-ordination of Engineering, a Statistical Service, a Documentation and Public Disclosure Service and an Administrative Support Service.
- The Board, in accordance with article 1, is composed of: the Secretary General of the Ministry of Industry and Commerce; the Secretary General of the Secretariat of Planning of the Presidency; Representatives of the Ministries of Transport, Mining and Energy and the Interior; and two representatives of business associations.

Total staff: 53

Professionally qualified staff: 23

Director or Chief Executive

Secretary General of the Ministry of Industry and Commerce: Marcos José Marques

Principal sources of finance

Budget of the Ministry of Industry and Commerce.

Publications

- Informativo CNICC (CNICC Information) every 3 months;
- Boletim Estatístico (Statistical Bulletin), every 6 months.

BRA.124

Laboratório L.A. Falcão Bauer

L.A. Falcão Bauer Testing Laboratory

Rua Aquinos 111, 05036 Sao Paulo, SP

Telephone: 2629833 (011)

Date of establishment

Commercial Committee of Sao Paulo State 178653, 15 April 1953

The Laboratory is a private institution with branches in the major states of Brazil. It performs services upon request.

Main fields of work

Laboratory Services:

Measurements, analyses, evaluations and testing in:

- civil construction materials such as concrete, cement, agregates, bricks, iron, steel, ceramics, etc.
- mechanical industries

in the areas:

- Chemical analyses
- Physical and Mechanical Testing
- Metallurgical Testing
- Weld Procedure and Welder Qualification
- Consumer products
- Nondestructive testings

Engineering Services:

- Expert Witnesses
- Foundation Analyses
- Failure Analyses
- Plant Surveys
- Pavement Analyses
- Portland Cement and Concrete Proportioning
- Recuperation of Buildings
- Equipments Expediting/Inspection
- Quality Assurance
- Vendor Surveillance
- Equipment Calibration
- Quality Assurance Program Development

Total staff: 387

Professionally qualified staff: 245

Director or Chief Executive

President: L.A. Falcão Bauer

Heads of department

Quality Assurance: Fernando Simões Souto

Engineering Department: Roberto José Falcão Bauer

Chemical Department: Marcia Laube

Metallurgical Department: Victor Andres Toscanini

Civil Construction: Fernando Jardim Mentone

Physical Department: Eduardo Bernasconi

Mechanical Department: Cristina Rolim Dias

Principal sources of finance

Services and research

Publications

Technical standards/recommendations are issued through the Brazilian Association of Technical Standards (ABNT). Catalogues are available on request.

BRA.125

NORIE Centre for Innovation in Building

Curso de Pós-graduaçao em Engenharia Civil, Universidade Federal do Rio Grande do Sul

Av Osvaldo Aranha 99, 3°andar, Porto Alegre, RG do Sul, CEP 90.000

Date of establishment

1974

A building research and teaching organization.

Main fields of work
Project and Construction of Innovative Building including school buildings and social amenities centres

Master of Science Programme in Building Construction comprising the following areas of teaching and research: Construction Management; Construction Economics; New Building Technologies; Systems Analysis and Operational Research in Construction; Environmental Performance of Buildings.

Total staff: 14
Professionally qualified staff: 10
Director or Chief Executive
Professor L.F. Heineck
Principal sources of finance
Through the National Council for the Development of Science Budget, US $280 000 per annum.
Publications
These and scientific papers are published regularly.

BRA.126

Technological Centre of Minas Gerais (CETEC)
Centro Tecnológico de Minas Gerais (CETEC)
Av José Cândido da Silveira 2000, Horto 30.000, Belo Horizonte, Minas Gerais
Telephone: (031) 461-7933
Telex: (031) 1031
Date of establishment
21 March 1972

The CETEC is a non-profit foundation linked to the Operational System for Science and Technology of the Secretariat for Science and Technology of the State of Minas Gerais.

CETEC provides services to private enterprises, R and D institutes, government departments and development agencies, mainly by request; CETEC also carries out technological studies, applied research development projects, analysis and tests either on contract or through its own initiative, aiming at improving industrial processes.

Main fields of work
- Architectural design
- Building technology
- Low-cost building
- Multi-purpose types of building
- Environmental engineering
- Research and studies for clients

Major considerations of regional trends, social and economical features of sectors of the economy, as well as the availability of resources, are the principal criteria setting the boundaries of the studies performed by CETEC in each of its fields of action; the latter are:

Diffusion of Technical Knowledge; Energy; Industrial Chemistry; Foodstuffs; Metallurgy and Mineral Treatment; Product Development; Construction Technology; Basic Sanitation; Ecology and Environmental Engineering; Natural Resources, Tests and Analysis; Wood Charcoal; Remote Sensing.

Total staff: 741
Professionally qualified staff: 350
Director or Chief Executive
President: Milton de Lima Filho
Heads of department
General Director: João Nelson de Senna
Director for External Affairs: Evaldo Abdala
Director for Research Development: Geraldo Queiroz Cançado Sobrinho
Director for Administration and Finance: Carlos Armando Gontijo César
Principal sources of finance
CETEC is financed to about 70% from allocations by the State and Federal Government and to about 30% from services to industry.

Publications
- Manual para Construção e Operaçao Fornos de Carbonização Carbonizacão
- Manual para Construção e Operação de Biodigestores
- Anais Sobre Eventos Técnicos: Enchentes Urbanas
- CETEC Notícias, bi-monthly
- Boletim do NIT, bi-monthly
- Proceedings of the Latin American Symposium: rational organization of building applied to low cost housing.

BRA.127 FM

Technological Research Institute of the State of São Paulo
Instituto de Pesquisas Tecnológicas do Estado de São Paulo (IPT)
Cidade Universitária Armando Salles de Oliveria, Caixa Postal 7141, 055008 São Paulo
Telephone: (011) 2682211
Telex: (011) 22831 INPT BR
Cables: TECNINST
Date of establishment
1934 e.g. Law 13.979

IPT is a non-profit government enterprise.

Main fields of work
- *Civil construction:* construction materials, dams, foundations, paving floors, roads, soils (airport infrastructure), structures, tunnels
- *Building technology:* acoustics, ecological and geographical planning, fire research, heating and ventilating, housing lighting and solar energy, psycho-sociological behaviour, building materials
- *Building:* education, materials, technology, acoustics, environmental planning
- *Housing materials:* standards, testing, technology, education
- Cement and concrete
- *Materials technology:* aggregates, bituminous materials, cement materials, ceramics, concrete, insulating materials, metal (corrosion, failure analysis, melting and casting, powder metallurgy, welding), oils, paints, plastics, refractory materials, rubber, surface treatment, textiles
- Tests and analysis
- Pilot plant experimentation
- Production of standard materials
- Technical assistance
- Teaching assistance
- Teaching programmes
- Participation in committees, councils and standardization activities
- Collaboration with technical societies
- Training programmes for adaptation of foreign technologies
- Development and production of scientific equipment
- Development and pilot production of industrial equipment.

Total staff: 2800
Professionally qualified staff: 1484
Director or Chief Executive
Director Superintendent: Henrique Silveira de Almeida
Technical Director: José Adeodato do Souza Neto
Research Director: Stefan Wolynec
Development Director: Waldemar Bon Júnior
Financial Director: Ruy Guilherme Granziera
Heads of Department:
Mining and Engineering Geology: Ricardo Fernandes da Silva
Civil Engineering: Manuel Martins
Building Technology: Roberto de Souza
Ocean Engineering: Toschi Ichi Tachibana
Mechanical Engineering: Walter Link
Industrial Electricity: Luiz Harold Dirickson
Matallurgy: Edson de Oliveira Pinto
Chemistry and Chemical Engineering: Tokio Morita
Forest and Woods: Amantino Ramos de Freitas

Ore Treatment: Vitor Alberto Mário Lo Ré
Economic and Systems Engineering: Fabio Mariotto
Fertilizers: Marco Guilietti
Pulp and Paper: Leopold Rodés
Textiles: Edson Bittencourt
Railroad Development: Fernando B. Gandara Mendes
Industrial Equipment Technology: Ricardo Kunihiko Yamaguishi
Housing Programme: Caio Fabio Attadia da Motta

Principal sources of finance
Government sector: Government of the State of São Paulo; Federal Government.
Private sector: through the most important clients: FINEP, FIPEC, STI/MIC, CNPq, SADESP/CESP/COSIPA/BNH. About 22% by government grant and 78% by income earned from services.
Principal related bodies
Secretary of Industry and Commerce, Science and Technology of the State of São Paulo directs part of its resources to the development of projects in IPT.
Publications
Numerous publications (lists are available).

CAN.128 **AM**

Canada Mortgage and Housing Corporation (CMHC)

National Office, Montreal Road, Ottawa, Ontario K1A 0P7

Telephone: (613) 748–2000

Telex: 053 3674

Date of establishment

Act of Parliament 18 December 1945

CMHC is responsible for administering the National Housing Act. It is a Crown Corporation, which acts as agent for the federal Government in the implementation of national housing policies and programmes.

Main fields of work

The main functions of CMHC are the provision of federal Government insurance on mortgage loans made by private lending institutions, and the administration of federal social housing programmes, many of which are implemented in partnership with provincial governments. In support of these activities CMHC undertakes or commissions technical, social and economic research and establishes standards governing the housing quality of projects receiving assistance under the National Housing Act. It has some 90 regional and branch offices throughout Canada.

Total staff: approximately 4100

Professionally qualified staff: approximately 75

Director or Chief Executive

President of CMHC: Robert C. Montreuil

Officer responsible for CIB Liaison: R.W. Anderson

Heads of department

Research Division: W. Griggs

Planning Division: B.J. Lacombe

Statistical Services Division: A. Stukel

Program Evaluation Division: J. Potter

National Office Support Centre: J.-L. Bolduc

Materials Evaluation: G. Walt

Principal sources of finance

Mortgage insurance fees; profit on Real Estate Portfolio; fees for services (inspections, material evaluations, agency fees, etc.)

Principal related bodies

Canadian Housing Design Council; Inter-governmental Committee on Urban and Regional Research

Publications

- *Journal:* HABITAT, quarterly
- *Technical standards:* Builder's Bulletins, giving details of CMHC's technical requirements; Materials Evaluation Reports, Advisory Documents on technical subjects
- *Research reports:* All research work carried out or commissioned by CMHC into all apsects of housing, for which a report is prepared, are available to research workers from CMHC's Canadian Housing Information Centre at CMHC National Office
- *Bibliographies/catalogues:* List of CMHC publications; Periodic listing of research projects financed by CMHC; Bibliography on specific housing topics.

CAN.129

Canadian Construction Management Development Institute (CCMDI)

Institut Canadien pour le Développement de la Gestion en Construction

5799 Yonge St, Suite 901, Toronto, Ontario M2M 3V3

Telephone: (416) 222-6449

Date of establishment

10 February 1981, incorporated under Canada Corporations Act

To serve those associations and their affiliated organizations as a national focus for the co-ordination and promotion of management development and training in the construction industry.

Main fields of work

Acting as a catalyst between the industry and the colleges and universities, the Institute's mandate is as follows:

- To monitor current and evolving needs
- To evaluate existing management programmes and improve industry access to these programs
- To encourage educators, through consultation with the industry, to develop courses designed to meet the specific needs of the construction industry
- To encourage industry itself to develop management development programmes
- To keep industry members informed of the availability of management programmes
- To promote the value of management development programmes to the construction industry as a means of improving management effectiveness
- To develop means by which national accreditation on behalf of the construction industry can be accorded the graduates of management development programmes.

Total staff: 3

Director or Chief Executive

President and Executive Director: David Judge

Principal sources of finance

Canadian Construction Association; Housing and Urban Development Association of Canada; Canadian Government (for first 5 years).

Principal related bodies

Board of Directors comprising representatives of the three funding organizations, the Association of Canadian Community Colleges and the universities.

Publications

Education for Management in Construction (catalogue of programs and courses available across Canada), revised edition January 1983.

CAN.130

Canadian Institute of Steel Construction (CISC)

Institut Canadien de la Construction en Acier (ICCA)

201 Consumers Road, Suite 300, Willowdale, Ontario M2J 4G8

Telephone: (416) 491-4552

Telex: 06-986547

Other addresses:

Atlantic Region, 1057 Barrington Street, 536 Roy Building, Halifax, NS B3J 2A1

Telephone: (902) 422 5954

Quebec Region, 1400 ouest rue Sauve, Bureau 224, Montreal, Quebec H4N 1C5

Telephone: (514) 334 2222

Ontario Region, 201 Consumers Road, Suite 300, Willowdale, Ontario M2J 4G8

Telephone: (416) 491-4552

Central Region, 213 Notre Dame Avenue, Suite 700, Winnipeg, Manitoba R3B 1N3

Telephone: (204) 943 5074

Alberta Region, 5923 Buckthorn Road NW, Calgary, Alberta T2K 2Y8

Telephone: (403) 274 3707

British Columbia Region, 1033 Davie Street, Suite 208, Vancouver, BC V6E 1M7

Telephone: (604) 681 0014

Date of establishment

Formed in 1930, granted Federal Charter in 1942

CISC is a non-profit national industry association representing the structural steel, platework and joist industries.

Main fields of work

CISC is both a technical and marketing organization promoting good design and safety together with efficient and economic use of structural steel, through uniform practices, engineering and technical research, as a means of expanding the construction markets for structural steel, steel joists and platework.

Members receive services covering information on statistics, marketing, accounting practices, business ethics, taxes, exports, industry public relations; CISC liaises with government and steel mills. Owners, developers, architects, engineers, educational institutions, government, member and non-member firms may all use CISC services including: design aids, computer programs, project cost analyses; information on codes, standards, fire protection, specifications, steel design; speaker services, films, slides, training courses, workshops, seminars, conferences and technical library.

CISC is the management agent of the Canadian Steel Construction Council. CISC develops and distributes puiblications relating to steel construction.

Total staff: 26
Professionally qualified staff: 16
Director or Chief Executive
President: Hugh A. Krentz
Heads of department
Engineering: M.I. Gilmor
Statistics: D.A. Carrick
Marketing: J.K. Ritchie
Public and Member Services: D.K. Turner
Principal sources of finance
Dues and fees from Fabricator, Honorary and Associate Mambers.
Principal related bodies
24-man Board of Directors: elected representatives, including Chairman, from 63 Fabricator Members. Associate and Honorary Members represent major Canadian steel mills, USA and British mills.
Publications

- *Technical standards/recommendations:* Design and detailing handbook, texts, tables, booklets and manuals (CISC); CISC Code of Standard Practice (CISC) (French and English); Fire Protection Bulletins – a series (CSCC); Standards for Quick Drying Primer and One-Coat paint for use on Structural Steel (CISC) (French and English)
- *Research reports:* Design Guidelines for Rectangular Steel Bins (CSCC)
- *Bibliographies* Proceedings Index (Index des comptes Rendus)
- *Others:* Canadian Structural Engineering Conference Proceedings (CSCC) 1968 to 1982; Computer and Calculations Programs (CISC/CSCC) – over 20 programs; A Project Analysis Approach to Buildings (CSCC); Fundamentals of Structural Shop Drafting (CISC); Steel Joist Facts Recommended Practice (CISC); Computer and Calculator Programs for Structural Steel Design (CSCC) publications brochure; Publications brochure (CISC/CSCC).

CAN.131

Canadian Steel Construction Council (CSCC)
Conseil Canadien de la Construction en Acier (CCCA)
201 Consumers Road, Suite 300, Willowdale, Ontario M2J 4G8
The CSCC can be contacted through the Regional Offices of the Canadian Institute of Steel Construction.
Telephone: (416) 491 9898
Telex: 06-986547
Date of establishment
Originally formed in 1960 as the Steel Industries Advisory Council, incorporated in 1965 as the Canadian Steel Industries Construction Council. The present name was adopted in 1980.

An industry association representing the basic manufacturers of steel used in construction, the fabricators of structural steel, open web steel joist, steel plate, corrugated pipe and sheet steel and the manufacturers of fasteners.
Main fields of work
To promote the manufacture, fabrication and use in the construction industry in Canada of construction steel (being structural, plate, sheet and stainless steel and related steel products). CSCC is the development of university and industry research in the areas of design improvement, building codes, specifications and standards through engineering and technical research. CSCC sponsors a Steel Fellow at the National Research Council of Canada in order to study the behaviour of steel in fire and calculation of fire endurance of structural members and assemblies for buildings.

CSCC specializes in Building Codes and Fire Protection and contributes to the work of various government code writing bodies at the national, provincial and municipal levels, and acts in an advisory capacity on codes and fire protection.

CSCC originate computer programs for use by the engineering profession in Canada to aid in the design of structural steel buildings.

The CSCC Project Analysis Division works with project planners on the comparison of costs of structures built with steel and competitive materials making extensive use of the CSCC computer programs.

CSCC Committees are established for special assignments or studies for improvement of the industry, one of which is the biennial Canadian Structural Engineering Conference, an international forum for the exchange of engineering knowledge related to structural steel.

Government liaison is another area in which CSCC represents the steel construction industry.

Total staff: 26
Professionally qualified staff: 17
Director or Chief Executive
Chairman: Hugh A. Krentz
Heads of department
Project Analysis Division: J. Keith Ritchie
Codes and Standards and Fire Protection: Stewart G. Frost
Research and Development: Michael I. Gilmore
Special Projects: Donald K. Turner
Principal sources of finance
Member contributions.
Principal related bodies
The Canadian Institute of Steel Construction acts as a management agent for the CSCC, sharing facilities and key staff members. Supervisory Board of Directors representing member companies and institutions: The Algoma Steel Corporation Ltd, Dofasco Inc., Sidbec-Dosco Inc., Stelco Inc., Canadian Institute of Steel Construction, Canadian Fasteners Institute, Canadian Sheet Steel Building Indsitute, Corrugated Steel Pipe Institute (Canadian Welding Bureau is an Associate Member).
Publications
CSCC publications are included with those for the Canadian Institute of Steel Construction.

CAN.132 AM

Centre des Etudes sur le Bâtiment (CEB)
Centre for Building Studies
1455 de Maisonneuve Blvd W, Montréal, Québec H3G 1M8
Telephone: 879-8551 (514)
Date of establishment
January 1977

A research and teaching centre with two main functions: to define, undertake, and execute programmes in research, development and information transfer in response to the needs of the building industry and to offer building engineering programmes (BEng, MEng, Combined and PhD.)
Main fields of work
The Centre offers an accredited undergraduate programme and an MEng and PhD programme. It conducts research in four areas of building engineering: Building environment, Building science, Building structures and Construction management.

The Centre provides extensive laboratory facilities to assist researchers in energy conservation and in refining design of building components. There are ten laboratories: two for building materials, one for mechanics of materials, one for thermal environment engineering, a thermo-aerodynamic laboratory with wind tunnel, one for building science, one for building enclosures, an acoustics laboratory, a two-storey prototype laboratory, a computer-aided design laboratory, including several support laboratories such as an instrumentation and calibration laboratory, a fabrication shop and roof-top exposure site. The Centre provides expertise of personnel and equipment to assist industry and government in problem solving and product development.

Total staff: 28

Professionally qualified staff: 17

Director or Chief Executive

Professor Dr Paul Fazio

Principal sources of finance

Concordia University, La Formation de Chercheurs d'Action Concertée (FCAC), industry in the form of contracts and the National Science and Research Council of Canada.

Principal related bodies

Advisory Board of approximately 12 members drawn from industry and government agencies.

Publications

- Proceedings of the World Congress on Space Enclosures, 1976
- International Symposium on Low-Cost Housing Problems, 1974
- List of publications available on request.

CAN.133

Centre for Research and Development in Masonry (CRDM)

Centre de Recherche et de Développement en Maçonnerie

105 4528 6A Street NE, Calgary, Alberta T2E 4B2

Telephone: (403) 277-3024

Date of establishment

1979

A non-profit organization, registered with Consumer and Corporate Affairs Canada for the purpose of promoting the advancement of the science and practice of masonry. CRDM is open for membership to individuals, companies, institutions and associations wishing to utilize some or all of the services offered or simply to keep abreast of technical developments in the field of masonry.

Main fields of work

- To undertake research, development and testing projects in all matters relating to masonry and to the design, quality and performance of masonry products and to undertake or assist in experiments, tests and research relating to materials, building and development of equipment, machinery and processes for masonry and its application to trade or industry.
- To prepare, print and publish books, papers, periodicals, etc. bearing upon masonry and its application to trade or industry and to establish, form and maintain a library of scientific data.
- To publish and to aid in the dissemination of information to assist in training in masonry.
- To encourage the discovery of, and investigate and make known the nature and merits of, inventions, improvements, processes, etc. affecting masonry and its application to trade or industry.

Principal departments are: Laboratory and Computing Facilities; Library with computerized information storage and retrieval systems; Education and Training; Materials and Structural Testing.

Total staff: 6

Professionally qualified staff: 6

Director or Chief Executive

Director General: Emlyn L. Jessop

Research Director: D.J.N. Warren

Principal sources of finance

Grants-in-aid of research, consulting and testing activity, sales of technical publications, industry funding and membership fees.

Principal related bodies

A Board of 12 Directors is elected from the membership.

Publications

- *Technical standards:* Specifying the Design and Construction of Masonry Fireplaces and Domestic Chimneys: Part 1, An Appraisal of North American Practice
- *Research reports:* Masonry Connectors: A State-of-the-Art Report; Thermal Performance and Energy Storage in Masonry
- *Bibliographies:* Masonry Research in Progress, 1982; Masonry Bibliography: Volume 1, 1830–1982
- *Others:* Fireplace Technology in an Energy Conscious World; Metric Design Handbook for Reinforced Concrete Masonry Elements (also available in French).

CAN.134

Clay Brick Association of Canada (CBAC)

Association Canadienne de Brique d'Argile Cuite

5218 Yonge Street, Willowdale, Ontario M2N 5P6

Telephone: (416) 225 7763

Date of establishment

1940

A non-profit trade association comprised of clay brick manufacturing companies.

Main fields of work

- To represent its member companies through interaction with Government, Code and other bodies
- To promote the use of clay brick to homebuilders, developers, homeowners, architects, engineers, contractors and the construction industry in general through advertising in the media and conferences, seminars and exhibitions
- To communicate the relative advantages of clay brick to these groups
- To educate the specifier and mason trades in the proper application and use of clay brick
- To encourage a high level of product quality and development for the continued profitability of the industry
- Encourage, through scholarships and awards at professional and trade schools, an awareness of and preference for clay brick
- Make available technical assistance and service support to clay brick manufacturers, specifiers and users.

Total staff: 2

Professionally qualified staff: 1

Director or Chief Executive

Managing Director: John F. Cutler

Principal sources of finance

Dues paid by clay brick manufacturers in Canada.

CAN.135 FM

Division of Building Research of the National Research Council of Canada (DBR NRC)

Division des recherches en bâtiment du Conseil national de recherches du Canada

Montreal Road, Ottawa, Ontario K1A 0R6

Telephone: (613) 993 2607

Telex: 053 3145

Cables: Research

Other addresses: British Columbia Regional Station, 3904 West 4th Avenue, Vancouver, British Columbia V6R 1P5

Prairie Regional Station, Saskatoon, Saskatchewan S7N 0W9

Atlantic Regional Station, 1411 Oxford Street, Halifax, Nova Scotia B3H 3Z1

Date of establishment

1 August 1947

The Division of Building Research is one of twelve research divisions of the National Research Council of Canada, which is a Federal Government institution. The work of the Division is carried out by seven Research Sections, a Codes and Standards Group, and a Technical Information Group, which includes a Publications Section and a Library.

The Division's facilities are available to all levels of government and to the public.

Main fields of work

The Division of Building Research provides a national research facility for the Canadian construction industry and Canada Mortgage and Housing Corporation. This facility is equipped to study problems that are unique to Canada because of its climate, geography and raw materials. The Division also provides research and administrative support for the National Building Code and the National Fire Code.

Research undertaken by DBR falls into three categories: basic studies, field or performance studies and standards development. Besides its laboratories in Ottawa, the Division operates Regional Stations in Vancouver, Saskatoon and Halifax. These Stations provide information and assist in identifying and solving regional building problems. The Division also operates a Field Station at Rogers Pass, British Columbia (for avalanche research) and a Fire Research Field Station, 65 km from Ottawa.

Total staff: 279

Professionally qualified staff: 102

Director or Chief Executive

Director: G. Seaden

Associate Director: L.W. Gold

Heads of department

Building Materials Section: V.S. Ramachandran

Building Performance Section: K. Latta

Building Structures Section: W.A. Dalgliesh

Building Services Section: R.C. Biggs

Fire Research Section: T.Z. Harmathy

Geotechnical Section: H.L. Palmer

Noise and Vibration Section: A.C.C. Warnock

Codes and Standards Group: R.A. Hewett

Technical Information Group: R.G. Turenne

Principal sources of finance

Annual Parliamentary vote of funds to the National Research Council. Some supplementary funding is received from other government agencies and from industry for specific projects. Income earned from testing services and sale of publications is credited to the National Research Council.

Principal related bodies

Canadian Committee on Building Research, formed by the National Research Council to provide an interdisciplinary forum for the stimulation and application of building science and technology in Canada with particular reference to the design, performance, and use of buildings.

Publications

Results of research are available as reprints of articles by DBR staff published in scientific or technical journals.

Separate series of publications have been established based on user requirements and nature of the information. Examples of such series: Canadian Building Digests, Building Practice Notes, and Special Technical Publications.

The Division is also active in the transfer through preparation of audio-visual presentations, building science forums (held each year in major centres across the country), lectures, inquiry services, and consultations.

CAN.136

Forintek Canada Corp

6620 NW Marine Drive, Vancouver V6T 1X2

Telephone: 604 224 3221

Telex: 04 508552

Date of establishment

1 April 1979

A private, non-profit organization serving the forest products industry (with the exception of the pulp and paper sector), and the federal and provincial governments. Forintek is Canada's national wood products research institute with laboratories in Vancouver and Ottawa, Canada.

Main fields of work

Basic and applied research; productivity improvement; codes and standards; and training and education are the broad areas of activity. Disciplines serviced include: Wood Engineering, Housing, Composite Products, Lumber Manufacturing Technology, Biotechnology, Wood Protection, Packaging, Seasoning, Wood Machining, Energy from Wood, Wood Science and the production of Food and Chemicals from Wood.

As Canada's national wood products research institute, Forintek provides a leadership role to ensure the continued national and international acceptability of Canadian forest products and to help strengthen Canada's international trade position. Towards this end, our staff actively participate in many national and international forest products associations and committees. Organizations such as the Truss Plate Institute of Canada and the Waferboard Association regularly invite our staff members to their Technical Committee meetings; this facilitates keeping up to date with technical problems in various forest products industry sectors in Canada. In some of these associations our staff have acted as President, Chairman, Co-chairman, Secretary and in other capacities.

Total staff: 200

Professionally qualified staff: 50

Director or Chief Executive

President and Chief Executive Officer: K.A. French

Heads of department

Eastern Laboratory, Ottawa, Director of Research: R.W. Stephens

Western Laboratory, Vancouver, Director of Research: Dr C.C. Walden

Principal sources of finance

Forest industry, provincial governments, federal government.

Principal related bodies

Board of Directors; Research Program Committees, to assist in the development of the most appropriate research programmes for the industry.

Publications

A list of publications is available free of charge.

CAN.137

McGill University, Centre for Minimum Cost Housing (MU CMCH)

School of Architecture, McGill University, 3480 University Street, Montreal, Quebec H3A 2A7

Telephone: (514) 392 8021

Telex: 5268100 MTL

Date of establishment

May 1970

The Centre provides a Minimum Cost Housing postgraduate programme for architects and planners who wish to pursue a career in housing in less developed countries. Two degrees are offered: a Master's of Architecture (2 year course of study) and a Graduate Diploma in Minimum Cost Housing (7½ month course of study). The Centre also conducts research into human settlements problems, financed by government grants and research contracts, and with the support of the University.

Main fields of work

Technological aspects of housing in less developed countries. Research into the following specific areas:

- Sulphur building
- Low-cost sanitation technologies
- Strategies for mass housing (sites and services, support structures, etc.)
- Alternative building techniques and materials.

Total staff: 3

Professionally qualified staff: 2

Director or Chief Executive
Director: Professor Witold Rybczynski
Principal sources of finance
McGill University; Canadian International Development Agency (project in progress); International Development Research Centre (project application); Previous projects: Agriculture Canada, United Nations Environment Programme, Department of Northern and Indian Affairs, Central Mortgage and Housing Corporation.
Principal related bodies
McGill University.
Publications
Research papers
Books: The Ecol Operation (English and French versions available), Stop the Five Gallon Flush (survey of alternative waste disposal systems), Use It Again Sam (investigation into the design of consumer containers for re-use as building components), Rooftop Wastelands (community roof-top gardening), Water Conservation and the Mist Experience (mist spraying for showering using 1 litre of water).

CAN.138 AM

Public Works Canada (PWC)
Ottawa, Ontario K1A 0M8
Telephone: (613) 998 8728 (Assistant Deputy Minister D and C)
Telex: 053-4235
Date of establishment
1868

The Department of Public Works fulfils many roles pursuant to the Public Works Act and other acts and regulations. These roles fall into three different areas of responsibility:

Provision and management of a major portion of the government's accommodation inventory, under the direction and control of the Minister of Public Works, required for federal government operations. In this role, the department not only provides operational management, but performs the role of trustee on behalf of the government. This role includes all aspects of property investment and operations management including the owner/investor role.

The administration and delivery of specific products on behalf of the federal government as follows: fire prevention and emergency preparedness; construction and realty technology development; marine transportation and other engineering; federal land management; municipal grants in lieu of taxes; and public works crown corporations.

The provision and administration of professional and technical expertise and services in support of the above, in support of other government departments and agencies.

Main fields of work
Public Works Canada is one of the largest realty operations in Canada and, indeed, one of the largest in the world. Through its six regions, it directs virtually all its resources to the support of other departments and agencies. It provides accommodation to federal departments and agencies through construction, purchase or lease, as well as architectural, engineering and related services. It constructs and maintains roads and bridges under federal jurisdiction, manages and develops federal lands and issues grants to municipal and other authorities in lieu of taxes on federal property.

The organization of the department is based on the concept of matrix management with functional groups at headquarters concerned with policies, priorities, strategies and standards, and operational groups in six regional centres, charged with implementation and delivery of departmental services. Conventional management support groups round out the organizational structure. Under this organization, Departmental operations are decentralized to a very high degree.
Total staff: 8972 (1981/83)
Director or Chief Executive
Minister of Public Works Canada: Romeo LeBlanc
Deputy Minister of Public Works Canada: J.A.H. Mackay
Assistant Deputy Minister Design and Construction Public Works Canada (Departmental representative to CIB): Guy Desbarats
Heads of department
Deputy Minister: J.A.H. Mackay
Assistant Deputy Minister, Operations: A.D. Wilson
Assistant Deputy Minister, Realty: A.J. Perrier
Assistant Deputy Minister, Design and Construction: G. Desbarats
Assistant Deputy Minister, Revenue Dependency: F.S. Currie
Director General, Departmental Planning and Co-ordination: D.J. Hartt
Director General, Personnel: H.A. Carswell
Executive Secretary: M.J. Desormeaux
Director, Public Relations and Information Services: A. Normand
Regional Directors General:
Atlantic Region: J.M. Dunphy
Quebec Region: G. Wolfe
National Capital Region: H.D. McFarland
Ontario Region: E.D. Manchul
Western Region: N.M. Hoyt
Pacific Region: R.B. Angus
Fire Commissioner of Canada: G.A. Hope
Director, Equal Opportunity and Official Languages Programs: V. Hambly
Principal sources of finance
Financed through an annual Parliamentary vote of funds to PWC. As well, some programmes are funded by other government departments.
Publications
Departmental Documents List is available through Information Research and Library Services.

CAN.139 AM

Université de Montreal, Faculté de l'Amenagement
BP 6128, Succ. 'A', Montréal, Québec
Telephone: (514) 343 6001
Date of establishment
1968

Provides professional training in architecture, landscape architecture, industrial design and town-planning, as well as a research oriented training at the Master and Doctoral levels. It undertakes research assignments subsidised and under contract in aspects of development.
Main fields of work
The research is channelled into the following main directions:

- Conserving and developing the landscape
- Planning, territorial and regional development
- Methodology of appropriation, policies and management of land
- Habitat, urban structures and quality of life
- Building science and technology (industrialization, innovation and energy)
- The relationships between man and the environment
- Structural topology
- The socio–technological aspects of adaptation (adaptation of the handicapped to the environment)
- The environment, within the context of the interplay of political forces and human life.

Director of Chief Executive
Dean: Colin H. Davidson
Heads of department
School of architecture: Professor Jean Ouellet
School of landscape architecture: Professor Ronald Williams
School of industrial design: Professor Roger Camous
School of town planning: Professor René Parenteau
Faculty of advanced study: Professor Iskandar Gabbour
Research: Professor Peter Jacobs, vice-dean
Principal sources of finance
University of Montréal.

Principal related bodies
Research Group 1F (research and publications concerning industrialization and technological innovations).
Research Group on structural topology (research and publications concerning topology and structures).
Research Group on the socio–technological aspects of adaptation (research and industrial design related to problems facing the handicapped).
Publications
Journal: 'Micro-1F' four times annually and 'Topologie Structurale' twice annually.
Research reports: 'Notes de recherche' and 'Cahiers de recherche' two series. Irregular publications dates.

CAN.140
Calgary University, Department of Civil Engineering
2500 University Drive NW, Calgary, Alberta T2N 1N4
Telephone: (403) 284 5821
Telex: 038 21545
Date of establishment
The University of Calgary became an autonomous institution in 1966. The Department of Civil Engineering existed from 1963 when the Department was part of the Division of Engineering of the University of Alberta.

The Department of Civil Engineering, University of Calgary is typical of Civil Engineering Departments on the North American continent. Members of each of the principal groups meet regularly to plan postgraduate teaching and research. Several research contracts have been obtained by research groups within the department instead of individuals.
Main fields of work
- Structures and Solid Mechanics, the largest group in the Department
- Materials Science, second largest group
- Transportation Engineering
- Geotechnical Engineering
- Water Resources Engineering.

Total staff: approximately 73
Director of Chief Executive
Professor and Head, M.A. Ward
Principal sources of finance
The Department is supported wholly by public funds in the Province of Alberta. In addition the Department receives approximately $800 000 annually in research grants and contracts to support its research programmes.
Publications
Since it is the University's policy not to become involved in confidential work, the results of the Department's research work are published regularly in the open literature. The Department does maintain its own research report series. Generally this is simply a mechanism to produce preprints well in advance of the availability of reprints from a particular publisher. Documents in this series are available at no charge, under reasonable circumstances.

CAN.141
New Brunswick University, Department of Civil Engineering (UNB CE)
PO Box 4400, Fredericton, New Brunswick E3B 5A3
Telephone: (506) 453 4521
Telex: 014-46-202
Date of establishment
The first engineering courses were instituted in 1854.

The department is in the Faculty of Engineering which consists of Civil, Mechanical, Electrical, Chemical, Surveying, and Geological Departments. The university is financed through grants from the provincial government, tuition fees, and contracts for research. The role of the university is to provide English Language training for students of the Maritime Provinces; however, students from other parts of Canada comprise a large proportion of the student body. A limitation on visa students is in place.
Main fields of work
Research is presently being undertaken into the behaviour of lightweight concrete in a marine environment; geotechnical studies include: the behaviour of peats and other soft soils, foundations; environmental and sanitary: the treatment of food waste; structures: the behaviour of local buckling of steel sections, connections, the interaction of steel frames with block masonry infill; hydrology and water resources: the behaviour of river bends, hydrological studies.
Total staff: 17
Professionally qualified staff: 17
Director or Chief Executive
Dean of Engineering: F.R. Wilson
Department Head, CE: R.M. Francis
Principal sources of finance
Financed from grants from the provincial government, tuition fees, grants from research contracting with outside organizations.
Publications
Publications are on a random basis covering all fields of engineering endeavour. Publications lists are published each year and are available on request.

CAN.142
Waterloo University, Waterloo Construction Council (UW WCC)
Department of Civil Engineering, University of Waterloo, Waterloo, Ontario N2L 3G1
Telephone: (519) 885 1211
Telex: 069-55259
Date of establishment
University 1957, Council 1973

Waterloo Construction Council is made up of individual companies representing contractors, consultants, owner clients and suppliers to the industry.

It began as the Construction Management Programme to attempt to provide a vehicle by which the performance of the construction industry is improved. The objectives are to transform existing management techniques through research in order that they can be applied to construction; and to educate professionals on the application of these techniques towards resolving the unique problems of construction.

There is a working partnership between the Council and the University which ensures that the programme is relevant by providing a mechanism for the two-way flow of ideas, information and personnel. The Council provides assistance in the provision of visiting lecturers, advice on the topics and guidance on research as well as a portion of the financial support. The University provides direct supervision and participation in the educational (lectures and workshops) and research (student and faculty) programmes.
Main fields of work
To develop a core of professional managers for the construction industry and to undertake research into problems of relevance to the industry.

Course titles offered by teaching on a co-operative basis are:
- Labour Relations in the Construction Industry
- Construction Economics, Accounting and Finance
- Organizational and Legal Responsibilities in Construction
- Administration of Construction Projects

- Construction Methods and Equipment
- Construction Planning.

Total staff: 9
Professionally qualified staff: 8
Director or Chief Executive
Director of Programme: Dr R.W. Cockfield
Principal sources of finance
Government funds and annual construction from the members of the Waterloo Construction Council.
Principal related bodies
Board of Directors
Publications
Research reports: (available on demand at cost)
Papers published in technical journals.
Others: Manual of Labour Relations with the Construction Trades 1981; Construction Equipment Management, 1978.

CAN.143
University of Western Ontario, Boundary Layer Wind Tunnel Laboratory (BLWT)
Faculty of Engineering Science, University of Western Ontario, London, Ontario N6A 5B9
Telephone: (519) 679 3307
Telex: 064-7134
Date of establishment
November 1965

The Boundary Layer Wind Tunnel Laboratory has pioneered new techniques for studying the effects of wind on structures, on the dispersion of pollutants, snow drifting and other phenomena. The established objectives of the Laboratory are to further research, to undertake innovative studies for industry and to provide opportunities for graduate studies. The contributions made by the Laboratory over the years to the study of wind engineering have demonstrated the important interdependencies between these three activities. Construction of a second generation boundary layer wind tunnel began in October 1982 and completion was expected in 1983.
Main fields of work
The field of wind engineering; the term defines a generic collection of engineering problems in which interaction with the wind is a governing factor. It generally involves the modelling of the earth's turbulent boundary layer over uniform and complex terrain in various conditions of atmospheric stability. It embraces the following subjects:

- *Wind effects on buildings and structures:* stresses, deformations and sway of major structures; pressures on the exterior glass and cladding; influence on heating and ventilation
- *Wind energy:* performance of wind energy machines in turbulent wind; influence of topography on siting; optimization of wind energy 'farms'
- *Environment:* dispersion of pollutants and chimney design and location; study of spills of liquid natural gas, study of wind conditions in city streets; influence of wind breaks in agriculture and forestry; drifting and deposition of snow
- *Transportation:* drag alleviation and fuel economy of vehicles; design of highway cuttings, e.g. wind breaks, snow drifting; ride qualities of trains and rapid transit during periods of strong winds

All the topics above are significant and several now have strategic importance. The list can be extended; one direction is the study of problems influenced by

- *Wind and wave interaction:* wind/wave interactions on off-shore structures; design of sailboats and wind powered vessels; mooring and behaviour of ships in stormy seas; behaviour of pack ice under wind.

Total staff: 27
Professionally qualified staff: 12
Director and Chief Executive
Professor and Director: Dr A.G. Davenport
Manager and Associate Research Director: Dr N. Isyumov
Associate Research Director: Dr D. Surry
Professor and Associate Research Director: Dr B.J. Vickery
Professor and Associate Research Director: Dr M. Novak
Principal sources of finance
NSERC research grants; industrially sponsored research projects.
Principal related bodies
Faculty of Engineering Science.
Publications
Numerous studies (list available upon request).

CHN.144

China Academy of Building Research

中國建筑科学研究院

9 Xiao Huang Zhuang, An Wai, Beijing
Telephone: 46 3631
Telex: 22477 CSCEC CN
Cables: 5912
Date of establishment
1956

The Academy is a comprehensive research institution affiliated to the Ministry of Urban and Rural Construction and Environmental Protection.

The Academy has eight research institutes: Institute of Building Structures, Institute of Foundation Engineering, Institute of Earthquake Engineering, Institute of Concrete, Institute of Air-conditioning, Institute of Building Physics, Institute of Building Mechanization and a Computer Centre.

Main fields of work
The main emphasis is on housing construction. Research covers the following fields:
- The key technical problems in the construction of industrial and residential buildings
- Compilation and administration of building codes, specifications and standards
- Testing and computation techniques
- Basic problems in applied sciences
- Technical bottlenecks in the key projects of the country
- Technical services and technical training.

Total staff: 1279
Professionally qualified staff: 407
Director or Chief Executive
President of the Academy: Xu Zhengzhong
Principal sources of finance
The Chinese Government.
Publications
Journal: Building Mechanization, bi-monthly.

CHN.145 **FM**

China Building Technology Development Centre

中国建筑技术发展中心

19 Che Gong Zhuang Street, Beijing
Telephone: 8992567
Cables: Beijing 2594
Date of establishment
May 1983

The Centre is an independent, non-profit organization under the leadership of the Ministry of Urban and Rural Construction and Environmental Protection. Its main task is to promote the advance of building technology and to contribute to urban and rural development.

Main fields of work
- Comprehensive studies and provision of materials according to the requirements of the Ministry for the Development of science and technology and technological transformation of the building industry
- Technical estimation and evaluation, display of building products, technical exchanges, technical training and consultation
- Compilation of building standards and regulations
- Research work on building standard design, rural development, building economics, building information and history of architecture
- Collection of books, periodicals and literature published in China as well as in other countries. Provision of documentation retrieval services
- Promotion and organization of international co-operation and exchange of building science and technology.

Total staff: 700
Professionally qualified staff: 370
Director or Chief Executive
Senior Engineer: Xu Ronglie
Heads of department
Division of Technology Development: Meng Lei
Division of Comprehensive Studies: Shen Caiwen
Institute of Building Standard Design and Research: Zhuang Guowei
Institute of Building Information: Wang Yangzu
Institute of Building Economics: Wu Luoshan
Institute of Rural Development: Jin Daqin
Institute of History of Architecture: Cheng Jingqi (Ms)
Exhibition Hall for Building Products: Meng Lei
Principal sources of finance
Ministry of Urban and Rural Construction and Environmental Protection.
Publications
- Heating, Ventilating and Air Conditioning (Chinese), quarterly
- Water and Sewage Engineering (Chinese), bi-monthly
- Construction Technique (Chinese), bi-monthly
- Building Structures (Chinese), bi-monthly
- Rural Development (Chinese), quarterly
- Studies on Building Economics (Chinese), bi-monthly
- China Building Selection (English), quarterly
- Synopses (English), irregular.

CHN.146

Ministry of Urban–Rural Construction and Environmental Protection, Bureau of Science and Technology

Bai Wan Zhuang, Westsuburb, Beijing
Telephone: 8992282
Telex: 22477 CSCEC CN
Cables: 2260
Date of establishment
May 1982

Governmental organization.

Main fields of work
Governmental administration in urban–rural development.
The principal divisions are:
- Division of Planning
- Division of Scientific Research
- Division of Scientific and Technical Development
- Division of Regulation

Director or Chief Executive
Director: Xu Ronglie
Principal sources of finance
Government.

CHN.147 **FM**

Shanghai Research Institute of Building Sciences

75 s Wan Ping Road, Shanghai, China
Telephone: 380803
Cables: 3295
Date of establishment
1958

Under the direct leadership of the Shanghai Municipal Government Construction Committee, the Shanghai Research Institute of Building Sciences is a basic comprehensive research institution for building sciences in the Shanghai region and an authorized testing station for checking, inspection and arbitration of building materials, structural elements and building products as well as construction quality. It also provides consultative technical and technical services.

Main fields of work
- Research on structures and structural elements of residential and high-rise buildings

- Research on various building materials, including concrete, special concrete (high-strength concrete, high-impermeable concrete, anti-corrosion concrete etc.), artificial lightweight aggregate, new types of deformed steel, etc.
- Research and development on the rational utilization of industrial wastes for building materials, such as steam-cured lime-fly ash blocks, panels, slabs etc., steel slag cement, concrete admixed with fly ash and artificial lightweight fly ash aggregate.
- R & D on decorative building materials, building coatings and building plastic products as well as their applied technology
- Research, design and practical trials of construction engineering machinery
- Research on building functions, R & D on materials for water-proofness, thermal and acoustic insulation and their test methods
- Research on applications of computer techniques in construction
- Research on building information systems and building economics.

Total staff: 559
Professionally qualified staff: 338
Director or Chief Executive
Director, Senior Engineer: Dr Wang Pu
Heads of departments
Research Department: Mr Wang Fu-yuan
Development Department: Mr Lu Ji-guang
Scientific and Technical Service Department: Mr Wang Xing-long
Principal sources of finance
Shanghai Municipal Government Construction Committee; Shanghai Municipal Government Scientific and Technical Committee; Ministry of Urban and Rural Construction and Environmental Protection; Ministry of Building Materials; Contracts with other corporations and institutions.
Principal related bodies
The Institute's Academic Council; Design and Scientific Research Department, Shanghai Municipal Government Construction Committee.
Publications

- Shanghai Building Science and Technology, two-monthly (GB).
- National Standards For Wall-Building Bricks Test; National Standard For Semi-unplasticized Polyvinyl Chloride Floor; National Standard for Unplasticized Polyvinyl Chloride Drain Pipes and Fittings For Civil Buildings; Designing Regulations For Reinforced Lightweight Aggregate Concrete.
- Carbonation Resistance of Steam-Cured Lime-Fly Ash Concrete; Analysis of the Core of Tube In Tube Subjected to Lateral or Torsional Loading by Standard Plane Frame Programe; etc.
- A Survey Report of Pressure Distribution and Displacement of Tall Building In Strong Winds; In situ And Wind Tunnel Investigation on Wind Pressure And Vibration of Bai-Yun Hotel.

CHN.148

Tianjin Fire Research Institute
PO Box 27 Tianjin
Telephone: 28812
Cables: 3194
Date of establishment
1964

A Government research institute subordinated to the Fire Bureau of the Ministry of Public Security, carrying out investigation and research projects for the Fire Bureau and some research sponsored by industrial firms or factories.

Main fields of work

- Petro-chemical fire protection
- Research into extinguishants
- Fixed fire extinguishing systems
- Basic research into fires
- Fire investigation, statistics and analysis
- Research into fire protection of buildings
- Building fire protection research and fire resistance testing
- Standards and fire codes.

Total staff: 293
Professionally qualified staff: 170
Director or Chief Executive
Director: Xu Bingyao
Heads of departments
Research Planning and Administrative Division: Li Bingjun
Petro-Chemical Fire Division: Sun Lun
Fire Theory Division: Song Yuchi, Master
Fire Fighting Agents Division: Jin Hongbin
Fixed Fire Extinguishing System Division: Xie Delong
Fire Investigation and Statistics Division: Wu Juying
Standardization Division: Yuan Xiaozhi
Fires in Buildings Division: Liang Jirong
Information Division: Yuan Jizheng
Principal sources of finance
Financed mainly by the Government and also by fees for services, testing and research from various industrial firms.
Publications
Fire Science and Technology News, non-periodical.

COL.149

Instituto Colombiano de Productores de Cemento (ICPC)
Colombian Institute of Cement Producers
PO Box 52816. Medellín, Colombia, or Edificio Camacol, Piso 6, Medellín
Other address: Carrera 6ª No. 26–35, Piso 7, Bogotá
Telephone: 94 30 02 11
Telex: 6697 CNARE
Cables: INCOCEMENTO
Date of establishment
18 July 1973

A private non-profit institution, created and maintained by the Colombian cement producers.

Main fields of work
Basic objectives: to improve and develop the production and uses of cement; to co-ordinate, harmonize and protect the activities of the associated enterprises; to promote study and research in the field of cement, concrete and construction, and to disseminate the results of these studies and research, through the technical cement department, the technical concrete department and the documentation centre (for cement, concrete, building and civil engineering).

Total staff: 20
Professionally qualified staff: 8
Director or Chief Executive
President of the Board of Directors: Juan M. Ruiseco
Executive Director: Germán Botero Arango
Heads of department
Technical Department: Germán Madrid
Industrial Department: Carlos Ossa
Documentation Center: Maria Eugenia de Restrepo
Principal sources of finance
Cement industry through percentage of total sales
Publications
- Technical papers and books related to cement and concrete technology of importance for or adapted to the Colombian situation
- Bibliographies on different specific subjects related to the defined field of work are available for purchase upon request
- Bulletin: Boletín ICPC, quarterly.

CUB.150 **FM**

Centro Tecnico de la Construcción y los Materiales (CTCM)
Technical Centre for Construction and Materials
Apartado 6180, La Habana

Date of establishment
Re-constituted in 1979 from the former Centre for Construction Experimentation and Research (CIEC)

Functions as the national research centre for the building and building materials industries under the Ministry of Construction.

Director or Chief Executive
Director: D. Rodriguez Cedrón.

CSK.151

Building Institute for Testing Materials and Structures
Technický a zkušební ústav stavební (TZÚS)
Konviktská 24, 113 33 Praha 1
Telephone: 222 698, 265 043
Telex: 122 531
Date of establishment
25 June 1953

A state public institution at the disposal of all state and municipal services, national enterprises and civil services in the Czech part of Czechoslovakia. The board of management is in Prague but there are branches in all regional towns.

The institution is governmental and the basic raison d'être is to carry out systematic and complex research and to test materials and structures in order to improve the quality of building materials and products; also to develop testing apparatus in the building industry.

Main fields of work

- To carry out forecasting, research, development and studies in the field with the goal of improving quality in the building industry
- To research and develop new testing and checking procedures and methods of evaluation for the building industry, with experimental verification and application of the new procedures and methods
- To draw up state standards for the testing of building materials, products and structures
- To carry out research into new apparatus for testing in the building industry
- To perform tests on building materials, products and structures and provide expertise and advice
- To bring about state inspection and to grade building materials and products in different quality classes
- To perform geological surveys for raw materials for the industrial production of building materials.

Total staff: 500
Professionally qualified staff: 300
Director or Chief Executive
Managing Director: Jirí Petrán, MSc
Principal sources of finance
Research is financed by the Ministry of Building Industry from the state budget. Other work is paid for by customers.
Principal related bodies
The Institute is managed by the Ministry of Building Industry.
Publications

- Bulletin of the Institute, quarterly
- Bulletin of New Testing and Evaluation Procedures, 10 times a year
- Defects of Structures, annually.

CSK.152 FM

Building Research Institute, Prague
Výzkumný ústav pozemních staveb Praha (VÚPS)
102 21 Praha 10, Hostivař, Pražská 16
Telephone: 752641–9, 752741–9
Telex: 122682 vups c
Date of establishment
27 March 1958 by Foundation Document of the Ministry of Building Industry

The Institute is a research and development organization of the building industry, and as a departmental institute it is directly controlled by the Ministry of Building Industry of the Czech Socialist Republic.

The role of the Institute is to ensure the technical and economic development of residential, public, industrial and agricultural buildings, including their reconstruction and modernization.

The Institute is principally engaged in:

- Research studies
- Detailed technical and economic predictions of scientific and technical development
- Preparing technical standards and technical bases for classification
- On the basis of economic agreements supplying building and design-and-planning organizations with the results of problem-solving and co-operating in development
- Manufacturing prototypes and delivering prototype elements, fragments and/or units for experimental verification as part of the Institute's research activities.

Main fields of work

- Research on the theory of building structures, theoretical and experimental research in the field of building heat engineering, building light engineering, building acoustics, building chemistry, building mechanics, technology of concrete production (including lightweight concretes) and fire safety of buildings
- Research into new structures, their elements, components and whole systems of buildings of residential, public, industrial and agricultural construction including new building technologies
- Research into new products, especially those for the finishing of buildings and building services as well as research into the techniques and technologies of reconstruction work, modernization and repair of buildings
- Problem-solving in relation to the mechanization of building processes in building construction, transport and handling of materials (including the care of building machines) and foundations of buildings
- Estimating the building industry's requirements for the development of products in other supply branches and participating in the preparation of these products
- Problem-solving and co-ordinating research into the organization of building processes and management of buildings.

Total staff: 685
Professionally qualified staff: 442
Director or Chief Executive
Director: Hanuš Austerlitz
Heads of department
Section of Technical Development of Building Construction: Zdeněk Nekolný
Section of Research into Residential and Public Buildings: Otakar Paulus
Section of Research into Industrial and Agricultural Buildings: Vladimír Reichel
Principal sources of finance
State Budget and finance obtained from boards of management of building construction branches on the basis of economic contracts for executing professional work.
Principal related bodies
Advisory Board composed of Institute's leading workers. Scientific Council composed of prominent experts of teh Ministry of Building Industry, Universities and building organizations.
Publications

- Bulletin-Synopses, 6 times a year
- Results of Research Activities, annually
- Anniversary Volume of Research Tasks, annually
- 25–30 standards every year from the field of designing and executing building structures, heat engineering, acoustics, finishing work, etc.
- Guiding Rules and Technological Instructions, 25–30 parts a year
- Advertising materials referring to the Institute's research activities.

CSK.153
Czechoslovak Building Centre
Československé středisko výstavby a architektury (ČSVA)
Václavské náměstí 31, 111 21 Praha 1
Telephone: 268 269
Telex: 122 056 csva c
Cables: Baucentrum Praha
Date of establishment
1 January 1968
State Commission for Technology by passing an Establishment Paper

Scientific research and development institute operating as a specialized research organization whose basic task is to influence the development and quality of materials and products for construction in accordance with the needs of the national economy.
Main fields of work
The basic task of the Centre is to obtain, process and put into practice scientific and technical information on the development of products for construction, to build information systems and to provide information services, to follow the technical and aesthetic standard of products for construction. The Centre controls and co-ordinates the development of the system of selected information for construction, applies the results of research in the sphere of mechanization and automation of information processes and solves the research tasks in this field. The Centre also promotes international co-operation as a member of the International Union of Building Centres (UICB), in the framework of a CIB Working Group and as a specialized Centre of the International Building Information System of the CMEA countries (Council for Mutual Economic Assistance or Comecon). It functions also as the State Testing Laboratory No. 239, participates in the co-ordination of the State Testing Programme and the testing of products for construction.
Total staff: 136
Professionally qualified staff: 109
Director or Chief Executive
Director: Miloslav Matašovský
Technical Deputy Director: Josef Plsek
Economic Deputy Director: Jaroslav Krampol
Heads of department
International relations and information service: Ing Arch Navrátil
Development of material base and agrément of quality of building products: Ing Obr
Development of information system on products for construction: Ing Arch Kališová
Polygraphy and reprography, editions, publications: Ing Holubec
Branch Office, České Budějovice (processing of catalogue information): Ing Zavadil
Principal sources of finance
Payments by users of the Centre's information services; economic contracts; partly State budget.
Principal related bodies
Ministry for Construction and Technology of CSR, Federal Ministry of Technical Development and Investment, Council for Mutual Economic Assistance (CMEA), Secretariat of the Cz. delegation in CMEA.
Publications
- Czechoslovak Building Catalogue, quarterly
- Index of products and producers, annually
- Legislative Regulations, twice a year
- Blok magazine, quarterly
- Catalogue on Typification issued annually with supplements
- Manuals, up to 10 a year.

CSK.154
Refractories Research Institute
Vyskumny ústav hutníckej keramiky (VÚHK)
Februárového vít'azstva 71, 832 49 Bratislava
Other address:
Establishment for the applications of refractory materials: 703 00 Ostrava, Vrázova 31
Telephone: 661 31, 662 87
Date of establishment
1951

Set up by the Czechoslovak Ministry for Siderurgy to serve industries producing refractory materials and also the metallurgical industries.
Main fields of work
- *Research:* undertaken in the laboratory on models
- *Development:* construction of technical installations with documentation for the project; elaborating documentation for factory construction and for the production of refractory materials
- *Standardization:* standardization of refractory materials
- *Patents:* preparing the necessary documentation
- *Co-operation:* at the national level within the framework of Comecon or the Council for Mutual Economic Assistance (CMEA).

Total staff: 125
Director or Chief Executive
Director: A Cermák
Deputy Director: E. Kubran
Principal sources of finance
The State and enterprises.
Publications
- Abstracts from world literature
- Reports on research results.

CSK.155 **AM**
Research Institute for Building and Architecture
Výzkumný ústav výstavby a architektury (VÚVA)
Letenská 3, 118 00 Praha 1, Malá Strana
Telephone: 539651–6
Cables: VÚVA PRAHA
Other addresses:
Brno, Jakubské nám. 3
Ústí nad Labem, Belehradská 29
Date of establishment
1952

A research institute within the Ministry of Construction and Technology of the ČSR (Czech Socialist Republic).
Main fields of work
- Research activities in the field of complex house-building construction, architecture, urbanization, physical planning
- Realization of research results by assistance to central authorities, national councils and other authorities involved in construction through studies, expert comments, analyses, etc., and organization of seminars, conferences and training for investing and project organizations

In addition to these main activities the Institute performs special work for the Ministry of Construction and Technology.

Research departments are: Residential and Civic Construction; Technical Equipment of an Area; Urbanization and Physical Planning; Living Environment of the North Bohemian Region; Exact Methods (Mathematics); Economics, Sociology and Construction Management; Theory of Architecture; The Living Environment; Architectural Experimental Projects; Urban Experimental Projects. There are also Information Centres.
Total staff: 244
Professionally qualified staff: 120
Director or Chief Executive
Director: Ing Arch Václav Kasalický
Deputy Director: Ing Irena Lérová

Principal sources of finance
State budget.
Principal related bodies
Scientific Council, Board of Institute's Heads of Departments, Operational Board of the Director, Qualification Commission (discusses questions relating to the Institute's researchers gaining higher scientific grades).
Publications
Journal: Building and Architecture, monthly.

CSK.156

Research Institute for Civil Engineering
Výskumný ústav inžinierskych stavieb (VÚIS)
815 37 Bratislava, Lamačská 8
Telephone: 429 41
Telex: (09) 344
Date of establishment
1951

A research and development institution within the Ministry of Building; controlled by the Ministry of Technical Development; under present form and name since the Ministry of Building Decree of 1963. Chief co-ordination institution in civil engineering research and development. Co-ordination centre for standardization in civil engineering branch. Documentation and information centre for civil engineering.
Main fields of work
- Research, development and consulting in the following fields: bridges, roads and motorways, hydraulic constructions (dams and others)
- Underground service conduits, underground structures, tunnels, soil mechanics, foundations, earthworks and structures, blasting works, concrete technology and concrete works
- Consulting and assistance in the practical realization of research results, modifications to technical prescriptions, technical and scientific documentation concerning civil engineering
- Expert work in the matter of licensing foreign technologies

There are three sections: Research; Technical policy; and Administration service. The Research Section has the following departments: Civil Engineering Structures; Geotechnics; Concrete; and Auxiliary Services.
Total staff: 400
Professionally qualified staff: 200
Director or Chief Executive
Director: L. Kompis
Principal sources of finance
The major enterprises which benefit from the results of commissioned research; and the State.
Publications
- Bulletin VÚIS, quarterly
- *Research reports:* brochures issued after the completion of research summarizing the results
- Annual review: contains a review of the Institute's activities during the past year.

CSK.157

Slovak Academy of Sciences, Institute for Construction and Architecture
Ustav stavebníctva a architektúry, Slovenská akadémia vied (ÚSTARCH-SAV)
Dúbravská cesta, 842 20 Bratislava
Telephone: 451 63
Date of establishment
1953

The Slovak Academy of Sciences was founded as an organic part of the Czechoslovak Academy of Sciences. In its framework the Institute of Construction and Architecture was organized as the institute for engineering research.
Main fields of work
The present goal of the Institute is to study the problems of linear and geometrically non-linear theory of two dimensional structures, the theory of deformation and failure of reinforced concrete, metallic and combined building structures, the problems of physico-chemical mechanics of building materials, particularly on the silicate basis. Other goals to study are the problems of building physics, namely of thermal performance of buildings, of daylighting and earthquake engineering, the theory of town and country planning and environment as well as the problems of the influence of social factors on planning and development.

The Institute consists of ten research departments:
- Theoretical and Experimental Mechanics
- Structural Dynamics
- Theory of Structures of Non-homogeneous Materials
- Mechanics of Structures of Homogeneous Materials
- Physico-Chemical Mechanics
- Rhealogy of Heterogeneous Materials
- Performance and Climate in Buildings
- Building Daylighting
- Town Planning Theory
- Research into Social Factors in Planning and Development

In addition to these scientific research divisions, there is a number of independently organized units within the structure of the Institute. They are: the Mechanical and Technological Laboratory, the Laboratory of Dynamic Strength of Structures, the Engineering Department including the mechanical and electric shops and maintenance service, the Department of Scientific Information and the Economic and Administrative Section.
Total staff: Approximately 150
Director or Chief Executive
Director: Rudolf Skrúcaný
Deputy Director: Svetozar Lichardus
Principal sources of finance
Slovak Academy of Sciences.
Principal related bodies
Institute Committee comprising the director, deputy director, scientific secretary and heads of department.
Publications
The publishing activities of the Institute are numerous. Its bibliography is classified by departments. The Institute is responsible for the publication of two periodicals: the Stavebnícky časopis (Building Journal), published since 1953 and the Architektúra a urbanizmus (Architecture and Town-planning), appearing since 1967.

DNK.158

Aalborg Portland Technical Advisory Services

Aalborg Portland Cementfabrikkernes tekniske Oplysningskontor (CtO)

Rørdalsvej 44, PO Box 165, DK 9100 Aalborg

Telephone: (08) 16 77 77

Telex: 69646 cemex dk

Date of establishment

1934

Main fields of work

Technical information on cement and on the production and control of concrete for consumers and authorities.

Total staff: 12

Professionally qualified staff: 9

Director of Chief Executive

P. Nepper-Christensen

Publications

- Beton-Teknik (series of occasional publications)
- Beton-Bogen (handbook of concrete technology)
- Betonteknologi: (courses compendium)
- Betonlitteratur: (documentation and library services).

DNK.159

Cement and Concrete Laboratory

Cement- og Betonlaboratoriet

Aalborg Portland, Rørdalsvej 44, PO Box 165, DK 9100 Aalborg

Telephone: 8/16 77 77

Telex: 69646 (cemex dk)

Cables: Cementcentral

Date of establishment

1976

The Laboratory is part of Aktieselskabet Aalborg Portland-Cement-Fabrik and is engaged in research development and control in relation to production and usage of cement.

Main fields of work

- Laboratory support in cement production (quality control of cement and materials received, trouble-shooting in cement production, etc.)
- Laboratory support in the marketing of cement (testing of materials from customers, publications and courses for customers, trouble-shooting in concrete production)
- Research and Development (new cements, new applications of cement, alternative use of company resources)
- Library and documentation services
- Authorized cement testing.

Total staff: 40

Director or Chief Executive

Manager, Products Development: L. Hjorth

Principal sources of finance

Aktieselskabet Aalborg Portland-Cement-Fabrik.

DNK.160

Danish Asphalt Industries' Laboratory for Road Construction

Asfaltindustriens Vejforskningslaboratorium

Stamholmen 91, DK 2650 Hvidovre

Telephone: (01) 78 08 22

Date of establishment

November 1962, Danish Asphalt Industries' Information Office for Road Construction.

Private research laboratory for the members of the Danish Asphalt Industries' Information Office for Road Construction.

Main fields of work

To carry out research related to road construction as far as the application of asphalt is concerned.

Total staff: 6

Professionally qualified staff: 2

Director or Chief Executive

Chief of laboratory: J.M. Kirk

Principal sources of finance

Financed through contributions from the members of the Danish Asphalt Industries' Information Office for Road Construction.

Principal related bodies

Board of Directors appointed by the Danish Asphalt Industries' Information office for Road Construction.

Publications

Research reports: Internal reports.

DNK.161 **AM**

Danish Building Centre

Byggecentrum

Vester Voldgade 94, DK 1552 Copenhagen V

Telephone: + 45 1 12 73 73

Telex: 1 60 56 build dk

Cables: Buildingcentre

Other address:

Training Centre and Exhibition, Hindsgavl Allé, DK 5500 Middelfart

Telephone: + 45 9 41 14 41

Date of establishment

1956

Byggecentrum is a non-profit private foundation. The aim of the foundation is to serve the building industry and property owners with information, services and documentation.

Since 1983 the Ministry of Housing and 22 trade and professional organizations have elected a council of 27 to supervise Byggecentrum and elect its board.

Main fields of work

- Research and documentation projects
- Publishing of manuals, text books, reference books and periodicals
- Postgraduate training courses, seminars and symposiums
- Winter work consulting and training
- General information and marketing services
- Product information and distribution
- Permanent exhibitions covering building materials, components and accessories (1000 m² indoor exhibition area: 7000 m² open air exhibition)
- Operation of a training centre (12 lecture rooms seating from 12 to 200 people; 65 single bedrooms and 6 double bedrooms).

Total staff: 70

Professionally qualified staff: 20

Director or Chief Executive

President of the board and Council: Jørgen Bagge Andersen

Managing director: Petor Houmark-Nielsen

Financial director: Erik Holst

Heads of department

Projects: S. Hald

Publishing: J. Skovgaard

Training: F. Løve

Winter work consulting: Axel Kofoed

General information and marketing: P. Bindsbøll

Exhibition: L. Böye

Training Centre: L. Horsfeld

Development: H. von Scholten

Principal sources of finance

Revenue from book sales, subscriptions, course fees, training centre hiring out, exhibition area hiring out, and marketing services to private companies: 80%; revenue from required research and documentation projects, general information campaigns and winter work consulting services for governmental bodies: 20%. Annual turnover: 25 m Dkr (1983).

Principal related bodies

Byggecentrum co-operates with governmental bodies related to construction and housing and with semi-public and public institutions related to research and development within construction.

Affiliated company: Byggecentrums Litteraturtjeneste A/S, Vester Voldgade 94, DK 1552 Copenhagen V.
Telephone: + 45 1 12 70 70
Book-shop and abstracts retrieval services.

Publications

In Danish:

- Byggedata: A 6000 sheet manual on building legislation, standards, codes of practice, contract matters, references on research reports, and product information, circulation 1200 copies. Subscription includes updating at subscriber's office 6 times a year.
- Byggeorientering: A periodical, circulation 3500 copies.
- Byg-Erfa: Construction failures sheets, circulation 4500 copies.
- Construction rules, 1982
- Heating Pumps, 1981
- Total Insulations, 1980
- Winter construction, 1980

In English:
The Danish Building Sector Guide, 1981

DNK.162

Danish Building Industry Development Board
Byggeriets Udviklingsråd
Lønganstræde 37A, 1468 Copenhagen K
Telephone: (01) 12 28 11
Date of establishment
19 May 1971

A governmental institution to promote measures with the purpose of increasing the quality and productivity of the Danish building industry, as well as its competitiveness on the international market.

Main fields of work

The Board, according to the Law of 19 May 1971, seeks to identify conditions for the future development and structure of the industry and for export, as well as to increase the productivity of all parts of the building process. The Board sponsors viable experimental and developmental projects which promote the building industry's development, and provides for the dissemination of the results of these projects for future utilization within the industry.

Total staff: 6
Professionally qualified staff: 4
Director or Chief Executive
Flemming Frydendal
Principal sources of finance
Government grants.
Publications
Various reports are published in Danish concerning the results of individual sponsored projects.

DNK.163 FM

Danish Building Research Institute
Statens Byggeforskningsinstitut (SBI)
Postboks 119, DK 2970 Hørsholm
Visitors: Dr Neergaards Vej 15
Telephone: (02) 86 55 33
Cables: BYGFORSK
Date of establishment
1947 by Law No. 123 of 19 March 1947

An independent public institute working for the Danish building sector and serving all parties from legislators to users.

Main fields of work

The Institute has the following research divisions: Acoustics; Building physics; Service installations; Structural engineering, Building and environmental design; Urban and regional planning; Indoor climate techniques; Farm building; Economics.

The technical divisions are supported by laboratories, also mobile laboratories for measurements of indoor climate, energy and climate in farm buildings. There is also a secretariat and a contact and publications division.

Total staff: 205
Professionally qualified staff: 95
Director or Chief Executive
Director: Hans Jørgen Larsen
Principal sources of finance
Ministry of Housing, 70%; Ministry of Agriculture and Ministry of Energy 20%; private and other clients, 10%. Total 50 m Dkr (1981).
Principal related bodies
Governing body: five members appointed by the Minister of Housing; Council (also council for the Building Industry Development Board): about fifty members, appointed by the Minister of Housing.
Publications:

- SBI publications (about 50 per year) are issued in series according to the type of contents:
- SBI Directions give the results of building research work in a form directly application in practice.
- SBI Reports cover specific research projects.
- SBI Bulletins are preliminary reports and accounts of preparatory tests and reports from conferences and the like.
- SBI Town Planning give the results of town planning research.
- SBI Farm Buildings correspond to Directions, but for farm buildings.
- SBI reprints are offprints of important articles from periodicals.

SBI publications (mostly in Danish) are obtainable through Danish booksellers or direct from SBI. The list of publications, Latest SBI Publications, is normally brought up to date twice a year and contains short summaries of publications from the last five year period. It is available in Danish free of charge.

DNK.164

Danish Concrete Association
Dansk Betonforening (DBF)
Vester Farimagsgade 31, 1606 København V
Telephone: 01 15 65 65
Cables: Difingeniør
Date of establishment
1947

Association of individuals and firms interested in concrete technology and structures (1100 members including 130 firms.)

Main fields of work

The aim of the Association is to increase and distribute the knowledge of the properties, the production and the use of concrete. This is performed in collaboration with other technical groups, institutions, firms and persons interested in the associated field of activity.

The Association arranges several meetings during the year, where lectures on relevant subjects are given.
Director or Chief Executive
President: Carl F. Møller
Principal sources of finance
Members' subscriptions.
Publications
Nordisk Betong, 6 issues per year.

DNK.165
Danish Corrosion Centre
Korrosionscentralen
Park Alle 345, DK 2600 Glostrup
Telephone: +45 2 63 11 00
Telex: 3 33 88 SVC DK
Date of establishment
1 January 1965

A public institution at the disposal of all state, provincial and municipal services, civil services, industry and private individuals.
Main fields of work
The Centre works on the physical (metallurgical) and chemical (i.e. corrosion) behaviour of metals, and to a large extent on the behaviour of metals in buildings. Examples are:
- Metals in concrete: electrochemical measurements of corrosion of reinforcement (NDT)
- Metals in water supply systems; also pick-up of heavy metals in the water
- Metals in heating systems: corrosion in flue gases
- Metallic coatings
- Painting as corrosion protection
- Secretariat for the Danish Committee for Approval of Oil tanks
- Secretariat for the Danish Hot Dip Galvanizers Association.

Total staff: 55
Professionally qualified staff: 22
Director or Chief Executive
Director: Erik Nielsen
Heads of department
Corrosion Department: Ernst Maahn
Metallurgical Department: P.B. Ludwigsen
Research Department: Hans Arup
Principal sources of finance
Income approximately 75%; state grants approximately 20%; other grants approximately 5%.
Publications
- *Journal:* Newsletter (Corrosion of Steel in Concrete), issued periodically
- Research reports (in Danish).

DNK.166
Danish Hydraulic Institute
Dansk Hydraulisk Institut (DHI)
Agern Alle 5, 2970 Hørsholm
Telephone: 45-2 +86 80 33
Telex: 37 402 dhicph dk
Cables: hydroinstitute
Date of establishment
February 1964

DHI is an independent non-profit consulting organization, affiliated to the Danish Academy of Technical Sciences, performing comprehensive research programmes (field surveys and mathematical and physical model studies) in relation to port, coastal, offshore, river, water resources and environmental engineering.
Main fields of work
- Studies of wave conditions, ship motions, sedimentation, environmental forces related to design of ports and other coastal structures
- Surveys, environmental data evaluations and scale model studies for marine pipelines and offshore structures
- Cooling water recirculation studies, investigations of oil spill and sewage discharges
- Oceanographic investigation, hydrographic and seismic surveys
- Irrigation and drainage, flood control and urban hydrology
- Hydrodynamic modelling, tidal and storm surge studies

Main emphasis is put on development and application of advanced mathematical models and scale modelling techniques. A three-dimensional hydrodynamic-numerical model is currently under development and a new test basin for offshore structures subject to natural short-crested waves has just been implemented.
Total staff: 155
Professionally qualified staff: 78
Director or Chief Executive
Director: Torben Sørensen
Heads of department
Ports and Marine Structures Department: Jens Kirkegaard
Environmental and Coastal Department: Peter Mortensen
Water Resources Department: Torkil Jønch-Clausen
Computational Centre: Gaele S. Rodenhuis
Offshore Department: Peter I. Hinstrup
Principal sources of finance
Internally financed with support from the Danish Council of Technology.
Principal related bodies
The Danish Academy of Technical Sciences has appointed the Board, consisting of ten members from government, science and private enterprise within the field of civil engineering.
Publications
- *Journal:* Danish Hydraulics, once, or twice per year
- *Research reports:* about 20 scientific papers are published annually.

DNK.167
Danish Illuminating Engineering Laboratory
Lysteknisk Laboratorium (LTL)
Bygning 325, Lundtoftevej 100, 2800 Lyngby
Telephone: (02) 87 39 11
Date of establishment
1 April 1964 by the Illuminating Engineering Society in co-operation with the Technical University.

An independent laboratory performing research, testing and other contract work. Affiliated to the Academy of Technical Sciences. Authorized by the National Testing Board. Approved by the Ministry of Industry as a technical service institute.
Main fields of work
Research, testing, calculation and product development related to illuminating engineering. Postgraduate courses in lighting and teaching at the Technical University. Main fields of research work: road lighting, reflection properties of road surfaces, work lighting, daylight and energy.
Total staff: 18
Professionally qualified staff: 10
Director or Chief Executive
Director: Bjarne Nielsen
Principal sources of finance
Council of Technology (Ministry of Industry) 40%; contract work 40%, other sources 20%.
Principal related bodies
Supervisory board of directors appointed by the Academy of Technical Sciences.
Publications
- *Journal:* Lampetten, in Danish, published in association with the Illuminating Engineering Society

- A series of research notes (LTLI-notater) is available on subscription (in Danish).
- Own series of technical reports: 34 reports published, one-third in English. List of reports available on request.

DNK.168

Danish Lime and Brick Laboratory
Kalk- og Teglværkslaboratoriet (K&T)
Teglbaekvej 20, 8361 Hasselager
Telephone: (06) 28 38 11
Date of establishment
1948

An independent institution, authorized by the Danish National Testing Board and approved as a Technological Service Institute by the Danish Department of Industry.

Main fields of work
The Laboratory works as a research and testing laboratory for manufacturers and users of materials and constructions connected mainly with the brick building trade. Besides equipment for physical and chemical tests of the most important raw materials of the industry (clay, lime and sand) and equipment for testing the manufactured products including their dimensions, water absorption, water and air permeability, suction rates, density, volume stability, compressive strength, pore conditions, frost resistance, etc., the Laboratory has a number of kilns for firing tests. More specialized equipment includes a van with equipment for measuring and determination of energy consumption by drying and firing processes at brickworks, a drying plant for investigating the drying properties of newly moulded products, apparatus for analysis of plaster mix and mortar, a fully automatic plant for tests of frost resistance, and a clamp plate for strength tests of storey-high wall units.

The Laboratory also undertakes surveys and evaluations of defects as regards building methods as well as education classes and lecture activities.

Total staff: 21
Professionally qualified staff: 15
Director or Chief Executive
C. Falk
E. Kjaer
Principal sources of finance
The expenses are covered by contract research, the Brick and Lime Association and the Danish Government.
Principal related bodies
The Board of Directors of the Laboratory is formed by members of the Brick and Lime Association and the brick building trade.
Publications

- Laboratory reports
- Reports submitted at national or international conferences and/or published in the two Danish periodicals Kalk- og Teglnyt (quarterly) and Tegl (quarterly) or in international journals.

DNK.169

Danish National Centre for Building Documentation
Byggeriets Studiearkiv (BSA)
Peder Skramsgade 2 D, DK 1054 Copenhagen K
Telephone: (01) 12 37 95
Date of establishment
Established in 1930 by the Royal Danish Academy of Fine Arts, School of Architecture.

Public information and documentation centre with a special library at the disposal of those in research or education, consulting architects and engineers, contractors, manufacturers, state, provincial and municipal services and private individuals.

Main fields of work

- All aspects of modern architecture, building, housing and planning. The information and documentation services cover:
- Retrospective searches and SDI-profiles on relevant national and international data bases; current awareness service, editing and abstracting Danish input to international data bases.
- Maintenance of files of abstracts and bibliographies; analyses of new building literature; compilation of subject bibliographies; editing of state-of-the-art reports; editing of guides and manuals; arranging courses, seminars and lectures.
- Research and development within building information and documentation.
- Library containing books, reports, periodicals; special collections of laws, bylaws, regulations, etc. statistics, standards, trade catalogues, contract documents – basic and samples, maps and drawings. Collection of audio-visual material is planned.
- Library services: document delivery, loans, photocopies. Other services: reprographic studio, offset printing.

Total staff: 24
Professionally qualified staff: 10
Director of Chief Executive
Chief Executive: Steen Estvad Petersen
Heads of department
Information and documentation service: Leif Jønsson
Library: Mogens Fischer
Publications and other services: Karen Simonsen
Principal sources of finance
Financed as a section of the Royal Danish Academy of Arts, School of Architecture by the Ministry of Culture.
Principal related bodies
Supervisory Council with members from the Federation of Danish Architects, Danish Association of Consulting Engineers, Chamber of Danish Trade and Crafts, the Danish Building Industry Development Board, the School of Architecture and the staff.
Publications

- *Journal:* Byggreferat, Scandinavian building abstract journal, monthly
- Occasional papers on various subjects
- State-of-the-art reports with select bibliographies
- Select subject bibliographies with introduction to the subject proper
- Subject bibliographies
- Information folder
- Accession list, quarterly
- Current periodical list, annually
- List of publications, annually.

DNK.170

Danish Society for Structural Science and Engineering
Dansk Selskab for Bygningsstatik (DSBy)
ABK, bygning 118, Technical University of Denmark, DK 2800 Lyngby
Telephone: (02) 88 35 11
Telex: 37529 dthdia dk
Date of establishment
1928

The object of the society is to arrange meetings with lectures and discussions.

Director or Chief Executive
Chairman: Professor M.P. Nielsen
Secretary: J. Nielsen
Principal sources of finance
Members' fees.
Publications
Journal: Bygingsstatiske Meddelelser, quarterly

DNK.171

Danish Society of Chemical, Civil, Electrical and Mechanical Engineers

Dansk Ingeniørforening (DIF)

29 Vester Farimagsgade, 1606 København V

Telephone: (01) 15 65 65

Date of establishment

1892

A professional society.

Main fields of work

DIF is a professional society of engineers who have graduated from the Technical University of Denmark and the Danish Engineering Academy. Through permanent and provisional committees the society aims at promoting technical development. DIF formulates codes of practice for different building methods and building materials.

Director or Chief Executive

President: Niels Hjorth, CivEng

Chief Manager: Hans-Ole Skovgaard, CivEng

DNK.172

National Institute for Testing and Verification (DANTEST)

Dansk institut for Prøvning og Justering (DANTEST)

Amager Boulevard 115, DK 2300, Copenhagen S

Telephone: +45 1 54 08 30

Telex: Fotex DK 16 600

Cables: Governtest

Date of establishment

Founded 1896; 1909 State Institute; Statutory document of 12 December 1979; 1980 Independent Foundation.

Public Test House and Metrology Institute accredited by the Danish National Testing Board and established as an independent foundation supported by public grants.

Main fields of work

Dantest is a modern centre for technical testing covering the main fields of: fire technology, mechanical, building, chemical engineering and calibration and verification of measuring instruments.

The primary objectives of the Institute are to offer services and aid to Danish commerce, trade and industries, and to offer advice and guidance to public authorities through unprejudiced technical expertise relating to:

- Technical testing (fire technology- mechanical, constructional and chemical engineering)
- Calibration (mass, length, volume, force and pressure)
- Pattern examination and calibration of measuring instruments liable to calibration or verification
- Technical Advisory Service and Consultation, concerning: testing techniques; fire prevention and fighting; quality control and surveillance

Through the wide scope of its activities, Dantest also contributes to the community's endeavours to promote (through testing and voluntary control measures) industrial safety, environment and consumer protection.

The Institute is directed by an Executive Board consisting of 11 members, representing the principal user interests within trade, commerce, consumer and public authorities.

Total staff: 175

Professionally qualified staff: 45

Director or Chief Executive

Director: E. Repstorff Holtveg

Heads of department

Verification Branch: B. Flensted Sørensen

Mechanical Branch: A. Klingenberg

Metrology Division: L. Rosenkilde

Building Materials Branch: N. Schøttz Andersen

Fire Technology Branch: Ejnar Danø

Chemical Technology Branch: Åse Skov

Principal sources of finance

Public grants (for industrial metrology): approximately 20%; revenues: balance.

Principal related bodies

Board of Directors: (Chariman: Professor Dr Tarras Wanheim); Advisory Committees: one each for: fire technology; chemical, constructional and mechanical engineering; metrology.

Publications

- DANTEST Annual Report
- *Technical standards:* DANTEST methods
- *Research reports:* DANTEST Technical Reports; DANTEST Technical Monographs
- *Others:* DANTEST brochures (in Danish and English); DANTEST: A Centre for Technical testing, Calibration and Verification, 1983.

DNK.173

Scandinavian Paint and Printing Ink Research Institute

Nordisk Forskningsinstitut for Maling og Trykfarver (Danish)

Nordiska Institutet för Färgforskning (Swedish) (NIF)

Agern Allé 3, 2970 Hørsholm

Telephone: (02) 57 03 55

Date of establishment

1969 by amalgamation of the Central Research Laboratory of the Danish Varnish and Paint Industry (founded 1946) and the Central Research Laboratory of the Swedish Paint and Varnish Industry (founded 1943).

A self-governing institution established in affiliation with the Danish Academy of Technical Sciences and the Royal Swedish Academy of Engineering Sciences.

Main fields of work

- Coating division: Interaction between pigments, binders and solvents. Dispersions in their production and stabilization. Application of paints and paint film testing for mildew, weathering (outdoor and in laboratory), testing of corrosion resistance.
- Painting on different building materials (concrete, plastics, wood, etc.) Dirt pick-up and outdoor durability of paint films. Investigations by scanning electron microscope. UV light stability and transmission.
- Mechanical and physical properties of paint (liquid) and paint films. Flow out defects. Internal stresses in paint films. Use of computer techniques in paint technology. Viscosimetry and rheology in general.
- Printing Ink section: Investigation of optical, rheological and printing properties and testing of printing inks. Measurements of colour, testing of pigments (both organic and inorganic grades). Environment questions, pollution, toxicology, etc.
- Analytical section: Chemical analysis of paints and printing inks in wet as well as in dry (film) conditions. Advanced techniques and instruments are used (IR-spectroscopy, gas chromatography, NMR, etc). Development of new analytical methods in the paint and lacquer field. A special field in analytical investigations of environmental problems arising from manufacturing and use of paint products. Forensic analysis.

Total staff: approximately 25

Professionally qualified staff: 0

Director or Chief Executive

Director: Charles M. Hansen

Deputy Director: Ago Saarnak

Principal sources of finance

Contributions from the Swedish, Finnish and Danish industries, grants from Swedish, Finnish and Danish Research and Technology Councils and funds paid by individual companies and organizations for contracted investigations.

Publications

- Technical reports to members (paint manufacturers in Denmark and Sweden)

- Research reports of an official character sent to the grant organization (e.g. The Swedish Building Council)
- Papers presented at international congresses
- Papers presented in foreign journals e.g. Jocca (England), JCT (USA)
- Papers presented in the Scandinavian journal Färg och Lack.

DNK.174 **UM**

School of Architecture, Department of Building Administration

Kunstakademiets Arkitektskole, Instituttet for byggeteknik, Afdelingen for Byggeriets Administration, Peder Skramsgade 1, DK 1054 København K

Telephone: 01 126860

Date of establishment

The Department of Building Administration was established in 1960.

The Department of Building Administration within the School of Architecture of the Royal Academy of Arts is occupied with the organization and structure of the building industry, building economics and finance, legislation relevant to building together with the tasks of the Architect in connection with the organization, planning, recording and control of building projects.

Main fields of work

The programme of the department comprises studies and seminars in design methodology, particularly emphasizing methods for design cost control and cost control during the building process proper. The courses of the department comprise inter alia drawing planning, specification, cost planning, quantity surveying, time planning, tendering procedures, contracts, site supervision, cost control, etc. Courses include computer-aided building administration.

The research programme of the department comprises management information systems for the building process, particularly for use with computers.

Total staff: 3

Professionally qualified staff: 2

Director or Chief Executive

Bjørn Bindslev, Arkitekt, docent, dr techn

Principal sources of finance

State financed.

DNK.175

Denmark Technical University, Building Materials Laboratory

Danmarks tekniske Højskole, Laboratoriet for Bygningsmaterialer (DtH LfB)

Bygning 118, 2800 Lyngby

Telephone: 45 2883511

Telex: 37529 DTHDIA DK

Date of establishment

1969

The Laboratory is part of the Civil Engineering Department of the Technical University of Denmark and perfoms testing upon request.

Main fields of work

Research and teaching in the field of building materials (primarily cement, concrete and wood).

Total staff: 14 of whom approximately half are support staff.

Director or Chief Executive

Professor Torben C. Hansen

Principal sources of finance

Government.

Publications

Papers published in professional or scientific journals.

DNK.176

Denmark Technical University, Department of Construction Management

Danmarks tekniske Højskole, Instituttet for Anlægsteknik (DtH IfA)

Building 115, DK 2800 Lyngby, Denmark

Telephone: 02 884 4200

Telex: 37529 DTHDIA DK

Date of establishment

1962

The Department is part of the Technical University of Denmark.

It conducts teaching and research (also upon request).

Main fields of work

Organization and working conditions in the construction industry. All major aspects of planning and execution of construction contracts, except the actual design of permanent structures. The theoretical background necessary for understanding the more basic economic and managerial problems encountered by firms in the construction industry.

Total staff: 16

Professionally qualified staff: 12

Director or Chief Executive

Head of Department: Kristian Kreiner

Principal sources of finance

Public.

Publications

- Theses
- Articles
- Reports.

DNK.177

Technical University of Denmark, Department of Environmental Engineering

Danmarks tekniske Højskole, Laboratoriet for teknisk Hygiejne (DtH LtH)

Bygning 115C, DK 2800 Lyngby

Telephone: (02) 88 42 00

Telex: 37529 DTHDIA DK

Date of establishment

1865

The Department is part of the Technical University of Denmark; it teaches environmental engineering and performs environmental investigations upon request.

Main fields of work

Water and wastewater treatment; urban run-off; soil pollution; solid waste.

Total staff: 25

Professionally qualified staff: 15

Director or Chief Executive

Head of the Department: Nils Juul Larsen

Principal sources of finance

Government.

Publications

Research reports: numerous: a list is available from the librarian of the Department.

DNK.178

Denmark Technical University, Department of Structural Engineering

Danmarks tekniske Hojskole, Afdelingen for Bærende Konstruktioner (DtH ABK)

Building 118, 2800 Lyngby

Telephone: (02) 88 35 11

Telex: 37529 DTHDIA DK

ABK is a department of the Technical University of Denmark which covers education and research in structural engineering. In addition, measuring methods and the reliability of structures are dealt with.

Main fields of work

The main field of work comprises all aspects of research, sponsored research and teaching of the stress, strength and deformation analysis of building materials and structural components subjected to static or dynamic loadings, of fibre concrete technology, of structural reliability and of structural design and construction.

Total staff: 60

Professionally qualified staff: 38

Director or Chief Executive

President of the Board: Professor Vagn Askegaard

Principal sources of finance

Government; to this is added some income from sponsored research.

Publications

- Reports containing results of major research, dissertations, etc. (Series R)
- Internal reports (Series I)
- Lecture notes (Series F)
- Numerous papers in professional or scientific journals
- List of summaries of papers published by the staff, annually (free of charge).

DNK.179

Denmark Technical University, Foundations Laboratory

Danmarks tekniske Hojskole, Laboratoriet for Fundering (DtH LF)

Maglebjergvej 1, DK 2800 Lyngby

Telephone: 02 88 44 44

Date of establishment

1955 when the Chair in Soil Mechanics and Foundation Engineering was created.

The laboratory is a department of the Technical University of Denmark, located at and co-operating closely with the Danish Geotechnical Institute.

Main fields of work

Soil mechanics and foundation engineering courses at the Technical University of Denmark. Research on strength and deformation properties, particularly of Danish soil types, and on methods of analysis concerning these properties and the design of foundations, earth-supporting structures, etc.

Total staff: 3, plus postgraduate students; and assistance from the staff of the Danish Geotechnical Institute

Professionally qualified staff: 3

Director or Chief Executive

Professor dr tech Bent Hansen

Publications

Lecture notes are currently being revised with the aim of establishing an up-to-date textbook system in soil mechanics and foundation engineering. Within the framework of collaboration with the Danish Geotechnical Institute, the Laboratory is contributing to the bulletin series of the DGI appearing at irregular intervals, to international conference proceedings. Publications also include postgraduate theses in soil mechanics.

DNK.180

Denmark Technical University, Heating and Air-Conditiong Laboratory

Danmarks tekniske Højskole, Laboratoriet for Varme- og Klimateknik

Building 402, DK 2800 Lyngby

Telephone: +45 2 88 46 22

Telex: 37529

Date of establishment

1935

Institute at the Technical University of Denmark. Higher education is covered and research is carried out. Testing and development work are performed for industry and others.

Main fields of work

Higher education and research at the Laboratory comprise heat production and distribution as well as the effect of indoor climate on man's comfort, health and work.

The field of activity can roughly be divided into heating and air conditioning:

Heating comprises: heat production by combustion of fossil fuel (combustion technique); construction of boilers and burners; heat distribution in tube networks; heat supplies to rooms; and air pollution by smoke emission from chimneys.

Air conditioning comprises: thermal and atmospheric demands on the indoor climate; heat, humidity and pollution loads in buildings; climatic processes and climatic systems; heat-recovery; air movements in rooms.

Both subjects comprise energy optimazation, automatic control and data simulation.

Building fire technique has a natural connection with combustion technique and is also within the field of activity.

Total staff: 25

Professionally qualified staff: 18

Director or Chief Executive

Assistant Professor Ole Albrechtsen

Principal sources of finance

Annual appropriation from the Government. Support from funds and from other institutions for fulfilment of projects.

Publications

Numerous papers are published in international scientific and professional journals, many of these are reprinted by the Laboratory and circulated to the most appropriate audience for the particular paper. Annually, the Laboratory together with other institutions at the University publishes its Research Programme.

DNK.181

Denmark Technical University, Institute of Building Design

Danmarks tekniske Højskole, Instituttet for Husbygning (DtH IfH)

Building 118, DK 2800 Lyngby

Telephone: +45 2 88 35 11

Telex: 37529 DTHDIA DK

Date of establishment

1857, reorganized 1970

A laboratory within the Civil and Structural Engineering Department of the Technical University of Denmark.

Main fields of work

Teaching, research and consultative activities within the fields of:

- Building Design and Technology in general Design and Technology of Prefabricated Building Systems
- Co-ordination of Architectural and Engineering Design
- Structural Fire Engineering and Fire Safety Design
- Resource-economizing Design of Building Structures.

Total staff: 19
Professionally qualified staff: 12
Director or Chief Executive
Head of the Institute: Professor Johs. F. Munch-Petersen (1982)
Principal sources of finance
Public.
Principal related bodies
Other laboratories within the Civil and Structural Engineering Department; Danish Building Research Institute (SBI).
Publications
Research reports: updated list available.

DNK.182

Denmark Technical University, Institute of Hydrodynamics and Hydraulic Engineering
Danmarks tekniske Højskole, Instituttet for Strømningsmekanik og Vandbygning (DtH ISVA)
Building 115, DK 2800 Lyngby
Telephone: 45 2 88 42 00
Telex: 37529 DTHDIA DK
Cables: Hydroeng
Date of establishment
1971

The Institute is part of the Technical University of Denmark.
Main fields of work
Hydrology, water resources planning, hydraulics, hydrodynamics, sediment transport, stratified flows, coastal engineering, wave mechanics, offshore engineering, harbour planning, ice problems.
Total staff: 36
Professionally qualified staff: 23
Director or Chief Executive
Director: Associate Professor Dan Rosbjerg
Principal sources of finance
Governmental funds.
Publications
- Progress Report (research notes), 3 times per year.
- Series papers (major research reports), irregular.
  Both are sent free of charge on an exchange basis.

DNK.183

Denmark Technical University, Thermal Insulation Laboratory
Danmarks tekniske Højskole, Laboratoriet for Varmeisolering (DtH LfV)
Bygning 118 D, DK 2800 Lyngby
Telephone: (02) 88 35 11
Telex: 37529 DTHDIA DK
Date of establishment
1959

LfV is a laboratory at the Technical University of Denmark. The laboratory teaches engineering, carries out research projects and performs testing (including testing upon request).
Main fields of work
Building physics, specifically concerning the building envelope as protection against the outside environment. Since 1973 the laboratory has also been investigating the application of solar energy for heating purposes, and in 1979 several projects on heat storage were initiated, now including seasonal storage.
The research projects, most of which are based on external funds, cover the following topics:
- Heat and moisture transport in building materials, e.g. insulation materials
- Low-energy houses (research and demonstration projects)
- Passive solar systems and buildings (research and computation methods)
- Active solar systems for heating buildings and domestic hot water (research, demonstration projects and computational aids)
- Testing of solar systems (development of test procedures for and commercial testing of total systems and single components)
- Short-term thermal storage systems (research, demonstration projects and testing)
- Seasonal thermal storage (research, computation methods and pilot demonstration projects)
- Energy consumption and thermal indoor climate (calculation methods and weather data for such calculations)
- Thermal comfort and discomfort (research and development of measuring instruments)
- Heating systems and heat sources for energy conservation houses (research and demonstration projects)
- Insulating window shutters.

Total staff: 68
Professionally qualified staff: 38
Director or Chief Executive
Study Guide: Professor Vagn Korsgaard
Director: Bjarne Saxhof
Principal sources of finance
Danish Ministry of Education; Danish Ministry of Energy; The European Community.
Publications
Research reports: within the period 1976–82 the laboratory issued about 100 publications, many of which appeared as papers for international congresses and symposia or as articles in international periodicals. Within the same period about 290 internal technical reports were printed.
List of publications available upon request.

EGY 184 **FM**

General Organization for Housing, Building and Planning Research (GOHBP)

El-Tahreer Street, Dokki, Giza, PO Box 1770, Cairo

Telephone: 711 564

Cables: HOUDPLAN, CAIRO

Date of establishment

January 1977 by Presidential Decree, by the amalgamation of the Building Research Centre (founded 1954), the General Directorate for Research and Central Laboratories of the Ministry, in addition to assuming the responsibilities of the Ministry in the fields of housing and physical planning research.

An independent Government research establishment which has a Board and a Chairman working under the supervision of the Minister of Housing and Reconstruction. 14 Departments cover building materials, construction, economics, housing, environment, acoustics, physical planning and research.

Main fields of work

GOHBP is the Ministerial body responsible for scientific research and technical development in the fields of housing, building and planning. Scientific Departments are:

- Regional and town planning
- Foundations and underground structures
- Strength of materials and testing
- Building materials
- Dwellings and public buildings
- Sanitary and hydraulic structures
- Financial and administrative management
- Building physics and environment

Principal areas of research are:

Housing: to study the architectural and structural designs of housing, services and public buildings with the aim of improving their standards and reducing expenses to meet the economic conditions of the country, and also to lay down measurements and norms for architectural components. To participate in researches on standardization of architectural constituents in similar constructions, components and prefabricated units that are parts of various constructions. Finally, to develop structural theories in design in order to economize on building materials and reduce construction costs.

Construction: to study the characteristics of various building materials and undertake the necessary researches to improve them, eliminate their defects and increase their efficiency while reducing production costs. To produce building materials from raw materials and industrial and agricultural residues with the aim of discovering substitutes for the currently used materials; to study the traditional methods of building and construction and endeavour to develop them; to study the methods for laying foundations appropriate for the nature of the soil; to participate in drawing up plans for applied researches and studies carried out by general organizations and institutions, companies or any other organs in the field of building materials manufacture.

Physical Planning: to undertake research related to physical planning in urban and rural communities; to study conditions and principles in respect of land use for different purposes; to specify the ratio of public services buildings in residential centres.

Training: to co-operate with the relevant institutions in the implementation of a training policy in scientific research.

Total staff: approximately 250, among whom approximately 60% are support staff.

Director or Chief Executive

Chairman: Dr Abov Zeid Rageh

Principal sources of finance

Largely financed through the annual Government fund for ministries, governorates and independent organizations. Substantial resources are also drawn from the construction industry for work done at the Organization. Another source of finance is joint research work with foreign institutions such as the National Science Foundation, Washington, international agencies such as UNESCO, etc.

Principal related bodies

Board of Directors; ad hoc Committees for Codes of Practice, Current Costsfof Construction, etc.

Publications

A number of papers have been published individually by the staff members in professional and scientific journals. Numerous problems have been investigated and the results were published in Arabic locally for circulation within the local building industry and within some of the Arab countries, but plans are under way to have publications in English as current papers. It is also intended to have a periodical in English which would report on the current activities here and abroad in the field of research.

ETH.185 **FM**

Faculty of Technology (South)

ቴክኖሎጂ ፋካልቲ /ደቡብ/

PO Box 518, Addis Ababa
Telephone: 447030
Telex: AA-UNIV 21205
Cables: BUILDINST

The Faculty was formerly known as the Ethio-Swedish Institute of Building Technology and was established under a bilateral co-operation agreement between Ethiopia and Sweden. At present it is part of the Addis Ababa University.

Main fields of work

Education, including a five-year degree course in architecture, a three-year diploma course in building technology and at times upgrading courses for construction foremen. Testing and research on indigenous building materials. Information, library and other services.

A number of staff serve as committee chairmen and members of the different Ethiopian Standard Institute's technical committees and of the building codes technical committees of the Ministry of Construction.

Total staff: 120

Professionally qualified staff: 30

Director or Chief Executive

Associate Dean: Dr Zawde Berhane

Heads of department

Department of Architecture and Town Planning: Mulugeta Metaferia

Department of Building Technology: Yohannes Hailu

Materials Research and Testing Department: Assistant Director: Makonnen Wondemagegnehu

Principal sources of finance

Mainly the Government and a small amount of fees for services.

Principal related bodies

Academic Commission of the Faculty of Technology- the Senate of the Addis Ababa University.

Publications

A number of research and investigation results have been published. A list of these can be provided on request free of charge.

FIN.186
Association of Finnish Civil Engineers
Suomen Rakennusinsinöörien liitto (RIL)
Meritullinkatu 16A, 00170 Helsinki 17
Telephone: 358 0 661 693
Date of establishment
4 April 1934

A professional, technical and scientific association of Finnish graduate civil engineers. (Members of RIL are Masters of Science in building and civil engineering.)
Main fields of work
The Association:
- Develops Finnish techniques and economy in civil engineering
- Publishes books, standards and two professional journals
- Participates in standardization in close co-operation with the public authorities
- Exercises quality control of certain building materials
- Makes proposals and issues expert opinions on matters within the scope of the Association
- Organizes high-standard postgraduate courses and seminars
- Holds design competitions
- Holds a competition under the title The Construction Engineering Work of the Year
- Co-operates with other national and international societies.

There are four departments: Publishing; Course; Research; and Management.
Total staff: 11
Professionally qualified staff: 5
Director or Chief Executive
Managing Director: Yrjö Matikainen
Principal sources of finance
Membership dues; publishing; courses; research and consulting activities.
Principal related bodies
Board of directors, having the highest authority; Divisions, auxiliary organs of the Board; Committees, working groups set up by the Board; Editorial committees, responsible for production publications.
Publications
- *Journal:* Rakennustekniikka, 8 issues per year.
- *Technical standards:* RIL series (publications of the Association of Finnish Civil Engineers); Standards and directives
- Member catalogues
- RIL series: Handbooks; Course books.

FIN.187
Brick Laboratory of the Finnish Brick Industry Association
Kimmeltie 3, 02110 Espoo 11
Telephone: (9)0 455 0755
Date of establishment
Decision in principle of the board of directors of the Finnish Brick Industry Association 29 November 1948: the first laboratory director 16 March 1949.

The Brick Laboratory is an interior department of the Finnish Brick Industry Association.
Main fields of work
- Testing and research of clays and clay masses as raw material in the brick industry
- Testing of bricks and clay pipes and other brick products

Tests are made according to standards SFS 2803 (RIL 85) and RIL 128–1981.
Total staff: 2
Professionally qualified staff: 2
Director or Chief Executive
Laboratory Director: Martti Romu
Principal sources of finance
Financed principally by membership fees paid by brickworks.
Principal related bodies
Board of directors of the Finnish Brick Industry Association.
Publications
Several technical recommendations for masonry structures in practice.

FIN.188 **AM**
Building Information Institute
Rakennustietosäätiö Bygginformationsstiftelsen
Lönnrotinkatu 20 B, SF 00120 Helsinki 12
Telephone: 90 645 615
Telex: 122303 stand sf
Date of establishment
1932 Helsinki Building Centre; 1942 SAFA Planning and Building Standards Institute; April 1972 Building Information Institute

A private non-profit foundation.
Main fields of work
The Institute promotes information services, research and standardization of planning and building construction. It maintains committees for information development, research programming and standardization. The Institute also grants funds for research and development especially where information activities are demanded.

The Institute is responsible for standardization in the field of building construction. The results are issued as national standards by agreement with the Finnish Standards Association.

The Institute is the author of a comprehensive information system called RT Building Information File. The file consists of
- 2000 pages of a Standards File: standard solutions of building elements, product standards, design guidelines and unified contracting terms
- 1000 pages of a Regulations File: continuously updated legislative regulations which may be needed during the building process
- 2500 pages of a Products File: technical data of available building products
- 500 pages of an Information Source File: institutions as information sources, updated references to periodicals and books, literature indices

Building Information File is published by Building Book Ltd.

The Building Book Ltd (Rakennuskirja Oy) acts as publishing and printing house for the Institute. It is totally owned by the Institute, but has a management of its own. Besides the Building Information File the company publishes different kinds of technical literature.

The Building Centre Ltd (Rakennuskeskus Oy) runs an over-desk information service in Helsinki. It has a building products exhibition of 1500 m² floor area in the centre of Helsinki. In connection with the exhibition there is free distribution of trade literature and a bookshop.

The Building Centre also maintains EDP data bases of building materials, organizations and literature. A focal point of the development in this area is job specification data base and CAD procedures for building design.
Total staff: approximately 60
Professionally qualified staff: 20
Director or Chief Executive
Director General: Professor Esko Lehti
Heads of department
Building Book Ltd: Managing Director: Timo Olkkonen
Deputy Managing Director: Martti Tiula
Building Centre Ltd: Managing Director: Juhani Keppo
Principal sources of finance
Sale of publications (60%); commissions (30%); government grants (10%).
Principal related bodies
Council (an advisory and policy-making body); Board (the body responsible for finance and activities); Board of Directors (management); Advisory and standardization committees for different purposes.

Publications
- RT-uutiset, 8 copies per year
- Muoto, quarterly
- Building Information File
- Publishing activities by the Building Book Ltd cover about 15 titles yearly: hand books, text books and research reports on the branches of architecture, planning and construction.
- Building Products Index
- Organization Index for the Construction Industry
- Building Literature Index.

FIN.189
Concrete Association of Finland
Suomen Betoniyhdistys, Finska Betongföreningen ry
Mikonkatu 18 B 12, SF 00100 Helsinki 10
Telephone: 358 0 651 411
Date of establishment
16 October 1925

A technological–scientific association, whose main function is to improve concrete technology.
Main fields of work
Standardization, advancement of research work, supplementary training, organization of congresses, seminars and meetings, etc.
Total staff: 6
Professionally qualified staff: 2
Director or Chief Executive
Managing Director: Heikki Kaitila
Principal sources of finance
Training courses, publications, membership fees, standardization fees.
Publications
- *Journal:* Nordisk Betong, 6 per year
- *Technical standards:* Codes of practice for structural design and construction of concrete structures; other unofficial recommendations
- *Research reports:* Nordic Concrete Research 1982; Annual report of the Nordic Concrete Federation
- *Others:* Training course publications.

FIN.190 **AM**
EKONO Consulting Engineers
EKONO Oy
EKONO Oy, Tekniikantie 4, 02150 Espoo 15
Mailing address: EKONO Oy, PO Box 27, SF 00131 Helsinki 13
Telephone: 358 0 46911
Telex: 124822 ekono sf
Cables: EKONO HELSINKI
Date of establishment
EKONO was founded in 1911 as the Finnish Steam Boilers Owners' Association

EKONO is a world wide independent consulting engineering company owned by The Finnish Energy Economic Society. Members of the society are owners of industrial and municipal boilers and buildings, totalling approximately 620.
Main fields of work
EKONO's research work covers the following fields in the building sector:
- Structural engineering
- Steel construction
- Industrial building
- Heating, sanitary and air conditioning systems for buildings
- Municipal energy supply systems
- Energy economy of buildings
- Calculation methods for constructions, energy balances, etc.
- Effect of climate on buildings
- Lighting systems for buildings

Type of research is close to practical applications. Typical projects include:
- Elaboration of design guidelines
- Product development
- Building system development
- Feasibility studies
- Experimental building projects
- Development of calculation methods
- Technology assessment studies.

EKONO is divided into three sectors of roughly similar size, sectors for building, energy and industry. The research work is mainly carried out in the technical departments, which are: Civil Engineering Department, Electrical Engineering Department and Heating, Air Conditioning and Sanitary Engineering Department. The Building Research Project Office co-ordinates all research projects in the building sector.
Total staff: approximately 650
Professionally qualified staff: approximately 400
Director or Chief Executive
President: Dr Sven O. Hultin
Vice President: Juha Gabrielsson, (Building Division)
Manager: Eero Erkiö, (Building Research)
Principal sources of finance
EKONO is a privately owned company. Research and development work is carried out partly with private funds and partly with governmental funds. A portion of the profit is used for its own research and development work.
Publications
Numerous papers are published annually in professional journals. Some of the research results are published by governmental organizations. EKONO also publishes a publication series of its own; it is available through the Finnish Energy Economy Society.

FIN.191
Helsinki University of Technology, Department of Civil Engineering
Teknillinen korkeakoulu Rakennusinsinööriosasto
Rakentajanaukio 4 A, 02150 Espoo 15
Telephone: Helsinki 460144
Telex: 12-1591 TKK sf
Date of establishment
1908

With 12 full and 6 associate professorships the Department is the largest of the three Civil Engineering Departments in Finnish universities.
Main fields of work
The Department educates engineers (MSc) for all principal civil engineering fields. The postgraduate programme gives the degrees LicTechn and DR Techn. The research projects are mainly connected with the master, licentiate and doctorate degrees, often in co-opeation with state and municipal organizations or private enterprise.
Total staff: 90
Professionally qualified staff: 50
Director or Chief Executive
Head of the Department: Professor S. Lyly
Heads of department
The following laboratories:
Building Technology: Professor P. Kanerva
Structural Mechanics: Professor M. Mikkola
Bridge Building: Professor H. Paavola
Concrete Technology (vacant)
Steel Structures (vacant)
Building Economics: Professor J. Kiiras
Soil Mechanics: Professor K.-H. Korhonen

Traffic and Transportation: Professor S. Lyly
Highway Engineering: Professor J. Hyyppä
Water Resrouces (vacant)
Hydraulic Engineering: Professor H. Sistonen
Water Supply and Sanitary Engineering: Professor E. Kajosaari
Principal sources of finance
Education: State; research: partly public, partly private sources.
Publications
Laboratories of the Department have their own publication series.

FIN.192
National Board of Public Building
Rakennushallitus
Postal address: PB 232, 00531 Helsinki 53
Street address: Siltasaarenkatu 18A, 00530 Helsinki 53
Telephone: 90 717 511
Telex: 12-2349
Date of establishment
1811

A governmental agency operating within the Ministry of Finance and whose task is to provide and maintain office facilities for the use of various governmental agencies.
Main fields of work
- To provide office facilities by buying, renting existing facilities, building new premises or repairing the old, and the maintenance of this accommodation
- To supervise the use of governmental building sites
- To issue reports on building projects of local authorities and public organizations receiving state aid.

Total staff: 4050 (3650 in the District Building Bureaux)
Professionally qualified staff: 280
Director or Chief Executive
Jaakko Mäki
Heads of department
Administrative Department: not available at time of going to press
Projection Department: Toimi Hämäläinen
Construction Department: Pekka Metsola
Real Estate Management Department: Pertti Tuominen
Maintenance Department: Jukka Liede
Principal sources of finance
State budget.
Publications
Standards and recommendations to be followed in governmental building projects such as: Directives for energy saving, Directives for projecting plumbing and ventilation and electricity, etc.
Series of reports are published yearly (in 1982 reports on 16 subjects); mainly they are preliminary studies and studies on various technical problems and possible solutions.

FIN.193
Nordic Committee on Building Regulations
Nordiska Kommittén för Byggbestämmelser (NKB)
Centrala sekretariatet, Eriksg 24 A,00100 Helsingfors 10
Telephone: 90 6946045
Date of establishment
1955 on the Recommendation of the Nordic Council

NKB is a joint committee for the national building authorities in Denmark, Finland, Iceland, Norway and Sweden.
The NKB authorities are:
- Ministry of Housing in Denmark
- Ministry of the Interior in Finland
- Directorate of Town and Country Planning in Iceland
- Ministry of Municipal Affairs and Labour in Norway
- National Board of Urban Planning in Sweden.

Main fields of work
The main object of NKB is the advancement of Scandinavian co-ordination of all technical regulations concerning the production of buildings and building components.
Total staff: about 50
Director or Chief Executive
Chariman: Esko Mononen, Ministry of the Interior (Finland)
Secretary: B.-G. Lindqvist, Central Secretariat (Finland)
Principal sources of finance
Financed partly by the national building authorities of various countries and partly by contributions from the Nordic Council of Ministers. Much work is also performed on a voluntary, unpaid basis.
Principal related bodies
Main Committee.
Publications
- Building regulations
- Guidelines
- Programmes of work.

FIN.194
Roads and Waterways Administration (RWA)
Tie- ja vesirakennuslaitos (TVL)
Pasila Office Centre, PO Box 33, SF 00521 Helsinki 52
Telephone: 90 1541
Telex: 124589 tvh
Date of establishment
17 December 1799

The RWA provides infra-structural services for society by attention to routes for road and water traffic. The Central Administration of the RWA directs and supervises the administration and work of the RWA. District administration of the RWA is carried out through 13 Road and Waterway Districts.
Main fields of work
The RWA attends to
- The planning, construction, improvement and maintenance of public roads
- The development of traffic conditions
- The planning, construction, improvement and maintenance of navigable routes and floating channels, canals and harbours
- Public canals
- The development of water traffic conditions.

Total staff: 720
Director or Chief Executive
Director General: Jouko Loikkanen
Director-in-Chief: Väinö Suonio
Heads of department
Administrative Department: Raimo Salmi
Economics Department: Erkki Koskinen
Design Department: Eero Hietanen
Construction Department: Anton Ortamo
Operations Department: Pekka Härkönen
Waterways Department: Jarkko Saisto
Principal sources of finance
Financed from public budget funds.
Principal related bodies
The Central Administration of the RWA is the central office subordinated to the Ministry of Communications.
Publications
The most important research reports are noted in the RWA's list of publications. In addition the offices have report lists of their own showing the research results they have produced.
Reports of importance are listed in the OECD's International Road Research Documentation (IRRD) scheme.

FIN.195 **AM**

Tampere University of Technology, Department of Civil Engineering

Tampereen teknillinen korkeakoulu, Rakennustekniikan osasto (TTKK R-osasto)

PO Box 527, SF 33101 Tampere 10

Telephone: 931 162 111

Telex: 22313 ttktr-sf

Date of establishment

Tampere University of Technology was established on 13 August 1965 as a university affiliated to Helsinki University of Technology. The university gained full independence in 1972.

The Department is one of four departments in Tampere University of Technology. It gives instruction in the construction field and performs research upon request.

Main fields of work

The Department gives instruction and performs research in eight laboratories. Research in the laboratories centres on the following fields:

- *Building Construction:* product development of building components (especially thin-walled steel structures, nailplates, concrete brackets, concrete blocks, timber–concrete slabs, steel–polystyrene elements, polyurethane-insulated concrete and timber elements).
- *Construction economics:* forecasts of quantity and quality of building construction and building stock and the demand of inputs in building construction and in operation of buildings (i.e. labour, energy, materials) and development of planning methods (value analyses, lifecycle costing, worth determination, planning systems for energy conservation measures, etc.)
- *Earth and highway construction:* transport services in rural and urban areas; lifetime of cars
- *Engineering geology:* utilization of waste products; properties of building materials; groundwater investigations
- *Geodesy and photogrammetry:* measurement technics in construction (especially photogrammetry and numerical methods)
- *Soil mechanics and foundation engineering:* arctic earthworks; groundwater and soil investigations in refuse tips
- *Theory of structures:* the theory of element and stone structures (stability, strength, contact problems, ADP programs)
- *Water supply and sanitary engineering:* aerobic, anaerobic and chemical waste water treatment; postgraduate courses and water research for developing countries; water treatment.

Total staff: 50

Professionally qualified staff: 23

Director or Chief Executive

Professor Dr Olli-Pekka Hartikainen

Heads of department

In charge of the laboratories are:

Building construction: Professor Lauri Mehto

Construction economics: Professor Raimo Salokangas

Earth and highway construction: Professor Olli-Pekka Hartikainen

Engineering geology: Associate professor Raimo Uusinoka

Geodesy and photogrammetry: Associate professor Hannu Salmenperä

Soil mechanics and foundation engineering: Professor Leo Keinonen

Theory of structures: Professor Herman Parland

Water supply and sanitary engineering: Professor Matti Viitasaari

Principal sources of finance

The university budget for 1981 totalled 38.5 million marks. An additional 11.8 million marks was contributed by research commissions and other outside sources, which made it possible to engage nearly 100 research workers and assistants. The share of the Department is approximately 15% of the total finance.

Principal related bodies

Board of the Department.

Publications

The Department of Civil Engineering and the laboratories publish report series, most of them in Finnish. The Department supplies these publications free of charge.

Many postgraduate course publications concern water treatment in developing countries. Some examples in 1981:

- Selection of settling basin for sediment removal
- Windmills within water supply
- Horizontal roughing filter in pretreatment of slow sand filters.

FIN.196 **FM**

Finland Technical Research Centre, Division for Building Technology and Community Development

Valtion teknillinen tutkimuskeskus (VTT) Rakennus-ja yhdyskuntatekniikan tutkimusosasto

Vuorimiehentie 5, 02150 Espoo 15

Telephone: 358 0 4561

Telex: 122972 vttha sf

Date of establishment

1942

A government research and testing institute with the purpose of maintaining and raising the level of technology in Finland and of meeting the research and testing needs of the public and private sectors.

Main fields of work

The Division for Building Technology and Community Development covers two main fields of work: building technology and community development.

Building technology, fields of research: soil and foundations, structural mechanics, building physics, concrete and silicate technology, wood technology, installations, fire technology, building maintenance and renovation, building economics.

Community development, fields of research: community planning, mapping and surveying, municipal engineering, road construction, traffic and transport, environmental protection.

The Technical Research Centre of Finland is a polytechnical research institute. Research and testing are carried out in five divisions: Division for Building Technology and Community Development; Division for Energy Technology; Division for Information Technology; Division for Manufacturing Technology; Division for Process Technology.

The Division for Building Technology and Community Development comprises ten laboratories:

Concrete and Silicate Laboratory, Betonimiehenkuja 5, 02150 Espoo 15
Research and development work on concrete and its constituents, mortars, masonry and other ceramic products, concrete structures and their reinforcements as well as masonry structures, together with the production techniques related to the above-mentioned matters.

Geotechnical Laboratory, Lämpömiehenkuja 2 B, 02150 Espoo 15
Research on building technological properties of the soil and rock and development of foundation methods for different structures and ground construction methods.

Laboratory of Heating and Ventilating, Lämpömiehenkuja 3, 02150 Espoo 15
Research and development on heating, ventilating and plumbing technology and questions related to air protection and municipal engineering in the field.

Laboratory of Land Use, Revontulentie 7, 02100 Espoo 10
Research related to community planning, including land surveying, planning, evaluating and real estate technology, community technology.

Fire Technology Laboratory, Kivimiehentie 4, 02150 Espoo 15
Research into fire safety properties of building materials, products, elements and buildings, including regional fire protection and fire extinguishing equipment and technology.

Forest Products Laboratory, Puumiehenkuja 2 A, 02150 Espoo 15
Structure and properties of wood and technology related to raw materials, manufacture, properties, further processing and use of forest products.

Laboratory of Structural Engineering, Betonimiehenkuja 3, 02150 Espoo 15

Research and development work concerned with physical properties and strength of structures, components and building materials.

Building Laboratory (Oulu), Postal address: Box 181, SF-90100 Oulu 10, Telephone: 358 81 346488

Research concerned with modernization, reparation, management, maintenance and administration problems of building construction, including economic investigations, and research of building materials, structures and methods required by the special conditions of Northern Finland.

Laboratory of Building Economics, Revontulentie 7, 02100 Espoo 10

Research directed to building and building planning, with the aim of clarifying the choice of solutions in the light of value and cost factors.

Road and Traffic Laboratory, Lämpömiehenkuja 2 A, 02150 Espoo 15

Research and development of the technical and economic questions related to the design, construction and maintenance of traffic routes and of traffic engineering, economics and safety.

Total staff: 680

Professionally qualified staff: over 200

Director or Chief Executive

General Director: Professor P. Jauho

Research Director of the Division for Building Technology and Community Development: Professor H. Poijärvi

Heads of laboratory

Concrete and Silicate Laboratory: Professor Asko Sarja

Geotechnical Laboratory: Professor Markku Tammirinne

Laboratory of Heating and Ventilating: Professor Erkki Äikäs

Laboratory of Land Use: Acting Director Hilka Lelektonen

Fire Technology Laboratory: Professor Pentti Loikkanen

Forest Products Laboratory: Professor Tuija Vihavainen

Laboratory of Structural Engineering: Acting Director Erkli Helanden

Building Laboratory: Acting Director Heikki Matala

Laboratory of Building Economics: Professor Per-Olof Jarle

Road and Traffic Laboratory: Professor Otto Wahlgren

Principal sources of finance

About 50% of the financing is covered by the State budget, the other half by external financing consisting of income from customer orders and research contracts.

Principal related bodies

Council of Technical Research Centre, Board of Technical Research Centre, Advisory Committees of Laboratories of Technical Research Centre.

Publications

The Centre publishes the following publications series:

- Technical Research Centre of Finland Publications
- Valtion teknillinen tutkimuskeskus, Tutkimuksia, Technical Research Centre of Finland Research Reports
- Valtion teknillinen tutkimuskeskus, Tiedotteita, Technical Research Centre of Finland Research Notes
- VTT Symposium (congress paper series).

FRA.197

Association des Industries de Matériaux, Composants et Equipements pour la Construction (AIMCC)

Association of Building Materials, Components and Fittings Producers' Federations

39 rue Rouelle, 75015 Paris

Telephone: 579 35 26

Date of establishment

1971

An inter-professional association which groups together industrial syndicates and federations concerned with construction (governed by Law of 1901). AIMCC is an official organization representing industries with the aim of solving common problems in the building and construction sector. Established at the initiative of the Ministry of Industry and Research, formerly known by the title: Association pour l'Etude des Matériaux et Composants pour la Construction (AEMCC). This was altered on 14 January 1976 to Association des Industries de Matériaux, Composants et Equipements pour la Construction (AIMCC).

Main fields of work

Objectives: to assist in defining an industrial policy for construction; to assist in strengthening industrial representation in construction.

Commission for technical studies; modular co-ordination; problems of industrialization; standardization problems.

Juridical Commission: problems common to industry in the building market.

Principal sources of finance

Subscriptions paid by the federations and syndicates; subventions for research.

FRA.198 **AM**

Association des Propriétaires pour l'Optimisation de la Gestion, de l'Entretien et de l'Exploitation (APOGEE)

Building Owners' Association for the Optimization of Management, Maintenance and Improvement

4 rue Cambon, 75001 Paris

Telephone: (1) 260 87 21

Date of establishment

Founded in 1978

APOGEE is a non-profit organization, governed by the law of 1901, which gathers about 60 of the major owners, managers and investors of housing units, from both public and private sectors, among them: the main organizations of housing professionals (owners, managers, real estate agents, insurance companies), CNAB, FNAIM, Union des HLM, Fédération des SEM, Groupement des SII, Fédération Française des Sociétés d'Assurances; owners, managers, public and private.

The purpose of APOGEE is to develop new methods for the management and maintenance of buildings, and to improve information in this field among its members and others working in the same field.

APOGEE makes surveys of maintenance costs (national observations, comparing maintenance and improvement costs between buildings), and helps its members to conceive and operate planning systems for maintenance. APOGEE has also developed methods for taking account of the 'life cycle cost' of buildings at the architectural and technical design stage.

Main fields of work

- Observation of maintenance and improvement costs for both housing units and office buildings
- General surveys of maintenance and improvement: energy saving; life cycle cost; connection between technology and maintenance; methods for maintenance planning and budgeting; social aspects of housing
- Development of software for computers (for practical work in the fields listed above), especially: GETAL and GETEQ: computerized maintenance log and planning for housing units and equipment
- Technical and financial surveys for assessment of housing units and office buildings, according to: maintenance costs (past and future); improvement costs (energy, management costs); long-term financial stability
- Information for members and the public.

Professionally qualified staff: 15

Director or Chief Executive

Director Michel Jouvent

Principal sources of finance

Contributions from members of the Association, and payments received for services rendered from both members and other organizations.

Publications

Journal: La Lettre d'APOGEE, every 2 weeks

Research reports: reports of working sessions (selection of some recent reports):

- Gestion immobilière locative et informatique
- Le gardiennage des immeubles
- Choisir et utiliser un micro-ordinateur
- Nettoyage des immeubles dans le tertiaire
- Grosses réparations et améliorations dans le tertiaire
- La réalisation des états des lieux dans le logement
- La loi Quilliot un an après
- Les tableaux de bord
- Annual reports of the Observations of improvements, for housing and office buildings
- Reports on Life Cycle Cost for the RAUC:
- Les couvertures
- Les menuiseries extérieures et occultations.

FRA.199 **AM**

Association for Initiation in Building Research

Association pour l'Initiation à la Recherche en Bâtiment (IRBat)

242bis Boulevard Saint Germain, 75007 Paris

Telephone: (1) 544 72 66

Date of establishment

1976

A private association supported by public authorities and professional organizations to provide newly graduated engineers or architects with an opportunity to become acquainted with building research in France or abroad.

Main fields of work

Building research.

Total staff: 3 permanent in Paris, 5 regional

Director or Chief Executive

President: G. Blachère

Secretary General: J. Baillon

Heads of department

Building Science and Technology: G. Blachère

Building Economics: J. Baillon

Principal sources of finance

Research scholarships, grants and contracts both public and private.

Principal related bodies

Board of Directors and General Assembly

Publications

Research reports: About 100 per year, many of them restricted.

FRA.200 AM

Bureau Securitas (BS)

Bureau Securitas (BS)

Les Quadrants, 3 Avenue du Centre, F 78182 Saint Quentin en Yvelines Cedex

Telephone: (3) 043 99 13

Telex: 698 684 F

Date of establishment

13 June 1929

Non profit-making association, governed by the law of 1 July 1901, Bureau Securitas is a private institution at the disposal of all public services and private building firms and organizations, except private individuals.

For 30 years, the main activity of the Bureau has been concerned with building pathology.

Main fields of work

The aim of the Bureau is to contribute towards improving techniques in the art of building and more especially those concerned with the safety of immovable constructions, the adaptation of their structures, their equipment and installations viewed from the aspect of functional requirements and the safety of the material used by the firm.

The principal methods used to achieve these objectives are:

- To undertake or arrange to have undertaken all theoretical or experimental studies which bear on those problems of general or specific interest likely to arise in the area of safety of structures both during and after execution
- To study and formulate any recommendations or rules which are judged to be relevant to these matters, especially when new methods or materials are used
- To undertake or assist with expertise in all construction works both during and after execution
- To undertake or initiate statistical and technical surveys into accidents or mishaps relating to structures with a view to establishing preventive recommendations or rules
- To arbitrate in technical disputes that may arise during construction projects
- To collaborate with other organizations having similar aims
- To execute or set in motion all operations, research projects or tests which are relevant to these objectives or which would aid in their realization
- To lend assistance in all aspects of education and improvement of technicians engaged in construction: this is effected through initiating numerous seminars or courses on the pathology of construction.

Total staff: permanent: 10

Professionally qualified staff: members of technical Commissions: 170

Director or Chief Executive

President: Pierre Laborde

General Director: Louis Logeais

Principal sources of finance

Members' subscriptions; subventions for research, contracts of research from the State; fees for seminars and conferences (especially about building pathology).

Principal related bodies

Administrative Council (15 members); Advisory Board (13 members).

Publications

- Recommendations concerning the Conception, Calculation, Execution and Control of Ground Anchors (TA 77)
- Recommendations concerning the underground cavities (CS 78)
- Pathology annals: Les Constructions en Béton Armé (1975); La Construction Métallique (1979)
- Books about pathology: La Pathologie des Fondations (Pathology of Foundations) (1982); La Pathologie des Murs de Soutènement (Pathology of Earth-retaining Walls) (1982)
- Numerous articles in French and foreign reviews.
- Statistical study about 10 000 cases of building failures (1979).

FRA.201 AM

Centre d'Etudes et de Recherches de l'Industrie du Béton Manufacture (CERIB)

French Precast Concrete Study and Research Centre

rue des Longs Réages, BP 59, F 28230 Epernon

Telephone: (37) 83 52 72

Telex: F 782.048

Date of establishment

5 January 1967

Technical Industrial Centre (Law of 22 July 1948) established by the Ministry of Industrial and Scientific Development (the Ministry of Research and Industry). Technical Industrial Centre for French precast concrete industry with collective research activity, technical assistance, quality control and standardization.

Main fields of work

- *Research activities:* concrete products manufacture; concrete products use;
- *Technical assistance:* assistance with quality control and replies to questions submitted, dissemination of information, data documentation); studies or surveys for fees; expert advice in matters involving litigation; advanced courses (for engineers and technicians); observatory for innovation
- *Quality control and standardization (25% of the activities of the centre) embracing:* the control of quality marks for AFNOR; the control of quality labels; co-operation with CSTB in matters relating to ministerial approval of sanitation ducts; participation in commissions engaged in the preparation of standards and regulations for concrete products and in numerous other bodies working in statistical methodology, in technical advice, in the preparation of codes of practice, etc.

Total staff: 107

Professionally qualified staff: 65

Director or Chief Executive

President: Maurice Gillet

General Director: Michel Darcemont

Heads of department

Administration Department: Pierre Challamel

Information and Diffusion Department: Michel Rollet

Quality Department: Michel Valles

Studies and Researches Department: Françis Dutruel

Principal sources of finance

Financed by: a parafiscal tax of 3.5% on the volume of business outside the levy on industry manufacturing concrete products; repayment of loans effected by the Centre; contracts for studies concluded with the State. Budget 1983: 30.6 million Francs.

Principal related bodies

Administrative Council: 22 members nominated by Ministerial decree (drawn from professional industrialists, the scientific world and representatives from the profession); Technical and scientific committee: 32 members nominated by the Administrative Council (professional industrialists, users of concrete products, public and private research bodies with an interest in concrete).

Publications

- Technical publications (nos. 1–76)
- Data documentation quarterly
- Monograph (nos. 1–10).

FRA.202

Centre EDF-GDF d'Information des Professionnels du Bâtiment (CEGIBAT)

Information Centre for Professionals engaged in Building (EDF-GDF)

42/46 rue du Rocher, 75008 Paris

Telephone: 522 31 41

Telex: 641 524 F

Date of establishment

1972

Communicating information from specialist engineers on the application of gas and electricity in construction.

Main fields of work

- Monthly publication (edition of 40 000 copies) which contains a synthesis of current developments in the building world and a selection of up-to-date information and articles.
- Documentation at the service of professionals.
- Initiatives. In co-operation with professional organizations and the manufacturers of materials, exhibitions are organized. The themes are derived from the various applications of gas and electricity in construction, both new and current. As part of these exhibitions discussion meetings are held to further explore the subject matter.

Other gatherings of various kinds are put on at the request of building professionals, e.g. meetings, extempore discussions, round tables, study groups, etc.

Director or Chief Executive

Head of Centre: Robert Mouchel

Principal sources of finance

EDF–GDF budget.

Publications

Journal: Cegibat-Relations, monthly publication (edition of 40 000 copies), which contains a synthesis of current developments in the building world and a selection of up-to-date information and articles.

Reports of discussions.

FRA.203

Centre Méditerranéen de Recherches et D'Etudes Expérimentales pour l'Industrie et la Construction (CEMEREX)

Mediterranean Centre for Research and Applied Studies in Industry and Construction

6ème rue no 14, 13127 Vitrolles, Marseille

Telephone: (42) 89 93 87

Telex: CEMEREX 402 687 F

Date of establishment

Founded in 1917, reorganized in 1971

CEMEREX is an institution where the former Marseille Chamber of Commerce and Industry and Centre Expérimental de Recherche et d'Etude du Bâtiment et des Travaux Publics (CEBTP) regional agencies are amalgamated. It is managed by CEBTP and placed under sponsorship of the Marseille Chamber of Commerce and Industry.

Main fields of work

Studies, analyses, tests and controls in the following fields:

- Soil mechanics, in situ investigations, laboratory tests, foundations and road engineering
- Building materials: concrete, aggregates, etc.
- Research of quarries and borrow pits
- Finishing materials
- Pathology of construction
- Non-destructive tests on pipes, pile foundations, beams, etc.
- Metals, metallography, corrosion, protection against corrosion
- Stress and strain analyses, extensometry
- Acoustics, vibrations
- Analytical vibrations

It provides technical assistance for French companies world wide: site laboratory management, quarries, crushing and screening plant, earth works and road control, concrete, admixtures, pouring in cold or hot weather.

Total staff: 35

Professionally qualified staff: 35

Director or Chief Executive

Manager: René Bertrandy

Heads of department

Soil Mechanics, Foundations and Road Engineering: P. Burguiere
Building Materials and Civil Engineering: J.P. Guillelmet
Physics, Metals, Corrosion: A. Vincens
Chemistry: Mrs D. Collart
Acoustics, Vibrations: J. Benazet

Principal sources of finance

Fees paid for work carried out; research grant from CEBTP.

Publications

- *Journal:* Contribution to professional publications: ITBTP, monthly
- *Research reports:* Aggregates and fillers; Marine concrete work; Lightweight concrete; Board use for building construction.

FRA.204 **FM**

Centre Scientifique et Technique du Bâtiment (CSTB)

Scientific and Technical Building Centre

4 avenue Recteur Poincaré, 75782 Paris Cedex 16

Telephone: 45 24 43 02

Telex: 610710 F

Cables: CESITEBA PARIS

Other addresses:

84 avenue Jean-Jaurès, 77420 Marne-la-Vallée
24 rue Joseph-Fourier, 38400 St-Martin-d'Hèrbes
11 rue Henri-Picherit, 44300 Nantes
BP 21 Sophia-Antipolis, 06562 Valbonne

Date of establishment

1947. Public institution with an industrial and commercial character since 1953.

Public organization under sponsorship of Ministry for Construction and Ministry for Research and Industry. Performs research, technical, sociological and economic projects for Government and/or private organizations. Performs tests and technical assessments for private industry. Publisher of building standards.

Main fields of work

- Construction technology: light frame structures; thermal insulation; precast concrete; reinforced concrete structures; masonry; windows; roofings, roof components; partitions; structure stability and safety; seismic engineering.
- Equipment: plumbing, sanitary, heating systems
- Heating and ventilating, energy management
- Solar energy in building
- Acoustics in building and in towns
- Fire safety in buildings
- Climatology and wind energy
- Aerodynamics of buildings
- Lighting
- Building materials: aging, durability, new materials
- Economics in construction and 'prospective economique'
- Sociology and psycho-sociology related to building production organization and building use
- Architecture

Total staff: 563

Professionally qualified staff: 223

Director or Chief Executive

President: André Laure
Director: Pierre Chemillier

Heads of department

Construction Technology: E. Farhi, J. Lugez, J. Mathez

Equipment: F. Perrier
Heating and Ventilating: M. Croiset
Solar energy: M. Rubinstein
Climatology, Aerodynamics, Lighting: J. Bietry
Acoustics, Materials: R. Josse
Economics: R. Urien
Sociology: M. Conan
Architecture: C. Queffelec
Library: U. Soulaille

Principal sources of finance

Ministry of Research and Industry: about 60% (through Ministry of Construction); other organizations and public agencies: 15%; private industries: 15%; publications: 10%.

Principal related bodies

Administrative Board; Research Advisory Committee.

Publications

- *Journal:* Cahiers du CSTB, monthly;
- *Technical standards:* REEF (Recueil des Eléments Utiles à l'Etablissement et à l'Exécution des Projets et Marchés de Bâtiments en France): includes standards and building codes of practice: DTU), quarterly
- *Research reports:* General Activity report, annually; some research reports are available in the Cahiers du CSTB.

FRA.205 **AM**

Centre Technique du Bois (CTB)

Wood Technical Centre

10 Avenue de Saint-Mandè, 75012 Paris

Telephone: (1) 344 06 20

Telex: 214 280 F CTBOIS

Experimental Station: 84 avenue J. Jaurès, 77420 Champs-sur-Marne

Date of establishment

1952

The CTB is an institution of public interest founded by the application of the law of 22 July 1948 concerning professional technical centres. It is managed by a board of directors including industry representatives and two government representatives (Department of Agriculture and Department of Industry).

Its role is to favour technical progress and encourage its application in enterprises in order to improve raw material recovery and product quality.

Main fields of work

- Studies and investigations (fundamental and applied): improved knowledge of wood material; development of use of home-grown species; raw material savings and adaptation of industrial processes; optimum capacity in usage of wooden products; development of products and end uses.
- Standardization and quality control: CTB is responsible for an office of standardization of wood and furniture; it also manages and controls, within AFNOR, the NF (French Standard) mark and CTB quality labels set up by CTB.
- Active collaboration with several international organizations: ISO, CEN, IUFRO, FAO.
- Technical assistance: CTB provides technical support to industry by means of: intervention in the factory; studies and tests; information and consultation; vocational training.
- Documentation and information: objectives are to assure the dissemination of knowledge through: publications; a documentation service and a library (4000 books, 260 scientific and technical journals, 30 000 references on file).

Total staff: 240

Professionally qualified staff: 130

Director or Chief Executive

President: Gérard Moulet
General Director: Pierre Malaval

Heads of department

Director of Studies and Research: Walter G. Kauman
Director of Industrial Interventions: D. Guinard
Documentation and Information Service: B. André
Standardization and Quality Control: B. Petetin
Physics and Mechanics Department: M. Avale
Biology, Chemistry, Preservation Department: M. Romeis
Industrial Processes Department: J.P. Lego
Wood in Building Department: Ph. Loiseau
Furniture Department: P. Parisot
Logging Department: J.M. Nierat
Sawmill and Conditioning Department: F. Mondy
Panel Department: J.L. Jaudon

Principal sources of finance

Subventions from National Forestry Fund; agreement for studies and research concluded with the State; sharing in fees for specific tests, studies, technical assistance, vocational training, publications.

Principal related bodies

Board of Directors; Technical Committee for Research; Professional Commissions; Department of Agriculture; Department of Industry.

Publications

Periodical publications:

- Courrier de l'Industriel du Bois et de l'Ameublement (CIBA), quarterly
- Courrier de l'Exploitant et du Scieur (CEFS), quarterly
- Bulletin d'Informations Techniques, quarterly
- Bulletin Bibliographique (abstract journal), 5 per year

Irregular publications:

- Cahiers
- Tirés à part (Reprints).

FRA.206 **AM**

Centre Technique Industriel de la Construction Métallique (CTICM)

Technical and Industrial Centre for Constructional Steelwork

20 rue Jean-Jaurès, 92807 Puteaux

Telephone: 776 41 31

Telex: 611677 F

Date of establishment

1962 (Law of 22 July 1948)

The CTICM was founded in 1962 in order 'to promote the progress of constructional techniques and improve the efficiency and quality in the steel construction industry'.

CTICM is under the control of the Ministry of Research. A Board of Directors, half of whose members are representatives of the steel constructional industry, orientates the main activities of the Centre and supervises its management.

In order to cover the various problems of constructional steelwork, the activities of the Centre are divided into several departments.

Main fields of work

- Technical assistance in various fields such as computer structural design, automatic drawing, management computer systems.
- Mechanics of structures: structural safety (research carried out in co-operation with the Bridge and Road Services); fatigue strength of offshore structures (in co-operation with the French Petroleum Institute (IFP))
- Investigations for difficult projects at design office level (aerial, industrial furnaces, complex buildings, etc.)
- Thermal problems: fire resistance tests of structural materials carried out at the Fire Research Station; programmes for the thermal design of buildings (heating, ventilation, insulation, etc.)
- Computer applications: development of computer codes for design, namely in the field of off-shore construction; use of micro-computers to solve problems of design or management (computer-aided design packages)
- Professional education, information and documentation
- Regular and international at both national and international levels

(assistance for the export trade by analysis of foreign rules and specifications)
- Experts' reports
- Fire research and testing station.

Total staff: 95
Professionally qualified staff: 47 engineers
Director or Chief Executive
Chairman: Henry Deschênes
Director: Lucien Wahl
Heads of department
Department of Research: Jacques Brozzetti
Department of Applications: Michel Cambon
Department of Standardization: André Pousset
Fire Research Station: Paul Arnault
Principal sources of finance
64% own resources (revenue from sales); 27% professional income; 9% public research grant.
Principal related bodies
Board of Directors under the control of the Ministry of Research.
Publications
- CTICM-Construction Métallique (technical industrial review with national and international recommendations), quarterly.
- Rapport d'Activité (general information about the Centre's work such as final activity reports), biannually available on request

Papers on theoretical and experimental research, available on request.

A list of publications of CTICM is available on request.

FRA.207
Comité Français de l'Isolation (CFI)
French Insulation Committee
4 rue Cimarosa, 75116 Paris
Telephone: 504 94 55
Date of establishment
1976

A non-profit association.
Main fields of work
- Co-ordination of study and research on sound and thermal insulation for housing, industrial buildings and those of the tertiary sector
- A great deal of information for professionals and for the public
- Professional training
- International co-operation with national counterparts: COMITA (Belgium), SKOOP (Netherlands), ANDIMA (Spain) and SIA (Great Britain).

Total staff: 5
Director or Chief Executive
Chairman: J.B. Grosborne
Director: J.F. d Roumefort
Principal sources of finance
Members' dues.
Publications
- Bulletin: Isolation Information, 10/12 issues per year
- Papers in specialized or general information reviews.

FRA.208
Institut de Céramique Francaise (ICF)
French Institute for Ceramics
6 Grande Rue, 92310 Sèvres
Telephone: 6269111
Date of establishment
Set up in 1917

Establishment for advanced free education, recognized as a Public Utility by the Law of 10 August 1920.
Main fields of work
- Basic training and on-going education for engineers, technicians and workers
- Studies of general interest in the area of materials and ceramic products and of articles in enamelled metal.

Total staff: 8 permanent staff workers plus temporary personnel
Director or Chief Executive
President: Robert Bouchet
Director: Jean Tuleff
ICF comprises two executive departments: Continuing Education; Laboratories.
Principal sources of finance
Subventions for professional training.
Principal related bodies
Administrative Council.
Publications
Bulletin Céramique-Cerindex.

FRA.209
Institut Français de l'Energie (IFE)
French Institute of Energy
3 rue Henri Heine, 75016 Paris
Telephone: (1) 524 46 14
Telex: 620 132
Cables: CENTRCHAUFF Paris
Date of establishment
1919, under present form since 1952

A private non-profit organization.
Main fields of work
- Training
- Studies and information: library for consultation of the documents collected at IFE or for a search for other sources of information; response to precise inquiries; access to the Documentation Centre of Human Sciences (CDSH) processing system into which part of the IFE documentation is integrated.

IFE manages a Training Department and the Studies and Information Centre. It administers the secretariat of the French Thermicians' Society.
Total staff: (Studies and Information Centre): 10
Director or Chief Executive
Chairman: Jean Couture
Director: Edouard Maintrieu
Director, Studies and Information Centre: Francine Brenière
Heads of department
Director, Training: Jacques Claudon
Director, Studies and Information Centre: Mrs Francine Breniere
Principal sources of finance
Financed by a few Members' dues and especially by consultancy fees.
Publications
- Actualite Combustibles Energie, monthly digest of the French and foreign publications and synthesis of topics of the day or of general interest
- Les déchets, source d'énergie (Waste material as a source of energy); Volume 1, Déchets animaux et végétaux (Animal and vegetable waste); Volume 2, Déchets industriels et urbains (Industrial and urban waste)

- Directory of Information Centres of the World
- Rational use of energy (Congress: Perpignan-Grenoble)
- Regulation– Financial aids (Rational use of Energy), quarterly
- Competitiveness of Firms and Energy
- District Heating
- Coal substitution

FRA.210

Laboratoire Central des Ponts et Chaussées (LCPC)

Road and Bridge Central Laboratory

58 boulevard Lefebvre, 75732 Paris Cedex 15

Telephone: (1) 532 31 79

Telex: 200 361 F LCPARI

Other address:

BP 19, 44340 Bouguenais, Nantes

Telephone: (4) 65 14 88

Telex: 710 805 F LCNANTE

Date of establishment

1831: Chemical laboratory of the Academy for Roads and Bridges

1949: External service of the Ministry (currently Town Planning and Housing)

The central laboratory of Roads and Bridges provides a service on behalf of the Ministry of Town Planning and Housing. It is a body engaged in scientific and technical research whose work is of interest to the Minister of Transport and the Secretaries of State for the Ocean, the Environment and the Quality of Life as to the Minister for Industry and Research. The scope of the research takes in the areas of civil engineering, planning and the environment.

Main fields of work

The research which is managed in consultation with the regional laboratories for Roads and Bridges, is undertaken hand in hand with activities for assessing its effectiveness; training, technical advice, special studies (including some executed abroad), technical co-operation, elaboration of technical regulations, industrial protection of processes under refinement, documentation, publications.

Total staff: 602

Professionally qualified staff: 282

Director or Chief Executive

Director: Jean Claude Parriaud

Scientific Director: Alain Bonnet

Director of Programmes and Applications: Jean Claude Chantereau

Technical Directors: Mme Marcelle Brachet; François Baguelin; Claude Bois; Jacques Bonnot; Marcel Rat

Heads of department

Eleven divisions, classified according to their objectives

Materials for roads: Fernand Perez

Structures and characteristics of roads: Paul Autret

Methods and materials for execution: Robert Baroux

Use, signalling systems, painting: Anne Marie Serres

Geotechnics, soil mechanics (Paris): Georges Pilot

Geotechnics, soil mechanics (Nantes): Jean François Corte

Geotechnics, ordnance survey rock mechanics: François Baguelin

Water and environment: Claude Bois

Materials and structures for construction works: Claude Bois

Tests on structural elements: Jean Paul Gourmelon

Inspection and pathology of construction works: Jacques Chatelain

Eight services classified by their specialized field:

Physics: Jean Billard

Chemistry: Igor Voinovitch

Mechanics: Bernard Halphen

Mathematics: Michel Fremond

Information: Serge Savoysky

Scientific and technical information: Jean Pierre Magnan

Materials: Michel Roger Moreau

International studies and research: Paul Autret

Principal sources of finance

Service du Ministère de l'Urbanisme et du Logement.

Principal related bodies

Conseil d'Orientation et de Gestion des LPC (Président: André Herzog)

Conseil Scientifique des LPC (Président: Pierre Piganiol).

Publications

Bulletin de Liaison des Laboratoires des Ponts et Chaussées (bimonthly: 6 nos per year)

- Recommendations
- Directives
- Notes d'information technique
- Research reports from the Central Laboratory
- Catalogues
- Guides
- Studies.

FRA.211

Laboratoire Central d'Hydraulique de France (LCHF)

French Central Hydraulics Laboratory

rue Eugène Renault 10, 94700 Maisons-Alfort

Telephone: 893 28 28

Telex: 240-985 F

Date of establishment

1939

A private group of consultants acting on behalf of public or private organizations who deal with hydraulics engineering or water management.

Main fields of work

- Maritime hydraulics: oceanography with a measurements-in-situ service
- Sedimentology
- River hydraulics, water resources (hydrology, hydro-geology, urban and agricultural hydraulics)
- Information processing

The laboratory has developed a special simulation device with which it is possible to reproduce the sound propagation (in three dimensions) in an urban area; this procedure also makes it possible to search for an overall plan that would reduce external noise at building facades.

The LCHF is very flexible in adapting to problems for study, acting according to requirements as follows:

- Making preliminary contact with any public or private organization having hydraulics problems to solve.
- Sending an expert: an LCHF specialist goes on site for a general examination of the problem

Actual studies may be carried out: from plans only; with the collaboration of various laboratories; with mathematical models, using computers; from nature or with scale models.

Total staff: 100

Professionally qualified staff: 50

Director or Chief Executive

General Director: P. Prudhomme

Deputy General Director: J. Fischer

Heads of department

Development: M. Berger

Maritime Hydraulics: M. Orgeron

Sedimentology: M. Migniot

River Hydraulics: M. Bouloc

Water Resrouces, Hydro-geology: M. Henry

Principal sources of finance

Omnium Technique Holding; Paribas Bank.

Principal related bodies

Geohydraulique, for water resources studies.

Publications

- Study reports for clients
- Internal reports

- Various publications in specialized reviews
- Commercial brochures (French, English and Spanish) on the general activities of the firm.

FRA.212 **AM**

Qualitel

Qualitel

136 Bd Germain, 75006 Paris

Telephone: 325 56 43

Date of establishment

1 April 1974

Qualitel is a non-profit-making association which provides the consumer with information regarding the technical quality of constructions.

Main fields of work

To disseminate information on the technical quality of buildings based on a list of criteria which are evaluated by independent experts.

The objective is to secure an improvement in the general quality of construction and the method used is subject to Government certification and control.

Director or Chief Executive

Managing Director: Claude Tréhin

Heads of department

Technical Service: Marc Weckstein

Information: Natalie Luquiens

M. Hélène Sergent

Principal sources of finance

Percentage of construction cost.

Principal related bodies

Consumer associations; architects and construction professionals.

Publications

- Qualitel Guide and Information Seminars for consumers
- Recommendations in the Qualitel Guide.

FRA.213

Recherche, Architecture Urbanisme et Construction (RAUC)

Research Centre for Architecture, Urbanization and Building

55 avenue Kléber, 75116 Paris

Telephone: 720 10 20

Date of establishment

1966

The RAUC is a centre for research into urban planning, architecture and construction.

It is an association governed by the Law of 1901 and was set up by: l'Office général du bâtiment et des travaux publics, le Conseil national de l'ordre des architects, l'Union nationale des syndicats français d'architectes, la Fédération nationale du bâtiment, l'Union technique interprofessionnelle des fédérations nationales du bâtiment et des travaux publics.

Main fields of work

To promote development in the built environment through theoretical and experimental research. Research projects are undertaken in the following fields among others: quality of dwellings; fire safety; acoustics; new towns; innovation in building; building economics.

Diagnostic method for assessing quality of building structures.

A Technical Committee comprising around 20 members representing architects and entrepreneurs is responsible for the conduct of research. It selects the subjects and there is a Library open to the public.

Principal sources of finance

Financed by subventions from architects and contractors and by public contracts.

Principal related bodies

Technical Committee.

Publications

The RAUC has published over 100 studies on a wide variety of subjects. Articles are also published in external journals (e.g. Le Moniteur, Editions Eyrolles, etc.) These look at construction as much from the architectural as the sociological or purely technical aspects. They are intended to be practical reference works, an aid for better construction. A list is available. Examples include: Approche d'évaluation d'effectif d'un chantier, Les micro-climats urbains, L'éclairage dans l'habitat, Guide sécurité incendie, Guide acoustique, Mémento des termes du bâtiment, Isolation et insonorisation des menuiseries, Prévention du bruit dans les installations sanitaires, La Protection acoustique des logements, Introduction à l'acoustique des bâtiments d'habitation, Les salissures de façades, Protection acoustique des bâtiments contre les bruits de circulation.

Audio-visual representations of activities are available for display.

FRA.214 **AM**

Syndicat National du Béton Armé et des Techniques Industrialisées (SNBATI)

Syndicates of Reinforced Concretes and Industrialized Techniques

9 rue la Pérouse, 75784 Paris Cedex 16

Telephone: 4720 10 20

Other addresses:

Centre de Formation et de Perfectionnement de Villejuif, 66 rue Guy Moquet, 94800 Villejuif

Telephone: 4720 10 20

Director of the Centre: Didier Picard

Direction de la Recherche (de la FNB)

Domaine de Saint-Paul, BP no1, 78470 Saint-Remy-les-Chevreuse

Telephone: 34052 92 00

Counsellor for Research and Development: Jacques René Kramer

Date of establishment

1903

Professional syndicate which groups together firms in the construction and public works sectors working with prestressed and reinforced concrete. Its aim is to stimulate innovation throughout the civil engineering industry (industrial techniques, prefabrication, products and new methods).

Main fields of work

The study and proposal groups are actually made up of 14 teams to initiate their work.

The Centre de Formation et de Perfectionnement de Villejuif has two categories of training, full-time technological education and on-going professional training:

- Technical and technical commercial assistance for the member firms concerning economic, juridical, commercial, scientific and technical documentation and courses, studies, seminars
- Adjustment of research programmes, follow up, publication of research results undertaken within the cadre of the Research Directorate of the Union Technique Interprofessionelle des Fédérations Nationales du Bâtiment et des Travaux Publics in the laboratories.

This research takes place in the following fields: knowledge of soils, quotations and materials; safety and durability (fire, pathology and prevention, safety-timber yards); structures (wooden, metallic mixed frameworks, reinforced concrete, prestressed concrete); linings (hygrothermic behaviour, acoustic isolation); eqiupment; regulating the methods and apparatus for testing.

- Formulating and putting at the disposal of firms programmes for calculation (technical and economic aspects)
- Active involvement in the field of technical regulations
- Representing the syndicate in professional affairs at the national level and in technical organizations within the profession, commissions for

technical regulations, advisory commissions for techniques and approvals.

Total staff: 12 permanent councillors to the Syndicate (plus the management and training personnel from the Centre du Formation at Villejuif)

Director or Chief Executive

President: Phillipe Levaux

General Delegate: Alain Kergall

Technical Counsellor: Jacques Mathivat

Principal sources of finance

Subscriptions from the member firms.

Principal related bodies

Deliberative groups (representing the firms); Administrative Council: Office (12 persons) and Council Members (12+40 persons); Study and Proposal groups (representing the firms and assisted by technical councillors, and permanent cadres from the Syndicate).

Publications

Results of tasks and surveys executed within or at the initiative of the syndicate:

These are published through:

- internal documents with limited circulation
- specific documents as required
- professional journals
- Annales de l'Institut Technique du Bâtiment et des Travaux Publics
- Le Bâtiment – Bâtir.

FRA.215 FM

Union Technique Interprofessionelle des Fédérations Nationales du Bâtiment et des Travaux Publics (UTI)

Interprofessional Technical Union of National Federations of Building and Public Works

6–14 rue la Pérouse, 75784 Paris Cedex 16

Telephone: 720 10 20

Telex: 611 975

UTI is grouped into three separate organizations:

- UTI itself which is a non-profit making association (Law of 1901)
- Technical Institute for Building and Public Works (ITBTP), association governed by the Law of 1901
- UTI Services, SARL and CACT

Main fields of work

The work of UTI is always orientated towards perfecting enterprises and furthering progress of the construction industry. Their tasks are developed in the fields of research information, technical assistance and training.

Research: this covers most of the sectors of building and a great many of public works, definition of actions, soils and foundations, the physico-chemical investigation of materials, structures and safety of works, fire security and the physics of building.

Information: this activity is undertaken by CATED which operates the bibliographical centre of ITBTP (12 000 volumes, 150 000 index articles) and the Documentation Centre of the Research Directorate (85 000 recorded documents).

Technical assistance: this activity is also undertaken by CATED (Telephone information service and Ariane system) which assists firms in finding solutions to their technical problems and those concerned with Regulations. Technical assistance relating to information applications (calculations) is provided by UTI and CACT; that relating to organization and marketing is provided by CDC. CEBTP also renders technical assistance.

Training: takes place at the Centre for Advanced Studies in Construction (CHEC); at the International College for Building Sciences at the Centre for Advanced studies in Thermics and Ventilation (CHETA). Courses and seminars are organized by the Professional Unions, CATED, CEBTP, and CHEC.

UTI was designed as a central organization responsible for bringing together the technical services of the National Federatioh of Building and for co-ordinating their programme of action. Within the group of three organizations as presently constituted, UTI proposes the general policy, determines the broad objectives and ensures control over all the Group's services. Responsibility for these tasks at group level belongs to the Director General and, at UTI level, the different functions are executed by:

- A Department charged with Management Control (Ms C. Gigomas) whose main task is the organization and implementation of the services;
- A Financial Department (L. Morvan) who is in charge of financial services, contracts, purchases, etc.

UTI comprises:

- The Regulations Directorate (M. Adam) which groups together: The Standardization Office and the Regulations governing civil engineering calculations.
- The Directorate for Research (J.C. Maldague) which is drawing up a three year plan for the orientation of research and deriving an annual programme. In addition it transmits research mandates to laboratories, follows through their execution and collates the findings. The laboratories responsible for research are the Experimental Station for Study and Research into Building and Public Works (CEBTP), which covers almost the whole field of construction, the laboratory of the Scientific and Technical Committee for the Heating and the Air Conditioning Industries (COSTIC).

The Director General of UTI has the same function for the two other organizations within the Group:

The ITBTP which is responsible for all the activities covering assistance to the firm and which comprise:

- The Centre for Construction firms managers CDC (J.C. Saunier)
- The Centre for Technical Assistance and Documentation, CATED (C. Mouroux), whose purpose is to bring together the items of technical and professional knowledge and to communicate them to industry: this is achieved principally by means of the Ariane information system. CATED also operates a bibliography centre of Proceedings (telephonic information service);
- An audio-visual department (J.M. Brouard)
- NTI Services and CACT provide computer services to the FNB group and to the firms (Staff: approximately 100).

Total staff: approximately 350

Director or Chief Executive

President: J. Brunier

General Delegate: J. Buchor

General Director: Y. Aubert

Principal sources of finance

UTI is financed by the Fédération Nationale du Bâtiment whose resources are derived from voluntary contributions of French firms engaged in construction and public works. However, payment received for services rendered to enterprises accounts for 50% of the resources of UTI.

Principal related bodies

Directorate; Professional Council for innovation and research; Technical commissions of national unions; Specialized technical committees.

Publications

- The proceedings of ITBTP which publishes records of the ITBTP Conferences and the results of research
- Le Bulletin Signalétique de la Direction de la Recherche Monographs
- Publications of the International College for Building Science (reports of seminars).

DDR.216 **FM**

Bauakademie der Deutschen Demokratischen Republik (BA der DDR)

Academy of Building of the German Democratic Republic

1125 Berlin, Plauener Strasse

Telephone: 3780

Date of establishment

1951

The Academy is the Central Research Institution of Building of the DDR and is subordinate to the Ministry of Construction. It is concerned with research, development, planning, design and consulting in architecture, buildings and civil engineering.

Main fields of work

The Academy can award to individuals the title of Doctor and of Professor.

The Academy comprises 12 Institutes with a total staff of 4000, Building Information Centre and Experimental Project Design Division included. Most of them are located in Berlin, with a few, however, in Dresden, Leipzig and Weimar.

The principal research areas of the Institutes (each of which has a staff of 200 to 600) are summarized below:

- *Institute for Town Planning and Architecture:* Theory of architecture; History of architecture; Long term future evolution of living conditions and human settlements; Planning of settlement systems and the regions around towns; Development of methodology of master plans of town planning; Reconstruction and reshaping of central city cores specifically as regards medium and also small size towns; Planning of new residential areas; Planning of industrial regions
- *Institute for Housing and Related Buildings:* Development of large panel systems; Executing experimental projects for housing; Development of systems for public buildings; Adaptation of industrialized methods for the purpose of inner city modernization projects
- *Institute for Industrial Buildings:* Methods for taking stock and evaluating the existing industrial buildings; Development of reconstruction methods; Development of assembly methods for power stations through the use of large size steel space units; Assessment of the level of technology in industrial constructions; Development of systems with a high potential of application in industrial constructions; Development of building systems; Development of programming and management of industrial constructions based on computers
- *Institute for Agricultural Buildings:* Buildings for raising cattle, poultry and growing vegetables ensuring up-to-date industrial production methods; Development of economical building systems and components ensuring adequate protection against fire, good climatic and lighting conditions, energy conservation; Development of unified prefabricated components to be used for different types of agricultural buildings; Establishing conditions for modernization of rural settlements and their building stock
- *Institute for Civil Engineering:* Inner city reconstruction of municipal services with use of existing pipe networks and cellars of buildings; Reconstruction methods of sewage pipe networks with the use of chemical processes and pipes made of plastics; Development of road construction and reconstruction technologies; Development of sewage treatment plants; Research on problems of building mechanics. In the field of acoustics, the Institute has been awarded contracts not only for performance halls in the GDR but also abroad.
- *Institute for Technology and Mechanization:* (staff of 500). Development of technological processes; Development of measurement and control techniques including the use of laser techniques; Development of organization and management methods with application of computers; Development of organization and management methods with application of computers; Development of robots and introduction of microelectronics in construction and prefabrication; Development of new methods to mix and cure concrete.
- *Institute for Design and Standardization:* Development of methodologies for up-to-date design work; Development of computer aided design methods, computer graphics and microfilms; Computer programmes to be used in design processes
- *Institute for Heating, Ventilation and Structural Theory:* (staff of 300). Energy conserving district heating systems; Systems for the recovery of heat; Development of alternative energy systems; Use of microelectronics in HVAC systems; Protection against corrosion; Improving sound insulation and creation of good acoustical conditions in performance halls; Investigations of heat and moisture properties of building materials and components and elaboration of design methods aiming at protection against heat and moisture
- *Institute for Building Materials:* Additives for concrete; Special concretes; Improvements in the economics and energy consumption in the manufacturing of building materials; Protection of reinforcements against corrosion in cellular concretes; Use of by-products in the production of building materials; Development of new building materials for the external envelope of buildings; Testing of building materials and control of durability of plastics; Control of fire resistance of building materials and structures, development of design methods taking into account fire
- *Institute for Economics:* Complex economic analysis of decision making, central management and planning in the building sector; Research into long term building needs and optimal use of the building stock; Development of methods to evaluate the efficiency of the work of enterprises; Methodology of improvements to increase productivity and rentability; Models for the further development of the building industry in individual regions
- *Building Information Centre* (staff of 270). Main areas of activity are: Literature research; automation of such research; advisory service on patents; Library services; providing designers with essential basic data; Exhibitions, services in graphics, photo, film, video; Printing, publishing and marketing publications
- *Experimental Project Design Division:* This division is engaged in the planning, design and construction of projects for export as well as experimental projects in the fields of building research and special projects in the sphere of community buildings. General planning for complex export projects in urban and residential development as well as community buildings; Elaboration of projects for pilot experimental buildings in building research; Undertaking special studies and designs.
- *Structure of the Academy:* The President, the First Deputy and the Vice-Presidents constitute the Presidium of the Academy. The supreme scientific body is the Plenum. Members of the Plenum are the Ordinary Members (numbering 30), the Candidate Members (35) and the Corresponding Members (slightly more than 70).

It is headed by the President who chairs the Presidium and the Academy's plenary sessions. The Director of Research of the Academy is his First Deputy, whereas the three Vice-Presidents represent him in complex fields of research work. At the same time they are Directors of one of the Academy's Institutes.

Principal sources of finance

Financed by fees paid for work performed for the construction industry and by funds from the Ministry of Construction.

DDR.217 **AM**

Technische Universität Dresden, Sektion Bauingenieurwesen, Sektion Architektur, Sektion Energieumwandlung

Technical University Dresden, Section of Civil Engineering, Section of Architecture, Section of Energy Transformation

Mommsenstrasse 13, 8027 Dresden

Telephone: 4630

Telex: 02278

Date of establishment

1828 Technical Educational Establishment; 1858 Polytechnic; 1890 Institute of Technology; 1961 Technical University

The Technical University is a governmental institute of advanced education in the social, natural and technical sciences. The research work is provided by order of the Ministry of Advanced Education, the Ministry of Building, the Academy of Building and the Building industry.

Faculties: Social Sciences; Natural Sciences and Mathematics; Mechanical Engineering; Electrical and Electronic Engineering; Building, Water and Forestry Sciences.

The Faculty of Building, Water and Forestry Sciences includes the following Sections: Civil Engineering; Architecture; Geodesy and Cartography; Water Engineering; Forestry.

Main fields of work

Regional and town planning, urban renewal; design of technical communications and supply in suburban areas; design of residential, social, industrial and agricultural buildings; design and testing of structures in concrete, steel and wood; building and soil mechanics structural analysis and synthesis; maintenance, modernization and reconstruction; building climatology, heat and moisture transfer; thermal loads and energy conservation, low temperature heating and heat recovery; planning methods for the building process, computer aided manufacturing; thermal treatment of concrete; building diagnostics.

Total staff: 180 (Sections of Civil Engineering and Agriculture, Department of Technical Building Equipment)

Professionally qualified staff: 100

Director or Chief Executive

Dean: Professor Kurth

Vice Dean: Professor Deutschmann

Heads of department

Section of Civil Engineering: Professor Schmidt, Director

Section of Architecture: Professor Papke, Director

Department of Technical Building Equipment in the Section of Energy Transformation: Professor Kraft

Principal sources of finance

Financed by the Ministry of Advanced Education. Research work is additionally financed by the Academy of Building and the building industry.

Principal related bodies

Council of the Society; Council of Science; Council of the Section.

Publications

Wissenschaftliche Zeitschrift der Technischen Universität Dresden

Papers concerning building research are published in the following professional journals: Bauzeitung, Bauplanung und Bautechnik, Betontechnik, Deutsche Architektur, Stadt- und Gebäudetechnik, Verlag für Bauwesen, Berlin.

DEU.218
Battelle-Institut eV
Battelle Institute
Am Römerhof 35, 6000 Frankfurt 90
Telephone: (0611) 79080
Telex: 041 1966
Date of establishment
1952

Private, registered contract research institute.
Main fields of work
- Physical technologies
- Product and process developments
- Laser and optics
- Materials technology
- Safety technologies
- Vehicle technology
- Information technology and automation
- Plant and process engineering
- Economics research
- Human affairs and technology

Total staff: 620
Professionally qualified staff: 350
Director or Chief Executive
Managing Director and Vice President: Dr rer nat Horst Haeske
Heads of department
R & D Departments: Director Dr H. Haeske
Quality and Inventions: Director Dr G. Schimmel
Marketing: Director K.W. Staehle
Internal Research: Director Dr G. Sandstede
Administration: Director Dr N. Petersen
Principal sources of finance
Contracts sponsored by industry and governmental agencies.
Publications
- *Journal:* Battelle aktuell, quarterly
- *Bibliographies:* List of published papers and articles, twice a year.

DEU.219
Beratungsstelle für Stahlverwendung
German Steel Information Centre
Kasernenstrasse 36, Postfach 1611, 4000 Düsseldorf 1
Telephone: (0211) 829-1
Telex: 08582286
Date of establishment
1927

A group organization of the German steel industry.
Main fields of work
All fields of the utilization of steel, especially in the building industry, machine construction, mechanical engineering, mining industry and agriculture.

Consultation of experts, students and other people concerned with problems related to the utilization and processing of steel.
Director or Chief Executive
President: G. Theodor Wuppermann
Director: Dipl Ing Hans Gladischefski
Principal sources of finance
Yearly contributions of the members.
Publications
Wide programme of publications; list available in one bibliographical survey.

DEU.220 **UM**
Betriebswirtschaftliches Institut der Westdeutschen Bauindustrie GmbH (BWI-Bau)
Institute for Business Administration of the West German Building Industry
Schillerstrasse 33, Postfach 17 13, 4000 Düsseldorf 1
Telephone: 0211 67 03-0
Date of establishment
1963

BWI-Bau is an institute of the employer's federation of the building companies in Nordrhein-Westfalen.
Main fields of work
- Business administration for the building industry
- Branch economics and national economics
- Problems of the building trade (including technical-economics questions): research work and advice; business comparative and branch statistical research; short and long term programmes for building trades; vocational training and advanced training; general and special consultancies for companies, organizations and public institutions.

Total staff: 20
Professionally qualified staff: 6
Director or Chief Executive
Dr K. Robl
Heads of department
Analysis and statistics of the building market and building companies: H. Tipp
Organization and data processing: D. Ohlert
Research and advanced training: Dr B. Refisch
Technical-economics questions: Bauassessor V.O. Tripcke
Principal sources of finance
Financed from assignments and members' fees.
Principal related bodies
Advisory board, whose function is supervision.
Publications
Bauwirtschaftliche Informationen: brief contributions on the activities of the Institute which appear several times yearly in an indefinite order

List of publications available on request.

DEU.221
Bundesanstalt für Geowissenschaften und Rohstoffe (BGR)
Federal Institute for Geosciences and Natural Resources
D 3000 Hannover-Buchholz, Stilleweg 2, Alfred-Bentz-Hans, Postfach 61 01 53
Telephone: (0511) 6468-1
Telex: 923 730 bfb (bgr) ha d
Cables: Geobund
Date of establishment
1 January 1959

The Bundesanstalt is the central institution for geosciences and mineral resources in the Federal Republic of Germany and is subordinated to the Federal Ministry of Economics.
Main fields of work
The BGR has the following responsibilities:
- Advising the Federal ministries on all questions relating to the geosciences and natural resources
- Participation in securing the supplying of the Federal Republic of Germany with mineral and energy resources, advising German industry, as well as carrying out prospecting and exploration projects at home and abroad, including geoscientific marine research
- Participation in activities of the Federal government involving the geosciences and natural resources; participation in Technical Co-operation projects with developing countries

- Participation in measures taken for geotechnical safety (e.g. final disposal of radioactive wastes, siting of nuclear power plants), as well as environmental protection problems involving the geosciences
- Development of methods and instruments in all fields of the geosciences, as well as their practical use
- International scientific co-operation: antarctic research monitoring of seismic events, geological maps, deep-sea drilling program; co-operation with foreign geological surveys.

Total staff: 740
Professionally qualified staff: 279 Geoscientists
Director or Chief Executive
President: Professor Dr Friedrich Bender
Heads of department
Economic Geology: Professor Dr Martin Kürsten
Technical Geology: Professor Dr Helmut Venzlaff
Engineering Geology and Geotechnics: Professor Dr Michael Langer
Geophysics: Professor Dr Hans-Juergen Duerbaum
Geochemistry, Mineralogy: Professor Dr Wolfgang Stahl
Administration, Library, Documentation, etc.: Karl Abraham
Principal sources of finance
Ordinary Budget of the Federal Ministry of Economy; also financial support from the Ministries of Economic Co-operation and Research and Technology.
Principal related bodies
Board of Trustees.
Publications

- *Journal:* Geologisches Jahrbuch, monthly
- Reports on the Commodities Situation of individual countries
- Studies on Supply and Demand for Mineral Commodities

A catalogue of publications is available.

DEU.222 — AM

Bundesanstalt für Materialprüfung (BAM)
Federal Institute for Materials Testing
Unter den Eichen 87, D 1000 Berlin 45
Telephone: 81 04-1
Telex: 1-83 261 bamb d
Cables: BAM BERLIN
Date of establishment
Founded in 1871 as the Mechanical-Technical Institute. After being combined with several other institutes, the BAM changed to its present name in 1956. By decree of the Federal Ministry for the Economy from 1 September 1964 the tasks were defined.

The BAM is a scientific-technological institute, aiming to promote the development of the German economy by carrying out materials research and materials testing, and by further developing chemical safety engineering. Within the framework of certain laws BAM has federal authority. Applicants for testing or research work are industry, ministries, courts and private individuals.
Main fields of work
The main fields of work can be seen from the following scheme:

- Department 1: Metals and Metal Structures: Physical Metallurgy and Metal Technology; Materials Mechanics and Behaviour of Structures; Corrosion and Protection of Metals; Inorganic Chemical Investigations
- Department 2: Building: Structural Safeguards; Secretariat for UEAtc Affairs; Mineral Building Materials; Load-Bearing Capacity of Building Structures; Geotechnics and Bituminous Building Materials; Preservation of Building
- Department 3: Organic Materials. Elastomers, Plastics and Paints; Textiles and Leather; Paper, Printing, Packaging; Special Tests for Organic Materials
- Department 4: Chemical Safety Engineering: Transport of Dangerous Goods; Heterogeneous Combustion Reactions; Special Fields; Technical Gases; Explosive Solids and Liquid Substances; Gas Equipment and Acetylene Installations
- Department 5: Special Fields of Materials Testing; Biological Testing of Materials; Rheology and Tribology; Surface Investigations and Electrochemistry; Colorimetry and Optical Properties of Materials
- Department 6: Methods Independent of Material Type; Measuring Techniques and Basic Experimental Engineering; Non-Destructive Testing; Nuclear Technology in Materials Testing and Radiation Protection; Joining Techniques.

Total staff: 1200
Professionally qualified staff: 1000
Director or Chief Executive
President: Professor Dr G.W. Becker
Heads of department
Metals and Metal Structures: Professor Dr D. Aurich
Building: Professor Dr A. Plank
Organic Materials: Professor Dr H. Feuerberg
Special Fields of Materials Testing: Professor Dr H. Czichos
Methods Independent of Material Type: Professor Dr E. Mundry
Principal sources of finance
Federal Ministry for the Economy: other Ministries through special research programmes; research-sponsoring institutions.
Principal related bodies
State materials testing institutes within the Federal Republic of Germany, testing institutes abroad, standardizing institutions.
Publications

- *Journal:* Official journal of BAM (Amts- und Mitteilungsblatt der BAM), quarterly
- Annual report, annually
- *Technical standards:* in co-operation with DIN (German Institute for Standardization) UEAtc agréments
- *Research reports* (Forschungsberichte), irregularly
- Conference reports, irregularly
- Documentation services: Biological materials testing, organisms and materials; Rheology documentation; Tribology documentation (wear, friction, lubrication); Colour, Urgent reports on colour; Measurement of mechanical quantities; Non-destructive testing; Welding and related processes.

DEU.223

Bundesenstalt für Strassenwesen (BASt)
Federal Highway Research Institute
Brüderstrasse 53, D 5060 Bergisch Gladbach 1
Telephone: (02204) 430
Date of establishment
1951

A technical and scientific agency of the Federal Government. Established 1951 by the Federal Minister of Transport; in 1970, the Federal Parliament voted to nominate the Institute additionally as the centre for research into traffic safety and accidents.
Main fields of work
Achievement of both technically sound and economical highway construction; improvement of the capacity of highways and highway networks; reduction of the environmental impact of highways; improvement of traffic safety.
Total staff: 370
Professionally qualified staff: 110
Director or Chief Executive
President: Professor Heinrich Praxenthaler
Deputy: Professor Karl Krell
Heads of department
Highway Construction Techniques Department: Professor Egil Nakkel
Traffic Engineering Department: Professor Karl Krell

Traffic Safety and Accident Research Department: Professor Karl-Heinz Lenz
Central Department: Professor Peter Canisius
Principal sources of finance
Federal Government Budget; Federal Minister of Transport.
Publications
- Serial publications
- Regular activity reports
- Special reports

Numerous papers are published in scientific and professional journals.

DEU.224
Bundesforschungsanstalt für Forst- und Holzwirtschaft (BFH)
Federal Research Centre for Forestry and Forest Products
Leuschnerstrasse 91, 2050 Hamburg 80
Telephone: (040) 73962 1
Date of establishment
1950 (1939)

Governmental research organization directly related to the Federal Ministry of Food, Agriculture and Forestry with the task of elaborating the scientific fundamentals for decisions of the Federal Government in the fields of forestry and wood industry.
Main fields of work
Research activities on the following subjects:
- Conservation of forests and improvement of their wood production to ensure a sufficient supply of raw material as well as environmental benefits for mankind
- Improvement of productivity, structure and economy in forestry and forest products industry as well as of the marketing of their products
- Conservation, improvement and standardization of the properties of timber and wood products to warrant supply of high-quality products to the population
- Development of advanced techniques for treatment and processing of wood and rationalization of processes in the wood industry in order to make better use of the raw material wood, to improve the quality of products, and to promote production technologies which are less harmful with respect to the environment
- Properties and use potentials of lesser-used wood species (especially of tropical and subtropical areas) to ensure a sufficient wood supply in the future by widening the basis of raw material
- Observation of market conditions as well as the improvement of markets for timber, semi-finished and finished wood products and pulp and paper
- Active co-operation in national and international standardization of timber and wood products
- Co-operative connection with the University of Hamburg.

Total staff: about 200
Professionally qualified staff: about 60
Director or Chief Executive
Executive Director: Professor Dr W. Liese
Heads of department
Institute for World Forestry: Professor Dr E. Brünig
Institute for Forest Genetics and Forest Tree Breeding Ltd: Director and Professor Dr G.H. Melchior
Institute for Work Science: Professor Dr G. Eisenhauer
Institute for Wood Biology and Wood Protection: Professor Dr W. Liese
Institute for Wood Chemistry and Chemical Wood Technology: Professor Dr H. Nimz
Institute for Wood Physics and Mechanical Wood Technology: Professor Dr D. Noack
Principal sources of finance
Federal Ministry of Food, Agriculture and Forestry.
Publications
- *Research reports:* Mitteilungen der Bundesforschungsanstalt für Forst- und Holzwirtschaft (Scientific originals issued as separate editions), about 6 per year
- *Bibliographies:* Schrifttumskarteidienst (Reference Card Documentation Services), quarterly; Computerized documentation services in the field of forestry and wood industry
- Publication of articles in related scientific journals

List of publications available on request.

DEU.225
Bundesforschungsanstalt für Landeskunde und Raumordnung (BfLR)
Federal Research Institute for Regional Geography and Regional Planning, D 5300 Bonn 2
Telephone: (0228) 826 1
Date of establishment
(1935, 1941) 1973 by ministerial decree

BfLR is a scientifically autonomous research establishment of the Federal Government under the jurisdiction of the Federal Minister for Regional Planning, Building and Urban Development.
Main fields of work
To provide the scientific and informative base for the implementation of government functions in the areas of physical and urban planning as well as spatially effective sectoral planning. This is done in collaboration with similar institutions in the Federal Republic and other countries. The BfLR observes, analyses and reports on present and future spatial developments. It supports and advises the Federal Government on projects of spatial or local relevance and advises the Minister responsible for regional planning on the formulation and further development of regional and urban policy.

The BfLR fulfils its task through three types of activity:
- Own research for the development of long-term foundations and suggestions on spatially effective policy
- Implementation of the Medium-Term Research Programme, Regional and Urban Planning of the Federal Ministry for Regional Planning, Building and Urban Development
- Provision of information and expertise (mostly in the form of ad hoc commissions for the ministries).

There are two departments: Research and Information.
Total staff: 130
Professionally qualified staff: 40
Director or Chief Executive
Director: Professor Dr W. Strubelt
Publications
- *Journals:* Informationen zur Raumentwicklung; Raumforschung und Raumordnung
- *Monographs:* Forschungen zur Raumentwicklung
- *Series:* Seminare-Symposien-Arbeitspapiere
- *Script series:* Landeskundliche Luftbildauswertung im mitteleuropäischen Raum
- *Literature and research documentation:* Referateblatt zur Raumentwicklung; Thematische Literaturanalysen; Forschungsdokumentation Raumordnung, Städtebau, Wohnungswesen
- *Maps:* Verwaltungsgrenzen der Bundesrepublik Deutschland
- *Thematic atlases:* Atlas zur Raumentwicklung
- *Map collections:* Geographische Landesaufnahme 1:200 000 – Naturräumliche Gliederung; Arbeitskarten zur Energie- und Wasserversorgung
- *Information leaflet:* BfLR-Mitteilungen.

DEU.226

Deutsche Gesellschaft für Mauerwerksbau
German Society of Masonry Construction
Kortumstrasse 50, 4300 Essen
Telephone: 0201 776248
Date of establishment
1978

Umbrella organization of the associations of the producers of masonry materials.
Main fields of work
- Collaboration in national and international standardization
- Collaboration in research projects
- Collaboration in all technical problems concerning masonry constructions
- Public relations on the above mentioned activities.

Total staff: 5
Professionally qualified staff: 3
Director or Chief Executive
President: Klaus Göbel
Vice-president: Roland Cordes
Secretary: Reiner Pohl
Principal sources of finance
Membership fees.
Publications
Technical standards: several recommendations.

DEU.227

Deutsches Krankenhausinstitut, Institut in Zusammenarbeit mit der Universität Düsseldorf
German Hospital Institute, Institute affiliated to the University of Düsseldorf
Tersteegenstrasse 9, D 4000 Düsseldorf 30
Telephone: 0211 434422
Date of establishment
1953

Private voluntary institute, sponsored by the associations of hospital self-administration and the University of Düsseldorf.
Main fields of work
- Research
- Education and professional training
- Documentation
- Information Service.

Total staff: 37
Professionally qualified staff: 21
Director or Chief Executive
Professor Dr Siegfried Eichhorn
Architect: Richard-Joachim Sahl
Heads of department
Research: S. Eichhorn
Education and Professional Training: G. Baugut, R. Hartwig
Documentation: G. Baugut
Principal sources of finance
Fees for services (research projects, seminars, consulting); Government.
Principal related bodies
Supervisory Board (Kuratorium); Managing Board.
Publications
- *Journal:* Beitrag in Das Krankenhaus, monthly
- *Research reports:* Medizintechnische Service-Zentren; Optimierung der Betriebsabläufe im Krankenhaus; Progressivpflege; Qualitätssicherung im Bereich der Krankenhausmedizin
- *Bibliographies:* Krankenhaus-Informationsdienst
- *Others:* Schriftenreihe des DKI mit 15 Bänden; 473 Veröffentlichungen in Zeitschriften; 92 Beiträge in Büchern.

DEU.228

Forschungsgemeinschaft Bauen und Wohnen (FBW) Zentrale Forschungsstelle Baden-Württemberg für Bauwesen, Städtebau und Raumplanung
Research Association for Building and Housing
Silberburgstrasse 160, 7000 Stuttgart 1
Telephone: (0711) 610 629
Date of establishment
1947

A public foundation established by the State of Baden-Württemberg. Financed mainly by subsidies from the State of Baden-Württemberg.
Main fields of work
To co-ordinate research projects and research works; to sponsor and to promote research projects; to execute, oversee and utilize research works; to stimulate and to lead working groups for investigations of relations in the field of work; to collaborate in other working groups which serve practitioners, especially in working groups for standardization; to plan in cases of examples, to collaborate in the erection of trial buildings and to utilize experience gained thereby; to publish new knowledge in planning and building practice; to give advice; to document: in the following fields:
- Building (housing infra-structure, buildings, building technique, economics in building, renewals)
- Town planning (town development, master planning, urban renewals, urban development schemes)
- Physical planning (regional policy, regional planning, sub-regional planning).

Total staff: 10
Director or Chief Executive
Chairman of the Board: J. Herkommer
Managing Director: Kl Brandstetter
Principal sources of finance
Subsidies from the State of Baden-Württemberg.
Publications
List available on request.

DEU.229

Forschungsinstitut der Zementindustrie, Verein Deutscher Zementwerke eV (VDZ)
Research Institute of the Cement Industry, Association of the German Cement Works
Tannenstrasse 2, Postfach 30 10 63, D 4000 Düsseldorf 30
Telephone: (0211) 45 78–1
Telex: 0858 4867
Date of establishment
1948

A private research institute, supported by the Association of the German Cement Works.
Main fields of work
The main tasks of the Research Institute concern the technical and scientific problems of the manufacture and use of cement.

Scope of activities: cement and concrete technology, cement chemistry and mineralogy, cement manufacturing processes, environment protection and works safety.

The results of research are published by means of lectures, meetings (e.g. the International VDZ Congress), advanced training seminars, training courses for foremen, or by means of readily accessible publications and information services.

Total staff: 124
Professionally qualified staff: 28
Director or Chief Executive
Director of the Institute: Professor Dr-Ing Gerd Wischers
Heads of department
Concrete Department: Professor Dr-Ing J. Bonzel
Cement Department: Professor Dr F.W. Locher
Principal sources of finance
Mainly financed by membership dues per ton of dispatched cement.
Publications
- *Research reports:* Schriftenreihe der Zementindustrie (Publication series of the cement industry), Düsseldorf: Beton-Verlag, from 1949, irregular; Betontechnische Berichte (Reports on technology of concrete) Düsseldorf: Beton-Verlag, from 1960, annually; from 1980, biennially
- Zement-Taschenbuch (Handbook of cement), Wiesbaden, Berlin: Bauverlag, since 1950, biennially
- Tätigkeitsbericht (Report on activity), since 1948, at present: triennially
- VDZ-Mitteilungen (Information circular), since 1963, issued three times a year, for members only
- Notes and technical directives, numerous papers in technical and scientific journals, mostly in Beton and Zement-Kalk-Gips, books and conference proceedings.

DEU.230
Forschungs- und Materialprüfungsanstalt Baden-Württemberg Otto-Graf-Institut (FMPA)
Institution for Research and Materials Testing
Pfaffenwaldring 4, D 7000 Stuttgart 80
Telephone: (0711) 685-1 or 685-3323
Telex: 07255 727 fmpa d
Cables: FMPA BW STUTTGART
Date of establishment
1884 at the Polytechnikum Stuttgart by Professor Carl v. Bach

The laboratory was part of the University of Stuttgart until 1979. Since 1980 it has belonged, as a public body, to the Ministerium für Wirtschaft, Mittelstand und Verkehr Baden-Württemberg and performs research work, testing and development upon request for the Government, industry and private individuals, in co-operation with the institutes of the University of Stuttgart.
Main fields of work
- Research, testing of materials and structures, consulting, development, control, inspection, expertise in all fields of civil engineering (building materials, structures, physics relating to construction, soils and rocks)
- Chemical and technical analysis and expertise
- New building materials and technologies
- Corrosion of materials, members and structures
- Fire resistance, fire protection
- Preservation of structures
- Foundation engineering, slope stability analysis.

Total staff: approximately 260
Professionally qualified staff: 50
Director or Chief Executive
Director: Professor Dr Gallus Rehm
Heads of department
Building Materials: Professor Dr W. Manns
Structures: Dr G. Völkel
Physics related to Construction, Preservation: Professor Dr W. Schwaderer
Soils and Rocks: Dr L. Wichter
Chemical Analysis: Dr C. Schiele
Administration: Dr E. Vordermeier
Principal sources of finance
Governmental (state of Baden-Württemberg); industrial and private orders; grants for research studies.
Principal related bodies
Ministerium für Wirtschaft, Mittelstand und Verkehr Baden-Württemberg (supervisory).
Publications
- *Research reports:* e.g. Schriftenreihe des Otto-Graf-Institutes, irregular
- *Bibliographies:* a booklet is available free of charge
- Numerous papers published in professional or scientific journals.

DEU.231
Forschungsvereinigung Kalk–Sand eV (FVKS)
Lime–Sand Research Association
Entenfangweg 15, D 3000 Hannover 21
Telephone: (0511) 79 30 77
Date of establishment
Founded in 1965, emerging from the research laboratories of the Association of the Sand, Lime, Brick Industry.

An independent institution within the Association of the Sand, Lime, Brick Industry (Hannover).
Main fields of work
Research and development in the field of calcium hydrosilicates, produced by autoclaving a mixture of lime and silica: basic research; technological research; applications; evaluation of research results (literature). Besides improving the mechanical and physical properties of the newly produced sand–lime bricks, there is a further aim of researching the development of new sizes of building stones and solving the manifold technological problems on applications (finish and calculation of masonry).

The principal departments are: General Administration, Building Research; Basics and Technology; Laboratories; Documentation.
Total staff: 9
Professionally qualified staff: 3
Director or Chief Executive
Dr R. Cordes, Dr U. Wittneben, S. Kreisl
Principal sources of finance
Subscriptions of the members and State subsidies.
Publications
Research reports.

DEU.232 AM
Fraunhofer-Institut für Bauphysik (FIBP)
Fraunhofer Institute for Building Physics
Nobelstrasse 12, D 7000 Stuttgart 80 (Vaihingen)
Telephone: 0711 6868 00
Telex: 7 255 167
Other address: Freilandversuchsstelle Holzkirchen, Free Field Station, Postfach 1180, D 8150 Holzkirchen
Telephone: 08024 5055
Date of establishment
1929

Research institute and testing institution.
Main fields of work
- Architectural acoustics and noise control (traffic noise and industrial noise)
- Heat and moisture protection of buildings
- Energy saving
- Indoor climate
- Weathering
- Development and test of constructions and structural elements.

Total staff: 87
Professionally qualified staff: 58
Director or Chief Executive
Director: Professor Dr Fridolin P. Mechel
Heads of department
Technical Acoustics: Dr Fuchs
Building Acoustics: Dipl -Ing Koch
Thermal Laboratory Tests: Dipl -Phys König
Thermal Numerical Simulation: Dr Plöger
Building Components: Dipl -Ing Bertsch
Free Field Station, Weathering: Dr Künzel
Principal sources of finance
Fees for research work and state grants.
Principal related bodies
Fraunhofer-Gesellschaft, Munich, society for the advancement of applied research, supported by the Federal Government.
Publications
- Veröffentlichungen aus dem Institut für Bauphysik, Stuttgart (Publications of the Institute for Building Physics, Stuttgart)
- IBP-Mitteilungen (IBP Information) contains summaries of research results.

DEU.233

Fraunhofer-Institut für Holzforschung, Wilhelm-Klauditz-Institut
Fraunhofer Institute for Wood Research, Wilhelm Klauditz Institute
Bienroder Weg 54 E, D 3300 Braunschweig
Telephone: (0531) 39 09 0
Telex: 952 942 wkibs d
Date of establishment
Established in 1946 by representatives of the Government and of industry. Since 1972 it has acted as an institute of the Fraunhofer-Gesellschaft for the advancement of applied research.

Institute of the Fraunhofer Gesellschaft, a registered non-profit-making association for the advancement of applied research.
The Institute advances applied timber research.
Main fields of work
Main research activities: optimization of production techniques for wood-based materials, development of new materials, design and setting up of process control systems, solutions to problems occurring in the field of wood utilization, pollutant emissions during production and daily use, development of glue resins, utilization of special wood species and assortments, surface properties of wood and wood-based products, surface coating, properties of practical use and long-term behaviour of wooden building components.
Main fields of research work: automation techniques in the timber industry, development of equipment and software, buildings construction, simulation techniques, techniques for optimization and testing of materials, quality control, properties analyses, anorganic-bound wood-based materials, chemistry of wood and of pulp and paper, chemistry of adhesives and of varnishes, environmental analytics, studies on the properties and the possible fields of application of raw materials, wood preservation, surface techniques.
Total staff: 74 (including 11 part-time)
Director or Chief Executive
Director: Professor Dr Gert Kossatz
Heads of department
Automation techniques: Obering L. Mehlhorn
Production techniques, wood-based materials: C. Harbs
Anorganic-bound wood-based materials: Professor Dr G. Kossatz
Chemical technology, environment protection: Dr R. Marutzky
Raw materials, coating: Dr P. Böttcher
Principal sources of finance
Orders from industrial and economic associations; funds from Ministries of the Bund and the Länder.
Principal related bodies
CEN, DFG, DGfH, DIN, FESYP, Gütegemeinschaft Spanplatten (German Association for the Quality of Particleboards), Gütegemeinschaft Sperrholz (German Association for the Quality of Plywood), IfBt, ISO, IUFRO.
Publications
- *Journal:* Holzforschung, bi-monthly, Ed. Dr G. Stegmann; publishers: De Gruyter, Berlin
- *Research reports:* periodical WKI Reports (published by the Institute).

DEU.234

Gesellschaft für Umweltplanung Stuttgart mbH (GUS)
Society for Environmental Planning
Johannesstrasse 71, D 7000 Stuttgart 1
Telephone: 049/(0)711/610765
Date of establishment
1971

The society is privately owned by the heads of five research (briefing, building, energy, costs), architectural and engineering (structure and HVEAC), private offices. The society has two associated partner offices (landscaping, traffic)
It has the following foreign associated partners:
- HOK-Hellmuth, Obata and Kassabaum, St Louis, USA, consultant
- Boorsma, Civil Engineering and Architecture, Drachten, NL, consultant

The work is performed on behalf of individual or frame contracts with private and public clients.
Main fields of work
- Regional and city planning: general development planning for housing, traffic, industry, education, professional training, leisure, sports; village development and renewal, agricultural development
- Project planning for buildings and landscape: preliminary planning and feasibility studies, briefing, project development, architecture, design of all environmental systems, executional planning, tendering procedures; building supervision
- Consulting: planning procedures; briefing and organizational planning; expertise for public subsidy programmes; expertise for energy management
- Research: study of basic planning data (handicapped, students, professional training, commercial and business activities, industrial plants); investigation of planning methods, (briefing, dimensional co-ordination, computer application, costs, energy); analyses of built projects or existing building, renewal modernization, gain of basic planning data; development and publication of planning aids.

Total staff: 85
Professionally qualified staff: 65
Director or Chief Executive
Manager, Head of Architecture Department: W.D. Breucha
Manager, Head of Research Department: F.O. Kerschkamp
Heads of department
Architecture: Büro für Bauplanung Stuttgart, W.D. Breucha, H.P. Klein
Structure: Spielkamp, Hauke, consultants
HVAC: Ingenieure für Technik i.Bau, P. Rentschler, F. Riedesser
Electricity: Unabhängiges W. Schwarz
Regional planning: Planungsbüro Plamper Stuttgart, S. Plamper SRL
Research: C. Anadere, H.W. Ingerfurth, K. Famira, F.O. Kerschkamp
Principal sources of finance
Fees for project planning, investigations, studies and research projects for private and public clients.

Publications

- *Technical standards:* Einrichtungen für Spiel- und Sport mit Behinderten, Geistigbehinderte SBL-Studien 56, 1981; Körperbehinderte, SBL-Studien 57, 1983; SÜA-Sachmittelstandards für überbetriebliche Ausbildungsstätten, BMBW, 1983
- *Research reports:* Kaiser, Kerschkamp, Modulare Bauwerksteilanschlüsse, IRB, 1984; Ingerfurth, Plamper, Stadtraum-Lebensraum: Wieder Platz für Menschen, in: Machensen, Umbach, Jung (ed) Leben im Jahre 2000 und danach, Perspektiven für die nächsten Generationen, 1984
- *Others:* Kerschkamp, u.a., Weiterverwendung bestehender Schulgebäude, Arbeitshilfen Schulbaumodernisierung, SBL Studien 69, 1983; Ingerfurth, Zur Planung überbetrieblicher Ausbildungsstätten, in: Bauwelt 33, 9/1983; Ingerfurth, Kerschkamp, Ansichten zur Schule der letzten 20 Jahre, in: Bauwelt 7, 2/1984

A catalogue is available on request.

Principal sources of finance
Work is financed by members' subscriptions, fees for commissioned work, as well as private and public grants.
Principal related bodies
Official Bodies of the Institute are the General Meeting, the Board of Governors, the Administrative Advisory Committee and the Directorate.
Publications

- Wirtschaftskonjunktur (Economic Review), monthly
- IFO Schnelldienst (IFO Flash Report), 3 per month
- IFO-Konjunkturperspektiven (IFO Business Perspectives)
- IFO-Spiegel der Wirtschaft, structure and business cycle in graphs and figures
- IFO-Studien (IFO Studies)
- IFO Schriftenreihe (IFO Research Series)
- Manufacturing Series: structural analysis of the major industrial sectors
- Other specific publications.

DEU.235

IFO-Institut für Wirtschaftsforschung
IFO Institute for Economic Research
Poschingerstrasse 5, D 8000 München 86
Telephone: (089) 92241
Telex: 522269
Date of establishment
1949

A non-profit and independent research organization; it has the legal status of a registered society. As the name implies, IFO's objective is to work in the field of empirical economics through information and research, and to analyse and publish the findings with a view to practical application.
Main fields of work

- Analysis and short-term forecasting of economic developments in the Federal Republic of Germany and abroad; analysis of cyclical developments in industry, construction and trade; the carrying out of the Business and Investment Surveys developed by the Institute as a new method of economic observation
- Medium and long term projections of gross national product and its components
- Structural studies in industry, construction, agriculture, wholesale and retail trade, transportation
- Preparation of studies designed to serve as a basis for economic policy decisions
- Research in and forecasting of regional developments of gross national product as well as the regional development in specific economic branches, particularly construction and transportation.

Total staff: approximately 200
Professionally qualified staff: 80
Director or Chief Executive
President: Professor Dr K.H. Oppenländer
Heads of department
Department of General Economic Research: Dr Krumper
Public Finance Department: Dr Leibfritz
Business and Investment Survey Department: Dr Nerb
Industry Department: Dr Bonhoeffer
Construction Department: Dr Russig
Marketing Department: Dr Batzer
Agricultural Department: Dr Lösch
Department of World Economic Studies: Mr Claus
Africa Studies Centre, Developing Countries Department: Dr Braun
Econometric and Data Processing Department: Dr Goldrian
Publications Department: Dr Laumer
Department of Structure and Growth: Dr Gerstenberger

DEU.236 FM

Informationszentrum RAUM und BAU der Fraunhofer-Gesellschaft (IRB)
Information Centre for Regional Planning and Construction
Nobelstrasse 12, D 7000 Stuttgart 1
Telephone: (0711) 6868-500
Telex: 7255167
Date of establishment
1977

An institution of the Fraunhofer-Gesellschaft, entrusted by the Federal Government with the functions of a central information agency in the fields of regional planning, town planning, housing and civil engineering. It collects and disseminates technical and scientific information in these fields (evaluations and recommendations are excluded).

The information services of the IRB are available to everybody, in so far as no restrictions are imposed by the authors. The IRB is committed to neutrality in its opinions.

Analysing and disseminating existing knowledge, the IRB pursues objectives that are articulated by the economy, planning, research, teaching, society and politics.
Main fields of work
Production of data bases in the fields of regional planning, town planning, housing and civil engineering. Literature data bases: RSWB (regional planning, town planning, housing and civil engineering); ORLIS (1974–78) (urban and regional planning); LINA (compiled bibliographies). Research project data bases: BAUFO (building research projects); FORS (research projects regional planning, town planning, housing). Factual data bases: BODO (building properties); SOFI (software information); VISDATA (Building Regulations information system).

Supply with planning-orientated scientific and technical information, in the fields of regional planning, town planning, housing and civil engineering. Information services: data base services (on-line services, magnetic tape services); services based on single orders (enquiries, searches, visitor's services, literature supply); Individual periodical services (individual profile services, INFIS, the individual journal information service); printed services (see Publications).

Operation of the CIBDOC system (CIBDOC agency).
Total staff: 96
Professionally qualified staff: 48
Director or Chief Executive
Dr W. Wissmann
Heads of department
Data: W. Coprian
Information transfer: F. Breckner

Literature: M.-L. Prankl
Information marketing: M. Herr
Computing operations: D. Bruns
IRB Publishing Department: D. Meincke
Staff Department: J. Acevedo-Alvarez

Principal sources of finance

Basically the annual expenditure of the IRB is financed by the Federal Government, and a smaller part by the Länder of the Federal Republic which contribute to the financing of the Fraunhofer-Gesellschaft.

Principal related bodies

The Centre works in close conjunction with more than 100 information and documentation centres in foreign countries and within the Specialized Information System (FIS Bau) as well as other information centres within the Federal Republic of Germany.

Publications

- Literature on civil engineering (Schriftum Bauwesen, SB), monthly
- Subject fields (standard profile services)
- Professional editions: Architect; Civil engineer; Eastern-language literature
- Literature on regional planning, town planning and housing (Schriftum Raumordnung, Städtebau, Wohnungswesen, RSW), monthly
- Professional edition: Town planning
- New books in the fields of SB and RSW (Fachbuchinfo RSWB) monthly
- Information service: housing and housing economics (Schriftum Wohnungswesen, Wohnungswirtschaft), quarterly
- Building research summaries (Kurzberichte aus der Bauforschung, KbF), monthly
- Research reports (according to cataolgue)
- Research documentation: regional planning, town planning and housing) (FRSW) once a year
- Catalogue of building research reports (KB)
- Catalogue of compiled literature references (LINA)
- Literature documentations
- Thematic documentations: No. 1 Bauschäden; No.2, Biologisches Bauen

DEU.237 UM

Institut für Arbeits-und Baubetriebswissenschaft (ifA) (Dr Gerhard Dressel GmbH & Co.)

Institute for Ergonomics, Organization and Management of Construction (Dr Gerhard Dressel GmbH and Co.)

Berliner Strasse 40, 7250 Leonberg

Telephone: 07152 207 0
Telex: 7 24129

A private institution which works in both research and advisory capacities, in addition to offering vocational training.

Main fields of work

Analyses and research are carried out for federal ministries and other developers. The Institute works in close conjunction with more than 200 contractors in both West Germany and Austria, exchanging work experience in periodic meetings and working committees.

Analyses and recommendations are passed on in reports specifically prepared for medium-size contracting firms.

Training courses are offered from all departments within the Institute to management, junior management, technical and economic personnel of construction firms.

Total staff: about 44

Director or Chief Executive

Dr Gerhard Dressel

Heads of department

Director: W. Schwendowius
Ergonomics: Dr H.-J. Fritz
Economics: W. Busch
Data Processing: J. Dressel
Personnel Development: G. Rothenberger

Principal sources of finance

Financed by fees for research work, advice and consultations for federal ministries and the construction industry.

Publications

- *Journal:* ifA nachrichten: a report on new developments within the Institute, monthly
- *Technical standards:* Final analysis reports, concerning the optimum use of construction equipment; reports gathering necessary economic and statistical data (Führungszahlen), and technical check list of functions in every field of construction (Funktionsliste Bauwesen)
- *Research reports:* Covering a wide range of construction and economic management concerns; improvements in the work peace on site; economic advantages of winter-time construction, etc.
- *Bibliographies:* ifA-Leistungen: an annual publication of services offered by the Institute.

DEU.238

Institut für Baubetriebslehre

Institute of Construction Management Research

Pfaffenwaldring 7, Postfach 80 1140, D 7000 Stuttgart 80

Telephone: 9711 685 6144/5
Telex: 07 255 445 univ

Date of establishment

1964

The Institute of Construction forms part of the Faculty of Civil Engineering of Stuttgart University. It is responsible for teaching and research in the field of construction management.

Main fields of work

- Construction management
- Cost planning
- Operations research in the construction industry
- Use of construction machinery.

Total staff: 9
Professionally qualified staff: 7

Director or Chief Executive

Professor Dr G. Drees

Principal sources of finance

Financed by the State of Baden-Württemberg and individual research grants.

Publications

Proceedings of the Institute of Construction Management Research (Schriftenreihe des Institut für Baubetriebslehre).

DEU.239 FM

Institut für Bauforschung eV (IfB)

Institute of Building Research

An der Markuskirche 1, D 3000 Hannover 1

Telephone: 0511/80 10 98
Other address: Gräfelfinger Strasse 128, D 8000 München 70

Date of establishment

23 May 1946

A registered non-profit organization.

Main fields of work

The Institute undertakes scientific research, the results of which are translated into practical advice.

The building research covers building site planning, building materials, building processes, etc., and is intended to provide advice and stimuli, thus serving town planning, building economy, housing users and the promotion of the building industry. To this end research is performed and scientific results and practical experience are gathered and evaluated. The knowledge thus acquired serves as information for all building parties as well as the basis for advice, further research and experiments.

The work of the Institute is solely and directly intended for scientific purposes on a non-profit basis.

The Institute is a non-profit organization with a public service. (The above is extracted from the statutes of the Institute).

The research and advisory work is spread over three departments: Building planning and productivity; Building physics and building construction; Building technique and experimental construction. Each research project is supervised by a responsible project leader, who is assured of scientific support from the different departments. Should the scientific capacity for a certain project be inadequate, free collaborators are engaged for the period required. Contracts are made in a similar way with other institutions.

Total staff: 23

Professionally qualified staff: 10

Director or Chief Executive

Director: Dr Herbert Menkhoff

Deputy Director: Gerhard Achterberg

Heads of department

Erichbernd Brocher

Karl Deters

Principal sources of finance

Fees for research work, advice, consultations for federal and regional ministries, housing companies, private institutions and individuals, etc.

Principal related bodies

The Assembly of Members, taking decisions about roles for the Institute; the Board, executing decisions for the Assembly of Members.

Publications

Journal: Wirtschaftlich Bauen, supplement to Bundesbaublatt, monthly

All reports on research work are published either by the source commissioning it or, with agreement, by the Institute

Summaries of all research works appear as a permanent appendix To Build Economically in the journal Bundesbaublatt (Bauverlag GmbH Wittelsbacherstr. 10, D 6200 Wiesbaden 1)

Results of research work are also disseminated at conferences and published in specialized journals.

DEU.240 FM

Institut für Bautechnik (IfBt)

Institute of Civil Engineering and Building Technology

Reichpietschufer 72–76, 1000 Berlin 30

Telephone: (030) 25 03 (1)

Telex: 185413

Date of establishment

1968

Public authority. Estabished by authority of the Law on the Institute of Building Technology. The foundation of the IfBt is based on a treaty of establishment and financing of such an institute between the 11 Bundesländer and the Federal Republic of Germany.

Main fields of work

The IfBt consists of five sections, each headed by a Director:

- Section P: administration, legal matters
- Section I: Civil engineering I: prestressed concrete, prestressing methods, ground anchors, nuclear engineering, reinforced concrete, metal construction, plastic for construction elements
- Section II: Civil Engineering II: concrete technology, building material chemistry, building physics, construction, scaffolds, reinforcing steel, prestressing steel, metallurgy, fire protection
- Section III: Services engineering: drainage, fire equipment
- Section IV: Basic problems: building research, international co-operation, quality control, documentation, public work, standardization

The IfBt is concerned with the homogeneous execution of transferred works in the field of construction supervision, i.e.:

- General approvals
- Approval marks (co-operation with UEAtc)
- Standardization
- Quality control
- Building research
- International co-operation
- Public work.

Total staff: 112

Director or Chief Executive

President: G. Breitschaft

Principal sources of finance

The Bundesländer and the Federal Republic, and fees for issuing general approvals and approval marks.

Publications

- *Periodical:* Mitteilungen des IfBt, published every two months by the Verlag Wilhelm Ernst & Sohn, Hohenzollerndamm 170, 1000 Berlin 31
- Lists of Approval Marks, published by the Verlag Erich Schmidt, Genthiner Str 30G, 1000 Berlin 30.

DEU.241

Institut für Betonstahl und Stahlbetonbau eV (IBS)

Institute for Reinforced Concrete

Landsbergerstrasse 414, 8000 München 60

Telephone: (089) 88 51 75

Date of establishment

10 April 1973

A registered association with its own laboratory and testing facilities, undertaking research work for IBS members and contract research for industry and private individuals.

Main fields of work

IBS research ranges over all aspects of reinforced concrete structures. Main aims of research: methods of analysis; development of reinforcing steels and elements; detailing of reinforcement; industrialization of reinforcement; bond, cracking, deformations; durability of reinforced concrete structures; contribution to establishing national and international principles and codes.

Total staff: 9

Professionally qualified staff: 3

Director or Chief Executive

Director Dr H. Martin

Heads of department

Materials: Dr Martin

Techniques of Reinforcement: Dr Schiessl

Principal sources of finance

Annual subscription of the members and contract research.

Principal related bodies

General Assembly of full members.

Publications

A list of research reports and publications in scientific journals is available upon request.

DEU.242

Institut für Fenstertechnik eV

Institute for Window Technique

Arnulfstrasse 13, 8200 Rosenheim-Aisingerwies

Telephone: (0 80 31) 65 245 or 65 296

Telex: 525746

Date of establishment

1966

A registered association.

Main fields of work

- Research division
- Quality control of manufactured windows at the factories
- Expertise and consultations
- Tests of windows and doors (mechanical tests on air and water-tightness)
- Tests of sound insulation of windows and parts
- Test of thermal insulation of windows and parts
- Literature
- Administration
- Workshop.

Director or Chief Executive

Director: Josef Schmid

Principal sources of finance

Financed by membership fees, income from research work, expertise and tests, and literature.

Principal related bodies

The Institue is a testing station for the Bundesanstalt für Materialprüfung Berlin (BAM). It co-operates with CEN (European Committee for Standardization) for standards and recommendations.

Publications

Publications concerning:

- Architects' problems with windows
- Problems of advertising wood, plastic and aluminium windows
- Moments of inertia for profiles
- Coating problems
- Glazing problems
- Tightness problems
- Modernization of old windows
- Classification of acoustic performance of windows
- Low energy window constructions.

DEU.243 **AM**

Institut für Stahlbau

Institute for Steel Constructions

Beethovenstrasse 51, 3300 Braunschweig

Telephone: 0531 391 3373

Date of establishment

1967

The Institute is a research institute of the Technical University of Braunschweig.

Main fields of work

Research and development in the following fields:

- Stability of steel structures, buckling, plate buckling
- Ductility of steel structures
- Connection techniques
- Cable structures, optimum design
- Bridges
- Dynamic response of structures due to earthquake and wind.

Total staff: approximately 40 of whom about 25 are support staff

Director or Chief Executive

Director: Professor J. Scheer

Heads of department

Department of Steel Constructions: Professor J. Scheer

Department of Mechanics Structures: Professor K. Hering

Principal sources of finance

Part of the technical university financed by the Ministry of Science and Arts of the State of Niedersachsen; research subventions; fees for services.

Publications

Research reports: Reports of the Institute.

DEU.244

Institut Wohnen und Umwelt GmbH (IWU)

Institute for Housing and Environment

Annastrasse 15, D 6100 Darmstadt

Telephone: 06151 26911

Date of establishment

July 1971

IWU is a private non-profit corporation owned by the State of Hessen and the City of Darmstadt. To some extent, IWU carries out contract research for public administrations.

Main fields of work

IWU develops applicable models and concepts for improving the housing of underprivileged population groups. Research results and proposals for amendments already available are investigated for their effects and applicability and further examined in specific projects. Where necessary, gaps in knowledge are filled by analytical and synthetic research efforts.

The Institute addresses itself to political decision makers, to the scientific community and to population groups affected by government policies.

Total staff: about 30 permanent

Professionally qualified staff: about 15

Director or Chief Executive

Dr Uwe Wullkopf

Principal sources of finance

Mainly grants given by the State of Hessen.

Publications

A comprehensive list of research reports, bibliographies and catalogues is produced.

DEU.245 **AM**

Normenausschuss Bauwesen im DIN Deutsches Institut für Normung eV (NABau im DIN)

Building Division of the German Standards Institution

Burggrafenstrasse 4–10, Postfach 1107, D 1000 Berlin 30

Telephone: (030) 2601 501

Telex: 184 273 din d

Cables: deutschnormen berlin

Date of establishment

1947 (but DIN has worked in the field of building standardization since 1917)

Registered association

Main fields of work

Through about 300 working committees NABau establishes DIN standards and regulations in the field of building, with the exception of parts of the fields 'Water supply and water drainage' and 'Engineering'.

The following fields of work exist:

Sub-Division 1: Basic and planning standards
Sub-Division 2: Unified technical building regulations
Sub-Division 3: Measuring standards
Sub-Division 4: Town and country planning
Sub-Division 5: Foundations
Sub-Division 6: Building components and building elements

Sub-Division 7: Concrete and reinforced concrete
Sub-Division 8: Steel construction
Sub-Division 9: Finishing
Sub-Division 10: Roads
Sub-Division 11: Highway bridges and foot bridges
Sub-Division 12: Industrial buildings
Sub-Division 13: Agricultural building
(sub-divisions as contained in the organization scheme of the NABau)

From the following interested associations and ministries, 4000 persons participate in the standardization work in an honorary capacity: Bund Deutscher Architekten BDA, Bonn; Zentralverband des Deutschen Baugewerbes, Bonn; Hauptverband der Deutschen Bauindustrie eV, Wiesbaden; Bund Deutscher Baumeister, Architekten und Ingenieure eV, Bonn; Deutscher Beton-Verein eV, Wiesbaden; Deutsche Bundesbahn, Zentralamt, Minden; Verband Kunststofferzeugende Industrie eV, Frankfurt/Main; Bundesvereinigung der Prüfingenieure für Baustatik, Stuttgart; Deutscher Stahlbauverband, Köln; Gesamtverband gemeinnütziger Wohnungsunternehmen eV, Köln; Verein Deutscher Zementwerke eV, Düsseldorf; Verband der Sachversicherer, Köln, as well as the Ministers of the Länder responsible for building control.

Total staff: 27
Professionally qualified staff: 12
Director or Chief Executive
President: Professor Dr H. Bub
Director: G. Lindemann
Principal sources of finance
NABau is promoted by the associations interested in standardization in the field of building, by the Ministers of the Länder responsible for building control and safety and by subsidies from DIN.
Principal related bodies
Advisory Council.
Publications
- News of standardization in the field of building, published at varying intervals
- *Technical standards:* DIN standards, Regulations, DIN pocket books (compilation of standards concerning particular fields of work)
- Catalogue of German standards and draft standards in the field of building, yearly.

DEU.246

Rechen- und Entwicklungsinstitut für EDV im Bauwesen eV (RIB)

Computing and Development Centre for Data Processing in Civil Engineering
Schulze-Delitzsch-Strasse 28, D 7000 Stuttgart 80
Telephone: 0711 7873-0
Telex: 07-255 580
Date of establishment
1960

RIB is an institution for research and software development in civil engineering. It has the status of an association. The members of the association represent industry, consultants, public authorities and science.

Main fields of work
RIB has been active in the development of portable standard software to support the design process in civil engineering. It is active in the following application fields:
- Computer aided design and automated draughting in building design
- Structural analysis, finite elements
- Highway design
- Quantity surveying, networking, cost evaluation

All programmes are designed in a modular way and can be integrated in program chains or with data base management systems.

Total staff: 34
Professionally qualified staff: 34
Director or Chief Executive
President of the Association: Volker Hahn
General Manager of the Institute: Wolfgang Haas
Heads of department
CAD, automated draughting: R. Gruschwitz
Structural analysis, finite elements: H. Hilber
Highway design, surveying: W. Völker
Quantity surveying, networking, cost evaluation: W. Leibbrand
Principal sources of finance
The Institution is financed by software development projects for private clients and public authorities and the marketing of standard software.
Principal related bodies
Institute for Software Development
Publications
- *Research reports:* For each research and development project a report is published. All reports deal with the EDP in civil engineering.
- Technical papers and user manuals of the standard software.

DEU.247

Rheinisch-Westfälische Technische Hochschule Aachen, Institut für Bauforschung (ibac)

Aachen Technical University, Building Research Institute
Schinkelstrasse 3, 5100 Aachen
Telephone: (0241) 80 51 00
Telex: 832704 thac d
Date of establishment
1948

Main fields of work
The Institute is connected with the chair of building materials science and the chair of metallic and organic building materials. The scientific staff is in charge of student training and research work.

Working groups: Admixtures and Masonry, Aggregates and Concrete; Steel and Plastics; Data Processing and Testing Techniques.

Research in progress: Deformation strength and fracture behaviour of concrete at short-time loading; deformation, strength and fracture behaviour of concrete at high sustained load; properties of fly ashes; deformation of masonry; rheological properties of fresh concrete; elastomers, polymers, measuring methods.

Total staff: 50 of whom 18 are scientific personnel
Director or Chief Executive
Director: Professor Dr K. Wesche
Deputy Director: Professor Dr H.R. Sasse
Principal sources of finance
The University, the Government and the building industry for the testing of materials.
Publications
ibac Mitteilungen, published yearly, at the beginning of the year, contains all publications of the current year.

Results of research work are published through lectures at meetings, in the technical press and by research reports.

DEU.248

Säurefliesner-Vereinigung eV

Institution for Research and Consultation on Tiles and Tilework

Im langen Feld 4, Postbox 12 54, D 3006 Burgwedel 1

Telephone: 051 39/30 61 + 30 62

Date of establishment

1942

An association of tile fixers for resistant floor and wall tiling (swimming pools, basins, etc.) with members in Scandinavia, Belgium, Netherlands, Luxembourg, Switzerland, Austria and the Federal Republic of Germany.

Main fields of work

- Research and examination of requirements of floor and wall tiles and tile work in usual and special conditions (ceramic and other materials with high physical and chemical resistance in different applications)
- Ceramic tiles (wall and floor)
- Split tiles
- Quarries
- Grout for tile work and joints
- Natural tiles
- Asphalt tiles
- Steel tiles
- Measurements of: water absorption; modulus of rupture; scratch hardness of surface; resistance to deep abrasion of unglazed tiles; resistance to surface abrasion of glazed tiles; linear thermal expansion; resistance to thermal shock; crazing resistance of glazed tiles; frost resistance; moisture expansion of unglazed tiles with a water absorption $E > 16\%$; resistance to staining of glazed tiles; resistance to household chemicals and swimming pool water cleansers; resistance to acids; resistance to alkalis
- Examination of slipperiness.

Total staff: 10

Professionally qualified staff: 6

Director or Chief Executive

Director of the Institution: Erhard Hopp

Principal sources of finance

Membership fees; examination charges and adrices.

Principal related bodies

Tile Council of America; British Ceramic Tile Council; Member of several working groups of DIN.

Publications

Research reports: A list of research reports and publications is available upon request.

DEU.249

Technische Hochschule Darmstadt, Fachgebiet Informationsverarbeitung im Bauwesen (THD, IVB)

Darmstadt Technical University Information Processing in Building and Civil Engineering Organizations

Petersenstrasse 13, 61000 Darmstadt

Telephone: (06151) 16 34 44

Telex: 419579

Date of establishment

1970

Main fields of work

Organization analysis, work studies, descriptions of the data structure in building projects, building contracting companies and building authorities; building planning and design research; development of procedures and methods for the storing and retrieval of documents and data in building project organization; planning methods of building processes; comparison and evaluation of software products, evaluation tests in the building industry; development of special calculation programmes.

Total staff: 3

Director or Chief Executive

Professor Dr Ing Heinz Schwarz

Principal sources of finance

Basic financing from the Federal State of Hessen within the framework of the budget of the University; research projects financed from German Research Community, Federal Ministries, etc.

Publications

A list of publications is available on request.

DEU.250 — AM

Technische Hochschule Darmstadt Institut für Massivbau (THD, IfM)

Darmstadt Technical University, Institute for Concrete Structures

Alexanderstrasse 5, 6100 Darmstadt

Telephone: (06151) 162144

Telex: 419579

Date of establishment

1939

The Institute is part of the Structural Engineering Department of the Technical University Darmstadt.

Main fields of work

Means of attachment for concrete elements; fracture mechanics models for plain and reinforced concrete; fatigue of reinforced concrete structures; probabilistic definition of earthquake loads; probabilistical investigations of prestressed constructions; crack width limitation; thermal stresses in reinforced concrete with particular reference to the design of LNG tanks; rail–gravel bank–superstructure interaction of railway bridges with high piers; non-linear behaviour of reinforced concrete structures under seismic loads; stability of reinforced concrete walls; three dimensional non-linear material model for reinforced concrete, analysis of reinforced concrete beams under torsion; economical optimization of high-rise buildings stabilized by cores; fatigue of reinforced concrete in uniaxial alternating tension–compression; non-destructive testing of concrete by acoustic emission, fracture toughness of plain concrete; behaviour of concrete under combined loadings such as moisture, temperature and tensile splitting stress; FE formulations for modelling non-linear material and bond properties, 2-D and shell elements; buckling of reinforced concrete beams under combined transverse and longitudinal loading.

Stress redistribution due to creep of concrete in prestressed folded plate structures.

Total staff: approximately 50

Professionally qualified staff: approximately 30

Director or Chief Executive

Director: Professor Dr Helmut Weigler

Professor Dr Gert König

Professor Dr Gerhard Mehlhorn

Professor Dr Horst Georg Schäfer

Heads of department

Director Laboratory for Plain and Reinforced Concrete Research: Professor Dr -Ing H. Weigler

Teaching and Research Divisions: Professor Dr -Ing G. König, Professor Dr -Ing G. Mehlhorn, Professor Dr -Ing H.G. Schäfer

Principal sources of finance

50% annual public funds; 50% research funds.

Publications

Research reports: Mitteilungen aus dem Institut für Massivbau; Forschungsberichte aus dem Institut für Massivbau

Numerous papers, published in professional and scientific journals as for example: Beton- und Stahlbetonbau, Bauingenieur, Betonwerk und Fertigteiltechnik.

DEU.251

Technische Universität Berlin, Institut für Baukonstruktionen und Festigkeit (IBF)

Berlin Technical University, Institute for Structural Design and Material Strength

Strasse des 17 Juni 135, D 1000 Berlin 12

Telephone: 030/314 3316/17/18

Telex: 1 84 262 tubln -d-

Date of establishment

1956

IBF is one of the five research institutes in the Department of Civil Engineering, which is part of the Technical University of Berlin. The organization structure is according to university law. Day-to-day management is carried out by the Director or Deputy Directors.

Main fields of work

IBF research ranges over all aspects of building and bridges (exceptions are foundations, hydraulics and roads). Some of the most important fields are the following: prefabricated homes; standardization licence; gas-foamed concrete; scaffolding; timber structures; steelwork construction; loadbearing concrete components; loadbearing roof members; components of scaffolding; light-metal constructions; joints in two stage weatherproof systems; constructions of glassfibre-reinforced concrete; lightweight concrete aggregates; heat loss in thermal insulation; loading capacity of prefabricated constructions; methods of measuring stress, strain, temperature and humidity of building constructions; physical problems in building constructions as thermal conductors; humidity and shrinkage; three dimensional stress distribution in concrete structures under concentrated loads; behaviour of plates under normal climatic conditions; problems of stability in reinforced concrete columns and in the reinforced concrete framework; problems of stability of steel columns and steel beams; load carrying capacity problems of connections in metal structures; trapezoidal steel and aluminium sheets. Research themes are from lectures given by the professors, resource requirements of the industry and court of justice, and the scientific research for regulations, standards and recommendations. The research work may be done by scientific calculations or by tests or in most cases by a combination of both.

Total staff: approximately 60

Professionally qualified staff: approximately 30

Director or Chief Executive

Director: Professor Dr M. Specht

Deputy Directors: Professor Dr J. Lindner, Professor Dr E. Cziesielski

Heads of department

Building Materials: Professor Dr K. Altmann, Professor Dipl H. Dombke

Concrete Structures: Professor Dr M. Specht, Professor Dipl P. Lorenz

Steel and Metal Structures: Professor Dr J. Lindner, Professor Dr D. Bamm

Physics of Civil Engineering and Wooden Structures: Professor Dr E. Cziesielski, Professor C. Scheer

Principal sources of finance

Mainly the university, but some from research bodies such as Deutsche Forschungsgemeinschaft. Some research activities are directly financed by industry.

Publications

Numerous papers are published in professional or scientific journals or printed in professional books.

DEU.252

Technische Universität Berlin, Hermann-Rietschel-Institut für Heizungs – und Klimatechnik (HRI)

Berlin Technical University, Hermann Rietschel Institute for Heating and Air-Conditioning Engineering

Marchstrasse 4, 1000 Berlin 10

Telephone: (030) 314 41 70; 314 26 18

Telex: 184262 tubln d

Date of establishment

1887

Established as a testing department for heating and air-conditioning equipment in the Royal Technical College, which later became the Technical University of Berlin.

Main fields of work

HRI is one of the eight institutes in the Department of Environmental Technology. The main task of the Institute is teaching.

Among its research activities are fundamentals of heating and air-conditioning engineering; the results of this research work often give the theoretical basis to standards such as DIN 4701, DIN 1946, VDI 2078.

Other research activities are the calculation of thermal loads and the energy consumption of buildings such as heating and cooling loads, the natural air flow through buildings from wind and thermal draft forces, the simulation of heating and air-conditioning equipment and systems, the influence of heating elements on thermal comfort, the optimization of the operation of heat pumps and the air-conditioning of hospitals.

HRI is a recognized DIN testing department for power measurements of heating elements.

Total staff: approximately 25 of whom 10 are support staff.

Director or Chief Executive

Managing Director: Professor Dr Horst Esdorn

Deputy Director: Professor Dipl Hubertus Protz

Principal sources of finance

Basically financed by the University; some of the investigations carried out are done on commission and paid for by institutions outside the University; some investigations are financed by industry.

Publications

- Various articles in scientific journals.
- Text-book for heating and air-conditioning engineering, whose first appearance dates back to Hermann Rietschel and is now available in the 15th edition (Rietschel/Raiss: Heiz- und Klimatechnik, Springer Verlag, Berlin).

DEU.253 **AM**

Technische Universitat Braunschweig, Institut für Baustof, Massivbau und Brandschutz

Braunschweig Technical University, Institute of Building Materials, Reinforced Concrete and Fire Protection

Beethovenstrasse 52, 3300 Braunschweig

Telephone: (0531) 391-5411

Telex: 9 52 698 ibstb d

Date of establishment

Established 1963 by amalgamation of a building materials laboratory, founded 1922, and the Institute for Building Materials, Science, founded in the early thirties.

The Institute is assigned to the Department of Civil Engineering. It performs educational and research work in the civil engineering sphere. It is headed by two Directors, one taking the chair of reinforced concrete structure, the other that of building materials science.

Main fields of work

Departments: Testing of building materials and structural elements including those of large dimensions; Technical acoustics and dynamics, vibrations; Chemistry and physics of building materials; Protection against fire; Special research area – Behaviour of structural members in fire.

Organization of the Institute is aimed at combining the work of civil engineers, architects, physicists and chemists in the various research spheres of civil engineering: safety of concrete constructions; basic physical properties of the strength and deformation behaviour of steel and concrete; development of new design methods in reinforced concrete building; behaviour of structures in fire; chemical analysis of hydraulic cements and the provision of basic knowledge and experience on the structure of building materials; heating processes in beams and plates; strength and deformation behaviour of concrete in fire. The Fire Section possesses a testing plant which is considered to be the most modern and capable of its kind in Europe. Current experiences and standard specifications dealing with the behaviour of structural members in fire, DIN 4102, are due in considerable measure to the research work accomplished in the Institute.

Total staff: about 200

Professionally qualified staff: approximately 40

Director or Chief Executive

Directors: Professor Dr Karl Kordina and Professor Dr F.S. Rostásy

Principal sources of finance

One third of the staff are sponsored by the regional government (Lower Saxony); all other costs, staff and expenses, must be paid for by research work (mostly contracts with the Federal Government), by scientific expertise and by commercial tests.

Publications

Numerous papers are currently published in professional or scientific journals as well as in a series edited by the Institute. Most of these papers are circulated free of charge to scientists and universities.

DEU.254 UM

Universität Essen, Bauphysik

Essen University, Institute of Building Physics

POB 68 43 D 4300 Essen 1

Telephone: 0201 183 2689 or 2695

Telex: 0857 0901 unie d

Teaching, science, practical research and developments in the field of building physics.

Main fields of work

- Energy conservation and thermal insulation of buildings
- Passive and hybrid solar technology
- Noise propagation in urban areas
- Moisture and humidity problems in structures and rooms
- Thermal and hygroscopic stresses of structures
- Environmental and microclimate research in urban areas.

Total staff: about 20

Director or Chief Executive

Professor Dr -Ing Karl Gertis

Publications

Publications in special periodicals. Institute publications number about 200.

DEU.255

Universität Stuttgart, Institut für Baubetriebslehre

Stuttgart University, Institute of Construction Management

Pfaffenwaldring 7, Postfach 80 1140, 7000 Stuttgart 80

Telephone: 0711/685 – 6144/45

Telex: 07 255445 univ

Date of establishment

1964 by order of the Ministry of Education

The Institute forms part of the University of Stuttgart.

Main fields of work

- Organization of building contracting firms
- Planning of building sites
- Management of prefabrication works
- Organization of stocks depots and stationery equipment in construction
- Tenders (calculation of construction costs)
- Management of large-scale projects
- Operational research application in construction
- Market research for contractors
- Optimal utilization of equipment
- Work study on building sites
- Energy studies for office buildings
- Network planning
- Cost control in construction and building
- Project management
- Value analysis in building

Total staff: 18

Professionally qualified staff: 7

Director or Chief Executive

Professor Dr G. Drees

Vice-director: Dr A. Bahner, responsible for Administration and Education

Principal sources of finance

The Ministry of Education, Federal State of Baden-Württemberg.

Publications

- Series of bulletins of the Institute: 23 volumes
- Numerous treatises on important problems, partly in co-operation with the building industry
- Books on problems of the construction industry: 10.

DEU.256

Universität Stuttgart, Institut für Baustatik

Stuttgart University, Institute of Structural Analysis

Pfaffenwaldring 7, D 7000 Stuttgart 80

Telephone: 0711 685 6123

Telex: 072 55 445

An institution within the University devoted to teaching and research in civil engineering.

Main fields of work

- Linear and non-linear analysis of structures, in particular geometrically and materially non-linear finite element analyses of thin structures such as three dimensional beams, slabs, shells
- Development of non-linear finite element solution techniques and computer programme NISA 80
- Stability and collapse analyses of thin-walled structures, especially shell structures with large displacements (local and global buckling, etc.) including the influence of geometrical imperfections, residual stresses, non-linear material behaviour
- Limit analysis of reactor containments (nozzle penetrations, etc.)
- Dynamic stability of shells

Total staff: 10

Professionally qualified staff: 10

Director or Chief Executive

Professor Dr -Ing E. Ramm

Principal sources of finance

Governmental.

Principal related bodies

The University and the Minister for Science and Art.

Publications

- Publications in scientific and technical journals
- Research reports of the Institute
- *Book:* Buckling of Shells (ed. E. Ramm), Springer-Verlag, 1982.

DEU.257 **UM**

Universität (TH) Karlsruhe, Lehrstuhl für Stahl- und Leichtmetallbau, Versuchsanstalt für Stahl, Holz und Steine

Karlsruhe University, Department of Steel and Aluminium Construction, Building Materials Testing Institute

Kaiserstrasse 12, D 7500 Karlsruhe 1

Telephone: 49-721 608 2205

Telex: 7-826521

Cables: UNI KARLSRUHE

Date of establishment

Building Materials Testing Institute 1921; Department of Steel and Aluminium Construction 1885

The Department of Steel and Aluminium Construction is part of the Building Materials Testing Institute within the faculty of Structural Engineering at the University of Karlsruhe.

The Building Materials Testing Institute is a public institution at the disposal of all state, provincial and municipal services, civil services, industry and private individuals.

Main fields of work

- Testing of materials, structural elements and complete structures in steel, aluminium, plastics, timber, boards and bricks (testing machines up to 50 000 kN available)
- Supervision and quality control, e.g. with respect to welders' approvals, suitability of welding procedures, approval requirements for all types of product
- Research projects comprising theoretical and experimental investigations within the above fields of activity. (Specialties: Thin-walled cold-formed structural elements, tubular structures, bolted and welded connections, shells, sandwich elements, glued laminated timber structures and connections)
- Product development and consulting work; preparation of documents for the approval of products.

Total staff: 65

Professionally qualified staff: 33

Director or Chief Executive

Directors of the Institute: Professor tekn dr Rolf Baehre (also Head of the Department); Professor Dr Jürgen Ehlbeck

Principal sources of finance

Fees from industrial contracts; research funds.

Publications

Berichte der Versuchsanstalt für Stahl, Holz und Steine der Universität Fridericiana in Karlsruhe (Contributions regarding development and research), at irregular intervals.

GHA.258 **FM**

Building and Road Research Institute (BRRI)
University PO Box 40, Kumasi
Telephone: (051) 4221/2
Cables: BRIGAS, KUMASITECH
Date of establishment
1952 as West African Building Research Institute; 1964 as Building and Road Research Institute

The BRRI is a wholly government financed research institution under the Council for Scientific and Industrial Research (CSIR). It is one of the ten research establishments under the Council. The primary objective of the BRRI is to help the building and road design and construction industries to be more efficient, safe and economical in providing the buildings and roads that the country needs.

Main fields of work
- Building design and planning: design, environmental physics, physical and regional planning, low cost and rural housing, building economics
- Materials: feasibility studies for the establishment of medium and small scale materials industries, e.g. brick and lime, kilns for burning bricks and lime. Studies into the use of local building materials from agricultural and industrial wastes, termite research
- Soil mechanics and foundation engineering: geotechnical properties and engineering behaviour of soils and bedrock materials, engineering geology and rock mechanics, highway and airfield pavements, foundations and earth structures
- Structures: structural system evaluation, industrialized systems analysis, design of structures utilizing structural steel, aluminium, reinforced concrete, reinforced masonry, plain masonry, heavy timber and laminated wood
- Traffic and transportation: highway feasibility studies, design of bridges and culverts, geometric design of roads and intersections, rural road planning and maintenance studies, traffic accident studies, land use, transportation studies, etc.

The Institute also undertakes the following activities:

- Consultancy and development: building design, physical planning, quantity surveying, land surveying, subsoil investigations, foundation design feasibility studies, etc.
- Construction: building and road construction, construction of culverts and bridges, establishment of medium and small-scale materials industries, brick and lime kilns, etc.
- Information dissemination: publications, editorial services, library, printing and photocopying, exhibitions, conferences, courses, general liaison and training.

Total staff: 496
Professionally qualified staff: 89
Director or Chief Executive
Dr M.D. Gidigasu
Heads of department
Administration: P.H. Manuh
Construction: K. Amoah-Mensah
Consultancy: I.A. Allotey
Geotechnical Engineering: J.K. Ayetey
Information: M.D. Mengu
Materials: A.A. Hammond
Structures, Building Design and Planning: Dr K. Amonoo-Neizer
Traffic and Transportation: J.D.N. Riverson
Principal sources of finance
Direct Government 85%; consultancy and others 15%.
Publications
- BRRI Newsletter, quarterly
- *Research reports:* 5 series: Information sheets; Current Notes; Current papers; Current reports; Special reports
- Miscellaneous Brochures
- Annual Reports
- Library Accessions List and Selected Bibliographies.

GRC.259

Agricultural College of Athens, Laboratory of Agricultural Structures

ΑΝΩΤΑΤΗ ΓΕΩΠΟΝΙΚΗ ΣΧΟΛΗ ΑΘΗΝΩΝ
ΕΡΓΑΣΤΗΡΙΟ ΓΕΩΡΓΙΚΩΝ ΚΑΤΑΣΚΕΥΩΝ

IERA ODOS 75, Botanikos, Athens (301)

Telephone: (01) 3460.025 3473.466

Date of establishment

Established in 1964 as a laboratory at university level dependent on the Ministry of Education.

The Laboratory is part of the Faculty of the Agricultural College of Athens and is mainly at the disposal of the University.

The role of the Laboratory is: education, research and development.

Main fields of work

- Courses at undergraduate level
- Courses and seminars for agricultural engineers employed by the Ministry of Agriculture, and for farmers
- Provincial and municipal services
- Research projects in the fields of: farm building and rural construction; farm management in livestock production; environmental control (research and applications) in farm and greenhouse buildings; energy saving and recovery systems; wastes management and biogas facilities; soil stabilization.

The Laboratory is supervised by the Faculty of the University.

Total staff: 7

Professionally qualified staff: 6

Director or Chief Executive

Director of Laboratory: Professor Sp. Kyritsis

Principal sources of finance

State grants mainly through the Ministry of Education; secondly, State research subventions.

Publications

- Piggeries
- Poultry houses
- Cow husbandry
- Silos for grain
- Sheep management
- Waste handbook management.

GRC.260

Aristotle Thessaloniki University, School of Engineering, Department of Civil Engineering, Division of Structural Engineering

Thessaloniki

Telephone: 99 2629/2604/ 2643/2642/2616

Telex: THESSALONIKI - GREECE

Date of establishment

1955

The Division of Structural Engineering is part of the Department of Civil Engineering, which is part of the University of Thessaloniki.

Main fields of work

The Division is active in teaching and research in the following areas:

- Static and dynamic analysis of structures
- Steel structures
- Reinforced concrete structures
- Building materials and techniques
- Experimental methods and testing of materials
- Earthquake engineering.

Total staff: 100

Professionally qualified staff: 40

Director or Chief Executive

Director of the Division: Professor G. Penelis

Principal sources of finance

The University.

Principal related bodies

The Assembly of the Department of Civil Engineering.

Publications

Numerous doctoral theses, papers and articles in conference proceedings and journals in the areas of interest mentioned above.

Research reports for Greek and international institutions.

GRC.261

National Technical University of Athens, Department of Theoretical and Applied Mechanics

5 Heroes of Polytechnion Avenue, Athens (624)

Telephone: (01) 7795575-7777056

Telex: NTUA 221682

Date of establishment

1920

The Laboratory is the main Construction Department of the National Technical University of Athens and performs testing upon request.

Main fields of work

In the Mechanics Section and in the Laboratory for Testing Materials, basic and advanced knowledge is imparted to the students, advanced research in theoretical and applied mechanics is carried out and certificates of quality control of materials are issued.

The Materials and Construction Section is concerned with the performance of building materials in use and with the development of new materials.

The main research in the Department is performed in the area of fracture mechanics using various analytical and experimental methods.

Total staff: 50

Professionally qualified staff: 20

Director or Chief Executive

Professor P.S. Theocaris

Principal sources of finance

The University.

Publications

Numerous papers and research reports (about 600) have been published in various international scientific journals as well as a number of books in the national and international press.

GRC.262 **AM**

Technical Chamber of Greece

Technikon Epimelitirion Ellados (TEE)

4 Karageorgi Servias Str., PO Box 673, Athens 125

Telephone: 32.54.591-9

Telex: 218374

Date of establishment

1928

TEE is a state-sponsored institution that registers all Greek graduate engineers (civil, mechanical, electrical, chemical, mining engineers, surveyors, architects and naval engineers).

Total membership is 35 000.

Main fields of work

- Scientific activity covering the fields of civil, mechanical, electrical, chemical engineering, etc.
- Research covering the environment, socio-economic development, regional planning, educational matters, etc.

The principal divisions are the Scientific Sections of: Civil Engineers; Architects; Electrical Engineers; Mechanical Engineers; and Chemical Engineers.

Total staff: 120

Professionally qualified staff: 50

Director or Chief Executive

President of Executive Committee: Nicolaos Dessylas

Secretary-General of Executive Committee: Stefanos Ioakimidis

Principal sources of finance
Members' contributions; sale of publications.
Principal related bodies
Executive Committee.
Publications
- *Journal:* Information Bulletin, weekly
- *Technical standards:* Fire protection; Ready-mix concrete; ETC
- *Research reports:* scientific publication, four yearly; individual publications covering many technical subjects
- Greek and international technical bibliography
- Publications covering the minutes of technical conferences, symposia etc.

GTM.263 **FM**

Universidad de San Carlos, Centro de Investigaciones de Ingeniería (CII)

San Carlos University, Engineering Research Centre

Ciudad Universitaria, Zona 12, Guatemala

Telephone: 760790–4, ext. 353,373

Cables: CII-GUA

Date of establishment

Established by Resolution 9-842 of 27 July 1963 of the University of San Carlos, Supreme Council.

A university and public institution attached to the School of Engineering of the University of San Carlos, which serves the needs of participating institutions and of government, municipal, and other public institutions, industry, private enterprise and all concerned. It was made possible by the amalgamation of the Materials Laboratory of the Bureau of Public Works (founded in 1952) and the Materials and Hydraulics Laboratories of the School of Engineering, University of San Carlos (founded in 1935). The water analysis laboratories of the Bureau of Public Works and the Municipality of Guatemala were incorporated in 1965; soil mechanics and structural engineering facilities were established in 1964; chemical, mechanical, and electrical engineering laboratories were established in 1967. The construction information centre (Centro de Información a la Construcción) started operations in 1979. It has been expanded to related information in the field of human settlements. In 1980, the Housing Research unit of the School of Architecture and the Building Technology unit of the CII, were organized as the Technology for Human Settlements, program co-ordinated by CII.

Main fields of work

- Educational laboratory services in several engineering branches at university and technical level to 5 universities and 3 Government technical vocational schools.
- Technical assistance and testing and analysis in the fields of materials of construction; structures; soil engineering; building technology in general; air and water quality evaluation.

Research activities in the following fields:

- *Materials:* properties and performance of building materials; development of materials from local resources. Current studies: Evaluation of pozzolanic materials, fibre reinforced normal and lightweight concretes; ferrocement applications in low-cost housing; technical properties and durability of native bamboo species. Evaluation of several forms of soil stabilization for building purposes.
- *Structures:* structural performance and testing of elements for walls and roofing. Simple dynamic testing of full-scale models of houses. Current studies: Evaluation of reinforced masonry walls and reinforced earth walls for low-cost housing.
- *Building methods, current studies:* Development of appropriate building techniques for earth construction in four selected regions of the country, using locally available resources.
- *Housing typology, current studies:* Analysis and evaluation of housing types in the country; climatic design recommendations for selected regions.
- *Non-conventional energy, current studies:* Development of mass production methods for making several models of wood-burning stoves, and organization of small industrial plants to produce them. Evaluation of two solar refrigeration models for domestic use. Development of small plants for fuel alcohol production in rural areas.
- *Sanitary engineering, current studies:* Contamination and eutrophication of selected lakes and rivers. Air and noise pollution in Guatemala City. Development and evaluation of solid waste and domestic waste-water disposal systems.
- *Standarization:* Adaptation and development of standards for the construction sector.

Principal research projects (staffed by research groups):

- Technology for human settlements: Housing Typology; Building Technology; Non Conventional Energy; Rationalization of the building process
- Sanitary engineering: The CII collaborates in the following research projects of the Central American Regional School of Sanitary Engineering: Rivers and lakes eutrophication and contamination; Environmental pollution in Guatemala City; Solid waste and domestic waste disposal for small settlements.

Note: Several other studies are made in the Laboratories, although not being part of a systematic programme, as those mentioned.

Total staff: 95

Professionally qualified staff: 17

Director or Chief Executive

Director: Emilio Beltranena Matheu

Heads of department

Administrative Department: Francisco Polanco
Concrete Laboratory: Aníbal Rodas
Materials Laboratory: Domingo Samayoa
Structural Engineering Laboratory: Juan Miguel Rubio
Building Methods Unit: Javier Quiñonez
Sanitary Chemical and Microbiological Laboratory: Alba T. de Abreu
Industrial Chemistry Laboratory: José A. López
Hydraulic Model Testing Laboratory: Alberto Solórzano
Construction Information Centre (CICON): Herbert Miranda

Principal sources of finance

Government (Bureau of Public Works) 17%; Municipality of Guatemala 9%; University of San Carlos 34%; Contracts, work paid for mostly by the construction sector 36%; international organizations' assistance 4%. Total budget for 1983: US$380 000.

Principal related bodies

University of San Carlos Research Co-ordination Board: the CII has a representative on the board, which defines the research policies, and co-ordinates research activities and development in the University; CII Advisory Committee of representatives of the participating institutions, which gives general guidance in relation to policies and development of the Centre.

Publications

- *Journal: Revista CII (CII Journal), bimonthly*
- *Technical standards:* Proposals for standards to the Guatemalan Standarization Commission, in the fields of Construction and Sanitary Engineering (at present, work is being carried out on 46 standards)
- *Research reports:* Technical monographs (in Spanish); Technical articles (in Spanish, some in English); Technical notes (in Spanish); Papers for short courses, seminars, symposia on building and housing (mostly in Spanish)
- Quarterly reference bulletin: Servicio documental de Información sobre la Construcción (Document information service for Construction)
- Monthly Bulletin of Information on journals and other periodicals received by CICON: Páginas de Contenido (contents pages of periodicals)
- Bibliographies on selected topics
- CII publications lists.

HUN.264

Central Research and Design Institute for the Silicate Industry
Szilikátipari Központi Kutató és Tervezö Intézet
Bécsi út 126/128, Budapest, III
PO Box 112, H-1300 Budapest.
Telephone: 804-311
Telex: 225269, 224367
Date of establishment
1953

Under the supervision of the Ministry of Building and Urban Development, it represents the R and D technical design and engineering organization of the silicate sector of the Hungarian building materials industry.

Main fields of work

- Research and development of the building materials industry: Research and qualification of raw materials, preparation of forecasts, utilization of wastes; Research activity directed to the production of new building materials, special cements and special ceramics, introduction of production on a laboratory level and in pilot plants; Performance of research and measurements related to heat power engineering, energetics, air- and dust-technique, as well as isotopes; Performance of up-to-date tests relating to chemistry, microscopy and microstructure.
- Technical design
- Prime contracting, engineering
- Complementary activities.

Total staff: 730
Professionally qualified staff: 290
Director or Chief Executive
Director: Jenö Simon
Research Deputy Director: Dr József Mátrai
Design Deputy Director: Gábor Ferencz
Business Executive Director: DrIstván Balog
Heads of department
Department of Brick and Insulating Materials: Dr Kálmán Toth
Department of Ceramics: Dr Endre Wagner
Department of Cement Research: Dr László Szatura
Department of Concrete Research: Dr Tibor Kausay
Department of Silicate Chemistry: Dr Tamás Träger
Department of Glass Research: Dr Lajos Londvay
Department of Automation: Dr Antal Fehér
Department of Special Materials: Dr József Gyorgy
Department of Long-range Planning: Dr Árpád Kunvari
Department of Energy Utilization: Dr Ferenc Szijj
Group for Geology: János Lingauer
Group for Organic Building Materials: Dr Lajos Nagy
Principal sources of finance
Fees from contracts with clients.
Principal related bodies
Hungarian and foreign building and building materials firms: research and development of new types of modern building materials; development of new technology.
Publications

- Scientific publications (Szikkti), periodical
- Monographs of silicate chemistry, periodical
- Home and foreign periodical reviews, irregular.

HUN.265

FTV Consulting Engineering
Földmérö és Talajvizsgáló Vállalat (FTV)
Reviczky 4, 1088 Budapest
Telephone: 130–260
Telex: 22–5194
Date of establishment
1950

A governmental institution at the disposal of all state, provincial and municipal services, civil services and industry.

Main fields of work

- Surveying
- Soil mechanics and foundation engineering
- Engineering geology
- Corrosion protection
- Water supply
- Environmental protection.

Total staff: 1100
Professionally qualified staff: 400
Principal sources of finance
Financed on a contract basis by the Ministry of Building and Urban Development, the Hungarian Committee for Technical Development (OMFB) and other state enterprises.
Publications

- Papers published regularly in scientific journals
- Quinquennial year-books on activities
- Project text-books
- Collection of Foundation Experiences (21 volumes)
- Prospectuses
- Technical Information Proceedings
- Information Booklet on technical literature.

HUN.266 FM

Hungarian Institute for Building Science
Épitéstudományi Intézet (ÉTI)
Dávid Ferenc u 6, H 1113 Budapest
Postal address: POB 71 H 1502 Budapest
Telephone: (36-1-) 852-544; (36-1-)668-883
Telex: 224285 ETI-H
Cables: HUNGBUILDSCHI BUDAPEST
Other address: ÉTI Technological Plant, Dózsa György u 26, H 2000 Szentendre
Date of establishment
Law of 12 270/1948

The central research institute of the Hungarian building industry, a governmental institute operating on a self-supporting basis under the supervision of the Ministry of Building and Urban Development. Research work is carried out partly for the Government, partly for the building industry and other branches of the national economy.

Main fields of work

- The Institute performs ground research for the building industry in the fields of building physics, theoretical mechanics, statics, hydrodynamics, building informatics and prognostics
- In the framework of industrial assignments it carries out research and development work, establishing new building structure methods and technologies
- It develops up-to-date, energy saving heating–ventilating–air-conditioning systems and takes part in developing and qualifying energy saving building materials and structures
- It undertakes work in the mechanization of building (with nationalization and the development of small building machines) and in automation with the planning of automatic technological processes
- Takes part in elaboration of up-to-date building structures and diagnostic procedures and in developing up-to-date structures and equipment for the maintenance and renewal of buildings.

Total staff: 665
Professionally qualified staff: 355
Director or Chief Executive
Director: S. Vasas
Managing Director: L. Béres
Secretary of Scientific Affairs: Dr P. Scharle
Deputy Directors: Technical: Dr F. Burkus; Finance: J. Huszár
Heads of department
Progress Analysis and Information: I. Révész
Building Systems and Structures: Z. Visy
Structural and Civil Engineering: Dr P. Lenkei
Environmental Engineering and Energetics: Dr Cs. Részegh
Building Mechanization and Technology: Dr V. Zoltánka
Computer Techniques and Automation: Dr Gy. Páti
Testing of Building Machines and Mechanization of Finishing Works: G. Szentiványszky
Technological Design: Dr Gy. Kercsmár
Development Secretariat: Dr K. Hamvay
Principal sources of finance
Financed partly by the Government in the form of research contracts amounting to approximately 25% of its total income; 75% is derived from research contracts with the building industry and other branches of the national economy.
Principal related bodies
Central Scientific Council; 4 specialized branches: their responsibilities include the evaluation of the Institute's research activities, qualifications, etc; Divisional Directors' Advisory Board; Board of Directors, Commission for Complex Research Projects, Hungarian Academy of Sciences.
Publications
- Építés, Kutatás, Fejlesztés (Building, Research, Development), quarterly, in Hungarian, with index in English, French, German
- ÉTI Yearbook, ÉTI Summary, in English, German, Russian
- ÉTI Bulletin, contains papers by researchers of the Institute, reprinted
- Centre for Standardization of Load Bearing Structures: Elaboration of standards concerning loadbearing structures and revision (super-revision) of all standards and recommendations, respectively, before they are put into force.
- Tudományos Közlemények (Scientific Proceedings), in Hungarian with abstracts in English, French, German and Russian, issued occasionally
- Számítástechnikai Közlemények (Proceedings on Computer Techniques), in Hungarian with abstracts in English, French, German and Russian, 4–5 issues yearly.

HUN.267

Hungarian Institute for Town and Regional Planning
Városépítési Tudományos és Tervező Intézet (VÁTI)
Krisztina krt 99, Budapest I, PO Box 20, H-1253 Budapest
Telephone: 359-960
Telex: 22-5134
Cables: VÁROSTERV
Date of establishment
1949, with the introduction of state-organized planning

VÁTI is Hungary's latest settlement planning body, and is under the supervision of the Ministry of Building and Urban Development. It acts as one of the focal points of the Ministry. However, VÁTI, in the form of a company, is a self-supporting institution.
Main fields of work
VÁTI focuses its activity on research and all kinds of physical planning work concerning the field of urbanology.

Research is carried out in order to give theoretical and methodological support to the physical planning process. It includes: research in the field of town planning; research preparatory to territorial development; research associated with the protection and development of the settlement environment.

Regional and settlement planning includes: regional plans and studies; regional development concepts of the branches (housing, transport, etc.), landscaping and planning of recreational areas; studies on the utilization of thermal waters; master plans for towns, rural communities, settlement groups; detailed development plans for large neighbourhood units.

Reconstruction and rehabilitation plans, urban design of areas with historic buildings and of town centres. Plans for the care and maintenance of historic and folk monuments, buildings.

Complex planning of industrial areas; location studies based on feasibility investigations and space suitability.
Total staff: 450
Professionally qualified staff: 220
Director or Chief Executive
Director: Sándor Kálnoki Kis
Heads of department
Department of Research: Miss Jolán Bakai
Department of Regional Planning: Etele Baráth
Department of Settlement Planning: Attila Molnár
Department of Monument Protection: József Kerényi
Principal sources of finance
The Institute is a self-supporting company which derives its income from Commissions given by the Ministry of Building and Urban Development, and County and Local Councils; and from professional consulting and planning activities at home or abroad.
Publications
Technical standards recommendations: Telepuléstudományi Közlemények (Settlement Planning Review), annual; Településtervezés (Settlement Planning), quarterly; Referátumgyüjtemény (Selected Literature), quarterly; Szakirodalmi Tájékoztató (Abstracts), quarterly.

Research reports: Teruletrendezési szabályok (Legal Rules); Társadalmi részvétel az általános és az összevont rendezési tery készitésénél (Public participation in the planning process: master and comprehensive plans); Társadalmi részvétel a részletes rendezési terv készitésénél (Public participation in the planning process: detailed plans); Településrendezési normativák és mutatószámok (Norms and index numbers for urban physical plans); Research reports on 19 themes of the National Medium-Term Research; Development Programme (marked B/4) which runs under the title A települések fejlödésének irányitását megalapozó kutatások (Research forming the basis to the direction of settlement development).

HUN.268 AM

Information Centre of Building
Építésügyi Tájékoztatási Központ (ÉTK)
Hársfa u 21, 1074 Budapest, VII
Telephone: 117–317
Telephone: ÉTÉKÁ 226564
Date of establishment
1950; under its present name from 1965

Scientific and technical information base of the Hungarian building sector. Profit organization.
Main fields of work
The largest Hungarian collection of building documentation is operated by ÉTK which collects, registers and processes the primary documents (books, periodicals, research reports, etc.) about building materials, construction works, the built environment, urbanization, building machines and equipment, finishing works, building elements, sanitary engineering and installations.

In accordance with the scope of its activities, it assists the scientific and practical work of building authorities, researchers concerned with building science, designers, investors, contractors, manufacturers and builders with information and documentation services from its databank, and with reproductions, films and photo services, a translation service and permanent exhibitions.

It organizes study tours and exchanges of experience in Hungary and abroad. It produces and distributes special films about building, propaganda and educational films and sound film strips and it operates the Eastern European Vision Habitat Distribution Centre. By the application of different forms of building information (e.g. consulting service) it promotes the private housing efforts of the population. It participates in the work of international special library and information organizations and maintains continuously broadening relations with 200 institutions in about 70 countries.

The principal divisions are: Central building library; Product information service; Library and informatics research, development, system organization; Building photograph, microfilm, film strip, and film information services; Exhibitions, professional study tours in Hungary and abroad; Publication of building books and journals, printing and photocopying services.

Total staff: 350

Professionally qualified staff: 180

Director or Chief Executive

Director: Dr István Tiszai

Deputy Directors: Dipl Ing Péter Hamvay; Dr Gyula Sütö

Principal sources of finance

Sale of information publications and services on the building information market (the users buy the services); self-supporting business enterprise.

Principal related bodies

Ministry of Building and Urban Development (charged with supervision); Technical Information Council (advisory council inside the ÉTK).

Publications

Journals: KWIC index – Architectural Design, monthly, in Hungarian; Technical and Economic Information on Building, monthly, in Hungarian; Informatory Publications on Hungarian Building, monthly, in Hungarian; World News about Building and Urban Development, weekly, in Hungarian; Hungarian Building Bulletin, quarterly abstracting journal, information in English on the Hungarian building sector for foreign countries; Accession List of Foreign Products Review, quarterly, in Hungarian; Selective Design Data Collection, 8–10 copies per year, in Hungarian; Illustrated Building News, monthly, original photographs of buildings in the building sector

Catalogues: Summary catalogue of building research activities, yearly, in Hungarian; Film catalogue and list of film strips, yearly, annotated list in Hungarian; Technical literature research, annually.

Research reports: Studies in information science, annually 1–2 volumes in Hungarian with English abstract.

ÉTK has a special computerized data-base from building literature resources. It contains 25 000 items of information. From the data-base ÉTK gives selective dissemination of information and a retrospective information service to the user.

HUN.269

Institute for Design Development and Systems Design

Tervezésfejlesztési és Tipustervezö Intézet (TTI)

Asbóth u 9–11, Budapest VII

Telephone: 226–240

Telex: 22–5129 TTI H

Date of establishment

1 January 1961

State institute for design and systems development (under the guidance of the Ministry for Building and Urban Development).

Main fields of work

- Drafting and forecasting the functions of residential and public buildings
- Unification of building constructions (standardization)
- Development of building systems with the application of steel, reinforced concrete and wooden constructions
- Development of design methods in building construction (with special regard to CAD, including software development)
- Co-ordination and information in the field of design in Hungary and abroad
- Real planning operations in Hungary and abroad (mainly residential and public buildings – groups of buildings and public institutions).

Principal divisions are: Architects' Department for the Development of the Housing Environment; Architects' Department for the Development of Building Systems and Constructions; Department of Technical Direction.

Total staff: 382

Professionally qualified staff: 237

Director or Chief Executive

Director: Miklós Bartók, Arch

Principal sources of finance

Ministries, councils, enterprises, co-operatives.

Principal related bodies

Ministry for Building and Urban Development (Architects' Department for the Development of Settlements).

Publications

- Technical information, monthly
- Technical aids, monthly.

HUN.270

Institute for Quality Control of Building

Építésügyi Minöségellenörzö Intézet (ÉMI)

Diószegi ut 37, 1113 Budapest XI

Postal address: Pf 69, 1502 Budapest

Telephone: 851–511

Telex: 22–6507 1 eemi h

Cables: 1113 Bp XI Diószegi ut 37

Date of establishment

Established in 1963 by the Minister of Building, on the basis of a decision taken by the Economic Committee of the Government

The Institute is a state research body charged with governmental work. Official, expert work is performed nation wide by a central organization and several regional quality control stations. The activities of the Institute are supported by laboratories for building acoustics, thermal physics, fire technics and elevators.

Main fields of work

- Quality control work connected with building activity and building materials (including the technical design of buildings) as determined by the legal rules for building supervision: controlling the quality and quality certification of building works, building materials and structures; studying the effects of quality control and detailing suggestions for the improvement of quality
- Acting as the standardization centre of the building branch, tasks connected with standardization and other technical regulations in building
- Tests of new products and building methods for applicability in building, issue of certificates for technical suitability and/or initiation of granting permissions
- Qualification of building products and building materials according to quality, domiciliation of quality certificates of imported products
- Fire technical tests and qualification of building products, building materials and structures
- Safety tests of machines for building and the building material industry and equipment for sanitary installation
- Expertise connected with the permission of use for elevators, supervision of putting into service and use
- Research (basic and applied) connected with work in the above items

Additional activities:

- Professional supervision of the quality control work of the organizations involved in building and the building materials industry
- Radiological tests of building materials, structures and buildings
- Expertise in all fields within the scope of the work
- Training of quality control experts for the organizations engaged in building and the building materials industry as well as training of experts for quality control of elevators.

Total staff: 460
Professionally qualified staff: 180
Director or Chief Executive
Director: Imre Borbély
Heads of department
Scientific Deputy Director: Milhály Heincz
Technical Deputy Director: Gusztáv Grofcsik
Principal sources of finance
State funds, feeds paid by firms, fees for expertise and other services.
Publications

- Epités-Minöség (Building Quality), 6 issues annually, contains general information on the tests and results of research achieved by ÉMI and foreign institutes
- ÉMI Tudományos Tájékoztató (Scientific Bulletin of ÉMI) gives information on the principal resupts of research
- List of publications on technical regulations in building
- Technical documents regularly issued: standards, technical guidelines, etc. in the field of building as well as state standards and provisions for building.

HUN.271

Scientific Society for Building

Épitöipari Tudományos Egyesület (ÉTE)
Kossuth Lajos tér 6–8, 1055 Budapest V
Telephone: 125–247, 113–250
Telex: 225–792
Date of establishment
2 February 1949

A voluntary, non-profit-making scientific society. A member society of the Federation of Technical and Scientific Societies.
Main fields of work
Organization of conferences, congresses, demonstrations, exhibitions, club meetings, debates, technical film shows, postgraduate professional courses, etc.; promotion of knowledge and use of building industry experiences in other countries; contacts and co-operation with federations with similar scope; delegation of members to international conferences of the building industry; contests, awards for scientific or practical solution of technical and economic problems, arrangement of competitions- dissemination of information to members about latest scientific results in the building industry.

Subjects: building, civil engineering, town planning, structural engineering and technology, heating and ventilating engineering, prefabrication, economic conditions in building, price and organization analyses. International relations.

In addition to the nine principal divisions, there are 29 regional sections in the main towns of Hungary.
Total staff: approximately 10 000
Professionally qualified staff: 100
Director or Chief Executive
President: Dr Rezsö Trautmann
Secretary General: Dr Miklós Márkus
Head of Secretariat: Magdolna Makra, deputy secretary general
Heads of department
Department of Building Construction and Technology: Arpàd Korda
Department for Town Planning: Làszl òVarga
Department for Sanitary Engineering: Tamàs Làszló
Section for Prefabrication: Tamas Làszló
Section for Engineering, Establishment and Public Building: Ferenc Hajdu
Section of Technology and Economy: Dr Tamàs Novàk
Section for Loadbearing Structures: Dr Lajor Gàdoros
Section of Law: Dr Jenö Gilyèn
Section of Economics: Dr Tamàs Kozma
Principal sources of finance
Financed by subscriptions from the 10 000 voluntary members and 218 collective members.
Principal related bodies
The Chair of the Federation of Technical and Scientific Societies supervises the activities of the Society.
Publications
Journal: Magyar Épitöipar (Hungarian Building Industry) monthly; Épületgépészet (Heating and Ventilating Engineering), bi-monthly; Városépités (Town Planning), bimonthly.

HUN.272

Technical University of Budapest, Faculty of Civil Engineering, Faculty of Architecture

Budapesti Müszaki Egyetem, Építömérnöki Kar, Építészmérnöki Kar
Müegyetem rkp 3 I em 16, 1111 Budapest
Telephone: 452–305 632274 632427
Telex: 225931 muegy h
Date of establishment
Founded in 1782 as the School of Surveying and Hydraulic Engineering; in 1846 it was merged with the secondary Joseph Industrial School; after 1856 it was named Joseph Technical University; after 1871 it functioned as an independent Civil Engineering Section of the former, later as the Faculty of Civil Engineering and Architecture. Finally it was divided into the Faculties of Civil Engineering and of Architecture.

Two faculties of the Technical University of Budapest, under the authority of the Ministry of Education.
Main fields of work
The two faculties are the only ones in Hungary to graduate certificated civil engineers and architects respectively.

Among the departments, institutes and laboratories of the two faculties, the following undertake specific research in construction and architecture: Department of Steel Structures; Department of Mineralogy and Geology; Department of Building Materials; Department of Geotechnics; Department of Building Construction; Department of Mechanics; Department of Road Engineering; Department of Reinforced Concrete Structures; Department of Hydraulic Engineering; Department of Descriptive Geometry; Department of Building Operations; Institute of Building Construction and Equipment; Department of Building Structures; Department of Sanitary Engineering; Institute of History and Theory of Architecture; Department of Industrial and Agricultural Architecture; Department of Public Buildings; Department of Housing; Department of Drawing and Composition; Department of Strength of Materials and Structures; Department of Town Planning; Laboratory of Construction.
Total staff: Faculty of Civil Engineering: 470; Faculty of Architecture: 248.
Director or Chief Executive
Dean of the Faculty of Civil Engineering: Professor Dr Ottó Halász
Dean of the Faculty of Architecture: Professor Dr György Deák
Principal sources of finance
Financed by the State budget through the Ministry of Education; an amount equalling a given percentage of annual faculty budgets is directly available from fees for commissions by ministries or by the building industry.

Publications

- Periodica Polytechnica, review of the Technical University of Budapest

The staff may publish papers in Hungarian special journals, e.g. Magyar Epitöipar, Mélyépitésitudományi Szemle, Ëpitöanyag, Magyar Építészet, Hidrológiai Közlöny, Közlekedéstudományi Szemle, etc., and in foreign ones or in Acta Technica, the review of the Hungarian Academy of Sciences, publishing in any of the universal languages

- Department Reports: Reports of the Department of Building Materials, in Hungarian, with English, German and Russian summaries and English captions; Reports of the Transportation Working Group of the Hungarian Academy of Sciences active at the Department of Railway Engineering.

ISL.273 **AM**

Building Research Institute

Rannsóknastofnun byggingarionaoarins (Rb)

Keldnaholt, 110 Reykjavík

Telephone: 91-83200

Telex: 2307

Date of establishment

1965

An independent Government agency for testing and research in the building and construction industry of Iceland, under the Department of Industry. Financed by Government funds and income for research projects and services.

Main fields of work

The Institute is the only one of its kind in the country. Its functions are divided in the following divisions:

- Concrete Technology and Structural
- Road Construction and Soil Mechanics
- Building Technology
- Cost Analyses
- Geology
- Information and Publications
- General Administration

In most of these fields materials testing and contract research are performed. The main fields of work are the first three listed.

Total staff: 42

Professionally qualified staff: 19

Director or Chief Executive

Director: Haraldur Ásgeirsson

Heads of department

Concrete Technology: Hakon Ólafsson

Road Construction and Soil Mechanics: Thorir Ingason

Building Technology: Jón Sigurjónsson

Cost Analyses: Pálmi Kristinsson

Geology: Thorgeir Helgason

Information: Gunnlaugur Pálsson

Principal sources of finance

Approximately 40% of the total budget is the Institute's own income and 60% from Government funds.

Publications

- Rb – technical recommendations
- 4–8 technical publications written in Icelandic with English summary, annually.

IND.274

Cement Research Institute of India (CRI)
M-10 South Extension II, New Delhi 110 049
Telephone: 660233; 660590; 666010
Cables: ANVESHAN
Other addresses:
NH 2, Delhi Mathura Road, Ballabgarh 121004, Haryana
Telephone: 88-260; 88-261; 88-305; 88-316; 88-318
Telex: 0343-234
Trichy-Ariyalur Road, Muduvathur, Kallakudi 621652, Tamil Nadu
Telephone: 69
3-5-874/1 Hyderguda, Hyderabad 500 001
Telephone: 221960
2/17 State Bank Officers Colony, Khajpura, Bailey Road, Patna 800 014, Bihar
Telephone: 24611
Date of establishment
August 1966

CRI is a national centre for providing intensive and co-ordinated technical support to the cement and allied industries and for boosting the related research and development activities. The Government of India, the entire cement industry, the cement machinery manufacturers and others participate in the work and activities of CRI.

Main fields of work
The R and D activities of the Institute cover the entire spectrum, with cement as the central theme, starting from the chemical elements available in nature and their exploitation towards making cement, through the manufacturing processes, engineering design and development of cement plant and machinery, to cement technology and concrete technology, ending in concrete structures where cement finds its final destination.

The Institute's multi-disciplined R and D activity also provides service and assistance to industries dealing with other building materials such as lime, asbestos, etc.

The principal divisions are as follows:
Geology, Mining and Raw Materials Technology; Chemistry and Technology of Silicates and Ceramics; Chemical and Process Engineering; Mechanical and Industrial Engineering; Electrical and Power Engineering; Electronics and Instrumentation; Civil Engineering: Concrete and Construction Technology and Architecture; Structural Engineering; Technology Planning; Technology Transfer Sciences.

Total staff: 332
Professionally qualified staff: approximately 170
Director or Chief Executive
Director General: Dr H.C. Visvesvaraya
Principal sources of finance
Ministry of Industry (Government of India) grant; Research and Development contribution; Committee of Direction of Cement Research (CDCR) project funding.
Publications
- *Journal:* CRI Abstracts, quarterly; CRI Current Contents, bi-monthly
- Special publications, research bulletins, monographs, bibliographies, seminar documents, etc.

IND.275 **FM**

Central Building Research Institute (CBRI)
Kendriya Bhawan Anushandhan Sansthan
Roorkee (UP), India, Pin: 247667
Telephone: 2243
Telex: 0597-203
Cables: BILDSERCH ROORKEE (INDIA)
Date of establishment
1947

An autonomous research institution within the Council of Scientific and Industrial Research (CSIR), New Delhi. Its objectives are:
- To assist the building industry in solving problems of planning, foundation, materials, design and construction with a view to achieving better comfort, functional efficiency, speed, economy and productivity in building construction
- To carry out research and investigation into problems of fire in buildings with a view to reducing loss of life and materials
- To adopt all possible means to bring results of research to the notice of potential users and to ensure through appropriate measures that they find expeditious and efficient application.

Main fields of work
- *Architecture and Physical Planning Division:* Space requirements, energy efficient designs, planning of different types of buildings to provide norms and efficiency indices for effective and functionally sound planning; socio-economic surveys, etc.
- *Building Materials Division:* Building materials assessment and production, their methods of use, compatibility of different materials, their standardization, development of new building materials from industrial and agricultural wastes, etc.
- *Building Processes, Plant and Productivity Division:* Building processes, building services, building economics, building plants and construction management
- *Efficiency of Buildings Division:* Thermal, acoustical, illumination and ventilation efficiency of building; performance of materials and building components under varying conditions prevalent in the country
- *Soil Engineering Division:* Soil mechanics and foundation engineering as applied to buildings and allied structures; laboratory work and model studies; helping construction agencies in soil exploration and testing
- *Information Survey and Planning Division:* Dissemination of technical information from the results of the Institute's work and contacts with sister organizations and the industry, technology transfer, industrial surveys, documentation from literature, and public relations
- *Fire Research:* Standard fire tests for buildings materials and components; fire protection and fire safety
- *Rural Buildings Division:* Development of new materials and new construction techniques for rural areas; rationalization of conventional techniques for rural areas; study of present practice and trends in rural areas
- *Development, Construction and Extension Division:* Experimental and prototype constructions; implementation of new techniques and materials in lime construction projects; design and consultancy on various construction projects where the Institute's techniques can be utilized; liaison and dissemination of information relating to processes and techniques developed at the Institute; project reports or feasibility reports for different construction agencies.

Total staff: 579
Professionally qualified staff: 220
Director or Chief Executive
Acting Director: T.N. Gupta
Heads of department
Architecture and Physical Planning Division: T.N. Gupta
Building Materials Division: Mohan Rai
Building Processes, Plant and Productivity Division: J.P. Kaushish
Efficiency of Building Division: B.K. Saxena
Soil Engineering Division: R.K. Bhandari
Information Survey and Planning Division: A.C. Banerjee
Fire Research: G.N. Badami
Rural Buildings Division: N. Verma
Development, Construction and Extension Division: S.K. Misra
Principal sources of finance
Financed by the Government of India very largely through the Annual Parliamentary Vote of Funds for the Council of Scientific and Industrial Research.
Principal related bodies
Executive Committee: approves all technical and administrative work of the Institute; Research Advisory Council examines and recommends all technical work of the Institute.
Publications
- Research studies are published in professional or scientific journals
- Popular articles on the activities of the Institute are disseminated

through trade journals, house magazines, popular science journals, newspapers, etc.
- CBRI Abstracts: informative abstracts of the Institute's papers published and/or communicated as well as book reviews, etc., quarterly
- Building Research Notes: summaries of information on selected building topics, irregularly
- Annual Report: progress on research activities during the calender year
- Process Literature: summarizes manufacturing processes, manpower, materials, machinery and financial requirements for selected building materials and components
- Research Programme: research projects for the current calendar year
- Proceedings of conferences and seminars are published and distributed as priced publications
- Special publications on special topics, bibliographies, translations, etc.

IND.276
Central Road Research Institute (CRRI)
Delhi-Mathura Road, PO CRRI, New Delhi 110 020
Telephone: 632173, 632325, 632376
Cables: ROADSEARCH
Date of establishment
16 July 1952

A national laboratory for roads having federal status. The major role is to carry out basic and applied research in line with the national priorities for investigation, design, construction and maintenance of different types of road and runway, including related materials such as aggregates, bitumen, etc. and related equipment, etc. with a view to effecting optimum economics. Study of traffic and transportation problems including roads network planning.
Main fields of work
- To conduct and carry out research on the utilization of locally available materials for construction and maintenance of roads and runways, economically
- To carry out research on the characteristics of different types of road under different climatic and traffic conditions
- To carry out research and development in all aspects of road traffic and transportation engineering, including study of accidents, development of road safety devices, psychology of road users and transportation economics, as related to different forms of transport
- To develop labour intensive methods and manual aids for construction of low-cost all-weather village roads for all-round economic development of rural and under-developed regions of the country
- To develop appropriate tools, machinery, equipment and instruments for adopting technologies as related to highway engineering and as relevant to the country to achieve indigenous self-sufficiency
- To render technical advice and consultancy services to various organizations in the related fields to avoid import of foreign expertise.
- To train road technologists through refresher courses/workshop and training programmes for efficient implementation of road development plans for the country
- To create and establish all the necessary infrastructure, both in equipment and expertise, in the various facets of highway and transportation engineering and allied subjects as and where needed to achieve judicious solutions for special problems
- Publication of scientific material by contributing scientific and technical papers and special publications to journals, symposia, conferences, etc. devoted to scientific and applied research and developments in the connected fields
- Demonstration and utilization of the results of the researches conducted by the Institute
- To enter into arrangements with the appropriate foreign agencies and institutions for exchange of scientists, study tours, etc., conducting joint projects, technical assistance, etc.

Total staff: 530
Professionally qualified staff: 170
Director or Chief Executive
Director: Professor C.G. Swaminathan
Heads of department
General Projects: Dr M.P. Dhir
Geotechnical Engineering: Shri T.K. Natarajan
Soil Stabilization and Rural Roads: Shri T.K. Natarajan
Bridges: Dr P. Ray Chaudhuri
Flexible Pavements: Shri Y.C. Gokhale
Rigid Pavements: Shri Y.R. Phull
Workshop: Shri S.S. Rup
Planning and International Relations: Shri R. Krishnamachari
Documentation and Library Services: Shri M.C. Raghavan
Information and Operational Research: Shri Gyan Prakash
Traffic and Transportation: Shri Joginder Lal
Instrumentation: Dr P.S. KrishnaMohan Rao
Principal sources of finance
Council of Scientific and Industrial Research
Principal related bodies
Executive Committee; Research Advisory Council – review and selection of R and D projects.
Publications
- *Research reports:* Annual Report; research papers: about 6 in Indian Roads Congress Journal every year and periodically in other journals
- *Bibliographies/catalogues:* supplied on demand to staff and other agencies.

IND.277
Highways Research Station (HRS)
PB No 2371, Guindy, Madras 25. Pin: 600025
Telephone: 414290
Cables: Highsearch
Date of establishment
1956

A Government research establishment within the Department of Highways and Rural Works. Established under the administrative control of the Chief Engineer, Highways and Rural Works Department, Government of Tamil Nadu.
Main fields of work
- *Soils Laboratory:* Soils tests, geotechnics, design of foundations for bridges and buildings, soil stabilization, pavement design, etc.
- *Concrete Laboratory:* Mix design, quality control of concrete for major bridges, chemical analysis of concrete samples, research pertaining to concrete technology
- *Structures Laboratory:* Structural problems of buildings and bridges, model tests, load tests connected with them, corrosion study, structural failures, etc.
- *Traffic Laboratory:* Traffic and parking surveys, island designs, lane marking, signal setting, etc.
- *Bitumen Laboratory:* Quality control of works, performance of experimental stretches and skid resistance tests
- *Library:* The library of HRS is equipped with 16 900 books on civil engineering and allied subjects and also has certain back issues of important Indian and foreign technical journals which are not available elsewhere

HRS also undertakes some of the Government of India's research projects regarding roads and bridges.
Total staff: 335
Professionally qualified staff: 268
Director or Chief Executive
Director: Thiru S.A. Latheef
Heads of department
Soils Laboratory: Thiru N. Raghavan
Bitumen and Traffic Laboratory: Thiru G. Ratnavelu
Concrete Laboratory and Administration: Thiru C. Sankaran

Principal sources of finance
Financed by the Government of Tamil Nadu. Some Research Projects are financed by the Government of India.
Publications
Numerous papers are published in professional or scientific journals. Many of these are reprinted by HRS.
HRS publishes research notes as and when a research scheme is completed.
Annually, HRS publishes a summary of Highways Research giving a detailed report of its activities in the year concerned. So far, 50 Research Records have been published by HRS.

IND.278 **AM**
Housing and Urban Development Corporation (HUDCO)
Hudco House, Lodhi Road, New Delhi 110003
Telephone: 699534
Cables: HOUSECORP
Date of establishment
April 1970

HUDCO is a national housing finance institution and is a fully owned Government of India enterprise.
Main fields of work
The main objectives of HUDCO are:
- To finance or undertake housing and urban development programmes in the country
- To finance or undertake the setting up of new or satellite towns
- To finance or undertake the setting up of building materials industries

The principal divisions are: Finance, Law and Secretarial Wing; Project Wing; Research, Development and Construction Wing; Consultancy Wing; Materials Wing; Economics Wing; Administration Wing.
Total staff: 202
Professionally qualified staff: 120
Director or Chief Executive
Chairman and Managing Director: H.U. Bijlani
Principal sources of finance
Equity from Government, Government Guaranteed Debentures, Life Insurance Corporation of India and General Insurance Corporation of India.
Publications
- Shelter, a biannual magazine of HUDCO
- Economy in Housing Projects
- Path Finder Projects, Vols 1–3
- Research and Development Series, Vols 1–6
- User's Handbook Matrix – Land Subdivision Utility Network and Affordability
- Computer Based Design and Analysis to Affordable Shelter – HUDCO Model
- Breaking new ground
- Financial Appraisal
- HUDCO Design Concepts
- HUDCO and Rural Housing
- Research reports: User Reaction Surveys.

IND.279 **AM**
Indian Institute of Technology, Kanpur (IIT, Kanpur)
IIT Post Office, Kanpur 208 016
Telephone: 40066
Telex: 0325–296
Cables: TECHNOLOGY
Date of establishment
1960

The Institute is one of the primary centres of technological education and research in India and was incorporated by an Act of Parliament in 1961 (No. 59 of 1961). The Act declared it to be an institute of national importance.
It offers BTech, MTech and PhD degrees in various branches of Engineering and MSc and PhD degrees in Pure Sciences. It also offers a PhD degree in Humanities and Social Sciences. Short-term intensive courses and seminars are undertaken to promote continuing education for engineers and academicians.
Main fields of work
The Institute consists of 11 departments and 3 inter-disciplinary programmes. The Department of Civil Engineering is one of the departments. In addition, there are advanced research centres for Materials Science, Electronics Systems and Laser Technology. The Institute is engaged in education and sponsored research in almost all departments. It also offers Industrial Consultancy.
The Department of Civil Engineering offers the following major fields of specialization and areas of research: Environmental Engineering; Hydraulics and Water Resources Engineering; Geotechnical Engineering; Structural Engineering; Engineering Geology; Survey and Photogrammetry; Transportation Engineering.
Total staff: approximately 2300 of whom 280 are members of the faculty
Director or Chief Executive
Director: Professor S. Sampath
Heads of department
Aeronautical Engineering: Dr N.G.R. Iyengar
Chemical Engineering: Dr M.S. Rao
Chemistry: Dr P.T. Narasimhan
Civil Engineering: Dr V. Lakshminarayana
Electrical Engineering: Dr S.S. Prabhu
Humanities and Social Sciences: Dr Rajendra Prasad
Mathematics: Dr R.K. Jain
Mechanical Engineering: Dr V. Soundararajan
Physics: Dr R.M. Singru
Metallurgical Engineering: Dr A.K. Jena
Computer Science and Engineering: Dr H.V. Sahasrabuddhe
Nuclear Engineering and Technology Programme: Dr A.R. Das
Materials Science Programme: Dr K.P. Gupta
Advanced Centre for Materials Science: Dr K.P. Gupta
Advanced Centre for Electronics Systems: Dr S.S. Prabhu
Advanced Centre for Laser Technology: Dr A.S. Parasnis
Principal sources of finance
Ministry of Education and Culture, Government of India.
Principal related bodies
IIT, Kanpur is one among the five Institutes of Technology set up by the Government of India under the Ministry of Education. The Institute enjoys close co-operation with the other four IITs at Bombay, Delhi, Madras and Kharagpur and the Institute of Technology, Banaras Hindu Uni (BHU).
Publications
- R and D Capabilities
- Annual Report
- Courses Bulletin

Numerous papers have been published by the faculty members of various departments/programmes/centres in national and international journals. These papers are listed in the Annual Report.

IND.280
Indian Institute of Technology, Madras (IIT, Madras)
Madras 600 036
Telephone: PABX (10 lines) 415342
Telex: TECHMAS MS 7362, Code No. 041
Cables: TECHNOLOGY, MADRAS 600036
Date of establishment
1959

To provide for instruction and research in such branches of engineering and technology, science and arts, as the Institute may think fit, and for the advancement of learning and dissemination of knowledge in such branches.

The Institute was established as one of the premier academic institutions in the country devoted to science and technology. Declared as an institution of national importance, the Institute is an autonomous body under the administrative control of a Board of Governors, appointed by the Goverment of India.

Main fields of work
The Institute conducts study programmes leading to the BTech; DIIT (Diploma of Indian Institute of Technology), MTech; MSc; MS and PhD conferments in the appropriate disciplines. Apart from teaching, the Institute is engaged in a wide spectrum of activities covering research and development, industrial consultancy, continuing education, curriculum and Faculty development. To supplement the activities of the Departments in the furtherance of the objectives, the Institute has established a few research centres, schools for educational development, and special units for interaction with outside organizations.
Total staff: 2009
Professionally qualified staff: 900
Director or Chief Executive
Director: Professor P.V. Indiresan
Heads of department
Aeronautical Engineering: Professor C.S. Ballal
Applied Mechanics: Professor P.S. Srinivasan
Chemical Engineering: Professor M. Ramanujam
Chemistry: Professor G. Aravamudan
Civil Engineering: Professor L.N. Ramamurthy
Computer Science: Professor R. Kalyanakrishnan
Electrical Engineering: Professor A. Kuppurajulu
Humanities and Social Sciences: Dr A.V. Krishna Rao
Mathematics: Professor S.D. Nigam
Mechanical Engineering: Professor D. Prithviraj
Metallurgical Engineering: Professor K.A. Padmanabhan
Physics: Professor J. Sobhanadri
Principal sources of finance
The Government of India.
Publications
- Annual Report and Handbook of Information, annually
- Journal of Mathematical and Physical Sciences, bi-monthly.

IND.281
Karnataka Engineering Research Station (KERS)
Krishnarajasagara 571607, Karnataka
Telephone: Belagula 23
Date of establishment
1944

The Station came into being as a very small wing of the PWD. It has gradually grown in strength and expanded by diversifying its areas of function. Today, the Station ranks as one of the pioneer institutes of its kind in the country. It is one of the zones under the Irrigation Department of the Karnataka State Public Works Department. The Station deals mainly with applied problems connected with irrigation and power projects, soil mechanics, concrete and pozzolana, chemical analysis, road research, ports and harbours and coastal erosion referred by the project authority. Fundamental and basic research problems sponsored by CBIP are also studied. In addition, in-service engineers of the State are trained through refresher courses in various disciplines of civil engineering.

Main fields of work
The Station has rendered service to the civil engineering profession in assisting the successful fruition of almost all irrigation and hydroelectric schemes in the State. Preliminary investigations, designs and physical tests on hydraulic models, lining of irrigation canals, gauging discharges, concrete and material testing, soil analysis for engineering properties, analysis of water samples for irrigation and municipal supplies, low-cost house, rock mechanics, identifying and suggesting remedial measures to erosion prone areas, minor irrigation studies are some of the many areas in which the Station is involved. In addition, a number of basic problems have been assigned under the sponsorship of the Central Board of Irrigation and Power (CBIP), Government of India, which allocates financial grants for the purpose.

Many important applied research problems have been studied on reference by project authorities. In addition, the Station has initiated investigations into some basic research problems. The Engineering Staff Training College attached to KERS has sucessfully completed 93 orientation courses for in-service engineers of the State.
Total staff: 254
Professionally qualified staff: 132
Director or Chief Executive
Director: Sri K.V. Narasimhamurthy
Heads of department
Hydraulics Division No. 1: Sri K. Chandrashekara Gupta
Technical Services Division: Sri B. Aswathanarayanasetty
Fundamental and Basic Research Division: Sri R. Narasimha Iyengar
Coastal Engineering Division: Syed Athaulla Bokhari
Special MI Division: Sri T.R. Chandrasekhar
Principal sources of finance
Government of Karnataka State; Government of India through the central Board of Irrigation and Power.
Principal related bodies
Works Department and Irrigation Departments of Karnataka PWD.
Publications
- Research Engineer: a bi-monthly bulletin
- Annual report of KERS
- Annual report of the Engineering Staff Training College.

IND.282
Maharashtra Engineering Research Institute (MERI)
Dindori Road, Meri Campus, Nasik 422 004, Maharashtra
Telephone: 2185 (O); 2534 (P)
Date of establishment
April 1959

A Government research institute under the Irrigation Department, Government of Maharashtra.

Main fields of work
The Institute mainly caters for the needs of projects of the Irrigation Department, Public Works Department, problems of ports in Maharashtra State, water supply and drainage schemes of the Water Supply and Sewerage Board.

It also deals with research schemes applied to river valley projects, testing of construction materials, hydraulic models, structural models, foundations or investigations for the material required for the construction of dams, including sub-grade and similar materials required for roadworks.

In addition to undertaking basic studies consideration is also given to applications in the field.
Total staff: 433
Professionally qualified staff: 200

Director or Chief Executive
Chief Engineer and Director: Shri P.K. Nagarkar
Heads of department
Hydrodynamics Division No.1: Shri K.A. Grampurohit
Hydrodynamics Division No. 2: Shri V.C. Shahane
Coastal Engineering Unit: Shri S.R. Fadanvis
Soil Mechanics Division No. 1: Shri M.K. Kulkarni
Soil Mechanics Division No. 2: Shri A.A. Javalekar
Soil Mechanics Division No. 3: Shri B.B. Tirth
Instrumentation Research Division: Shri R.E. Rane
Materials Testing Division: V.C. Khole
Structural Research and Physics Division: B.D. Konshirssagi
Highway Research Division No. 1: A.V. Dabholkar
Highway Research Division No. 2: A.M. Joshi
Environmental Engineering Research Division: Dr J.N. Kardile
Vibration Engineering Research Division: Shri A.Y. Daphale
Mechanical Division (workshop, installations, maintenance, etc.): Shri K. Chinayya
Principal sources of finance
Mostly financed by the State Government of Maharashtra; grant-in-aid to some extent is given by the Government of India through the Central Board of Irrigation and Power and Ministry of Transport for carrying out research. Some private testing and consultancy work is also undertaken.
Publications
- *Journal:* STHAPATYA, quarterly
- Revision of PWB Handbook chapters for the use of State engineers
- Numerous papers are published in journals of various institutions, such as the Institution of Engineers, the Indian Water Works Association
- Some of the important papers on research work carried out at the Institute are printed in the form of Technical Memoranda.
- Annual Report
- Numerous research papers are presented in national and international conferences and symposia.

IND.283 **FM**
National Buildings Organization (NBO)
राष्ट्रीय भवन निर्माण संगठन
'G' Wing, Nirman Bhawan, Maulana Azad Road, New Delhi 110011
Telephone: 388784
Cables: NATBUILDO
Date of establishment
July 1954

The Organization is an office attached to the Ministry of Urban Development, Government of India. It is also UN Regional Housing Centre for ESCAP.

NBO is a research and development institution engaged in the co-ordination of research and transfer of technology. It works in an advisory capacity. In its capacity as UN Regional Housing Centre, it promotes technical co-operation amongst developing countries.

NBO is spread out in the country through 12 Rural Housing Wings – Bangalore, Chandigarh, Howrah, Jodhpur, Madras, Ranchi, Srinagar, Simla, Trivandrum, Vallabh Vidyanagar, Varanasi and Gauhati and six Regional Liaison Cells – Bhopal, Bombay, Calcutta, Shillong and Srinagar. These carry forward the recommendations of the Organization by implementation of the research, co-ordinating and promoting experimental projects and undertaking liaison work with other construction and allied agancies in their region.
Main fields of work
NBO is not charged with the responsibility of carrying out original research of its own but it maintains a close collaboration with research institutions in the country and co-ordinates the results of research for wider dissemination among construction agencies, besides maintaining liaison with the building materials industry. It sponsors projects to research organizations. Important recommendations have emanated from projects, studies and surveys formulated, sponsored or conducted by NBO for improving planning, design and construction of buildings and houses, particularly low-cost housing, production and utilization of building materials and adoption of progressive and rational sociological measures for human settlements.

NBO achieves its objectives through building a repository of knowledge in the form of a well-stocked and documented library of technical literature and films, organizing a series of seminars, get-togethers, conferences and training courses, publishing technical literature and producing technical films on construction techniques and housing. It has established a chain of Display Centres throughout the country for the benefit of engineers, architects and others connected with the building and housing industry.

It performs its international role as the UN Regional Housing Centre by maintaining close links with the ESCAP Secretariat and member countries of ESCAP through the liaison officers appointed by them. It provides experts and consultants to ESCAP and other UN Specialized Agencies and also caters for the technical enquiries made. Training courses are held at international level for the benefit of the member countries of ESCAP and other developing countries. NBO has been maintaining close liaison with building research and development institutions in various countries (Denmark, France, Australia, Canada, the Netherlands, UK, Finland, Sweden, USA and other institutions in Asia, Africa and the Middle East).
Total staff: 150
Professionally qualified staff: 50
Director or Chief Executive
Director: G.C. Mathur
Heads of department
Design Division: K.S. Srinivasan
Materials Division: Dr A.V.R. Rao
Socio-Economic Division: Dr. J.N. Mongia
Publications Division: Sunil Bery
Principal sources of finance
Annual parliamentary Budget under the funds of the Ministry of Urban Development.
Principal related bodies
Government of India, Ministry of Urban Development for policy and implementation of housing and building programme; Planning Commission of India for preparation of Five Year National Plan and programmes in housing and the building sector.
Publications
- The Journal of NBO, twice a year
- Publications on subjects of topical interest to the building industry
- Handbooks
- NBO Information Briefs
- NBO Impact Report
- Directories
- Catalogues
- Leaflets.

IND.284
National Institute of Construction Management and Research (NICMAR)
Walchand Centre, Walchand Terraces, Tardeo Road, Bombay 400 034
Telephone: 891581/890453
Cables: INSCONMAR Tardeo Bombay
Date of establishment
March 1984

NICMAR has been established by the construction industry and was founded as a society by 30 of the major contracting organizations in India. Public sector undertakings are also members of the Institute.

The Institute is at the disposal of all Central and State Governments;

local and municipal bodies; the construction industry in its widest sense and individual contractors.

The Institute has the full support of the Builders Association of India and Overseas Construction Council of India, which is the export promotion council for the construction sector set up by the Commerce Ministry of the Government of India.

Main fields of work

The Institute has been set up to promote the managerial capabilities and technical competency of the Indian construction industry. It seeks to increase the availability of professionally trained technical manpower at all levels and to upgrade the managerial skills of those already employed in the construction industry.

Activities and Services:

- postgraduate degree diploma level academic courses for engineers and others
- Short-term in-service courses for middle level personnel
- Week-end courses for top executives of construction companies, government departments and public authorities
- Seminars and symposia for policy-makers
- Surveys on various aspects of the construction industry
- Research projects to develop and adapt various management techniques and skills
- Co-operation with similar institutions in India and abroad
- Consultancy services and in-house training for individual construction companies on request
- Research and consultancy to improve the delivery system of rural infrastructure such as approach roads, drinking water, sanitation, housing, etc.
- Surveys and research on safety practices in construction, workers' health and welfare at work sites and to inform on statutory measures
- Research and consultancy on the management aspects of contract documents and evaluation of educational needs.

The Institute has developed a flexible organization which can adapt to the requirements of the construction sector. A number of Functional Centres are planned, each of which enjoys freedom of activity within the overall financial and administrative control of the Director and Governing Council.

A UNIDO/UNDP Project for 'strengthening of construction industry management capabilities and technical service to industry' is under way.

Total staff: 10
Professionally qualified staff: 5
Director or Chief Executive
Honorary Executive Director: A. Nagabhushana Rau
Heads of department
Director, Centre for Construction Management Education: Professor K.N. Vaid
Centre for Consultancy: R.K. Joshi
Centre for Technician's Training: V.D. Pai
Documentation Centre and Library: B. Ganesh
Principal sources of finance
Contributions by the members.
Principal related bodies
Builders Association of India, Overseas Construction Council of India, Indian Institute of Foreign Trade, National Productivity Council, supporting the activities of NICMAR.
Publications

- Proposed NICMAR News, monthly
- *Research reports:* Construction, Development and Civilization
- Bibliography on Construction Management.

IND.285 **FM**

Structural Engineering Research Centre, Madras (SERC, Madras)

CSIR Campus, Taramani, Madras 600 113
Telephone: 412139; 412343
Telex: 041-363 CSIR IN
Cables: SER CENTER
Date of establishment
10 June 1965

The broad scope and functions of the SERC are:

- It acts as an information bank for all the latest available knowledge and knowhow relating to the design and construction of housing, building and other structures such as transmission line, television and microwave towers, chimneys, cooling towers, machine foundations, electricity poles, pipes, railway sleepers, nuclear containment pressure vessels, ship structures, power plant structures, off-shore structures, etc.
- It undertakes application oriented research on all aspects of structural engineering in consonance with national priorities.
- It accepts research and development work from other organizations in the public and private sectors, especially developing complicated structural designs involving the use of digital computation tests on models and prototypes.
- It undertakes turn-key consultancy assignments in India and abroad, particularly those involving the design of industrial and institutional complexes and residential colonies.
- It organizes from time to time short-term specialized courses in structural engineering with a pronounced practical bias for the benefit of practising engineers, to acquaint them with new advances and developments in design and construction.

Main fields of work

- Development and evaluation of techniques for low cost housing for rural and urban areas
- Computer aided design and interactive graphics and software development
- Investigations on transmission line towers and other tower-like structures
- Investigations in analysis and design of ship structures and power plant structures
- Analysis and design of off-shore structures
- Dynamic response of structures and machine foundations
- Development of concrete composites and applications
- Development of concrete construction techniques
- Laser holography and optical techniques for experimental stress analysis
- Analysis, design and construction of wind and disaster resistant structures
- Studies on RCC, prestressed concrete and limit state design.

The Centre conducts three to four advanced courses every year for the benefit of middle and senior level engineers, designers, builders and architects.

SERC, Madras is equipped with PRIME-750 computer and graphic terminals. A modern Tower Testing station has been set up with assistance from UNDP, near Pallavaram, Madras.

Total staff: 205
Professionally qualified staff: 79
Director or Chief Executive
Director: Dr M. Ramaiah
Heads of department
Computer-aided Design and Interactive Graphics: N.V. Raman and Dr. A. Rajaraman
Non-linear Analysis, Ship and Power Plant Structures: Dr Appa Rao
Concrete Construction Techniques: Zacharia George
Housing and Buildings: Dr A.G. Madhava Rao
Wind Engineering: Dr B. Venkateswarlu
Off-shore Structures: Dr M.N. Keshava Rao
Limit State Design: V.S. Parameswaran
Structural Dynamics and Machine Foundations: Dr P. Srinivasulu

Experimental Stress Analysis: Dr R. Narayanan
Concrete Composites: Dr B. Venkakeswarlu
Tower Testing and Research: P.R. Natarajan

Principal sources of finance

The Government of India through the Council of Scientific and Industrial Research (CSIR).

Publications

- Designs for Low Cost Houses, vols 1 and 2
- Small Capacity Ferrocement Water Tanks – 2 (free)
- Journal of Structural Engineering, quarterly, published from Madras, carries original research papers written by authors from India and abroad.
- Numerous papers are published in professional and technical journals in India and abroad
- Annual Report, brings out the research and development activities, infrastructural activities (library, publications, etc.) including research and technical staff and budget
- SERC also issues priced publications written by its scientists.

IND.286

University of Roorkee, Earthquake Engineering Department (EQD)

Roorkee, UP, 247667

Telephone: 2482

Date of establishment

1958

A department of the University of Roorkee.

Main fields of work

- Education and training in all their aspects
- Research studies of earthquake engineering and engineering seismology; development and installation of seismic instruments and data processing
- Sponsored research and consultancy
- Dissemination of knowledge; symposia; contributions to the formulation of codes of practice and design in earthquake engineering and seismic instrumentation.

Total staff: 79

Director or Chief Executive

Professor L.S. Srivastava

Principal sources of finance

The Uttar Pradesh State Government, the University Grants Commission, Government of India, and through sponsored research and consultancy services.

Publications

- Numerous publications in scientific journals and conferences
- Annual report, summarizes the activities of the Department from 1 July to 30 June and lists titles of papers published.

IDN.287 AM

Directorate of Building Research (DBR) and UN Centre for Research and Development on Human Settlements
Direktorat Penyelidikan Masalah Bangunan (DPMB)
84 Jalan Tamansari (Tromol Pos 15), Bandung
Telephone: (022) 81082/81083
Telex: 28327 DBR BD IA
Cables: REHOCE
Date of establishment
16 September 1953

The Directorate of Building Research is a government research institute within the Directorate General of Housing, Planning and Urban Development of the Ministry of Public Works. The Directorate carries out research and development on housing, building materials, structure and human settlements. The research and development contribute to the programme in the field of Cipta Karya, consisting of urban and rural planning, housing, sanitary engineering, building materials and structure. Besides carrying out national programmes, DBR also contributes to international work by means of being the regional centre of the ESCAP region for research on human settlements in tropical humid areas, named the Regional Centre for Research on Human Settlements (UNRCRHS), and is also involved in the Training Co-operation Among Developing Countries Programme (TCDC).

Main fields of work
Being a research institute, the main function of the Directorate is to evaluate buildings, houses, building materials and structure and human settlements from technical, economic and functional considerations by continuously developing, organizing and undertaking studies and research necessary to support government programmes in the field of housing, building, planning and urban development. It is also responsible for the dessemination of relevant information for national and international audiences.

Besides research activities, the Directorate also conducts training courses, seminars and workshops to improve the skill of technicians, usually in collaboration with foreign or international bodies.

The tasks of the different departments are:
- *Administration Division:* planning, programming, budgeting and supervising of the Directorate's activities; administration of finance and personnel; executing correspondence, filing, archiving and internal and foreign affairs of the Directorate
- *Sub-Directorate of Building Materials Development:* to undertake research and development activities on building materials and construction techniques and on construction tools and machinery for the production of building materials
- *Sub-Directorate of Human Settlements:* to collect data, to carry out research and surveys on the planning, administration and development of human settlements; to carry out literature searches, collect information and formulate systems, norms and standards on human settlements
- *Sub-Directorate of Development:* to establish, supervise and develop building materials factories, construction techniques and systematization of building operations; to evaluate and appraise the implementation of research results
- *Sub-Directorate of Documentation, Analysis and Information:* to undertake library and scientific documentation services, generally in the field of technology, and particularly in the field of human settlements; to carry out information services, training courses, seminars, workshops and generally develop technicians in the field of human settlements.

Total staff: 283
Professionally qualified staff: 61
Director or Chief Executive
Director: Karman Somawidjaja
Heads of department
Administration Division: Sahat Mulia Ritonga
Sub-Directorate of Building Materials Development: Ruland Benyamin Tular
Sub-Directorate of Human Settlements: Djauhari Sumintardja
Sub-Directorate of Development: Zulkarnaen Aksa
Sub-Directorate of Documentation, Analysis and Information: A. Hariman
Principal sources of finance
The Government of Indonesia (mainly); foreign technical assistance.
Principal related bodies
Directorate General of Housing, Planning and Urban Development, Ministry of Public Works.
Publications
- *Journal:* Masalah Bangunan, in English, quarterly; Abstract Bangunan, in Indonesian, monthly; Infobang, monthly.
- *Technical standards:* various
- *Research reports:* various
- Library bulletin, quarterly
- Seminar and Workshops reports.

IDN.288

Forest Products Research and Development Centre (FPRDC)
Pusat Penelitian dan Pengembangan Hasil Hutan
Jalan Gunung Batu, PO Box 84, Bogor
Telephone: 24032
Cables: PUSSILHUT BOGOR
Date of establishment
1913

To conduct research and development, technical and scientific counselling in the field of forest products.

The Forest Products Research and Development Centre (formerly the Forest Products Research Institute) is an organ of the Agency for Forestry Research and Development, the Ministry of Forestry.

Main fields of work
Field of forest products technology:
- Biology and preservation: wood anatomy; mycology and marine organisms; entomology; wood preservation
- Chemical technology: wood chemistry; biomass energy; minor forest products; fibre technology
- Utilization: wood physics and engineering; wood drying; wood working and sawmilling; composite products

  Field of forest logging and economics
- Forest logging: forest products harvesting; forest engineering; forest products transportation; forest ergonomics
- Forest economics: forest management; forest products marketing; forestry socio-economics.

Total staff: 267
Professionally qualified staff: 8
Director or Chief Executive
Director: Abdurahim Martawijaya
Heads of department
Operational Management of Research and Development: Nana Supriana, PhD
Field of Forest Products Technology: Toga Silitonga, PhD
Field of Forest Logging and Economics: Sutarso Priasukmana, PhD
Research Group Leaders:
Biology and Preservation: Abdurahim Martawijaya
Chemical Technology: Toga Silitonga, PhD
Utilization: Suparman Karnasudirdja, MSc
Forest Logging: Ishak Sumantri, MSc
Forest Economics: Sutarso Priasukmana, PhD
Principal sources of finance
The Government budget (development and routine budget) and World Bank Loan.
Principal related bodies
Agency for Forestry Research and Development.
Publications
- *Research reports:* Reports;Communications, approximately 30 titles per year, at irregular intervals
- Proceedings of Seminars, Symposia, etc., at irregular intervals.

IDN.289
Indonesian Road Research Institute (IRRI)
Direktorat Penyelidikan Masalah Tanah dan Jalan (DPMJ)
Jalan Raya Timur No. 264, PO Box 298, Bandung
Telephone: 78251, 78252, 78253
Telex: 28377 dpmj Bd.
Cables: DPMJ
Date of establishment
Law of 1929 (Netherlands' East Indies Law)

A Government research establishment in the Directorate General of Highways within the Department of Public Works.

Main fields of work
To develop technological input for the development of roads and bridges, by carrying out necessary activities, for example, testing, investigations, analysis and evaluation, and by collecting and classifying all documents, data and information, for the preservation and improvement of technology.

Total staff: 600
Professionally qualified staff: 300
Director or Chief Executive
Director: H.A.B. Hasibuan
Heads of department
Highway Engineering Division: Anas Aly
Geotechnics Division: Siti Sulastri
Structures Division: Rosyid Sastramihardja
Principal sources of finance
The Government of the Republic of Indonesia, Ministry of Public Works.
Principal related bodies
Directorate General of Highways (IRRI is one of its directorates), Department of Public Works. (Indonesian Ministry of Public Works).
Publications
- *Journal:* Informasi Teknologi: abstract journal, monthly
- *Technical standards:* BINA MARGA Manual for Road Materials, etc.
- *Research reports:* Study of the Use of Natural Rock Asphalt (Asbuton) in Road Pavements; Study of AWCAS Low Cost Roads in Indonesia; Road Drainage Investigation; Study of Pavement Characteristics; Study of Partially Prestressed Concrete Bridge Components, etc.

IDN.290
Institute of Hydraulic Engineering
Direktorat Penyelidikan Masalah Air (DPMA)
J1 Ir H. Juanda 193, Bandung
Telephone: 81607, 84553, 84554
Telex: 28283-BD
Date of establishment
24 August 1966

A government establishment within the Ministry of Public Works. Established by the amalgamation of the Institute of Hydraulic Structure Engineering, the Institute of Hydraulics and the Division of Hydrology.

Main fields of work
The Institute provides basic engineering and design services, applied research, hydrological information, inventory and monitoring of water quality and the investigation and testing facilities which are essential to the rehabilitation and extension of water resources development projects and hydraulic engineering works mainly for agriculture throughout the country. The Institute also prepares and disseminates manuals and guide-books which will assist projects officials who are responsible for the design and operation of hydraulic engineering works, making available to them the achievements and experience elsewhere in the field of water resources development.

Total staff: 495
Professionally qualified staff: 281
Director or Chief Executive
Director: Rachmat Tirtotjondro
Heads of department
Office Business Management: Willy Haryono
Division of Hydrology: Muhadi
Division of Hydraulics Structures: L. Taulu
Division of Hydraulics: Mochamad Memed
Division of Documentation and Data Processing: Mohamad Yahya
Principal sources of finance
The Ministry of Public Works.
Principal related bodies
The Directorate General of Water Resources Development (DGWRD) under the Ministry of Public Works is in charge of the study, planning, design and supervision of construction of irrigation and flood control projects. DGWRD consists of six directorates, of which DPMA is one.
Publications
- *Technical standards:* Manual Klimatologi (In Indonesian)
- *Proceedings:* Seminar on the Safety of Dams, Seminar on Dam Foundation Treatment; Seminar on Problems in the Design and Construction of Dams; Roving Seminar on Conceptual Models for Operational Hydrological Forecasting; Pertemuan Pembahasan tentang Pengaruh Degradasi Dasar Sungai dan Gerusan Lokal Terhadap Bangunan Utama.

IRN.291 **FM**

Building and Housing Research Centre (BHRC)
Markaz-e-Tahghighat Sakhteman va Maskan
PO Box 1882, Tehran
Telephone: 639942-6
Telex: 212117 MHUD IR
Date of establishment
1974

BHRC is the research branch of the Ministry of Housing and Urban Development. The Centre is run by the Director and a High Council consisting of 7 Ministers of the Iranian Cabinet. It includes 2 Research Departments, an Administration Department and an Information and Documentation Department.

Main fields of work
BHRC intends to co-ordinate, perform and implement research programmes concerning building, housing, materials, construction technology and methods and building regulations, as well as to provide an advisory service to both public and private sectors in procurement and industrialization of housing and building products suitable for local conditions.
Total staff: 105
Professionally qualified staff: 40

Director or Chief Executive
M. Yazdi
Heads of department
Building Research Department: H. Najafi
Housing and Architectural Department: S.J. Mir Latifi
Administration Department: A. Etemadi
Information and Documentation Department: F. Roshanbin
Principal sources of finance
Government budget confirmed by Parliament; fees for advisory and consultancy services offered by the centre.
Principal related bodies
High Council consisting of Ministers of Housing (Chairman), Industries and Mines, Plan and Budget organization, Economics, Labour and Social Affairs, Agriculture and Rural Affairs.
Publications
- Various research papers and reports resulting from research projects, published in Farsi
- Bibliography on earthquakes and earthquake engineering
- Bulletin consisting of the new documents of the Information and Documentation Department, annually.

IRQ.292 **FM**

National Centre for Construction Laboratories (NCCL)

المركز القومي للمختبرات الانشائيه

Mousa Bin Naseer Square, Tell Mohammed, Baghdad
Telephone: 98632, 97256, 96846
Telex: 212704 IK

Date of establishment

Law of National Centre for Construction Laboratories in November 1976

Governmental institution attached to the Ministry of Housing and Construction at the disposal of all State, provincial and municipal services, civil services and industry.

Main fields of work

- To carry out laboratory tests for quality control for all types of construction materials; sub-soil investigations together with recommendations pertaining to foundation design and construction and all necessary tests on construction works carried out by the Ministry's subsidiary organizations and directorates and by the ministries and organizations of the Government sector against fees
- To endeavour to enhance the scientific and standard technical performance of its personnel and to arrange extensive training courses and international seminars. The centre also undertakes to train personnel of other organizations, including the Iraqi technical institutes and university students
- To carry out applied research on construction materials and works in order to determine and realize the best methods for the construction of roads and structures in co-ordination with the organizations concerned
- To compile the research reports issued by the universities and scientific, national and foreign organizations relating to the specialization of the Centre in order to prepare them and place them before the various establishments

Besides quality control, the technical bodies approach the Centre for advice on problems in executing different types of project. The Centre then undertakes to prepare studies and provide recommendations through the consulting committee.

The principal directorates are the following: Directorate of Research and Technical Affairs; Directorate of the Middle Region Laboratories; Directorate of the North Region Laboratories; Directorate of the Middle Euphrates Laboratories; Directorate of the South Region Laboratories.

Director or Chief Executive

Director General: Dr Mufid A. Samarai

Principal sources of finance

Government of the Republic of Iraq.

Publications

NCCL Standards: Sulphur as partial substitue for asphalt for roads; study on suitability of stone in building construction and concrete products; Effect of sulphate salts in sands and soils on concrete; Objectives and activities of NCCL; Sulphur sand mixes as building material; Effect of high temperature on properties of fresh and hardened concrete: a review; improved asphalts containing reclaimed rubber for paving purposes: a review.

International conventions held by NCCL: Non-destructive Testing for Concrete Structures, 1978; Quality Control of Construction Materials, 1979; On the effect of Temperature on Concrete and Asphaltic Concrete, 1982.

IRL.293

Institute for Industrial Research and Standards, Construction Industry Division (IIRS)

Ballymun Road, Dublin 9
Telephone: (01) 370101
Telex: 25449 IIRS EI
Cables: Research Dublin
Date of establishment
1961 by an Act of Parliament: Industrial Research and Standards Act 1961.

IIRS is a statutory body located within the Department of Industry and Energy; its Board is appointed by the Minister of that Department.

85% of the Division's clients come from the private sector construction industry and the remainder from government, public utilities and local authorities. The Division has laboratory facilities in Dublin and Cork.

The Division has the following practical objectives:

- The effective application by the construction industry of the best of relevant known and proven technology
- To provide technological support and leadership at sector level to the construction industry
- To provide technological support to Government in the formulation and implementation of technical policies affecting the construction industry.

As part of IIRS which is a national, technological institute, the Division benefits from the availability of a comprehensive range of technological specialization and laboratory facilities operating co-operatively under one management structure. As construction makes use of a wide range of technology the Division is in a position to mobilize comprehensive support to the industry from the whole Institute.

Main fields of work

The Division offers laboratory based services to the operating sectors of the construction industry and to associated government departments, local authorities, and other institutions. These services which are provided in the laboratory, on the construction site and in the production plant include technological research, product and process development, product and process evaluation, technical investigation, technical consultancy, quality assurance and inspection, certification, technical advice and information.

In addition, one department is devoted exclusively to research programmes and specialist technical consultancy in support of the developing Irish forestry and timber industry, with particular emphasis on assisting the application of Irish timber in construction. This includes the classification of the performance characteristics of Irish timber, the development of processing techniques and the implementation of quality assurance procedures in the industry.

The technologies covered include: building materials, civil engineering, concrete technology, construction technology, energy conservation (including alternative energy, environmental engineering, foundation engineering, timber technology, timber engineering and structural engineering).

In all the above services emphasis is placed on joint project work with industry and with government agencies.

Total staff: 77
Professionally qualified staff: 24
Director or Chief Executive
Director, Construction Industry Division: S.F. Dunleavy
Heads of department
Concrete Technology Department: B.A. Roche
Construction R and D Department: Acting Head, S.F. Dunleavy
Construction Technology Department: Dr W. Crowe
Forest Products Department: P. Colclough
Principal sources of finance
Government Grant 45%; fees for services 55%. Buildings and equipment are funded by a State Capital Grant.
Principal related bodies
National Board of Science and Technology: responsibility for national programme and budget for science; National Institute for Physical Planning and Construction Research (An Foras Forbartha), environmental research agency attached to the Department of the Environment.
Publications

- *Journal:* Technology Ireland, monthly; Building Progress, bi-monthly
- *Technical standards:* IIRS is also the national standards body and publishes all Irish Standards
- *Research reports:* Almost all research reports are confidential to the client
- Papers to professional, scientific and trade journals
- Articles for the national press
- Lectures to professional, scientific and trade organizations
- Contributions to congresses, symposia and seminars at home and abroad.

IRL.294

Irish Cement Ltd, Technical Advisory Department

P.O Box 101, 19 Lower Pembroke Street, Dublin 2
Telephone: Dublin 765851
Telex: 25168
Cables: IRECEM DUBLIN
Date of establishment
1959

The Department was established to provide users of Irish Cement Ltd's product with technical information, advice and help free of charge.

Main fields of work

The Department is an information and advisory body set up to supply information to engineers, architects, builders and contractors on the current use of cement and concrete in all forms of building and civil engineering work.

The main activity is founded on the provision of immediate impartial information as the result of enquiries received. Secondary fields of activity include:

- Lectures to universities, technical institutes, builders and agricultural groups
- Publication of technical articles in national trade magazines relevant to the concrete industry
- Site visits including advice on construction techniques and trouble-shooting.

Total staff: 3
Professionally qualified staff: 3
Director or Chief Executive
Manager of Department: T.M. Duggan
Principal sources of finance
Financed totally by Irish Cement Ltd, sole manufacturer of Portland cement in the Republic of Ireland.

IRL.295 **FM**

National Institute for Physical Planning and Constuction Research

An Foras Forbartha (AFF)
St Martin's House, Waterloo Road, Dublin 4
Telephone: 01 602511
Telex: 30846
Date of establishment
1964

A state environmental research institute established by the Government and operating under the aegis of the Minister for the Environment.

Functions of the Institute are to undertake research and provide specialist advisory/testing and training/information services for the Department of the Environment and local authorities in particular, other government departments and development organizations, state sponsored bodies and the construction industry. In addition, the Institute undertakes overseas projects in developing countries.

The objective of the Institute is to undertake the research and provide the specialist technical support necessary for the development and implementation of national policies, legislation and plans, relating to the physical environment. Specifically the Institute's programmes aim at furthering economic and social developments by ensuring minimum environmental impact and securing the maximum efficiency from infrastructural investment.

Main fields of work

Programmes: the Institute's programmes are focused in particular on environmental quality and infrastructural development with particular emphasis on: physical planning (national, regional and local); preservation and improvement of environmental quality, including pollution control; maintenance and development of the national physical infrastructure.

Specialist Advisory/Testing Services: The Institute provides a comprehensive range of support services to local authorities, regional development organizations, the Department of the Environment, other government departments and the construction industry. These services include: Planning Advisory Service; Water Management Advisory Service; Industrial Effluent Advisory Service; Road Construction Advisory/Testing Service; Road Traffic and Safety Advisory Service; Building Products Advisory Service; Housing Cost Control Advisory Service. The Specialist Advisory/Testing Services provide users with immediate access to the wide range of expertise accumulated in the Institute from nearly twenty years of continuing research and to the extensive information compiled in the Institute's national data bases on: water (quantity and quality, effluent treatment and dispoasl); housing (needs, conditions, costs/prices, energy use); roads (traffic and accident statistics, skid resistence and strengthening requirements); national heritage (natural and man-made environment); building products (supply, quantity, import content).

Training/Information: A prime responsibility of the Institute is to provide training in, and advance knowledge of matters relating to the physical environment. In pursuit of this role the Institute undertakes comprehensive training and information programmes, e.g. Road design and construction; Water supply and Waste disposal; Water resources monitoring and assessment; Development plan preparation; construction design and management; Environmental studies for teachers.

Total staff: 200

Professionally qualified staff: 100

Director or Chief Executive

Chief Executive Officer: L.M. McCumiskey

Chief Technical Adviser: P.T. Piggott

Heads of department

Planning Division: G. Walker

Water Resources Division: P. Kenny

Construction Division: P.T. Pigott

Roads Division (acting): P.T. Piggott

Information and Training Division Centre acting: S. Smyth

Principal sources of finance

Grant in aid from the Department of the Environment; commissioned projects for other organizations; fees for routine services.

Principal related bodies

Construction Industry Federation, Department of the Environment, Economic and Social Research Institute, Industrial Development Authority, Institute for Industrial Research and Standards, National Road Safety Association, National Board for Science and Technology, An Taisce (The National Trust).

Publications

Journal: Irish Journal of Environmental Science, twice yearly

Research reports: Application of Modular Co-ordination in Building, 1982; Aspects of Design and Performance of a Precast Concrete Facade with Open Drained Joints, 1981; Thermal Performance and Energy Use in Housing, 1978; Efficiency and Energy Conservation Potential of Domestic Heating Systems, 1981; A study of Foreign Project Work of the Irish Construction Industry, 1981; National Standard Control Procedures for House Construction Costs, 1979; Proceedings of CIB W-65 Third Symposium on Organization and Management of Construction, 1981; Proceedings of CIB W-67 Third International Symposium on Energy Conservation in the Built Environment, 1981.

IRL.296 AM

Society of Chartered Surveyors in the Republic of Ireland

5 Wilton Place, Dublin 2

Telephone: 01-765500

The Society of Chartered Surveyors in the Republic of Ireland is a constituent body of the Royal Institution of Cartered Surveyors UK.

Director or Chief Executive

Administrator: A. Smith

IRL.297

University College Dublin, School of Architecture

Richview, Clonskeagh Road, Dublin 14

Telephone: 01-697111

Telex: 25278 UCD EI

A department within the Faculty of Engineering and Architecture of University College Dublin, a constituent College of the National University of Ireland.

The primary concern of the School of Architecture is the teaching of undergraduates pursuing courses of study leading to the BArch Degree. Postgraduate Degrees include MArch, MArchSc and PhD.

Academic staff are encouraged to engage in research, and a number of research groups operate within the School.

Main fields of work

The School of Architecture provides facilities for postgraduate studies and research over the whole spectrum of architecture, but with emphasis on certain areas related to interests and expertise of individual staff members – architectural design, including housing, computer applications, solar energy, architectural technology, history and theory.

Research is undertaken on an individual basis and in co-operation with and on behalf of private and public organizations, State and international organizations.

Total staff: 43 (including 5 full-time)

Professionally qualified staff: about 38

Director or Chief Executive

Professor Cathal O'Neill, BArch, MSArch, MRIAI, RIBA

Heads of department

Computer Design Workshop: Professor Dermot O'Connell

Housing Research Unit: Philip Geoghegan

Energy Research Group: J. Owen Lewis

Principal sources of finance

Financed through the University, and by endowments and grants for research work in particular.

Principal related bodies

Faculty of Engineering and Architecture.

Publications

- *Research reports:* Occasional publications by individual members of staff and research groups.
- Annexe: an annual publication of Studio C at the School of Architecture.

IRL.298

University of Dublin, Trinity College

Dublin 2

Telephone: 772941

Date of establishment

1591

Main fields of work

The School of Engineering adopts an engineering science approach, that is to say attempts to cover a large part of the field of engineering science: structural engineering, engineering soil mechanics, and engineering photogrammetry and surveying, environmental thermal properties, and materials. Possibly as much as about 20% of the effort could be said to relate to building.

Structure: Board of the College; Council; Faculty of Engineering and Systems Sciences; Graduate School of Engineering Studies.

Total staff: approximately 800, of whom around one-half are support staff.

Director or Chief Executive

The Provost: Professor W.A. Watts

Heads of department

Dean of Engineering and Systems Sciences: Dr R.C. Cox

Head of the School of Engineering: Professor W. Wright

Principal sources of finance

Financed almost entirely by the State.

Publications

Papers are published in professional, technical and learned journals.

ISR.299 FM

Building Research Station (BRS)
Hatahana L'Heker Habnia
Technion – Israel Institute of Technology, Technion City, Haifa 32 000
Telephone: 04-292242; 04-292243
Date of establishment
1953

BRS was established in 1953 within the framework of the Technion – Israel Institute of Technology, and it performs today all the functions of a national research institute in the area of building sciences.

BRS engages in the following activities: applied research; immediate assistance to the building sector in problems requiring special expertise or experimental facilities; quality control and assurance; participation in standardization committees; dissemination of knowledge.

Main fields of work

- *Building Materials and Technology:* active in research of fly ash, oil shale ash and other wastes and by-products utilization, improvement of cement, concrete, fibre reinforced concrete properties, solution of various problems in application of prefabrication technologies, industrialization of finish works in building, adaptation of materials to special uses, durability of building materials
- *Structures:* Active in analysis of earthquake effects, performance of concrete and steel structures, prefabricated systems, ferrocement elements, shelters and various special purpose structures
- *Construction Management and Economics:* Active in development of scheduling and cost control methods in building sites and prefabrication plants, computerized information systems, rapid cost assessment methods, economic analysis of building policy alternatives and construction methods, adaptation of managerial control to inflation conditions, analysis of productivity in building trades
- *Climatology:* Active in analysis of thermal performance of components and building systems, development of design criteria for solar energy buildings with the aid of physical and computer models, analysis of ventilation problems and wind effects in a built neighbourhood, solution of special problems in the area of thermal performance and energy saving
- *Acoustics:* Active in analysis of acoustic performance of components and building systems, improvement of acoustic insulation of floors, windows and exterior walls, improvement of noise insulation of residential areas
- *Performance of Buildings and Systems:* Active in development of performance in area of comprehensive building systems (residential, industrial and public) both massive and lightweight, evaluation of utility rating of building systems, analysis of special properties of buildings and components, such as durability, moisture penetration, etc.
- *Materials and Components Testing Department:* Active in quality control and quality assurance in plants for fabrication of building materials, components and systems, testing of concrete reinforcing steel and other building elements on site and off site, testing of insulation, and special performance problems of buildings under construction.
- *Research:* The main areas of applied research are: preparation of infrastructure for standards in the field of structures, building materials, climatology and acoustics; analysis and development of new building materials and components; development of managerial techniques and systems in the areas of estimating, scheduling, cost control, financial control and contracting; economic analysis of methods, building systems and building equipment; evaluation of long range policy alternatives in the development of the building sector; theoretical and experimental evaluation of physical performance of buildings; preparation of performance specifications for various types of building.
- *Immediate Assistance to Building Sector:* The assistance comprises mainly the following: experimental and theoretical analysis of models of structures to be erected; evaluation of performance of building components and systems; professional opinions in special problems of design or construction.
- *Quality Control and Assurance:* The Testing Department of BRS performs an extensive programme of quality control and assurance in the following areas: quality assurance on the building sites; quality control and assurance in the prefabrication industry; quality assurance in plants manufacturing building materials such as tiles, blocks, steel bars, etc.

Total staff: 135
Professionally qualified staff: 50
Director or Chief Executive
Head of BRS: Dr A. Warszawski
Deputy Heads
Dr A. Bentur (Research)
Dr U. Korin (Quality Control and Testing)
Principal sources of finance
Ministry of Construction and Housing; Cement Foundation; the building industry.
Principal related bodies
The Faculty of Civil Engineering – Technion. The building industry.
Publications

- *Journal:* In the Field of Building (B'sde Habnia), monthly (in Hebrew)
- *Technical standards:* Performance Specifications; Preparation of drafts for national building standards (in collaboration with the Israeli Standard Institute
- *Research reports:* These are issued for all the research projects performed at BRS.
- *Bibliographies/Catalogues:* Bi-annual summary of research reports and related publications (in English), List of Publications (updated annually)

Most of the Master and Doctor of Science dissertations in the Department of Civil Engineering of the Technion, IIT, in Building Materials, Construction Management and Building Sciences Programs, are carried out at the Building Research Station.

ISR.300

Transportation Research Institute (TRI)
Technion City, Haifa 32000
Telephone: 4-292901
Date of establishment
January 1977

The TRI, which was established by the Senate of the Technion – Israel Institute of Technology, began its operations in October 1977. Incorporated within the TRI was the Road Safety Centre, a university research unit under the auspices of the Ministry of Transport. The TRI serves as a centre and framework of co-operation for the research activities of faculty members of the Technion's various academic units which operate in the broad spectrum of transportation research.

Main fields of work

The research programme of the TRI is divided into four departments.

- Transportation planning and land use (public transport, bus route models, transportation planning models, car ownership models)
- Road and traffic engineering (vehicle flows, geometric design of roads, environmental elements on traffic conditions, properties of asphalt mixtures, pavement design)
- Vehicle engineering and energy (optimal loads of heavy vehicles, transport of passengers in commercial vehicles, cooling system designs)
- Road safety (information systems, data analyses and processing; traffic and infrastructure characteristics in urban areas and interurban roads; vehicle safety; road users)

The TRI carries out research and conducts surveys; publishes research reports and articles in professional journals in Israel and abroad; holds seminars and lectures for professional people; participates in working groups and advisory committees appointed by the pertinent safety-related authorities.

Total staff: 52
Director or Chief Executive
Director: Professor Moshe Livneh
Heads of department
Traffic and Highway Engineering: Dr A. Polus
Road Safety Centre: Dr A.S. Hakkert
Vehicle Design and Energy: Professor A. Stotter
Transportation Planning and Land Use: Dr J. Prashker
Principal sources of finance
Ministry of Transport; Department of Public Works.

Publications
- Research report series, mostly in Hebrew with English abstract
- Publications series, mostly in Hebrew with English abstract
- The Institute publishes a quarterly bulletin containing a report of its activities and local and international news related to the transport field
- The Institute also publishes a bi-annual Activity Report and an annual research programme report.

ITA.301

Alluminio Italia SpA., Istituto Sperimentale dei Metalli Leggeri (ISML)

Research Institute for Light Metals

Via G. Fauser 4, 28100 Novara

Telephone: 0321 – 24701

Date of establishment

1939

A research establishment within the state-controlled body EFIM (Ente Partecipazione Finanziamento Industrie Manifatturiere), regrouping the greater part of Italian aluminium activities.

Main fields of work

ISML is the only centre in Italy in the field of aluminium research and technology. Its research ranges over all aspects of aluminium and aluminium alloy production, working and application.

ISML studies both materials and their surface treatments for architectural applications, including both the principles and the methods of analysis that will help engineers in designing light-alloy structures.

Total staff: 98

Professionally qualified staff: 38

Director or Chief Executive

Director: Professor Ing Franco Gatto

Principal sources of finance

Financed through the annual EFIM funds and through external research activities.

Publications

- Numerous papers are published in specialized national and international journals
- The Institute publishes Monographic booklets dealing with general problems of aluminium production, technology and application in which the Institute has acquired considerable knowledge
- Annual directory of available publications.

ITA.302

Associazione Italiana Condizionamento dell'Aria Riscaldamento e Refrigerazione (AICARR)

AICARR Italian Association of Air Conditioning, Heating and Refrigeration

Via Sardegna 32, 20146 Milano

Telephone: 02/490185

Date of establishment

24 May 1960 No. 1971/678 Repertorio dr. Federico Porro (Registered in Abbiategrasso on 25 May 1960 No. 1019 Vol. 76)

An association of air conditioning, heating and refrigeration techicians. For promotional purposes we organize meetings and technical visits. The Association has about 1100 members.

Main fields of work

Promotion of sciences of air conditioning, heating and refrigerating; Organization of meetings referring to this specific technical field.

Principal sources of finance

Association fees; fees for admittance to meetings.

Principal related bodies

Board of Directors; Cultural Committee.

Publications

- *Journal:* Il Condizionamento dell'aria, monthly
- Proceedings of the various meetings.

ITA.303

Associazione Italiana Tecnico Economica del Cemento (AITEC)

Italian Technical and Economic Cement Association

Via di S. Teresa 23, 00198 Roma

Telephone: 864 714; 856 137

Telex: 611321 AITEC I

Other address: Via Borgonovo 12, 201 21 Milano

Date of establishment

1959

A free category association exclusively for cement manufacturers having the object of protecting and promoting the collective technical and economic interests of the national cement industry, and developing the economy of this field.

Main fields of work

The Rome headquarters office consists of the following departments: secretariat, statistics and studies, press and public relations, administration. The periodicals department is engaged in managing and editing the periodicals and special publications of the Association. The Milan Office has a managing function and a technical secretariat with the task of promotion and information activity.

The fields of work are two:

- Activities and public relations in the interest of the Associates; studies and information of an economic, statistical, standardizational and industrial character; relations with private and public bodies connected with the construction industry;
- Work devoted to the promotion of the application of cement, comprising the following: edition and diffusion of periodical publications, monographic publications (for sale), free instructional brochures, diapositives (colour slides), films; participation in specialized industrial exhibitions; advice.

Total staff: Headquarters Office, Rome: 14; Editorial Office, Rome: 14; Milan Office: 5

Director or Chief Executive

President: Dr Ing Aldo Aonzo

Director: Mario Manicardi

Edihor of the magazines 'L'Industria Italiana del Cemento' and 'Il Cemento'; Dr Ing Gaetano Bologna

Principal sources of finance

Financed by annual contribution of the associated manufacturers, proportional to their production; the amount of the proportion is fixed by the Assembly.

Publications

- *Journals:* L'Industria Italiana del Cemento, monthly magazine specializing in the field of concrete structures; for 50 years this publication has illustrated the structures of the best Italian and foreign designers, giving each structural detail; Il Cemento, quarterly, devoted to physics, chemistry and technology of cement and concrete
- AITEC special editions, for sale
- Educational brochures, for free distribution: on cement and concrete technology (in Italian); for road construction and agriculture: Technical Notes, informative documents on concrete technology and on its most important applications; Documentation Service and Bulletin of Bibliography (for Associates only), information on technology of cement production.

ITA.304

Centro di Ricerca e Sperimentazione per l'Industria Ceramica (Centro Ceramico, Bologna)

Research and Experimentation Centre for the Ceramic Industry (Ceramic Centre, Bologna)

Via Martelli 26, 40138 Bologna

Telephone: (051) 53 40 15

Telex: 510891 CENCER I

Date of establishment

Convention: 5 May 1976; official acknowledgement: DPR 10 April 1978, No. 806

The aims of Centro Ceramico, Bologna are the following:

- To promote and perform research, studies and experiments on: ceramic raw materials, products, plant and processes; non-polluting materials, processes and technology for the ceramic industry
- To disseminate the most recent knowledge in the different sectors of the ceramic industry, as well as the results of research and studies performed, by means of publications, meetings, courses, seminars, etc.
- To promote activities for the professional education of personnel employed in the ceramic industry and related fields.

Main fields of work

- *Research:* Materials and products. Most research deals with building products (ceramic tiles, heavy clay bricks, roofing tiles, sanitary ware, etc.); Industrial processes, with particular regard to drying and firing processes, and to energy saving technology; Air and water pollution from ceramic processes. Pollutant monitoring and control; waste disposal and reuse.
- *Technical aid to industry:* Chemical and physical laboratories, for analyses and tests on raw materials and products, with particular regard to building products; Chemical laboratory and mobile laboratory for pollutant sampling and analysing.
- *Didactics:* Education and training courses for personnel employed in the ceramic industry; University course on Chemistry and Technology of Ceramic Products for the Faculty of Engineering of Bologna; Specialization Course in Ceramic Engineering, for graduate engineers; Scientific management of the Journal: La Ceramica
- *Organized activities:* SIMCER: International Symposium on Ceramics. 6th SIMCER on 26–28 September 1983, with special sessions on building ceramics; Meetings and seminars on various topics.

Total staff: 25

Professionally qualified staff: 23

Director or Chief Executive

President: Antonio Camellini

Director: Professor Carlo Palmonari

Principal sources of finance

Self-financed from technical aid to industry; contributions by member organizations.

Publications

- *Journal:* La Ceramica, bi-monthly
- *Research reports:* Results of researches are published in Italian and international journals, and sometimes in books edited by Centro Ceramico, Bologna.

ITA.305 **AM**

Centro Edile SpA

Building Centre

Via Rivoltana 8, 20090 Segrate, Milano

Telephone: 02 7530951

Telex: CEDILE 1 320412

Cables: Centredile-Segrate

Date of establishment

1976

Centro Edile is a society operating in the field of technical information for the building industry. Its aims and objectives are to promote technological innovation in the market for building products. Centro Edile has obtained the patronage of the Lombardy region for the activities it carries out.

Main fields of work

Services provided by Centro Edile are as follows:

- *Technical Information:* Centro Edile collects and distributes technical information on building products using the most up-to-date system available for the collection and filing of information from producers and technical consumers. Centro Edile also operates a computerized data bank which is accessible to visitors.
- *Research and Consultation:* Centro Edile carries out research to define new instruments for technological information and provides consultation on the qualification and utilization of products in the processing sphere of the building market. The annual research programme is laid down by the Centro Edile Scientific and Technical Committee. The above activities are also carried out for public bodies and private contractors and the Centre lends its support to other projects.
- *Promotion:* Centro Edile provides public relations assistance for the organization of displays (meetings, expositions, products, conventions, etc.) for the promotion of entrepreneurial and productive activities in the building sector, taking into account the level of the market and quality control as expected by both public and private consumers.
- *Exposition and Exhibition:* Centro Edile holds a permanent exhibition of building products at its base in Segrate, from where it supplies technical information and information on the building market. In collaboration with the E.A. International Fair of Milan the Centro Edile annually organizes SICOMAT/EDILIZIA, a specialized review of systems, components and materials for the building industry.

Total staff: 20

Professionally qualified staff: 6

Director or Chief Executive

President: Arch Luigi Chiara

Managing Director: Umberto Sportelli

Director: Dott Giovanna Alini

Heads of department

Information: Ing Paola Albini

Exhibition: Sandro Pulella

Public Relations: Valeria Mugnai

Researches: Arch Emilio Colombo, Dott Giuseppina Limentani, Arch Alfredo Zappa. Supervisor: Arch Luigi Chiara

Principal sources of finance

Private.

Principal related bodies

Ministero Lavori Publici (Public Works Ministry); Regione Lombardia (Lombardy Region).

Publications

- *Journal:* Edilizia Informazione, quarterly
- *Technical standards:* Technical cards of building products
- *Research reports:* Feasibility study for a national catalogue of building products
- Official catalogue of SICOMAT/EDILIZIA review: proceedings of meetings and conventions.

ITA.306 **AM**

Centro Italiano dell'Edilizia (CIE)

Italian Building Station

Via Lazzaro Spallanzani 38, 00161 Roma

Telephone: 850248 - 853547

Date of establishment

1962

CIE is a technical centre so organized as to be able to execute research and studies, information, publications and documentation and public relations.

Main fields of work
The main areas of research are:
- Technical standardization
- Industrialization
- Protection and safety
- Technical and legal assistance to members.

Director or Chief Executive
President: Pietro Padula
Vice President: Pietro Catalano, Antonio Simoncelli
Director: Silvano Panzarasa

Principal sources of finance
Financed by members' subscription, by subventins from public authorities and by miscellaneous sources.

Publications
- Results of research
- Proceedings of the assembly.

ITA.307 FM

Consiglio Nazionale delle Ricerche – Instituto per la Residenza e le Infrastrutture Sociali (IRIS)
National Research Council Institute for Housing and Social Infrastructures (IRIS)
Via Re David, 200 – 70125 Bari
Telephone: 080/227898
Date of establishment
2 October 1980

Public research institute financed by National Research Council.

Main fields of work
- Building refurbishment, maintenance and preservation
- Building materials and components
- Evaluation of building failures and environmental parameters
- Advanced testing techniques and data processing in the same field
- Town planning for building refurbishment
- Economic aspects of building refurbishment and rehabilitation
- Technical legislation.

Total staff: 6
Professionally qualified staff: 3
Director or Chief Executive
Director: Professor Giovanni Tortorici
Principal sources of finance
National Research Council.
Principal related bodies
The Scientific Committee: Professor Umberto Di Cristina; Professor Giovanni Fuzio; Professor Marcello Grisotti; Professor Benedetto Colaianni; Professor Almerico Realfonzo.
Publications
- 'Quaderni dell'Instituto per la Residenza e le Infrastrutture Sociali' – annual
- Articles in scientific and professional journals and proceedings.

ITA.308 FM

Istituto Centrale per l'Industrializzazione e la Tecnologia Edilizia (ICITE)
Italian Building Research Station
Via Lombardia 49, Frazione Sesto Ulteriano, 20098 San Giuliano Milanese
Telephone: 02 9880282/3/4
Date of establishment
1966

Public Research Institute of the National Research Council (CNR), ICITE is the only institute of CNR working in the building field. Its principal aims are the standardization and control of building component requirements and performance. It also performs qualitative control of building components and systems in agreement with the aims and directions of the UEAtc organization.

Main fields of work
- Building technology
- Standardization
- Energy
- Industrialization of building
- International co-operation (UEAtc, ISO, RILEM, CIB)
- Research
- Building data bank
- Agréments
- Tests
- Quality marks for cement.

Total staff: 65
Professionally qualified staff: 20
Director or Chief Executive
Director: M. Ferri
Heads of department
Energy: W. Esposti
External envelope: A. Moretti
Internal partitions: G. Bedotti
Data processing centre: D. Bertola
Sanitation plants: F. Medici
Structures: F. Cantoni
Agréments: A. Franchi
R ad D: R. Vinci
Quality mark for cement: G. Nicolai
Plastic durability: F. Carlile
Principal sources of finance
The National Research Council (CNR).
Principal related bodies
Scientific Council: gives scientific approval to activity plane.
Publications
- Articles in specialized journals
- Monographs.

ITA.309

Istituto Sperimentale Modelli e Strutture (ISMES)
Experimental Institute for Models and Structures
Viale G. Cesare 29, 24100 Bergamo BG
Telephone: 935-243043
Telex: 301249 ISMES I
Date of establishment
1945

ISMES is a joint stock Company operating in the field of structural and civil engineering, in support of firms in the sector.

Main fields of work
By exploiting its geotechnical and structural capabilities, ISMES mainly operates in the following fields:
- Energy
- Construction
- Transportation
- Industry

Further new promising fields where ISMES has been operating in the last decade are:
- Structural restoration
- Software
- Education.

Total staff: 246
Professionally qualified staff: 133
Director or Chief Executive
President: Dr Morando Dolcetta Capuzzo
Managing Director: Professor Alfredo Marazio
Co-Directors: Dr Aldo Castoldi and Dr Roberto Riccioni
Heads of department
Statics: Mr Ruggeri
Dynamics: Dr Casirati
Instrumentation and Systems: Mr Anesa
Mathematical Models: Dr Peano

Geomechanics: Dr Rossi
Geophysics: Dr Carabelli
Geotechnical Laboratory: Dr Baldi
In-Situ Geomechanics: Dr Bruzzi

Principal sources of finance

ISMES invoices its services and its research works.

Publications

- *Journal:* ISMES News (2 issues per year), available both in Italian and English, free of charge
- *Research reports:* I Quaderni Dell' ISMES: technical bulletins published by ISMES. Since 1982, 172 issues have been published, the majority written in English, available free of charge.

ITA.310

Istituto Sperimentale per l'Edilizia SpA (ISTEDIL)
Building and Civil Engineering Tests and Research Institute
Via Tiburtina Valeria km 18 300, 00012 Guidonia Montecelio
Postal address: PO Box 7237, 00100 Roma Nomentano
Telephone: (0774) 378346; 378374-5-6
Date of establishment
1972

Established on the initiative of the Italian Association of Building Contractors (ANCE). Authorized by the Ministry of Public Works to perform building control tests and to deliver official certificates. The ANCE and local contractors organizations are the shareholders of the Association. Eventual profits are invested in laboratory equipment and research activities.

Main fields of work

- Tests for studies, technical assistance and control in the field of building and civil engineering
- Research in the field of building technology
- Technical agréments for building materials and components
- Courses and seminars
- International co-operation
- Technical publications.

Total staff: 50
Professionally qualified staff: 23
Director or Chief Executive
President: Dr Arch Franco Pesci
Deputy President: Dr Ing Pietro Favini
General Manager: Dr Ing Manfredo Manfredi
Heads of department
Geotechnics: Dr R. Croce
Concrete, aggregates, metals: Dr Ing S. Spina
Structures: Dr Ing O. Orazi
Masonry: Dr Ing G. Lapolla
Building materials and components: Dr R. Bianchetti
Thermoacoustics: Dr Ing E. Giraud
Chemistry: Dr P.M. Magarò
Tests for building sites: Dr Ing M. Andreoli
Research: Dr Ing R. Zari
Bituminous materials: Dr Ing E. Cipolloni
Principal sources of finance
Tests and applied researches performed upon request.
Publications

- *Research reports:* La resistenza all'usura delle pavimentazioni civili ed industriali; Progetto di fattibilità per la creazione di un complesso di laboratori tecnologici regionali e/o interregionali; La determinazione sperimentale delle dispersioni attraverso i ponti termici e loro correzione; L'invecchiamento delle pareti multistrato con elevate caratteristiche di coibenza termica
- *Bibliographies/Catalogues:* Prontuario per il calcolo del cemento armato (Modern manual for the calculation of reinforced and prestressed concrete); Prevenzione dei danni nell'edilizia civile – Impermeabilizzazione degli edifici – Vol. I, II, III; Repertorio prove integrato da commento.

ITA.311

Politecnico di Torino, Facoltà di Architettura, Dipartimento di Scienze e Tecniche per i processi di Insediamento
Polytechnic of Turin, Faculty of Architecture, Department of Science and Technology related to Human Settlements
Castello del Valentino, Viale PA Mattioli 39, 10133 Torino
Telephone: (011) 688.861
Telex: 220646 POLITO
Date of establishment
The Department was established in 1981 with the reorganization of the School.

The Department of Settlement Sciences is one of 18 departments of the Turin Engineering Polytechnic, organized following public law No. 382/1980 for the restructuring of all the Italian Universities.

The Department covers the general field of sciences and techniques for human settlement both on the side of what is called 'soft' technology and on the side of 'hard' technology. The main disciplinary competences which operate in the Department are: sociology, land use and planning, architectural design, structural design, building technology, environmental integration of building, building and land survey.

The Department operates as a research institution within the charter of the Italian University.

Main fields of work

Sciences and techniques for human settlement, environmental sciences, land use and planning, architectural and structural design, building construction technologies, components performance testing.

The disciplinary fields covered are: sociology, socioeconomics, land use and planning, architectural design, structural design, building construction technology, land survey, building survey, environmental design

Current research programmes and contracts:

- Regional public health organization: hospital services and their interaction with land use and planning
- The dynamic evolution of the city of Turin in the next three decades: main trends, the form of the city, pathological events and trends. The 'image' of the city and its evolution
- Energy saving on the built environment, energy saving 'retrofit' and current maintenance
- Passive and low energy technologies for building in the various climates
- Sociological acceptance of new energy models.

Total staff: 42
Professionally qualified staff: 42
Director or Chief Executive
Director: Professor Arch Lorenzo Matteoli
Heads of department
Socioeconomics: Professor Detragiache, Professor Bertuglia, Professor Mela
Architectural and structural design: Professor Donato, Professor Mosso
Technology: Professor Mattone, Professor Peretti
Environmental design: Professor Matteoli, Arch Grosso
Survey: Professor Bassi, Professor Debernardi
Principal sources of finance
Ministry of Public Education; Consiglio Nazionale delle Ricerche; Ministry of Public Building and Works; private contracts and contracts with Local Authorities.
Principal related bodies
Consiglio di Amministrazione del Politecnico di Torino.
Publications

- *Journal:* bi-monthly bulletin
- *Research reports:* under specific contracts
- *Bibliographies/Catalogues:* available upon request on specific topics.

ITA.312

Università degli Studi di Firenze, Facoltà di Architettura, Dipartimento di Costruzioni

Florence University, Faculty of Architecture, Construction Department

Piazza Brunelleschi 6, 50121 Firenze

Telephone: (055) 298 587; 215 351

Date of establishment

Construction Department according to the Law of 3 January 1983, DR No. 5; formerly Structural Engineering Institute according to the Law of 16 November, 1939 DL No. 2228/35

The Construction Department is part of Florence University Faculty of Architecture.

Main fields of work

The academic and research activities (research projects, thesis projects, etc.) lie within the competence of the following courses: structural theory, structural engineering, structural restoration, masonry structural analysis, structural computer analysis, bridge and large structures engineering, seismic engineering, mineralogy and geology, geotechnics and foundation engineering, experimental stress analysis on building materials.

The experimental activity, for research as well as for the private sector, is based upon the official laboratory for tests on building materials and structural models.

Total staff: 31

Professionally qualified staff: 20

Director or Chief Executive

Director: Professor Dr Salvatore Di Pasquale

Principal sources of finance

Board of Education; National Research Council; income from official tests on materials.

Principal related bodies

Official Laboratory for tests on building materials.

Publications

Research reports: Technical reports, Research papers. Many of the papers produced are available free of charge and are exchanged with other Institutions or researchers.

ITA.313 **AM**

Università degli Studi di Firenze, Facoltà di Architettura, Dipartimento di Processi e Metodi della Produzione Edilizia

Florence University, Department of Processes and Methods in Building Production

Via Cavour 82, 50129 Firenze

Telephone: (055) 283580

Telex: 572660 Unifi I

Date of establishment

1983

The Department belongs to the University of Florence (Faculty of Architecture). It is an academic institution involved with research for both public and private bodies. The Department was previously known as the Instituto di Ricerca Architettonica.

Main fields of work

- The development of studies and research on methods for the control of building processes at different stages
- Studies and research on problems concerned with fabrication and construction techniques, materials and systems for industrialized building and the design of industrialized products.

Total staff: 38

Director or Chief Executive

Director: Professor Dr Arch Romano Del Nord

Principal related bodies

Department Committee, Department Council.

Publications

- Department Bulletin, bi-monthly
- Articles in specialized journals
- Monographs.

ITA.314

Universita Degli Studio Di Napoli, Facoltà di Architettura Dipartimento di Configurazione e Attuazione dell'Architettura

University of Naples – Faculty of Architecture

Department of Architecture and Building

Via Monteoliveto 3 – 80134 Naples

Telephone: 081/321027 – 324415

Date of establishment

The Faculty was established in 1935

The Department was established, 1 January, 1985

University Department with several Chairs and a specialised library.

Main fields of work

The Department has an experimental laboratory for testing models, components, materials and building instruments. The Laboratory also carries out design experiments and group research. Testing is not done on request but only to the extent that it promotes the objectives of research.

Education and research is carried out in the international field in the areas of architectural and industrial design, graphic representation, control of materials and building procedures.

Total staff: approximately 30

Director or Chief Executive

Head of the Department: Professor Virginia Gangemi

Heads of department

Architectural Design: Professor M. Angrisani

Industrial Design: Professor R. Mango

Architetural Drawing: Professor A. Sgrosso

Building Technology: Professor V. Gangemi

Building Science: Professor E. Russo Ermolli

Other Chairs: Structural design of large structures
Building industrialization
Structural typology
Building components morphology
Cultural assessment in technology and design

Principal sources of finance

Ministry of Public Education funds through University and Faculty research contracts.

Principal related bodies

Department Council (all professors and researchers), Department Board (Head, 6 professors and 2 researchers).

Publications

- Study and research reports
- Publications in specialised, scientific and professional reviews
- Proceedings of national and international symposia
- A Department Review is scheduled, as soon as Department starts its functions and is financed.

ITA.315 **AM**

Università degli Studi di Roma La Sapienza, Dipartimento di Disegno Industriale e Produzione Edilizia

Rome University, Department of Industrial Design and Building Production

53 Via Antonio Gramsci, 00197 Roma

Telephone: 06/870936 – 870721

Date of establishment
Established in 1970 as Instituto di Tecnologia dell'Architettura; since 1983 under its present name.

University department engaged in research and didactics.
Main fields of work
In addition to instructing students in the field of building technology, the staff is involved in the following research groups:
- Quality in Building and New Technology
- Monument Rehabilitation and Restoration Technology
- Standard, Regulation, Planning and Management of Health System Building
- Environmental Industrial Design.

Total staff: 56
Professionally qualified staff: 44
Director or Chief Executive
Director: Professor Eduardo Vittoria
Heads of department
Technological innovation in building (postgraduate courses): Professor Massimo d'Alessandro
SfB Agency, Italy: Professor Giuseppe Vetriani
Research groups:
Quality in Building and New Technology: Professor Giuseppe Morabito
Monument Rehabilitation and Restoration Technology: Professor Paolo Marconi
Standard, Regulation, Planning and Management of Health System Buildings: Professor Roberto Palumbo
Environmental Industrial Design: Professor Corrado Terzi
Principal sources of finance
Basic accommodation and staff provided by University funds. The research projects are financed by University funds, Ministry of National Education, National Research Council (CNR), contracts with public or private organizations.
Publications
- Research results are published as:
- Articles in professional and scientific magazines
- Congress reports
- Books.

ITA.316 **AM**
Università di Bari, Istituto di Fisica Tecnica ed Impianti Termotecnici
Bari University, Institute of Technical Physics and Thermotechnical Plants
Via re David 200, I 70100 Bari
Telephone: 3980 228036/228740
Telex: 810598 UNIVBA I
Date of establishment
1 November 1965

The Institute is part of the Faculty of Engineering of the University of Bari.
Institutional activities beyond education are: research in the field of thermal and energy problems; research in the field of applied acoustics; testing upon request in the fields mentioned.

Main fields of work
- Meteorology for the purposes of thermal plant design, reduction of energy consumption and solar energy development
- Solar energy applications mainly in the field of building
- Analysis of the thermal performance of buildings and building components. Development of performance and test criteria for buildings and participation in the work of national and international organizations on standardization
- Qualification and testing of thermal insulating material for building and industrial use
- Applied acoustics. Acoustical transmission and insulation for building and industrial use
- Noise induced hearing defects. Recent and future work on impulsive noise.

Total staff: 15
Professionally qualified staff: 7
Director or Chief Executive
Director: Professor Ing Paolo Bondi
Heads of department
Unofficial working groups:
Buildings: Professor Ing Paolo Bondi
Meteorology: Professor Ing Albino Buizza
Acoustics: Professor Ing Ettore Cirillo
Thermal Insulation: Professor Ing Guiseppe Talucci
Principal sources of finance
Education Ministry; National Research Council; industry.
Publications
Research reports: Thermal performance of walls. Italian studies and researches (in English); Il clima di Bari (The climate of Bari: a statistical analysis) (in Italian), both free of charge.

JAM.317 **FM**

Jamaica Building Research Institute

Medi Centre Building, 34 Old Hope Road, PO Box 505, Kingston 5

Telephone: 92-96757/8

Telex: 2245

Cables; UNDEVPRO

Date of establishment

April 1982

A public institution (Division of the Ministry of Construction (Housing)) at the disposal of government agencies, the building construction industry and private individuals on request.

Main fields of work

- *Stone Technology:* Location of stone deposits, identification of deposits, exploration activities, analysis of surface features; Architectural and structural design of prototypes and demonstration units; Extraction of slabby materials and boulders for the construction of prototypes and demonstration units; Processing of stone for prototypes and demonstration units; Construction of prototypes and demonstration units; Evaluation and studies of comparative costs; Preparation of self-help manual and audio-visual module on stone technology; Extension work to disseminate technology; Preparation of stone catalogue
- *Stabilized Soil Technology:* Selection of soils, stabilization using cement, lime, asphalt, emulsion and other stabilizers; Design of building products, analysis of manufacturing options, production of prototypes; Introduction of reinforced masonry concept in stabilized soil techniques, e.g. design of vertical and horizontal reinforcement, studies on possibility of using spoilt bamboo as reinforcement, prototype construction of sections of walls and intersections; Static, dynamic and laboratory testing; Economic analyses
- *Clay Technology:* Assessment of brick clays; Analysis of available inputs, e.g. clays, labour, energy, etc.; Tests on clays (mineralogical examination, chemical analysis, particle size determination, etc.); Selection of production technologies; Pre-feasibility studies for different technologies; Pilot production
- *Experimental Development for Demonstration (unfired and fired bricks and blocks, floor tiles, roofing tiles):* Design of product for experimental development; Design and manufacture of production equipment, Production of prototypes; Structural analysis of reinforced masonry techniques (in relation to building codes); Construction of functional units. Intersections; Testing; Economic analyses; Demonstration programme; Dissemination of information
- *Lost-cost Roofing:* Fibre-cement roofing tiles and sheets; Jointing techniques in roofing for earthquakes and hurricane risk; Use of secondary wood species in roof structures; Pre-cast roofing units.
- *Concrete Components:* Concrete block technology; Precast building components; Ferro-cement components
- *Use of Indigenous and Waste Materials (Agricultural and Industrial) for Construction:* Building lime, Gypsum; Utilization of alumina red mud as building material; Pozzolana; Building boards from agricultural wastes
- *Low-cost Infrastructure Development:* Stone pavement; Pre-cast concrete units for pavement; Non-conventional water supply systems; Improvement in drainage systems; Sanitation in urban and rural areas; Use of new and renewable sources of energy in infrastructure and services development. Solar energy. Eolic Energy. Biogas.

Total staff: 15

Professionally qualified staff: 8

Director or Chief Executive

Director and Chief Executive: Keith Gilfillian

Heads of department

Construction Techniques and Design: Veitch Daley

Information and Documentation: Stephen Hodges

Administration and Finance: Whitsun McDonald

Building Materials and Components: Dr Arthur Geddes

Principal sources of finance

Government of Jamaica and UNDP.

JPN.318 AM

Architectural Institute of Japan (AIJ)

社団法人 日本建築学会

26–20, 5-Chome, Shiba, Minato-Ku, 104 Tokyo
Telephone: Tokyo 456-2051
Date of establishment
7 April 1886

A private institution which consists of Regular Members (individuals and corporations), Junior Members (students) and Patron Members (Corporations). Total membership in November 1983 was 31 095.

Main fields of work

This is the main scientific organization devoted to the advancement and development of the science, technique and art of architecture through mutual co-operation among its members. Members from public and private universities, research institutions, as well as such fields as designing, construction, production and governmental administration, co-operate with and inform each other at the Institute as individuals through investigation, research and such activities as dissemination of their findings.

Headquarters are in Tokyo and there are 9 regional offices throughout the country.

Total staff: 74

Director or Chief Executive
President: Professor Takuji Kobori

Principal sources of finance
Membership fees and publication of architectural books.

Principal related bodies
General Assembly; Board of Councillors; Board of Directors.

Publications
Journal: AIJ Journal Journal of Architecture and Building Science, monthly; AIJ Transactions, monthly, for members.

JPN.319 AM

Building Centre of Japan (BCJ)
Nihon Kenchiku Senta
Mori Bldg 30, 2–2 Toranonion 3-choine, Minato-ku, Tokyo 105
Telephone: 03 434 7165
Telex: J 23650
Date of establishment
25 November 1964

A comprehensive building and housing information centre, divided into public and private sectors. The public sector was established in 1965 by approval of the Ministry of Construction, entrusted with technical appraisal services under the Building Standard Law and other related Orders. The private sector was established in 1964, invested by representative building associations and building contractors, etc. in Japan, aiming at the provision of accurate and unbiased building information services.

Main fields of work

- *Evaluation Department:* Technical evaluation is the service initiated in compliance with the demand of the Ministry of Construction and includes the prior technical evaluation of the various building methods covering high-rise buildings, building equipment and fire protection materials for approval by the Minister of Construction under the Building Standard Law and what is required for approval by the Minister of Construction as industrialized housing performance. The Centre carries out evaluations upon request from the administrative agencies and the results are utilized effectively as important data for certification of the building in the country. For such purpose, evaluation conferences and committees are formed by more than three hundred top scholars in the field of building in the country. As at the end of the fiscal year 1981 the cases of evaluation exceeded 9500.
- *1st Business Department:* Product explanation meetings, movie shows, lectures on newly revised regulations and short training courses for technical experts are organized and sponsored by this department as a means of information services.
- *2nd Business Department:* For the exhibition service, the Centre operates an exhibition centre having indoor and outdoor display area. At present, about 210 corporations display nearly 2000 different items indoors, and 11 corporations demonstrate 14 model houses outdoors. Specialists provide the visitors with counselling services.
- *International Department:* As a member of the International Union of Building Centres (UICB), the Centre exchanges publications with overseas building centres and also subscribes to overseas technical magazines to collect building information from all parts of the world. It plans, organizes and sponsors regular Japanese–French, Japanese–Swedish and Japanese–Danish conferences for building modernization. The Centre also invites survey missions and trainees from abroad to its exhibition centre. In addition, it offers useful information to Japanese people going abroad for survey activities.
- *Building Information Research Institute:* The Centre collects books, periodicals, government reports, catalogues and other documents relating to the modernization of building construction. These publications are indexed and processed into catalogues of literature and performance appraisal sheets which are offered for public perusal or used to answer enquiries under the information exchange system.
- *Building Technology Research Institute:* In the field of investigation and development, the Centre is commissioned by the government and public organizations as well as by commercial enterprises to formulate standards and specifications and to undertake surveys concerning modernization and rationalization of the building and urbanization problems. At the same time, the Centre is carrying out independent research work.

Total staff: approximately 50
Professionally qualified staff: 10

Director or Chief Executive
President: Mitsufusa Sawada
Executive Directors: Kohzoh Sakao, Yujiro Kaneko
Director: Takeshi Gotoh

Heads of department
Planning Department: K. Fujikawa
General Affairs Department: Takahiro Gotoh
Evaluation Department: Takeshi Gotoh
1st Business Department: Y. Fukda
2nd Business Department: T. Wakayama
International Department: O. Taketani
Building Information Research Institute: H. Matsuoka
Building Technology Research Institute: Y. Kaneko

Principal sources of finance
Fees for technical evaluation; entrusted research; exhibition and publication services.

Principal related bodies
Ministry of Construction; all the associations related to housing and building associations in Japan.

Publications
Journals, trade books, committee documents and evaluation data sheets. The journals include The Building Letter, monthly, which has a number of subscribers. The trade books include annotations of building laws and regulations, lists of building manufacturers and dealers and guide books for prefabricated houses, while the useful documents of the committees are printed and distributed to the circles concerned. A list of publications (in Japanese) is available upon request.

JPN.320 AM

Building Contractors Society (BCS)
Kenchikugyo Kyokai
2–5–1 Hatchobori, Chuo-Ku, Tokyo 104
Telephone: Tokyo (03) 551-1118
Date of establishment
1 April 1959

A juridical Foundation authorized by the Ministry of Construction, which undertakes research and investigation upon request from both Government and member companies.

The Society comprises fifty of the largest building contractors in Japan.

The Society selects committee members from the affiliated companies and forms about forty committees to perform research and investigation in the required fields of study.

Main fields of work

- Annual award of BCS Prizes to outstanding buildings of the year in Japan
- Provision of scholarships for students in building engineering
- Financial support for education and training of skilled workmen
- Accumulation, analysis and circulation of domestic and international information concerning the building industry
- Investigation and research on technology, machinery, energy, materials and construction techniques in building engineering
- Investigation and research on management and labour in the building industry
- Investigation and research for safety and health related to the building industry.

Total staff: 13

Director or Chief Executive

Chairman: Hajime Sako, President of Taisei Corporation
Vice Chariman: Kazuaki Fujita, President of Fujita Corporation

Heads of department

Executive Director: Teizo Higashi
Administrative Director: Yoshihiro Matsumura and Kensuke Ouchi
General Affairs Department: Chuhei Masano
Research Committee Department: Kentaro Nishina

Principal sources of finance

Contributions by member companies.

Principal related bodies

Japan Federation of Construction Contractors, Inc. (Nikkenren).

Publications

- *Journal:* BCS, quarterly
- BCS Prize Works published each year since 1960. Details of prize-winning works are published in Japan as well as overseas through Embassies
- *Research reports:* The results of work by the research committees are published as technical reports by the Society
- Energy Conservation Buildings in Japan (English edn), edited and published by BCS in September 1982.

JPN.321 FM

Building Research Institute, Ministry of Construction (BRI)

建設省 建築研究所

Kensetsu-Sho Kenchiku Kenkyujo
1 Tatehara, Oho-Machi, Tsukuba-Gun, Ibaraki-Prefecture
Telephone: 0298 64 2151

Date of establishment

April 1946

The Institute is the research organization of the Ministry of Construction of the Japanese Government. The research results are used to give scientific and technical support for new policy, to ensure public welfare and to contribute to the international research community.

Main fields of work

The major objectives of BRI can be summarized as:

- The prevention of disasters
- The improvement of the living environment
- The rational organization of building production and the development of new building techniques
- The effective use of energy and resources
- The promotion of international research co-operation

To accomplish the above aims, both basic and applied research works are conducted. The research theme covers almost all aspects of building, housing and urban planning, building economy, building materials and elements, fire safety, structural engineering, seismology and earthquake engineering, building production techniques and environmental design and engineering.

Total staff: 179

Professionally qualified staff: 90

Director or Chief Executive

Hiroshi Takebayashi

Heads of department

Housing and Building Economy Department: K. Matsumoto
Material Department: Dr Eng S. Koizumi
Structural Engineering Department: Dr Eng M. Fuzumatsu
Production Department: S. Okamoto
Environment, Design and Fire Department: Dr Eng T. Wakamatsu
Urban Planning Department: I. Twakawa
International Institute of Seismology and Earthquake Engineering: Dr Eng Sa S. Hattori
Testing and Evaluation Department: K. Asano

Principal sources of finance

About 90% of BRI's budget is provided by the Ministry of Construction and the remainder by other agencies.

Principal related bodies

Japan Association for Building Research Promotion roles: inspection of materials for construction work; promotion of research results; execution of investigation upon request.

Publications

- *Research reports:* BRI Research Paper (in English); Report of Building Research Institute (in Japanese)
- *Bibliographies/Catalogues:* BRI (Information on BRI).

JPN.322

Hazama-Gumi Ltd, Technical Research Institute

Kabushiki Kaisha Hazama-Gumi Gijutsu Kenkyujo
17–23 Hon-Machi Nishi 4-Chome, Yono, Saitama 338
Telephone: Japan 488 54 3111
Telex: HZMGMTYO J 25898; 1 242 2764 HAZAMA J
Cables: HAZAMAGUMI TOKYO

Date of establishment

Established in April 1946 and recognized on 1 November 1975

The Technical Research Institute of Hazama-Gumi Ltd is one of the divisions of Hazama-Gumi Ltd, general contractors, whose task is to conduct research and development in construction technology in the fields of civil and building works.

Main fields of work

Research work in:

- Ocean engineering
- Foundations and underground construction
- Methods and management of construction
- Construction materials
- Environmental engineering
- Structural and seismic analysis
- Energy-related research
- Consultancy services.

Total staff: 157

Professionally qualified staff: 87

Director or Chief Executive

Director: Keiichi Fujita

Heads of department

Building Research Department: Tetsuo Kiuchi
Dynamic/Structural Engineering Department: Yoshihide Miyazaki
Construction Equipment Research Department: Kensuke Sumita
Civil Engineering/Soil Mechanics and Foundation Department: Yasuki Yamaguchi

Principal sources of finance

Hazama-Gumi Ltd.

Publications

- *Journal:* Technical Research Report of Hazama-Gumi Ltd, annually
- *Bibliographies/Catalogues:* The Technical Research Institute of Hazama-Gumi Ltd.

JPN.323 **FM**

Housing and Urban Development Corporation (HUDC)
Jutaku Toshi Seibi Kodan
14–6, Kudankita 1-chome, Chiyoda-ku, Tokyo 102
Telephone: 03 263 8111
Telex: J34562 BCJAPAN
Date of establishment
1 October 1981 (Law of 22 May 1981) Japan Housing Corporation (JHC) (1955–1981) and the New Town Development Public Corporation of Japan (1975–1981) were consolidated, and HUDC was established

The HUDC is a public organization having a wide range of functions to create a complete living environment, to promote the supply of good housing and residential land in metropolitan areas, and to promote urban redevelopment as well as to provide a comprehensive system of city parks.
Main fields of work
In order to achieve its goals the following tasks are undertaken.
- Construction, administration including lease and sale of housing
- Development, administration including lease and sale of land
- Construction, administration including lease and sale of public, shopping and service facilities
- Implementation of land adjustment projects, new residential urban area development projects, etc.
- Implementation of urban renewal projects, land readjustment projects, etc.
- Management of regional railroad and street transport
- Construction and administration of pay park facilities in national parks as well as construction of city parks based on requests from local governments
- Provision of technical assistance based on requests from local governments.

Total staff: 5318
Director or Chief Executive
President: Yoshihıto Maruyama
Vice President: Naoki Fiyii Hitoshti Kunıgo
Heads of department
Personnel Affairs; Principal Planning and Co-ordinating; General Affairs: Koji Yoshida
Finance: Masanon Shiga
Housing Planning; Urban Renewal: Tomoaki Kurashige
Estates Management: Uinoni Kyosu
Architecture and Construction: Yukio Yabuucti
Urban Development Business; Ken Daı
Urban Development Operation: Teruo Yoshioka
Related Public Facilities and Transportation: Takasti Kishida
Parks and Green Spaces: Syogaku Tanabe
Inspector's Office: Takumi Tange
Principal sources of finance
Capital; Government loan; private loan; business income; corporation bonds; subsidies.
Publications
- *Journal:* Town and Living, quarterly
- *Technical Standards:* General Specification of Construction, annually; Special Specification of Construction, annually
- *Research reports:* Report on Research; Quarterly Report on Research.

JPN.324

Kajima Institute of Construction Technology (KICT)
Kajima Kensetsu Gijutsu Kenkyūsho
19–1, Tobitakyu 2-chome, Chofu-shi, Tokyo 182
Telephone: (0424) 85-1111
Telex: Kajima Corporation: 0242-2467 KAHIMA. J
Date of establishment
1949

Technical research laboratory of Kajima Corporation, architects, engineers and contractors.
Main fields of work
- *Structures:* rationalization of the design of structures and analyses of their resistance to earthquake, wind and waves; development of concrete structures and precast concrete structures
- *Production and Construction:* research on construction control, production control, construction equipment, etc.; research on improvement of working environment at construction sites
- *Soil, Foundation and Rock Mechanics:* development of construction methods for improvement of soil foundations, chemical grouting, shield tunnelling, cast-in-place diaphragm wall by slurry trench method, shoring, etc.; survey and investigation on large-scale earth-moving projects; rock mechanics and development of rock excavation techniques
- *Materials and Construction Methods:* research on construction methods for concrete, quality of materials and performance; development and improvement of finishing materials for buildings and of building elements
- *Environment:* research on factors affecting the living environment such as acoustics, thermal insulation, moisture-proofing and ventilation; research on air-conditioning equipment; research on schemes for the prevention of public nuisance such as noise, vibration and water pollution
- *Hydrography:* research on the design and construction of river, offshore and other structures in water.

Total staff: 309
Professionally qualified staff: 222
Director or Chief Executive
Director: Dr Yasuhior Kameda
Principal sources of finance
About 10% of Kajima Corporation's after-tax profits are invested each year in the Institute.
Publications
- Annual Report, in Japanese with English abstracts
- KICT Reports, in English.

JPN.325

Maeda Construction Co. Ltd Technical Research Institute
Head Office: 10–26, Fujimi 2-chome Chiyoda-ku Tokyo
Telephone: Tokyo 03 265 5551
Telex: 232-4033 MAECON J
Cables: MAEHONSHA TOKYO
Other address:
TR Institute: 39–16, 1-chome, Asahi-cho, Nerima-ku, Tokyo
Telephone: Tokyo 03 977 2241
Date of establishment
Technical Research Institute: April 1963

The Institute, as the leader of Maeda's technological improvement promotes fundamental and applied technical research and studies to meet various social needs, and undertakes the task of solving various construction problems and improving the quality of construction works.
Main fields of work
Contracting, programming and designing for the following works:
- Civil Engineering: Dams and power stations (hydro, thermal and nuclear); Railroads, subways, tunnel and subterranean works; Highways and bridges; Harbour, river and levee works; Reclamation, dredging and land development; Water supply, drainage and sewage disposal systems; Irrigation and water control systems; Airports
- Architecture: Schools, hospitals and office buildings; Residences and houses; Stores; Athletic and entertainment facilities; Factories and Warehouses.

Total staff: 90
Professionally qualified staff: 30
Director or Chief Executive
Director: Hiroyuki Kitamura

Heads of department
Structures: Yasuharu Maeda
Materials: Kazuie Yamada
Soil and Geological Engineering: Masao Itoh
Construction Methods: Takeo Ikeda
Environment: Masayuki Shirmua
Principal sources of finance
Maeda Construction Co, Ltd.
Principal related bodies
Ministry of Construction, Japan; Tokoy Electric Power Company, Japan; other civilian firms, Japan.
Publications
- *Journal:* Journal of Technical Research Institute, annually
- *Research reports:* Ak occasion demands
- *Bibliographies/Catalogues:* These are issued periodically.

JPN.326
Misawa Homes Institute of Research and Development
Misawa Homes Sogo Kenkyujo
2-4-5, Takaido-Higashi, Suginami-ku, Tokyo 168
Telephone: 03-332-5111
Telex: 2322289 MISAWA J
Cables: MISAWA
Date of establishment
10 January 1973

A privately owned and operated research establishment of Misawa Homes Company Ltd.
Main fields of work
The Institute undertakes a varied series of projects either independently or in collaboration with Universities and the Government. Recent achievements feature mainly the House 55 system and PALC (precastable autoclaved lightweight concrete) technology.

Eight sections belonging to the Technical Development Division execute fundamental research related to energy consumption, building materials, manufacturing techniques and durability survey and full-scale tests.

The principal divisions are: Technical Development; Socio-Cultural Development; CI and Graphic Communication; General Affairs.
Total staff: 79
Professionally qualified staff: 64
Director or Chief Executive
President: Tamotsu Takahashi
Vice President: Tadashi Hanaoka
Chief Executive Officer: Chiyoji Misawa
Principal sources of finance
Misawa Homes Company Ltd: financed through its annual funds.
Publications
- Technical Reports: issued periodically in both English and Japanese
- House 55 Misawa System and PALC Technology: in English, Jan 1983.

JPN.327 AM
Nippon Telegraph and Telephone Public Corporation, Building Engineering Bureau
日本電信電話公社建築局
Nippon Denshin Denwa Kosya Kenchiku-kyoku
5–3, Ote-machi 1-chome, Chiyoda-ku, Tokyo 100
Telephone: 81 3 509 4300
Telex: 222–5300
Date of establishment
NTT was incorporated on 1 August 1952 by the Nippon Telegraph and Telephone Public Corporation Law.

NTT is the sole organization furnishing domestic telecommunications services in Japan. The Building Engineering Bureau is one of the bureaux of NTT Head Office.
Main fields of work
Activities of the Bureau include: the acquisition of land; preparation of contract documents for architectural, structural, mechanical, sanitary and electrical works; supervision of construction; maintenance and remodelling of buildings as need be.

The Bureau has carried out research and development on various aspects of building engineering. They include telecommunications building prefabrication, building maintenance technology, air-conditioning systems for future telecommunications equipment, earthquake and wind-resistant design, DEMOS (Dendenkosha Multi-access On-line System) building engineering programs, an product design of new terminal equipment.

The Bureau has rendered technical support in the design and construction of various telecommunications facilities in developing countries.
Total staff: 2100 architects and building engineers working in Head Office, 11 Telecommunications Bureaux and 80 Telecommunications Divisions throughout Japan.
Professionally qualified staff: 1500
Director or Chief Executive
Director General: Noboru Iwatsuki
Deputy Directorsvgeneral: Kiyoshi Morishita and Toshiyasu Koizumi
Heads of department
Building Engineering Bureau Design Standard Office, Head: Makoto Ohara
Integrated Design Office, Head: Kiyoshi Morishita
Musashino Electrical Communications Laboratory Building Engineering Section, Chief: Ayao Ishigami
Structural Engineering Section, Chief: Yuji Sato
Principal sources of finance
NTT is financed through annual Parliamentary Vote as a government-subsidized organization.
Publications
- *Journal:* NTT Architecture, bi-monthly
- *Technical standards:* Standard Details for Telecommunications Buildings; Standard Specifications for Telecommunications Buildings; Structural Design Guide for Telecommunications Buildings; Design Guide for Air Conditioning Works.

JPN.328
Ohbayashi-Gumi Ltd
株式会社 大林組
3, 2-chome, Kanda Tsukasa-cho, Chiyoda-ku, Tokyo 101
Telephone: 03 292 1111
Telex: 222-4091
Date of establishment
1936. The first research laboratories were established in 1948; The new Technical Institute was established at Kiyose in 1965.

Main fields of work
Research activities cover all fields of building engineering, Special omphasis is given to structural dyanmics, soil dynamics, soil chemistry, artificial improvement of soil stability, model-testing technology of acoustic and air-conditioning engineering, analytical technique by electro-analogy, etc.

The principal divisions are : Civil Engineering; Construction Techniques; Structural Materials; Finishing and Materials; Structural Engineering; Soil and Foundation of Buildings; Seismology and Vibration of Buildings; Environmental Technology; Acoustic Engineering; Chemical Engineering.
Total staff: 280
Professionally qualified staff: 140
Publications
Report of the Technical Research Institute Ohbayashi-Gumi Ltd, bi-annually.

JPN.329

Osaka University, Department of Architecture, Faculty of Engineering
Osaka Daigaku
2–1, Yamadaoka, Suita, Osaka
Telephone: (06) 877-5111
Date of establishment
1966

Buildings and their surroundings are deeply related to human life. They must be designed for good health and human well-being, and planned with precautions against earthquakes, typhoons and other disasters.

In the curriculum are lectures, exercises and practice in various fields of science which extend from engineering to cultural and social science. It includes the plastic arts and landscape architecture as well as mathematics, dynamics, experiments in materials, etc. Much time is allotted to exercises in architectural design.

The student is required to become a person of high culture and flexible thought as well as a technical engineer.

Main fields of work
- Structural mechanics, Soil mechanics, Theory of earthquake resistant structures
- Theory of building structures, Structural materials
- Steel structures, Building construction
- Architectural planning, philosophy of architecture, Town planning, Exercise of architectural design
- Human engineering, Design method of architecture, Exercise on architectural design
- Environmental engineering in architecture, Building equipment, Building materials.

Total staff: 20
Professionally qualified staff: 20
Principal sources of finance
National expenditure.
Publications
Research reports: Technology reports of Osaka University.

JPN.330

Sato Kogyo Co, Ltd, Central Technical Research Institute
Chuo Gijutsu Kenkyusho
47–3 Sanda, Atsugi-shi, Kanagawa Prefecture
Telephone: (0462) 41-2171
Date of establishment
The Sato Kogyo Co Ltd, which was founded in 1862, has a 120 year history as a construction company. In 1963 a Laboratory was founded in the Technical Research Department of the Company and was developed in scale into the present Central Technical Research Institute in 1972.

The Institute plays a key role in conducting intensive research into technology development for the construction industry.

The Institute staff are currently conducting research and studies into basic techniques in connection with essential construction technology. At the same time the Institute staff are also active in identifying the industry's needs and in taking them up as their research theme to develop new technology.

Main fields of work
- Subsoil and its correlated construction technology: research and studies, tests, and analyses in connection with the condition and character of subsoil and bedrock; development of various rational and economical construction methods
- Labour-saving methods and mechanical methods: research and studies, tests, analyses, and development of new technology in connection with building construction works
- Research and various studies, tests and analyses in connection with geotechnic resistance design: a wide range of design technology for earthquake resistance engineering in relation to the movement of subsoil, earthquake resistance elements of structures, and their interaction
- Quality control of concrete: development of technology in connection with research analyses and quality control for concrete which is a major construction material
- Energy saving technology: development of new systems as well as research and studies on air-conditioning systems from energy saving aspects
- Pollution-proof technology: Research and studies, tests, analyses and technology development in connection with industrial wastes, noise, vibration and other public nuisance in the light of environmental preservation and harmonious execution of constructions.

Total staff: 87
Professionally qualified staff: 39
Director or Chief Executive
General Manager: Taihei Ogawa
Acting General Manager: Yoshitake Sato
Administrative Manager: Shoji Omachi
Heads of department
Civil Engineering: Toshio Kurokawa
Building Construction: Tamio Kuwabara
Geotechnic Resistance Engineering Research Group: Kanehiro Ishihara
Laboratory: Chikazo Kokaji
Principal sources of finance
Self generated.
Principal related bodies
Building Contractors Society.
Publications
Journal: Kenkyusho Ho (the Institute Bulletin), annually since 1964.

JPN.331

Shimizu Construction Co Ltd, Research Institute
Shimizu-kensetsu Kenkyujo
4–17, Etchujima 3-chome, Koto-ku, Tokyo 135
Telephone: 643-4311
Date of establishment
1944

The Research Institute forms part of a department of the Shimizu Construction Co Ltd. Its primary function is to sustain the company's creative technological knowledge, which is directed towards the development of new construction and engineering technology, and also guaranteeing the quality of the buildings and structures which the company constructs.

Main fields of work
- Research on properties and applications of various kinds of concrete and finishing materials; labour-saving construction methods; energy-saving building systems
- Research on properties of reinforced concrete and concrete structures; soil–structure interaction and design of earthquake-resistant structures; wind pressure and environmental effects of wind around buildings
- Research on mechanical properties and behaviour of soils and rocks; in situ depth mix soil improvement method; interaction between piles and soil–structure; dewatering method for excavation works
- Research on construction methods, construction planning, scheduling and work analysis; analysis of prestressing system; ultrasonic testing methods for welding metal and quality control
- Research on computer applications for the control of building equipmentnsound control techniques to secure desirable acoustic conditions for building; counter-measures against noise; analysis of client's requirements for architectural space and designer's works
- Research on energy-saving and thermal insulation systems; purification systems for contaminated air and offensive odours; water purification systems and water recycling systems
- Research on ocean survey systems and data processing methods; impact of wave, winds and tides on offshore structures; improvement

and utilization of port facilities and wave absorbers; power generating systems utilizing wave power.
Total staff: 128
Professionally qualified staff: 10
Director or Chief Executive
General Manager: Toshihiko Ohta
Deputy General Manager: Akira Ohira
Heads of department
Research Planning Department: Toshio Maruichi
Administration Department: Haruhiko Furukawa
Materials Division: Shigeru Morinaga
Structural Engineering Division: Osamu Isobata
Soil and Foundation Division: Yoshinobu Inoue
Construction Method Division: Minoru Nonaka
Planning Division: Kihachiro Kubota
Environmental Engineering Division: Eiji Miyaji
Ocean Engineering Division: Hiroshi Kobayashi
Publications
- Reports of the Research Institute of Shimizu Construction Co Ltd, bi-annually, in Japanese (free of charge)
- Special Report of the Research Institute of Shimizu Construction Co. Ltd, at irregular intervals in Japanese (free of charge)
- SHIMIZU Technical Research Bulletin, annually, in English (free of charge).

JPN.332 AM

Society of Heating, Air-Conditioning and Sanitary Engineers of Japan (SHASE Japan)
Kuki Chowa Eisei Kogakkai
Nakajima Bldg, 8–1, 1-chome Kita-shinjuku, Shinjuku-ku, Tokyo
Telephone: 03 363 8261
Date of establishment
28 July 1917

A legal corporation under the jurisdiction of the Ministry of Education. Established as the Society of Heating, Ventilating and Refrigeration, then for some time known as the Society of Domestic and Sanitary Engineering and then under its present name since 1962. A non-profit-making organization whose benefits are returned to the members.
Main fields of work
Meeting of Technical Sessions: a meeting and seven local meetings with presentation of papers are held yearly to promote expertise in the specific field in research and practice. Two training courses are held yearly for educational purpose. In addition symposia and seminars are held several times a year.

Activities of committees: In order to improve techniques in the Society's specific field, the following committees are operating: committees managed by the President (Long Range Planning; Society Awards; Building Engineer SHASE); committees managed by the Director of Planning (Planning; Codes, Publication Handbook of Air Conditioning and Sanitary Engineering); committees managed by the Director of Science and Technology (Science and Technology; Financial Support for Research; International Relations; CIB-W62; Screening Papers; Library Service; Plumbing Code; Air-Conditioning Standards; Solar Heating, Cooling nad Hot Water Supply); committees managed by the Director of Compilation (Publishing the Society's Journal and Transactions of SHASE Japan)

SHASE Awards: In order to raise the level of science and engineering in the Society's specific field, SHASE Awards have been made yearly since 1963 to the members who accomplished excellent work in the following three categories: technical paper, design and installation of building facilities and others not included in the former categories.

Building Engineer SHASE: SHASE established a system for building engineers in 1955 to recognize highly qualified engineers and to upgrade the general technical level. By 1976, over 12 000 applicants had passed the annual qualifying tests and were designated as Building Engineer SHASE.

The Society is devoted to the development of science and culture in promoting intercommunication of studies in its specific field, which includes building services and environmental engineering such as heating, ventilating, air-conditioning, water supply, drainage, sanitary engineering and other related areas. Through these activities, the Society has contributed to progress of the related industrial circle.
Total staff: 19 069
Professionally qualified staff: 17
Director or Chief Executive
President: Shoichi Fujii
Principal sources of finance
Financed by membership fees and income from activities.
Publications
- *Journal:* Journal of the Society of Heating, Air-Conditioning and Sanitary Engineers of Japan, monthly
- *Technical standards:* Handbook of Air-Conditioning and Sanitary Engineering; Heating Air-Conditioning Sanitary Standards (HASS)
- *Research reports:* Transactions of the Society of Heating, Air-Conditioning and Sanitary Engineers of Japan; Proceedings of SHASE Technical Conference; Proceedings of Joint Technical Conference on Air-Conditioning and Refrigeration.
- Bibliographies on Water Supply and Drainage Services
- Seminar Texts
- Glossary of Air-Conditioning and Sanitary Engineering

All publications are written in Japanese. However, the technical reports of the Journal are accompanied by English abstracts. The Transactions papers are written either in Japanese or English. Japanese papers are accompanied by English abstracts.

JPN.333

Taisei Corporation, Technical Research Institute
大成建設技術研究所
344-1, Nasemachi, Totsuka-ku, Yokohama 245
Telephone: Yokohama (045) 812-1211
Date of establishment
1 June 1958

The Institute forms part of the Technology Division of Taisei Corporation. In close co-operation with such divisions and departments of the head office as are related to project management, business promotion, design planning and project operation, the Institute carries out the following services:
- Research and development of new techniques to meet economic and social needs
- Solution of various technical problems encountered in design and construction
- Collection, accumulation and presentation of technical information.

Main fields of work
Research, development and consulting on
- Construction engineering
- Building materials
- Structural materials
- Environmental engineering
- Acoustic engineering
- Structural engineering
- Foundation engineering
- Earthquake engineering
- Hydrodynamics
- Geotechnics

Total staff: 250
Professionally qualified staff: 117
Director or Chief Executive
Manager: Shoji Kondo
Heads of department
Planning and Information Department: Mitsuya Tazawa

Building Engineering Department: Akio Ikeda, Hideharu Kaga
Civil Engineering Department: Takashi Tsubahara
General Affairs Department: Tatsuhiko Saito

Principal related bodies

Technology Division of Taisei Corporation

Publications

Research reports: Reports of the Technical Research Institute, Taisei Corporation, annually.

JPN.334

Takenaka Komuten Co. Ltd, Technical Research Laboratory

株式会社　竹中工務店　技術研究所

5–14, 2-chome, Minamisuna, Koto-ku, Tokyo

Telephone: Tokyo (03) 647-3161

Date of establishment

January 1953

The Technical Research Laboratory is the nucleus for all advanced technological research and development programmes carried out by Takenaka Komuten Co Ltd and Takenaka Doboku Co Ltd. Through thie R and D programme, the company has been able to meet the changing requirements of society.

Main fields of work

- Ground improvement, soil dynamics and earthquake response analyses
- Geotechnical engineering in foundation and excavation works
- Geotechnical engineering in civil engineering structures
- Earthquake engineering
- Reinforced concrete/steel structural engineering, tests and analyses
- Building system and construction management
- Concrete material engineering and construction technology
- Building materials, fire engineering
- Air-conditioning, heating and plumbing
- Acoustic, wind and vibration engineering
- Offshore engineering, waste water treatment technology
- Construction equipment, mechanical engineering
- System design.

Total staff: 250

Professionally qualified staff: 180

Director or Chief Executive

Director and General Manager: Motoki Kondo

Heads of department

Manager, Management Department: Akito Niina

Principal sources of finance

Takenaka Komuten Co. Ltd; Takenaka Doboku Co. Ltd; Engineering Advancement Association of Japan; Mechanical and Social System Foundation.

Publications

Takenaka Technical Research Report, twice yearly, (in Japanese with English abstracts).

JPN.335

Toda Construction Co. Ltd
Toda Kensetsu Gijutsu Kaihatsu Senta

戸田建設株式会社

7–1, Kyobashi 1-chome, Chuo-ku, Tokyo 104

Telephone: 03 562 6111; 03 274 4378 (Direct)

Telex: 2224747 TODACO J

Cables: TODACONST TOKYO

Date of establishment

10 July 1936

Founded at Myoshi-cho Koto-ku Tokyo in 1958; moved to Matsudo Chiba in 1967; R and D Centre established in 1971; another Laboratory established at Tsukuba Gakuen City Ibaragi in 1982.

Research Laboratory of Toda Construction Co. Ltd, a company with 100 years history, and one of the leading construction companies in Japan.

Main fields of work

Research and development of construction works.

There are two principal departments:

- Technical Development, with project teams on Computerization of office work; Save energy; NATM technique (tunnel); PRC technique; Decreasing-pressure silo
- Research: Soil, Foundation and Bedrock; Aseismic design and structural analysis; Materials and construction; Construction methods and machines; Environmental design.

Total staff: 148

Professionally qualified staff: 143

Director or Chief Executive

Representative Director and Vice President: Takehide Ohmori

Principal sources of finance

Mitsubishi Bank Ltd, Nihonbashi Branch; DAi-Ichi Kangyo Bank Ltd, Kyobashi Branch; Bank of Tokyo Ltd, Yuraku-cho Branch; Sanwa Bank Ltd, Kyobashi Branch.

Principal related bodies

Todo Road Co. Ltd, Seiken Construction Co. Ltd, Construtora Toda do Brazil, Toda Construction of California Inc., Toda and Fuller Contracting Inc., Toda Construction (Nigeria) Ltd, Chiyoda Land and Building Co. Ltd, Shipko Industries Ltd, Toda America Inc., Sun Belt America Inc.

Publications

- *Journal:* Toda Technical Report, quarterly; R and D Centre Report, monthly
- *Technical Standards:* Toda's Cutter Knockdown Technique; TS Sleeve Joint Technique; RPC Technique; Toda System Housing; Toda's Site Cast RC Diaphragm Wall as Sheeting Technique; Underground Oil Stockpiling Technique; TS-H Caisson Dock Technique; Precast Subway Method; Earth Design Automatic Design System (TEDS SYSTEM); Computerization of Construction Design; Developed Total Computer System
- *Research reports:* Underground Working; Piling Work; Concrete Work; High Scraper Building System; HPC Housing System; Toda's Shielding Technique
- *Bibliographies/Catalogues:* various.

JPN.336

Toda Construction Co. Ltd, Technical Research and Development Centre

Toda Kensetsu Gijutsu Kaihatsu Senta

7–1 Kyobashi 1-chome, Chuo-ku, Tokyo 104

Telephone: 03 562-6111

Telex: 2524747 TODACO J

Cables: TODACONST TOKYO

Date of establishment

January 1958

The organization is a department of Toda Construction Co. Ltd and is engaged in research and development of construction technology, mainly for the company and its clients.

Main fields of work

- *Soil, Foundation and Bedrock:* Method of improving soil quality of soft ground; Soil pressure during excavation; Low temperature characteristics of soil and bedrock; Dynamic properties of soil and ground
- *Aseismic Design and Structural Analysis:* Experimental and analytical studies of dynamic properties of structures; Seismic waves, wind pressure; Dynamic analysis of earthquake resistant structures; Structural analysis of bedrock tunnels; Structural analysis of low temperature tanks; Elastic and plastic stress analysis of continuous structures
- *Materials and Construction:* Improving construction materials and working methods in respect of weight, strength, cracks, waterproofing, fire resistance, durability, etc. Re-utilizing industrial wastes; Materials for dam embankments

- *Construction Methods and Machines:* Technology for construction productivity and production system; Improvement of working quality and pollution-free working; Mechanization of working; Demolition method for reinforced concrete construction
- *Environmental Design:* Architectural acoustics; Noise, vibration, wind disaster and other pollutions; Energy saving building.

Total staff: 75
Professionally qualified staff: 55
Director or Chief Executive
Director: S. Maruyama
Heads of department
Toda Institute of Construction Technology (for research)
Soil, Foundation: T. Ambiru
Aseismic Design, Structural Analysis: R. Fukuzawa
Materials, Methods: T. Kemi
Environment: T. Kemi
Bedrock, Earth Structure: H. Hikage
Principal sources of finance
Toda Construction Co. Ltd
Publications
- Toda Technical Research Report, annually
- *Bibliographies/Catalogues:* Company Outline, Toda Construction Co. Ltd; Toda Institute of Construction Technology.

JPN.337

Tohoku University, Faculty of Engineering, Department of Architecture

東北大学　工学部　建築学科
Aramaki-Aoba, Sendai 980
Telephone: 0222-22-1800
Date of establishment
1 April 1951

The Department forms part of the Engineering Faculty of Tohoku University, which is governmental. The main role of the Department is education, including graduate courses and research in the field not only of architecture but also of building engineering.
Main fields of work
The Department has eight chairs, four in architecture and four in building engineering. The main areas in which the groups are especially active are as follows:
- *Architectural History and Design Group:* Restoration of historical architecture; Ornaments of traditional Japanese architecture; Influence of Asian architecture in Japan; Ancient Japanese castles
- *Urban Planning Group:* Urban affairs analysis; Planning of urban land use; Urban facilities, traffic systems and urban greens and parks
- *Architectural Planning Group:* Regional planning of medical and welfare facilities; Housing policy and planning; Structure of supply and demand for houses; Urban district planning
- *Environmental Engineering Group:* Heat transfer through walls, Indoor air movement and air pollution; Functional estimation and design of household equipment; Passive solar analysis and design
- *Building Materials Science Group:* Durability of concrete and mortar; Efficiency of concrete aggregates; Sensory evaluation of the texture of building materials
- *Structural Mechanics Group:* Structural dynamics and earthquake wave analysis; Design of anti-seismic structures; Source mechanism of earthquake wave; Mechanical behaviour and fracture mechanics of concrete
- *Structural Engineering Group:* Inelastic behaviour of reinforced concrete structural members; Anti-seismic design of buildings; Characteristics of strong ground motion; Seismic risk analysis of urban facilities
- *Disaster Proof Engineering Group:* Elastic and plastic stability of steel structures; Design of steel structures; Theory and experiment in static and dynamic stability of shells and suspensions.

Total staff: 48
Professionally qualified staff: 35
Heads of department
Architectural History and Design: Professor T. Sato
Urban Planning: Associate Professor H. Katsura
Architectural Planning: Professor K. Kakehi
Environmental Engineering: Professor F. Hasegawa
Building Materials Science: Professor K. Hirai
Structural Mechanics: Professor M. Izumi
Structural Engineering: Professor T. Shiga
Disaster Proof Engineering: Professor K. Uchiyama
Principal sources of finance
Educational Ministry of Japanese Government.
Publications
Architectural Reports of Tohoku University, annually.

JPN.338

Tokyo Institute of Technology, Department of Architecture and Building Engineering

Tokyo Kogyo Daigaku
O-okayama, Meguro-ku, Tokyo 152
Telephone: 03 726 1111
Telex: 2466360 TITECH J
Cables: TITECHOOKAYAMA
Date of establishment
26 May 1881, founded by the Ministry of Education

A national institution for education and research in engineering and science.
Main fields of work
Architectural history, planning and design of buildings, urban planning, structural and foundation engineering, environmental engineering and building materials.

JPN.339

University of Tokyo, Department of Architecture, Faculty of Engineering

東京大学工学部建築学科
7-3-1 Hongo, Bunkyo-ku, Tokyo
Telephone: (03) 812-2111
Date of establishment
1873

Department of the National University.
Main fields of work
Fields of specialization include research and education in planning and design, history, structural and construction engineering, environmental engineering, etc.
Total staff: 34
Professionally qualified staff: 34
Principal sources of finance
National.
Principal related bodies
The Architectural Institute of Japan.
Publications
Journal: Journal of the Faculty of Engineering, University of Tokyo, Series B, quarterly.

JPN.340

Tokyu Construction Co. Ltd, Technological Research Centre

Tokyu Kensetsu Gijutsu Kenkysho
13–9, Miyazaki 2-chome, Miyamae-ku, Kawasaki City 213
Telephone: 044 854 2111
Telex: J 25834 TOQCONST
Cables: TOQCONSTRUCTION
Date of establishment
1 February 1968

Centre operating under the supervision of the general manager of the Construction Works Division of the company. The main tasks of the Centre are: to perform the necessary research and development in the latest construction technology in order to meet social demands; to support the architects and the civil design engineers by solving various problems and by furnishing necessary data and information to accomplish their design works successfully; to assist the field construction engineers by solving various problems and by supplying necessary information regarding construction methods and materials during the construction period; and to publish and demonstrate new inventions, discoveries and improvements in construction methods, technology, materials, etc. to contribute to the academic and scientific societies and also to promote sales for the company.

Main fields of work

- To research and develop aspects of soil and foundations in civil engineering construction works, civil engineering structures and civil engineering construction materials.
- To research and develop building construction methodology, architectural construction materials and building foundations and structures.
- To research and develop wind effect problems, oscillations and vibrations, acoustic matters and other problems pertaining to the environment.

Total staff: 36
Director or Chief Executive
General Manager: Kunio Nakamura
Deputy General Manager: Hideo Maiye, Yasushi Fukui
Heads of department
Civil Engineering Department: Yasushi Fukui, Manager
Architectural Department: Hideo Maiye, Manager
Environment Department: Kuino Nakamura, Manager
Planning Department: Tsugio Kishi, Manager
Principal sources of finance
Tokyu Construction Co. Ltd
Publications
Tokyu Construction Technical Report, annually.

KEN.341 **AM**

Housing Research and Development Unit

University of Nairobi, PO Box 30197, Nairobi
Telephone: 27441, ext. 212
Cables: Varsity–HRDU
Date of establishment
1976. Memorandum of Understanding Signed by the Ministry of Housing, National Housing Corporation and University College, Nairobi

A multi-disciplinary unit within the Faculty of Architecture, Design and Development, serving the main housing research needs of the Government.

Main fields of work

- Low cost housing standards
- Building systems: prototype designs and demonstrations
- Construction materials and techniques
- Training
- Housing policy
- Housing co-operatives
- Dissemination of information.

Total staff: 19
Professionally qualified staff: 10
Director or Chief Executive
Director: T.S. Chana

Principal sources of finance
Ministry of Works and Housing; University of Nairobi; sale of publications.
Principal related bodies
Joint Advisory Board (University and Governmental Representatives); Faculty of Architecture Design and Development Board.
Publications

- *Research reports:* Rural Housing Conditions in Kisumu District, Kenya, 1982; Rural Housing Conditions in Kisii District, Kenya, 1981; Foundations for Low-Cost Structures in Poor Soil Conditions, 1981 (complete list available)
- Bibliography on Human Settlements with Emphasis on Households and Residential Environment – Kenya, 1982 (one of 3 published by the Swedish Council for Building Research)
- Foundations for Low-Cost Structures in Poor Soil Conditions, 1981
- Low Cost Housing Building Cost Index, 1981
- Case Studies of Site and Service Schemes in Kenya, 1980
- Construction of Farm Buildings, 1980
- Progressive Annuity Tables, 1979
- Urban Housing and Community Development Course, 1982
- Construction Guidelines for House Building in Site and Service Schemes, 1982.

KOR.342 **AM**

Architectural Institute of Korea (AIK)

대한 건축 학회

No. 2–7, 2-Ka, Myung-Dong, Chung-Ku, Seoul 100

Telephone: 776-0809, 776-3124

Date of establishment

1945

A public organization for academic activities on architecture under the supervision of the Government.

Main fields of work

Architectural Design and Planning, Structural Engineering, Building Materials, Building Construction, Building Law, History of Architecture, Housing Technology and Policy, Architectural Environment, Architectural Education, Urban Planning, Building Safety, Prefabricated Building, Building Installation, Energy, Rural and Sea Village Planning; Computer-aided Design, Architectural Industry Standard, Structural Standard, Foundation Structure, Architectural Psychology.

Total staff: 124

Director or Chief Executive

President: Keun-Duck Kim (Professor of Architecture, Yonsei University, Seoul)

Heads of department

General Affairs: Duck-Jai, Kim (Professor of Architecture, Chungang University)

Researches: Kyung-Hoi, Lee (Professor of Architecture, Yonsei University)

Operations: Do-Keun, Yun (Professor of Architecture, Hongik University)

Principal sources of finance

Commissioned research and services (68%); membership fees (10%); Government subsidy (2%); Contributions (20%).

Principal related bodies

Korean Association of Registered Architects (an association of practising architects); Korean Institute of Architecture (an association of architects).

Publications

- *Journal:* Architecture, bi-monthly
- *Technical Standards:* Reinforced Concrete Structure Standard; Steel Structure Standard; Building Construction Standard Specification, etc.
- *Research reports:* Study on Building Performance and Electrical Equipment Criteria; Study on Design Briefing of Electrical GIS substation in Housing District, etc.
- *Bibliographies/catalogues:* AIK's Achievements, 1982
- Korea, Industry Standard Setting, Architectural Terminology, Setting.

KOR.343 **FM**

Korea Institute of Construction Technology (KICT)

한국건설기술 연구원

377 Mansoo-Dong, Nam-Gu, Incheon 160-01

Telephone: 422-0667/7052

Telex: KICTEX K28878

Date of establishment

26 May 1983

KICT is a private non-profit organization sponsored by Korean Construction Companies.

The purpose in establishing KICT is to carry out research and development of construction technologies in a systematic and multidisciplinary manner and to contribute to building, civil and plant engineering fields by introducing advanced technologies and promoting widespread dissemination of such technologies to the Korean engineering and construction industry.

Main fields of work

- Research on Construction Standards
- Research and development of new construction methods
- Research and dissemination of advanced construction technologies
- Research and development on construction equipment and materials
- Research and development in construction design and construction management
- Technical service and consulting activities on construction research
- Other research activities to promote the construction industry.

Total staff: 150

Professionally qualified staff: 118

Director or Chief Executive

President: Dr Kyung-Mo, AHN

Heads of department

Planning and Information Division: Dr John Kyungyoon Park

Civil Engineering Division: René S. Rhie

Water Resources and Environmental Engineering Division: Dr Sung-Wan Hong

Architectural and Engineering Division: Won K-Kim

Plant Engineering Division: Dr Ko Suk Moon

Principal sources of finance

Funds from government-invested organizations, bodies and construction companies; funds and subsidy from the Promotion Funds for Overseas Construction Works.

Principal related bodies

Founder: His Excellency Chun, Doo Hwan, President of the Republic of Korea. Overseas Construction Association of Korea; Constructors' Financial Co-operative; Members: Leading Korean Contractors. These bodies are the sponsors of KICT research works.

Publications

- *Journal:* Resarch News, monthly
- Research Theses, periodical
- Annual Research Report
- First Anniversary Transactions
- Detailed bibliographies and catalogues are produced.

NLD.344 **AM**
Algemene Bank Nederland N.V. Building Department
Vijzelstraat 32, Postbus 669, 1000 EG Amsterdam, Building Department (2.D.00)
Telephone: 020 299111
Telex: 16780 abnbz-nl
Date of establishment
1824

ABN Bank is the largest international bank of the Netherlands. It maintains more than 700 local branches including those of Bank Mees & Hope N.V., and Hollandsche Bank-Unie N.V. Moreover, it has established an international network of over 200 branches and affiliates in 43 countries on five continents. Assets amounting to over 132 billion guilders put it high on the list of the largest European banks, and only very few banks in the world can offer their clients a comparable on-the-spot service in so many financial centres all over the world.
Main fields of work
The ABN is engaged in all banking activities on a global scale.
The Building Department is concerned with new office buildings and with renovations and maintenance programs of the buildings.
It is also responsible for the quality of the offices both in the Netherlands and abroad although it is more involved with the home offices.
Other spheres in which the Building Department is active include technical and security installations, energy conservation and also costs and profits.
Director or Chief Executive
Chief Manager: P.A. van der Heiden
Principal Officer: Th. A. Heijne

NLD.345 **FM**
Bouwcentrum
Stichting Bouwcentrum
Weena 700, P O B 299, 3000AG, Rotterdam
Telephone: 010 309911
Telex: 22530 bouwc nl
Date of establishment
1946

An independent, non-profit-making institution dealing with research, consultancy and information in the field of building, housing, and the residential environment.
The Bouwcentrum's aim is to work on the improvement and renovation of the built environment. Within this framework research is carried out, advice given, and data and experience collected from disseminated to all those involved in the process of building and living.
The Bouwcentrum is occupied both under the instructions of third parties and on its own initiative with research, development and rating, but equally with consultancy, teaching, exhibitions, publications and training.
In addition it functions as a meeting place for industry, authorities, organizations and occupants on the occasion of congresses, lectures, brainstorming days, product presentations and the like.
Main fields of work
In the Bouwcentrum some three hundred men and four women are employed under the charge of a general manager and six managers. The managers conduct five groups amongst which the different activities have been distributed.
The division into six specialized groups was introduced in 1981 to show clear-cut ideas to those availing themselves of the services of the Bouwcentrum. Within the organization structure each of the six teams works for one or more specific groups of users or customers. In other words, each team serves an own-market segment. Each team bears a name that makes it clear that the team is a member of the Bouwcentrum family, but has a task of its own with its special clientele. The six teams are:
- Bouwcentrum-Infraplan: Management and research consultancy
- Bouwcentrum-Techno: Technical and physical research and consultancy
- Bouwcentrum U-bouw: Research and consultancy for building projects
- Bouwcentrum-Woningbouw: Research and consultancy for dwellings
- Bouwcentrum-Vakinfo: Information and training group
- Bouwcentrum-Wonen: Advisory and exhibition centre in the field of building and living

Detachments: The Bouwcentrum serves a number of institutes in the building industry in a special way. In such instances collaborators – male or female – are detached to an institute. They then compose the executive staff and work under the control of the governing bodies and boards of management of the relevant independent institutes. Within this framework the institute places available working capacity to those institutions that are engaged in the National Insulation Programme, the education on behalf of and in developing countries, school building and the building industry.
AKB: For subjects requiring specialized knowledge in the domain of mathematics, operations research and information theory, the Bouwcentrum has a daughter body: the Consulting Agency for Quality Control and Operations Research (AKB). It is established in the Bouwcentrum building at Rotterdam and also accommodates the computing centre of the Institute. The agency carries out commissions both on behalf of the Bouwcentrum and for third parties.
Total staff: 300
Professionally qualified staff: 180
Director or Chief Executive
Chief Director: Ir M.J. Spanraft
Heads of department
Bouwcentrum-Infraplan/Woningbouw: Dr J.J. Botman
Bouwcentrum U-Bouw/Techno: Ir H.C. van den Berg
Bouwcentrum-Kwaliteitsbeheer: Ir Drs L. van der Graaf
Bouwcentrum Bouwcentrum-Ir R.J.M.G. Peutz
Bouwcentrum-Info: Ir P. Kerkhoven
Principal sources of finance
Research grants, consultancy fees, training instruction, publications, exhibitions.
Principal related bodies
Board
Publications
- Bouw, bi-monthly, Bouwmarkt, monthly
- *Research reports:* Bouwbaak; Research Reports as separate publications, or published in the series Blauudrukken or Informatieve Uitgaven
- Bibliographies on various (technical) subjects.

NLD.346 **FM**
Building Research Foundation
Stichting Bouwresearch (SBR)
Weena 740, Postbus 20740, 3001 JA Rotterdam
Telephone: 010 12 35 28; 117276
Date of establishment
1959

Private foundation, established by organizations in the building industry.
Main fields of work
Research work mainly on:
- Business organization: organization of contracting firms; problems of organization on site; co-operation between partners in the building industry; economic structures
- Building technique: structures; materials; construction methods; failures, quality, safety; installations
- Building planning and design: city planning; building products; environmental problems; planning and building methods; general.

Total staff: 10

Director or Chief Executive
General Director: Ir J.J. de Bruijn
Principal sources of finance
Financed by the Fund for Education and Research of the Building Industry; payments by contracting firms on the basis of staff wages.
Publications
- Approximately 100 publications in book form
- Approximately 180 preliminary publications

All publications available through Bouwcentrum Boekhandel, Rotterdam.

NLD.347 UM
BV Articon, Bureaux for Architecture, Technical Installations and Constructions
BV Articon, Bureaus voor Architectuur, Technische Installaties en Constructies
Berkenweg 7, 3818 LA Amersfoort
Telephone: 033 630524
Date of establishment
14 May 1947

BV Articon is a subsidiary company of the Netherlands Railways. It performs only limited direct research activities but in consequence of the nature and volume of its work (design of integrated projects) it applies research results and carries out technical development in its own field.

Approximately 75% of its work is executed for the Netherlands Railways, the remainder being performed for other clients (bus companies, PTT, banks and other private firms).
Main fields of work
A multi-disciplinary approach to design and supervision during the construction of integrated projects: railway lines, bridges, viaducts, stations, workshops, depots, terminals, industrual areas, office buildings, residential buildings, recreational facilities.

Research in the field of architecture and civil engineering. For instance: noise abatement (railways); standardization of glass dimensions for window frames.
Total staff: 157
Professionally qualified staff: 16
Director or Chief Executive
Director: Ir J.W. Kruithof
Deputy Managing Director: H.Th. van Schaik
Heads of department
Bureau for Architecture: Ir M.F.M. Reichert
Bureau for Civil Engineering No. 1: Ir H. Kullberg
No. 2: Ir P.S. Noyon
No. 3: Ir B.W. van Eede
Bureaux for Heating, Lighting and Vertical Transport: Ing A.C. Lute
Principal sources of finance
Fees for services rendered
Publications
Papers are published in professional journals.

NLD.348
Centrum Hout (CH)
Timber Centre for Research, Information and Education
Lambertus Hortensiuslaan 76, 1412 GX Naarden
Postal address: PO Box 401, 1400 AK Bussum
Telephone: 2159 48704
Date of establishment
In existence since the early fifties, re-established January 1982

A co-ordination office for research, development, training and information on wood, timber and timber structures in the Netherlands. Research is partly financed by industry, depending on the subject of the research, and partly by governmental authorities. Training and information are mainly financed by the trade and industry. Also use is made of subsidies from various other institutes, e.g. Stichting Bouwresearch, Stichting Pronorm, Stichting KOMO, TNO, etc. as well as of university research.
Main fields of work
The aim of the CH is to initiate, stimulate and co-ordinate activities in the field of research, development and dissemination of knowledge in order to achieve the suitable application of wood in building structures. This aim is pursued through a number of activities:
- To gather all possible knowledge on wood, timber and timber structures
- To initiate research into material properties and the application of the results
- To set up small committees for specified research projects
- To transfer know-how by information and educational means
- To seek out failures in the use of timber and timber structures
- To co-operate with similar organizations, institutes and persons working in the same fields
- To stimulate interest in the goals of the Centrum Hout.

Total staff: 12
Professionally qualified staff: 3
Director or Chief Executive
Managing Director: Ir H. van Putten
Heads of department
Information Department: Ing L. van Diest
Research Department: Ir H. van Putten
Education Department: Ir A. Overbeek
Principal sources of finance
The timber trade and indstry; sales of publications; income realized through research and educational activities.
Principal related bodies
General Board; Nederlandse Houtbond; Nederlandse Bond van Timmerfabrikanten; Federatief Verbond Nederlandse Houtindustrie; information centres of Canada and Scandinavia; universities; TNO.
Publications
- Regular editions of research results and promotional publications (list available)
- *Research reports:* Reports on research work.

NLD.349
CIAD Vereniging voor Computertoepassingen in de Ingenieurspraktijk
CIAD Association for Computer Applications in Engineering
PO box 74, 2700 AB Zoetermeer
Office: Breedewater 6a, Zoetermeer
Telephone: 079 21 93 24
Date of establishment
9 February 1968

Main fields of work
At the end of 1984 CIAD had 93 members, representing firms, industries, agencies, government departments, hardware and software suppliers and the like. Membership grows by some 20 annually. The activities of CIAD are widespread. Firstly there are the 'free of charge' activities, including the regular conferences on certain themes (8 times a year), the publication of CIADSCOOP (6 times a year) and the KNIPSELKRANT (12 times a year), in which the most important articles on CAD/CAM appearing in the Dutch newspapers are clipped. Besides these, information on computer applications in engineering available at the CIAD Bureau (library, video tapes, magazines) is free of charge.

The Project Groups occupy a very important place in the activities of CIAD. Some of the reports, resulting from these project groups, have been published in English as well through the international FACE Organisation, of which CIAD is a member. In total there are about 15 project groups in existence, such as on digitizing techniques, information analysis, computer aided management, automation in organising the production and expert systems.

CIAD is also involved in actions with other Organisations. The Conferences on CAD CAD/CAM (CAPE '83, CAPE Netherlands '85) and several lectures and courses should be mentioned. In the autumn of 1984 a course of 16 lessons started on Dutch television about CAD/CAM, in which CIAD participates through the Director and two other workers of the CIAD-bureau.

CIAD is an association with General Meetings and an Executive Board, its affairs are administered by the Bureau.

Total staff: 10

Professionally qualified staff: 10

Director or Chief Executive

Director: J.C. le Clercq

Principal sources of finance

Membership Fees. Contributions to the project groups by the participating parties (members) (until August 1985 subsidised by the Government).

Publications

- Reports from the Project Groups, Monographs (A list is available)
- 'CAD/CAM in Bedrijf', 'CIADSCOOP', 'KNIPSELKRANT'.

NLD.350

Stichting Computertoepassing Bouwwezen (CTB)

Computer Applications for the Building Industry

Galvanistraat 1, 6716 AE Ede

Telephone: 0838 30137

Telex: 37074

Date of establishment

Ede, 1 January 1977

A computer service bureau for housing and road construction.

Principal sources of finance

Financed through the participants paying for computer services.

Principal related bodies

Co-operation with the Dutch Security for the Building Industry.

NLD.351

Delft Hydraulics Laboratory (DHL)

Waterloopkundig Laboratorium (WL)

Rotterdamseweg 185, PO Box 177, 2600 MH Delft

Telephone: 015 56 93 53

Telex: 38176 hydel nl

Cables: hydrolab delft

Other addresses:

De Voorst Laboratory, Voorsterweg 28, 8316 PT Marknesse, Noordoostpolder

PO Box 152, 8300 AD Emmeloord.

Telephone: 05274 2922

Telex: 42290 hylvo nl

Wageningen Branch: De Nieuwlanden, Nieuwe Kanaal 11, 6709 PA Wageningen

Telephone: 08370 82776, 82891, 82786

Haren Branch: c/o Institute for Soil Fertility, Oosterweg 92, 9751 PK Haren

Telephone: 050 346541

Date of establishment

1927

A semi-governmental non-profit-making foundation. Annual turnover approximately Dfl 65 000 000 (1981).

Main fields of work

Services: mechanical and electronics engineering workshops, instrumentation laboratory, mathematics divisions, on-line computer facilities for model control and data processing, photography section, library and documentation service and sediment analysis laboratory. The Laboratory acts as a consultant in the field of hydraulics, hydrodynamics and hydrology. The work comprises applied and fundamental scientific research on contract and basic research on its own account to broaden the consultancy capacity.

Scope of activities: theoretical studies, mathematical investigations, hydraulic model investigations, simulation model investigations, site investigations; basic and applied studies; providing data for the design of hydraulic structures, river and coastal engineering works; supplying design criteria for industrial circulations; calculating hydrodynamic phenomena; carrying out surveys in situ; processing and interpreting observations. Fundamental research, in co-operation with educational and other scientific research institutes, as a support for consultancy work. Consulting and research activities organized in specialized advisory branches.

Total staff: about 600

Professionally qualified staff: 150

Director or Chief Executive

Director: J.E. Prins

Deputy Director: J.G.H.R. Diephuis

Head of the Delft Laboratory: A. Paape

Head of the De Voorst Laboratory: J.J. Vinjé

Principal sources of finance

Financed by proceeds from consultancy activities on contracts in the Netherlands and abroad and by various revenues from special services.

Publications

- Official publications and scientific periodicals
- *Journal:* Hydro Delft, quarterly
- Documentation service, fortnightly
- Congress papers.

NLD.352

Delft Soil Mechanics Laboratory (DSML)

Laboratorium voor Grondmechanica (LGM)

Stieltjesweg 2, PO Box 69, 2600 AB Delft

Telephone: (0) 15 569223

Telex: 38234 soil nl

Date of establishment

1934

DSML is an independent consulting and research institute in the field of soil mechanics and foundation engineering. Since its foundation more than 26 000 assignments have been executed all over the world.

DSML forms with the Delft Hydraulics Laboratory the foundation Stichting Waterbouwkundig Laboratorium. DSML has international offices in the United Kingdom, Singapore, Canada and Belgium.

Main fields of work

The activities of DSML comprise in general geotechnical site investigation, geotechnical laboratory testing, project-orientated consultancy, applied and fundamental research, development of new measuring techniques, new calculation methods and computer programs, environmental geotechnics. DSML provides the following facilities: field testing equipment, a submersible working chamber capable of working in water depths of 200 metres, laboratory testing equipment, computerized data acquisition and data processing, model studies both in the field and in the laboratory, computer facilities, development of new field and laboratory testing equipment, development of mathematical models and computer codes, a library. Moreover, DSML has the disposal of a wave flume for soil mechanics–hydraulics research and a geotechnical centrifuge (operational in 1985).

Total staff: 260

Professionally qualified staff: 150

Director or Chief Executive

Managing Director: Ir W. Bokhoven

Deputy Managing Director: Ir E.H. de Leeuw

Assistant Director: Ir W.J. Heijnen

Heads of department

Operations Division: Ir E.H. de Leeuw

Development Division: Drs J.D. Nieuwenhuis

Foreign Division: Ir M.F.J.M. Kroezen

Principal sources of finance
Geotechnical field investigations; geotechnical laboratory testing; consultancy; research
Principal related bodies
The Board of the foundation Stichting Waterbouwkundig Laboratorium exercises supervision in the management of DSML.
Publications
- LGM-Mededelingen, twice a year; Delft Soil, twice a year
- LGM Abstracts, bi-monthly.

NLD.353
Delft University of Technology, Department of Civil Engineering
Technische Hogeschool Delft, Afdeling der Civiele Techniek (AfdCT THD)
Stevinweg 1, Postbox 5048, 2600 GA Delft
Telephone: 015 785919
Telex: 38070
Cables: BITHD
Date of establishment
1842

A department of Delft University of Technology engaged in education and research in civil engineering. The Stevin Laboratories form part of the Department of Civil Engineering.
Main fields of work
Training for civil engineers and research relating to civil engineering:
- On Structural analysis, building materials and construction elements (wood, steel, concrete and other materials)
- In the field of environmental engineering, with special emphasis on the purification of drinking water and the treatment of sewage and industrial wastes
- On traffic behaviour and traffic flow
- Static and dynamic loadings on road and railroad constructions
- In the field of fluid mechanics, river, harbour and coastal engineering.

Total staff: 435
Professionally qualified staff: 38 professors
Director or Chief Executive
Manager of the Department: Dr G.J. van Alphen
Principal sources of finance
The Dutch Government.
Publications
- Annual report on the research activities of the department
- In co-operation with IBBC Institute TNO for Building Materials and Building Structures, Rijswijk: Delft: Periodical Heron; individual: reports, publications.

NLD.354
Delft University of Technology, Institute for Town Planning Research
Technische Hogeschool Delft, Instituut voor Stedebouwkundig Onderzoek (ISO)
Berlageweg 1, Delft
Telephone: 015 781088
Date of establishment
1959

A research institute, one of the three research organizations within the architecture Department of Delft University of Technology (the two other organizations being Centrum voor Architectuuronderzoek and RIW-Instituut voor Volkshuisvestingsonderzoek). Established as a joint venture of the Department of Architecture and the Department of Civil Engineering.
Main fields of work
Various fields of physical planning, from the aspects of both the technical and social sciences. Some examples: sociological research in different urban areas, analysis of statistical data concerning housing and changes in population structure, studies on parking problems in town centres and on the methodology of transportation planning, investigations in relation to physical aspects (wind, noise) of town planning.
Total staff: 13 (7 scientific from several disciplines).
Director or Chief Executive
Director: Dr J. den Draak
Principal sources of finance
Financed through funds of the University (Ministry of Education and Sciences).
Publications
There are four series of publications:
- Reports (to date 44)
- Memoranda (to date 24)
- Reprints (to date 26)
- Periodic information about new publications, mostly twice a year.

NLD.355 FM
Directorate General of Housing
Directoraat Generaal van de Volkshuisvesting
Boerhaavelaan 5, Zoetermeer
Postal address: Postbox 3001, 2713 KA Zoetermeer
Telephone: 079 279111
Telex: 34429 VORO NL
Date of establishment
1946

The National Housing Agency of the Ministry of Housing, Physical Planning and Environment. Part of the national governmental administration.
Main fields of work
- Preparation and evaluation of housing policy
- Implementation of housing legislation
- State Inspectorate of Housing
- Research and development Housing
- State aid for housing.

The Directorate has branch offices in each of the provinces.
Total staff: approximately 1340
Director or Chief Executive
Director General/Inspector General of Housing: Ir J.M. Koopman
Heads of department
Office of the Inspector General: Dr A. Kampschöer
Directorate of Internal Management: H. Scheerder
Directorate of Policy Coordination and Finances: Dr R. de Haan
Directorate of Dwelling Production and Urban Renewal: A.H.A. Lensen
Directorate of Research: Dr C.M.I. Richter
Directorate of State Aid and Rent Policy: Dr A. Steensma
Principal sources of finance
State Budget.
Principal related bodies
Other parts of the Ministry such as: Directorate of Building (DCB); Directorate General of the Environment (DGMH); Directorate General of Physical Planning (RPD); Central Planning Agency; Netherlands Standardization Institute; Housing Experiment Advisory Board; Research Advisory Board (URA OGO)
Publications
- early housing planning and budget
- Annual report (Jaarverslag)
- *Research reports:* at irregular intervals on various items
- Ministry Bibliographies/Catalogues.

NLD.356

Dutch Cement Industry Association
Vereniging Nederlandse Cementindustrie (VNC)
Postbus 3011, 5203 DA, 's-Hertogenbosch
Telephone: 073 401286 or 073 401287
Telex: 50185
Date of establishment
October 1982

VNC takes care of the common interest (in non-competitive areas) of its members.
Main fields of work
Market development, promotion, dissemination of information on the applications of cement and concrete.
Total staff: 21
Director or Chief Executive
C.F.A. Wolterbeek
Principal sources of finance
Financed by ENCI and CEMIJ, the two Dutch cement manufacturers.
Publications
- *Journal:* Cement, monthly; Betoniek, monthly; Betonwegennieuws, quarterly; Agrabeton, quarterly; Brandbestand, bi-monthly
- The publishing department publishes books on concrete technology, design and construction
- The Library, which covers the same subjects, publishes an accession list.

NLD.357 **AM**

Economic Institute for the Building Industry
Stichting Economisch Instituut voor de Bouwnijverheid (EIB)
Cronenburg 150, 1081 GN Amsterdam
Telephone: 020 429342
Date of establishment
1956

An independent institute for scientific research. Established as a private foundation.
Main fields of work
According to the Articles of Association the aim of the EIB is to promote, in an independent and scientific way, the knowledge of economic and social problems and phenomena, occurring in, or of importance to, the building industry in the Netherlands, in the broadest sense. Consequently the EIB conducts research into economic and social aspects of building production and the building industry, also paying attention to the relevant technical development. In several cases research is done in co-operation with other institutions. In the field of work the following topics can be distinguished, each of which offers a number of subjects for research: The building industry in the national society; The structure of the building industry; The building market; Costs and prices; Building production; The built environment; Public authorities and building; Other subjects.
Total staff: 36
Professionally qualified staff: 22
Director or Chief Executive
Director: Dr A.P. Buur
Heads of department
Department of the Building Market and Housing: Dr J.B.S. Conijn
Department of the Labour Market and Organizations: Dr J. Veen
Computer Department: Dr A.J. Grootenboer
Principal sources of finance
Financed partly from a Fund for Education and Development in the Building Industry, partly by other subsidies and by payment for work carried out on behalf of interested parties.
Publications
- *Journal:* Bouw/Werk, bi-monthly
- A list of publications is available: most of the publications are in Dutch only.

NLD.358

Eindhoven University of Technology, Department of Architecture, Building and Planning, Materials and Structures Laboratory
Laboratorium voor Materialen en Constructies
PO Box 513, 5600 MB Eindhoven
Telephone: 040 479111
Telex: 51163
Date of establishment
1971, area of investigations increased January 1983.

The Laboratory is part of the Building Construction Group of Eindhoven University and performs research and contract research.
Main fields of work
Materials investigated: stones, masonry, timber, metals, concrete, synthetic materials, binders, waste materials.

Specialities: concrete mechanics; behaviour of concrete – cold formed steel – and wood constructions; mechanics of structures; chemical analysis; corrosion; model tests; fracture mechanics – testing equipment – offshore structures – environment, pollution, recycling, physical analysis, surface chemistry/physics; development of new constructions and building products.

All research is bundled under the head: optimization, innovation and re-use of building structures and materials.
Total staff: 30
Professionally qualified staff: 30
Director or Chief Executive
Professor Ir P.C. Kreijger; Professor Ir A.J. van Neste
Principal sources of finance
Government.
Publications
- *Technical standards:* Co-operation with Dutch Standards, CEN and ISO
- *Research reports:* 3–8 per year, partly in professional or scientific journals and for symposia and congresses; in addition there is a series of research reports for external use and a second series of reports for internal use.

NLD.359

Foundation for Architects' Research
Stichting Architecten Research (SAR)
Building 0, Eindhoven University, Postbus 429, Eindhoven
Telephone: 040 43 36 16
Date of establishment
1965

Operations and research office.
Main fields of work
- Built environment with emphasis on housing
- Development of design methods with emphasis on opening up and restructuring the decision making process
- Main aim: to study measures in the decision making process and the production process that bring about user control

Much effort by the research office goes into promoting its ideas and methods, trying to create a better understanding and a better context. The SAR represents a movement as well as research. The significance of SAR is that its research fits into a distinct ideology: bringing about user control. This gives the research results a large degree of coherence.
Total staff: 5
Director or Chief Executive
Chairman of the Foundation: Mr C.M. van den Hoff
Director of the Research Office: Ir J.C. Carp
Principal sources of finance
Sponsors (architects, builders, manufacturers, clients and Eindhoven University): one half budget; Government: one third budget; consultancy, courses, committees etc.: one sixth budget.

Publications
Some are published in English.
English language quarterly: 'Open House'.

NLD.360

Foundation for Paint Application
Stichting Verftoenpassing (SVT)
Postbus 71, Wassenaar
Telephone: 01751 7 80 44
Date of establishment
1956

Foundation created by the Bedrijfschap Schildersbedrijf, BS (organization of painters and painters – contractors) and the Vereniging van Verf en Drukinktfabrieken, VVVF (Organization of paintmakers).
Main fields of work
To promote and finance research by specialized laboratories on paint application.
Total staff: 2
Principal sources of finance
Financed by BS and VVVF and subsidized by other interested bodies.
Publications
In professional journals and circular letters to members of the organizations co-operating in the Foundation.

NLD.361

Foundation for Quality Statements, Organization for Building Materials and Building Components (KOMO)
Stichting Kwaliteitsverklaringen-Organisatie voor Materialenen Onderdelen voor de bouw (KOMO)
Sir Winston Churchill-laan 273, Postbox 240, 2280 AE Rijswijk ZH
Telephone: 070 994241
Telex: 32480
Date of establishment
1967

An independent institute, co-operating with testing institutes on certification and attestation on behalf of building and civil engineering and at the disposal of all parties concerned.
Main fields of work
KOMO activities range over all materials, products, components and constructions for building and civil engineering. The aim is to promote the availability of objective quality statements (Certificates, Agréments, approval documents, etc.). These documents inform the parties involved of the results of controls, assessments, tests and other activities based on standard requirements and criteria, which have been executed, on behalf of KOMO, by approved testing institutes under the supervision of quality control committees. The rules governing the application for, the issuance of and the maintenance of a quality statement are laid down in the Regulations for Quality Statements.
Total staff: 25
Director or Chief Executive
Chairman of the Board: Dr C.J. Verplanke
Director: Ir F. Wagenmaker
Deputy Director: Ing H.J. Verlinden
Heads of department
Administration: B. Ramrattan
Public Relations: A. van Holst
Quality Statements: Ing H.J. Verlinden
Secretariat: Mrs A.A. van Schaik
Principal sources of finance
Fees paid by industry for quality statements; fees paid by purchasers for batch inspection on demand; sale of publications; government subsidy
Principal related bodies
Board and Executive Board; Central Committee for Quality Statements: co-ordinates all activities concerning the realization of quality statements; Advisory Groups: each co-ordinates the activities on a certain group of materials; Board of Appeal – Assay committee
Publications
Journal: KOMO Guide: Part 1 contains general information on organization, quality statements and activities, published once a year; Part 2 contains specified summaries of quality statements issued, published three times a year; KOMO News: A bulletin with information on new activities, committees, publications, standard requirements and criteria, published three times a year.
Technical standards: Requirements other than official NEN standards and directives; Criteria.

NLD.362

Stichting Instituut voor Studie en Stimulering van Onderzoek op het gebied van verwarming en luchtbehandeling (ISSO)
Weena 740, Rotterdam
Postal address: PO Box 20740, 3001 JA Rotterdam
Telephone: 010 14 61 16
Telex: 22530 bouwc nl
Dates of establishment
20 April 1974

ISSO is a private institute whose aims are: to publish guidelines for designers and installers of heating, ventilating and air-conditioning systems as far as necessary in relation to building design; to stimulate and co-ordinate and carry out studies to make publication possible.

It aims to carry out research on problems in the field of systems providing indoor climate by heating, cooling, ventilating or air-conditioning treatment and related techniques in the widest sense. Apart from co-ordinating and stimulating the necessary studies and research projects the most important goal is to publish technical guidelines, design rules and/or standards for use in practice by designers, installers, consulting engineers and industry.
Main fields of work
- Calculation methods for the design of heating and cooling systems: cooling load calculation for buildings (developing a Dutch standard); heating demand calculation for buildings
- Design of heating systems: floor-heating systems (design, calculation method, guidelines for installations); pipelines for distance heating; air-heating systems (design, calculation, guidelines, etc.); central heating systems for space heating and domestic water; radiators and convectors
- Design of solar systems for domestic waterboilers
- Design of heat recovery systems for buildings
- Ventilation of buildings: minimum fresh air supply
- Indoor climate in buildings; standard indoor conditions
- Energy consumption of buildings: development of energy calculation methods for building, e.g. houses, office buildings, schools, hospitals, etc.
- Building properties and outdoor climate: to provide useful information for the above calculation methods, attention is paid to: *U*(k) and *R*-values of building constructions; ZTA-values of window systems
- Solar data of the Dutch climate
- Short reference year of weather conditions of the Dutch climate for energy calculations.

Total staff: 3
Director or Chief Executive
Director Ir J. Hogeling
Heads of department
Board of Technical Advisers:
TVVL: Ir W.H. Knoll (President)
Director ISSO: Ir J. Hogeling (Secretary)

Technical University of Delft: Professor Ir A.W. Boeke
TNO Institutes: Ir J. Claus
ONRI (Organization of Consulting Engineers): Ir J.H.L. Crooijmans
Technical University of Eindhoven: Dr Ir J.T.H. Lammers, Ir H. Meijer, Professor Ir J.K. Nieuwenhuizen
Technical Board T V V L: Ir W. Schuringa
Technical Dept. ACI: Ir E.J. Wagenaar
Installer CBVI: Ing C. van Vliet

Principal sources of finance

The three founders of ISSO: Algemene Vereniging voor de centrale verwarmings- en luchtbehandelingsindustrie (ACI); Stichting Centraal Bureau voor Verwarmingsindustrie (CBVI); Nederlandse Technische Vereninging voor Verwarming en Luchtbehandeling (T V V L); subsidies from governmental resources, industry and private institutes for particular research projects.

Principal related bodies

ACI, CBVI and T V V L have founded ISSO and are responsible for the policy of ISSO. The Board of the foundation is responsible for policy. The Board of Technical Advisers is responsible for technical work.

Publications

- *Technical standards:* ISSO publications (14)
- *Research report:* (1)

A list of ISSO publications is available.

NLD.363 AM

Institute for Housing Studies BIE (IHS)
PO Box 20718, 3001 JA Rotterdam

Telephone: 010 116181
Telex: 24548 ihs ne
Date of establishment
1958

Initially called Bouwcentrum International Education, IHS was established as a centre aiming at the transfer of housing techniques and technology rooted in the Netherlands and other developed countries to professionals from the Third World.

In the early 1970s, BIE began to move gradually towards approaches adapted to particular conditions in developing countries, with a focus upon the housing problems of low-income groups.

In 1983 BIE became an independent institute under the name of the Institute for Housing Studies (IHS).

Main fields of work

Each year IHS runs two 5 month courses in Rotterdam with different workshops covering the fields of Housing, Planning, Policy, Design and Technology for low-income families. Fellowships for the courses are granted by the Netherlands Ministry of Foreign Affairs, the EEC, the United Nations, the World Bank and other institutions. Many of IHS's activities, however, take place in Third World countries where IHS assists governments and institutions in setting up action-orientated housing programmes, and carries out and promotes research and advisory services.

The establishment of training and research centres (to date in Indonesia, Sri Lanka, Tanzania, Thailand and Colombia) has helped to increase the relevance and scope of IHS's work. These programmes focus directly on the housing problems of the countries and regions in which they are located. This has resulted in a better match between course content and specific training needs, and a move away from the generality of approach implicit in a course taught in Rotterdam.

Indeed the Rotterdam courses are revitalised by the experience and new knowledge gained by staff members in regional projects. The organization and support of local research networks in Third World countries has also contributed to this improvement.

Total staff: 25
Professionally qualified staff: 13
Director or Chief Executive
Director: Ir Cor Dijkgraaf

Heads of department

Regional Educational Programmes: Paul Baross
International Course on Housing, Planning and Building: Henk van den Born
Research Programmes: Edgardo Martinez
Consulting Services: Michael Rodell

Principal sources of finance

Dutch Ministry of Foreign Affairs.

Principal related bodies

Board.

Publications

- *Journal:* IHS Bulletin: Blue Series; Red Series, at irregular intervals
- IHS Staff Papers
- IHS Forum
- Annual Report.

NLD.364

National Housing Association
Nationale Woningraad (NWR)
Markenlaan 1, Postbox 50051, 1305 AB Almere

Telephone: 03240 91911
Telex: 76391 NWR H NL
Date of establishment
1913

The NWR is a master organization of housing associations. The number of members is 700 and these associations together own about 1.1 million dwellings.

Main fields of work

The head office of the NWR is situated in the new town of Almere. There are another five offices in different areas of the country. The NWR serves the interests of the member local housing associations where the national government is concerned. The Association exerts influence in many situations by conferring with other organizations. The main fields for advising and serving the member housing associations are: management and organization development; financial aspects of social housing; automation; technical aspects.

The NWR arranges training programmes for the employees, etc. of the member housing associations to enable them to function accurately.

Total staff: approximately 500
Director or Chief Executive
General Director: Dr B.G.A. Kempen
Director: N. van Velzen

Principal sources of finance

Financed through subscriptions from the housing associations and through fees for commissions.

Publications

- Woningraad, a newspaper which appears twenty times a year
- Woningraad-Extra, in edition which contains extensive information about topics in tho field of social housing

Various other publications.

NLD.365 **AM**

NBD Product-Information Systems BV

Stromarkt 17a, PO Box 23, 7400 GA Deventer

Telephone: 05700 10844

Telex: 49295 kludv

Date of establishment

1976

The NBD is a division of the publisher group Kluwer NV, and publishes the Netherlands Building Documentation system, annually distributed, updated and free of charge among architects, contractors and other bodies concerned with the choice of building products.

The information system consists of directly comparable general and product information of the building industry on three connected information levels that correspond to the phases of the construction planning process.

Main fields of work

The goal of the NBD is to be an intermediary between the producer and the user of building products.

The principal divisions are the editorial and sales departments.

Total staff: 17

Professionally qualified staff: 6

Director or Chief Executive

Director: F.R. Thedieck

Principal sources of finance

Participants whose product information is published in the NBD system.

Principal related bodies

Union of Dutch Architects (BNA) which through an editorial board authorizes the technical texts before publication.

Publications

- NBD system, consisting of 13 binders and an index
- Product information in connection with building projects for physically handicapped persons
- Holland Products: published annually, with product information about Dutch industrial goods and services suitable for export. Each product is presented on two pages. The left page gives a visual presentation and the right is text containing technical and commercial information about the displayed products. Holland Products will be distributed via the official Dutch representatives abroad (such as Embassies, Chambers of Commerce and Banks) in more than 100 countries. The first edition consists of 10 000 copies.

NLD.366

Netherlands Committee for Research, Codes and Specifications for Concrete (CUR–VB)

Stichting voor Onderzoek, Voorschriften en Kwaliteitseisen op het Gebied van Beton (CUR–VB)

Bredewater 26, Postbox 61, 2700 AB Zoetermeer

Telephone: 079 219313

Date of establishment

CUR–VB was formed in 1979 as a result of the amalgamation of CUR (Netherlands Committee for Concrete Research), which had been concerned with scientific investigations of concrete since 1952, and the CVB (Committee on Codes for Concrete), which had been responsible, in various forms of organization, for the preparation and drafting of the Netherlands codes and specifications relating to concrete since 1912.

The CUR–VB is an independent organization which operates under the joint responsibility of the Department of Building and Hydraulics of the Royal Institute of Engineers and the Netherlands Concrete Society, to which it is an affiliated body.

Main fields of work

The activities of the CUR–VB have two important aspects: the promotion and effecting of technical and scientific research of concrete and the drafting of codes, guidelines and specifications for concrete.

For the research projects, the following groups can be distinguished: Theoretical problems, Materials, Practical applications, Behaviour of the structure and Economy of concrete constructions.

Work on codes comprises developing national basic codes and standardizing specific items in the fields of concrete, such as concrete bridges, piles, etc.

The drafting of codes is undertaken in close collaboration with the Netherlands Standards Institute (NNI). When a draft code of practice has been completed, it passes through an approval procedure, in the course of which there is opportunity for critical comment. After this it is published as a Netherlands Standard (NEN) by the NNI. The method of working is usually through research or code committees made up of experts in the relevant area.

In recent years there has been a marked expansion in research projects connected with the offshore industry.

In 1983 a total of 30 research and code committees were operative, amounting in all to 300 members. For the greater part these members gave their services without renumeration.

The activities of CUR–VB are directed by a seven-member Executive Committee whose policy is carried out by a professionally staffed office comprising technical, scientific and administrative personnel.

The Executive Committee is assisted by an Advisory Council while financial policy is monitored by a Supervisory Council comprising representatives of the financing bodies.

Director or Chief Executive

Director: A.C. van Riel

Secretary: L.L. Kamminga

Technical Secretary: H.P.J. Vereijken

Principal sources of finance

Finance is provided partly by industry (of which a major share comes from the cement industry) and partly by government.

Principal related bodies

Betonvereniging (Netherlands Concrete Society): CUR–VB is an affiliated body; NNI (Netherlands Standards Institution): the body through which codes prepared by CUR–VB are published as Dutch Standards (NEN); Royal Institute of Engineers, Department of Building and Hydraulics, under whose joint responsibility CUR–VB operates.

Publications

- *Research reports:* The results of each research project are published as a CUR–VB report. Well over 100 have already appeared. Each is accompanied by a lengthy summary in English and the majority also have German and French summaries
- In general work on Codes results in the publication of Dutch Standards (NEN)
- On occasion certain new developments are not considered to have reached a level justifying the issue of a standard. In such cases they form the subject of CUR–VB Recommendations and are published in the trade press.

NLD.367

Netherlands Institute for Physical Planning and Social Housing

Nederlands Instituut voor Ruimtelijke Ordening en Volkshuisvesting (NIROV)

Van Speijkstraat 25, 's-Gravenhage

Telephone: 070 46 96 52

Other address: Library: Willem Witsenplein 6, 's-Gravenhage

Telephone: 070 24 69 96

Date of establishment

1917

A private professional organization for physical town planners and all those who are concerned with planning and housing in general (4000 members).

Main fields of work

Aim: to improve by all means physical planning and social housing conditions. The most active members are organized in sections. NIROV has many representatives and advisors in working groups, ministerial commissions, boards of universities, etc.

The Institute gives information to members and non-members about planning and housing, and arranges congresses (4/5 per year), educational courses, research, post-academic activities, discussion groups via the sections and has dealings with the relevant Ministry.

Total staff: Around 14 persons 50% of whom have a scientific background.

Director or Chief Executive

President: Ph. J.I.M. Houben

Secretary-director: Dr A. Everts

Principal sources of finance

30% by contributions; 70% by State subventions.

Publications

Reports on research work.

NLD.368 **AM**

Philips International BV Building Design and Engineering Division

Philips International BV Architecten- en Ingenieursbureau (AIB)

Cederlaan 4, 5616 SC Eindhoven

Telephone: 31 40 734500

Telex: 35000 PHTC/NLUEVBO – OAN 4a

Other addresses:

Inter Engineering BV, Cederlaan 4, 5616 SC Eindhoven

Intarco (International Architects and Consultants) BV, Cederlaan 4, 5616 SC Eindhoven

Date of establishment

1925; the present organizational form dates from 1958.

A division of Philips International BV employing architects, civil, mechanical, electrical, chemical and pollution control engineers, quantity surveyors and draughtsmen.

Main fields of work

The Division provides fully integrated designs (architectural, civil, mechanical and electrical) for all industrial plant buildings, laboratories and office buildings.

To keep abreast of the latest developments on the work-environment in factories and office buildings its specialists carry out the appropriate investigations and research.

Through its subsidiaries the same services can be offered to third parties; these services range from making preliminary studies, consultancy and project management to contracting turnkey projects in civil, mechanical and electrical engineering. The two subsidiaries of the Division: Intarco BV and Inter-Engineering BV provide the same professional services including the contracting of turnkey projects to third parties. The head office is in Eindhoven, the Netherlands; there are branch offices in Paris, Jakarta, Sao Paulo and Hong Kong.

Total staff: approximately 325

Director or Chief Executive

J. Hollander and H.J. Bordes

Principal sources of finance

Financed by fees received for orders and services rendered.

Publications

Papers are published in professional journals. These papers may describe novel designs or deal with developments and research in the field of work environment in factories.

NLD.369 **FM**

Planning Committee for Building Research TNO of the Netherlands Organization for Applied Scientific Research

Plancommissie Bouwresearch TNO van de Nederlandse Organisatie voor Toegepast Natuurwetenschappelijk Onderzoek (PCBR TNO)

PO Box 238, 2600 AE Delft

Telephone: (0) 15 60 60 00

Telex: 38270 ibbc nl

Date of establishment

1956

The Planning Committee co-ordinates the research activities of those TNO divisions and institutes partly or entirely engaged in building and living research which comprise not only dwellings and buildings but also civil engineering, offshore construction, environmental aspects, energy issues and interactions with traffic, working conditions, physical planning and economy.

The Committee's aims are:

- To present an overall insight into the joint TNO activities in building and living research
- To co-ordinate the tasks
- To ensure that building research is put at the service of society
- To ensure that expertise, knowhow and capacities present within TNO are utilized in an efficient way
- To deal with financial aspects of projects and working groups

Since 1977 the Committee has had at its disposal a small office, which acts as a focal point for access to all TNO's building research activities.

Main fields of work

- Institute of Applied Physics TNO–TH: thermal problems, noise insulation, acoustical problems, glass
- Research Centre for Physical Planning TNO: research into planological problems
- TNO Institute for Building Materials and Building Structures: building materials, building construction, building physics, concrete structures, experimental mechanics, steel structures, software engineering, dynamic problems, Technical Centre for Fire Prevention
- Division of Technology for Society, Heat and Refrigeration Engineering Department: heating and refrigeration techniques, ceramics, wind problems
- Forest Products Research Institute TNO: timber technology, materials science relative to timber
- Plastics and Rubber Research Institute TNO: plastics and rubber technology, development of products and structures in plastics
- Metal Research Institute TNO: welding technique, mechanics of brittle fracture corrosion and corrosion protection, surface technology
- Paint Research Institute TNO: paint applications, surface technology
- TNO Research Institute for Environmental Hygiene: indoor climatology, noise, lighting, sociological and socio-psychological research in the field of living

- Netherlands Institute for Preventive Health Care TNO: psychohygienic problems related to housing
- Prins Maurits Laboratory TNO, Institute for Chemical and Technological Research: physics of explosions, risk analysis.

Director or Chief Executive
Chairman: F.K. Ligtenberg
Vice-Chairman: P. Vermeyden
Secretary: H.C. Kortenhoff
Members: J. Claus, J.V. Witteveen, F.B. de Walle, J.W. Oldeman, F.P.M. Vonk, G.J. Kleinhoonte van Os and H.J. Pasman
Head of the Office: G.H.C. Kortenhoff
Heads of department
Institute of Applied Physics TNO–TH: J. Delcour
Research Centre for Physical Planning TNO: F.P.M. Vonk, J. Witteveen
TNO Institute for Building Materials and Building Structures: F.K. Ligtenberg
Division of Technology for Society, Heat and Refrigeration Engineering Department: J. Claus (Deputy Director)
Forest Products Research Institute TNO: G.J.N. Oldeman
Plastics and Rubber Research Institute TNO: L.C.E. Struik
Metal Research Institute TNO: B. Dane
Paint Research Institute TNO: H.J. de Vries
TNO Research Institute for Environmental Hygiene: F.B. de Walle
Netherlands Institute for Preventive Health Care TNO: W.M. J. van Duyne
Prins Maurits Laboratory TNO, Institute for Chemical and Technological Research: A.J.J. Ooms
Principal sources of finance
A government subsidy determined each year by the Minister of Finance subject to approval by Parliament; grants and contributions from research organizations and industry; contract research.
Principal related bodies
Institutions co-operating in terms of specific building matters: Delft Hydraulics Laboratory; Delft Soil Mechanics Laboratory
Publications
- *Journal:* Heron, a technical journal issued by the TNO Institute of Building Materials and Building Structures; Plastica, issued by the Plastics and Rubber Institute TNO
- Numerous papers published in professional and scientific journals
- Numerous research reports
- Reports published by co-operating research associations, e.g.: Committee for Research of the Concrete Association (CUR–VB); Building Research Foundation (SBR); Structural Steelwork Association (SG); Foundation for Paint Application (Stichting Verftoepassing); United Pipeline Administrators (Verenigde Pijpleiding Beheerders).

NLD.370 AM

Progresbouw
Vereniging Industriële Woningen Utiliteitsbouw
Beneluxweg 8, 2411 NG Bodegraven
Postal address: PO Box 252, 2410 AG Bodegraven
Telephone: 01726 19244
Telex: 39534
Date of establishment
1955

To promote a market for, and supply of, products for house building, other buildings (including factories, plants, schools, offices, public buildings, hospitals, sport and entertainments) and to assist its members in their activities in relation to these.
Main fields of work
- To promote the production of efficient building products with a high level of quality and relatively low costs
- To influence the Government so that its building policy contributes to efficient production in the building industry
- To promote unification and harmonization of building regulations in such a way that higher quality can be obtained at limited costs
- To promote rationalization of building components by modular co-ordination
- To influence international developments in the building sector on behalf of the interests of members
- To make clear the advantages of industrialized building products, especially as regards good quality and costs
- To supply information about industrial developments in the building industry and building market for the products of members of the association.

Total staff: 8
Director or Chief Executive
Chairman: Ir N.A.J. Hooijmaaijer
General Secretary: Dr P.M. van Onna
Publications
- ProgresProjecten, twice yearly
- ProgresExpres, six times a year
- Annual review
- Research reports
- List of research publications.

NLD.371

Research Centre for the Sandlime Brick Industries
Stichting Research Centrum Kalkzandsteenindustrie (RCK)
Ambachtsweg 10, Postbus 145, 3770 AC Barneveld, 3771 MG Barneveld
Telephone: 03420 13643
Telex: 47873
Date of establishment
1 November 1960

Research and development of building materials.
Main fields of work
Research centre for developing new products, mechanical handling and application of building materials. There are three principal departments: Mechanical; Physical; and Chemical.
Total staff: 14
Director or Chief Executive
Deputy Director: Ing A. Noorlander
Principal sources of finance
Research and other specially commissioned reports.
Principal related bodies
Advisory Committees; Standards Committees.
Publications
Internal publications only and specially commissioned reports.

NLD.372

Research Institute for Labour Economy in the Building Trade (SAOB)
Stichting Arbeidstechnisch Onderzoek Bouwnijverheid (SAOB)
Stationsweg 89, 6711 PM Ede
Telephone: national 08380 16421; international 0031
Telex: 35442 TTS NL SAOB
Date of establishment
1 May 1954

A private institute with 100 members: contractors, architects, manufacturers, installation engineers, designers, government departments, educational establishments, housing corporations, maintenance departments and consultants.
Main fields of work
- For the benefit of the building industry as a whole: to analyse the

production process on the building site and to gather basic times standards and cost selection data for every stage in the building process (execution research)

- To supply work study data and technical job execution data in order to assist the design of buildings and testing of designs against the possibilities of execution (product development)
- To supply the information for analysing the building process from its initial conception to final completion and acceptance, to establish the relevant decision-making criteria and to formulate rules for decision-making so as to enable the building process to be controlled (process study).

The principal departments are: Investigation of Productivity; Research and Innovation; Dissemination of Knowledge.

Professionally qualified staff: 10

Director or Chief Executive

Director: C.P. Verschuren

Principal sources of finance

Annual fees of the members; investigations commissioned by members, the Ministry of Housing, Building Research Institute, the Concrete Association, housing institutes and others.

Publications

- *Journal:* SAOBJECTIEF, quarterly, free of charge
- *Technical standards:* Standard Times for Building and Installation Operations, vols 1–5 plus supplements

A List of Publications is available.

NLD.373

RIW Institute for Housing Research

RIW Instituut voor Volkshuisvestingsonderzoek

Nieuwelaan 76, 2611 RT Delft

Telephone: 015 783049, 783046

Date of establishment

February 1969

The RIW is an intrinsic part of the Faculty of Architecture of the Delft Technical University. The Institute is active as a research group as well as a teaching group. The aims are:

- To research, as much as possible from the standpoint of the dweller, fundamental and applied, in the field of housing
- To publish the results of the research, to participate in the scientific and social discussions with regard to building and living
- To work on, and to maintain, forms of co-operation with similar institutes, persons or groups, who are working in the same scientific field
- To integrate the results of the research in the teaching programme of the Faculty of Architecture, Delft Technical University, especially in the programme for the degree in Housing.

Main fields of work

Research on four main themes:

- Quality and costs of building in social housing
- The housing situation of specific groups: special interest for elderly people, ethnic minorities and group housing
- The price of housing and national policies on this point: the subject is related to the rent and subsidy policy, and the division of the costs of housing over different income groups
- The Dutch housing system: this subject has in recent years focused on forms of tenure, local housing market problems, decentralization.

Total staff: 22

Professionally qualified staff: 17

Director or Chief Executive

Director: Dr P.P.J. Houben

Principal sources of finance

The normal flow of money for institutions of academic education; subsidies from institutions for whom the RIW is executing research programmes.

Principal related bodies

Faculty of Architecture and several working groups of that Faculty. The Faculty appoints the Board of the RIW, the Board consisting of five members, two from the Faculty, and three from the Institute. The Director of the RIW is adviser to the Board, and executes the decisions.

Publications

Articles, brochures, research memoranda.

Dokumentatie woningbouw (P. Groetelaers, H. Priemus en F. Wind red.), Aanvulling 29, P. Groetelaers, Samson, Alphen a/d Rijn 1981, 68 blz., RIW – Instituut voor Volkshuisvestingsonderzoek Delft

Dokumentatie Stadsvernieuwing (P. Groetelaers, H. Priemus en F. Wind red.), Aanvulling 9, P. Groetelaers, Samson, Alphen a/d Rijn 1981, 68 blz., RIW – Instituut voor Volkshuisvestingsonderzoek Delft

Groetelaers, P., H. Priemus, H.M.J. Bonsang en F.H. van Gigch, Wooneenheden: woonkwaliteit en woonlasten. Analyse en dokumentatie van huisvesting voor alleenstaanden en tweepersoonshuishoudens, Staatsuitgeverij, 's-Gravenhage 1980, 432 blz., RIW – Instituut voor Volkshuisvestingsonderzoek Delft

Groetelaers, P., Een toekomst voor zeggenschap van bewoners bij cascobouw, Cahier nr. 1 van de Initiatiefgroep Doe-het-zelf-huisvesting, Amsterdam 1982

Lijbers, R. en S. Pasteuning, Vervangende Nieuwbouw, Vakgroep 8, Delftse Universitaire pers, 1982

Houben, P.P.J., G.C. Jense, W. Lans, P.H.F.M. Moeskops, H.W.M. Toneman en F. Wind, Ouderen en hun woonbehoeften, Delft 1981, 124 blz., RIW – Instituut voor Volkshuisvestingsonderzoek Delft

H. Moeskops, F. Wind, P.P.J. Houben en G.C. Jense, Deskundigen over woonbehoeften van ouderen, RIW, Delft 1981, 101 pag.

F. Wind, H. Moeskops en P.P.J. Houben, Opzet fase 3 en volgende van het landelijk onderzoek naar behoeften van ouderen aan woon- en andere voorzieningen, RIW, Delft 1982

Houben, P., H. Moeskops en F. Wind, Beeldvorming van het wonen van ouderen, Delft 1982

RIW, Buurtonderzoek volkshuisvesting etnische minderheden, Delft, Februari 1981

W. Lans. Onderzoeksvoorstel buurtonderzoek volkshuisvesting etnische minderheden, pilot study, RIW, Delft, September 1981

Lans, W. en L. van Duyvendijk, Locatiekeuze buurtonderzoek volkshuisvesting etnische minderheden, RIW, Delft 1982, 27 blz

Brabers, P., H. Moeskops en H. Westra, De achterkant van Van Dam, Delft 1982

J.H. Kroes, Funding housing through local Government paper for Housing seminar on structural revisions in government's housing role van het Lincoln Institute/Wayne State University te Detroit (USA), Detroit 1981

Kroes, J.H. en F. Ymkers (red.), Nieuwe beheersvormen, Delft, 1980, 112 blz., RIW – Instituut voor Volkshuisvestingsonderzoek Delft

Centrale Directie van de Volkshuisvesting, Directie Onderzoek, Nieuwe beheersvormen, verkorte versie, Ministerie van Volkshuisvesting en Ruimtelijke Ordening, 's-Gravenhage 1981

J. van der Schaar, Non-profit verhuur- en volkshuisvestingsbeleid in: Sectie Volkshuisvesting van het NIROV, De non-profit huursector, nu en in de toekomst, 's-Gravenhage 1981

T.K. Hübner en J. van der Schaar, De functie van de non-profit huursector op de woningmarkt, in: Sectie Volkshuisvesting van het NIROV, De non-profit huursector, nu en in de toekomst, 's-Gravenhage 1981

RIW, Verslag van de NIROV–RIW studieconferentie 80 jaar Woningwet, volkchuisvestingsbeleid en verzorgingsstaat, Delft 1982, 223 blz.

Kroes, J.H. en F. Ymkers, Buitenlandse vormen van woningbeheer, RIW Delft 1982

Ymkers, F., Buitenlandse vormen van woningbeheer, Verslag studieconferentie, RIW, Delft, 62 pag.

Groetelaers, P., H. Priemus, J. van der Schaar en H. Westra, Woningbouw en volkshuisvesting (2 delen: collegedictaat en plandocumentatie), Delft, 1982

Schaar, J. van der, Housing costs in the Netherlands: trends in the period 1950–1980 and the distribution of housing costs in 1977, Delft 1982

Brouwershaven, R.P. en D. Reitsma, De Vijf Vinger Theorie en Doorstroming (scriptie), RIW – Instituut voor Volkshuisvestingsonderzoek en Geografisch Instituut Rijksuniversiteit Utrecht, Delft 1982

Kroes, J.H. en L. Vulperhorst, Bouwen voor de eigen woningbehoefte in de provincie Utrecht, RIW, Delft, 298 pag.
Schaar, J. van der, Woningbezit in Dordrecht, Een studie van eigendomsverhoudingen en van onroerend-goedtransacties in vooroorlogse delen van de woningvoorraad in Dordrecht, RIW, Delft 1982, 166 blz.
7 'Het wonen'.

NLD.374

Road Engineering Division
Wegbouwkundige Dienst (WBD)
Van der Burghweg, Delft
Telephone: 015 569307
Telex: 38043
Date of establishment
Decree of the Minister of Transport and Public Works of 13 February 1981. The State Road Laboratory (Rijkswegenbouwlaboratorium) was discontinued on 16 February 1981. At the same time the Dutch Minister of Transport and Public Works instituted the Road Engineering Division (Wegbouwkundige Dienst).

A Government R and D and control establishment within the Ministry of Transport and Public Works. Established as a branch of the Chief Directorate of Public Works.
Main fields of work
- R and D work in the following fields: the road-designing process; the structural design of pavements; materials technology; road surface characteristics; environmental impact of roads; relevant measuring techniques
- Assistance in maintaining quality control for materials, mixtures and constructions for roads and hydraulic works by means of laboratory research and measurements on roads.
- Advising on: road design and the application of mathematical methods and computerized systems in the road designing process; the structural design of pavements; the selection and processing of materials; methods of carrying out road building projects; rational road maintenance; environmental problems (noise nuisance; air pollution); the physical environment and landscapes.
- Assistance in preparing projects to be carried out by the Public Works Department (Rijkswaterstaat).
- Transfer of knowhow and information by giving the practical experience component of sandwich courses and the like, and by means of working visits and publications.

Total staff: 240
Director or Chief Executive
Director: Ir C. van de Fliert
Heads of department
Main Section of Economics and Domestic Concerns: Dr N. Parlevliet
Main Section of Road Design and Environment: Ir L. de Borst
Main Section of Materials: Ir H. Schram
Main Section of Structures: Ir D. Hogervorst
Technical – Scientific Bureau: Ir J.A.C. Th Brouwers
Personnel Section: K.J. Renema
Secretary to the Director: E.R. Tijssenraad
Principal sources of finance
Funds acquired through the annual Parliamentary Vote of funds for the Ministries.
Principal related bodies
Chief Directorate of Public Works, The Hague (Hoofddirectie van de Waterstaat, Den Haag)
Publications
- Annual Report (Dutch)
- Leaflets on several subjects are published as WBD–Wegwijzer (in Dutch)
- Reports on several subjects are published as Rijkswaterstaat Publications or as Rijkswaterstaat Communications (in English)
- Papers by Road Engineering Division co-operators are regulary published in periodicals
- Numerous papers are published in professional or scientific journals
- Lectures and papers, given during congresses and at universities.

NLD.375

Society for the Study of Precast Concrete Constructions
Studiekring tot Ontwikkeling van het Geprefabriceerde Beton (STUPRE)
Secretariaat Stupré: Raamweg 16 Postbus 85512, 2508 CE 's-Gravenhage
Telephone: 070 46 95 55
Telex: 31361

An association of engineers from the Department for Governmental Building, consulting engineering firms, construction departments of large construction firms and of precast concrete industries; professors and staff members of technical universities and the building institute of TNO.
Main fields of work
Four out of the seven members of the Board form a steering group, which decides on the type of studies to be initiated and institutes study groups to handle them.
Research subjects are:
- Principles of precast connections
- Principles of design in precast concrete
- Regulations and code of practice for construction in precast concrete
- Connection details
- The future of precast concrete.

Total staff: 86 members
Director or Chief Executive
Chairman and Managing Director: Ir F. Treffers
Secretary: Ir E.J. Apol, Chief of Technical Information Department, NCH Nederlandse Cement Handelmaatschappij BV
Principal sources of finance
The general activities of the association are financed by the annual fees of the members. For specific studies subsidies are obtained either from CUR or from Stichting Bouwresearch.
Principal related bodies
Board; study groups.
Publications
- Reports on research work
- Various articles on specific subjects published in the periodical of the Dutch Concrete Institute, Cement.

NLD.376

Study Centre for Road Construction
Studie Centrum Wegenbouw (SCW)
14 A Jansbuitensingel, 6811 AB Arnhem
Telephone: 085 425741
Date of establishment
December 1957

The SCW is an independent research institute, founded by the Netherlands Department of Transport, Waterways and Highways, and by the Dutch Association of Road Building Contractors.

The results of research are available for all state, provincial and municipal services, civil services, road building industry and private individuals.
Main fields of work
The statutory objectives are:
- The promotion of research and development in connection with pavement design and construction, and the co-ordination of such research
- The promotion and propagation of knowledge concerning pavement design and construction.

The scope of activities is: conducting, sponsoring, co-ordination and promotion of research in highway pavements including embankments, bases, sub-bases, road verges and encouraging the application of research findings.

The programme at present comprises design and construction of flexible, rigid and block pavements, pavement management systems, pavement rehabilitation and maintenance procedures, road embankment design and construction, roadside management, application of waste materials and industrial by-products in highway construction. The programme is carried out by some 20 working groups, comprising about 150 engineers, scientists and university teaching staff members, who give their services free of charge.

Total staff: 6
Professionally qualified staff: 4
Director or Chief Executive
Executive Director: K. Wester
Principal sources of finance
The institute is sponsored by governmental agencies (about 50%) and by the road building industry (about 50%). The total budget is about $800 000 annually.
Principal related bodies
State Road Engineering Department (WBD) Delft, Soil Mechanics Laboratory Delft, Regional Laboratories of the Association of Road Building Constractors Apeldoorn, Shell Research Laboratory Amsterdam, Institute of Applied Physics Delft, University of Delft, University of Wageningen, Netherlands Foundation for Lightning Engineering, private consultancy services.
Publications
- *Journal:* Articles in monthly road-building magazine Wegen
- Recommendations for practice (3)
- Research Reports (52)
- Publications (13)
- Records in English (5)
- Reports on partial research results (13)
- Annual Progress Reports, with a directory of all publications in each report
- Audio-visual presentation on Pavement Management
- Courses
- Information sessions.

NLD.377
Technical Centre Waalsteen
Stichting Technisck Centrum Waalsteen (STCW)
Nassausingel 4, PO Box 551, 6500 AN Nijmegen
Telephone: 080 234041
Telex: 48124
Date of establishment
13 October 1955

A private institution (non-profit organization) at the disposal of its participants which are brickmaking companies in the Netherlands, the United Kingdom and Belgium.
Main fields of work
- R and D in the brick industry
- Process and quality control
- Operational research
- Reporting in the technical field
- Instruction and education
- Recommendation and application of new processes and working methods
- Management consultancy
- Working for developing countries through Dutch or foreign Government or private enterprises
- Technology: clay; fuels and energy; process equipment; machinery; transport; electronics, measuring; maintenance
- Standardization
- Finance and economy of brick and building industries
- Statistics
- Co-operation with University of Technology and National Research Centre TNO

Total staff: 4
Professionally qualified staff: 3
Director or Chief Executive
Managing Director: Dr Ing P.C.F. Bekker
Heads of department
Supervisory Board of Directors:
W.J.C. van Wijck, Chairman
C.F.L. Tijsse Klasen, Secretary
N.H.J.J. Bots, Treasurer
Principal sources of finance
Levies from participants; from third parties due to external orders; government and UNO institutions (UNIDO, World Bank)
Publications
- Publications on Lifetime Theory and Maintenance
- Numerous papers are published in professional or scientific journals
- Lectures and papers, given during congresses and at universities.

NLD.378
Testing and Research Institute of the Netherlands Water Undertakings KIWA Ltd
Keuringsinstituut voor Waterleidingartikelen (KIWA NV)
Sir Winston Churchill Laan 273, Postbus 70, Rijswijk 2280 AB
Telephone: 070 90 27 20
Telex: 32480 Groningenhaven 7
Cables: KTWA NL 1072 3430BB
Other addresses:
Subsidiary research laboratory, Postbus Nieuwegein
Telephone: (03402) 60860
Sales: Treubstraat 8, Rijswijk
Date of establishment
1948

Limited company, in which shareholders are the Dutch non-private water undertakings.
Main fields of work
Two main departments:
- Testing Department (metals and corrosion protection; plastics and rubber; concrete and asbestos-cement; appliances and fittings; testing laboratory and working)
- Research Department (analytical chemistry; corrosion; ground water; hyperfiltration; microbiology, virology; toxicology; process technology).

The field of work ranges over two main subjects:
- Research for the water industry, dedicated to catchment, purification, transportation, distribution and the use of water
- Testing of different kinds of materials (for the water industry, building industry and partly for the gas and oil industry).

In the Testing Department KIWA operates two certification schemes:
- KIWA–KEUR (KIWA–Approval)
- KIWA–ATTEST (KIWA–Agrément).

Total staff: approximately 230
Director or Chief Executive
Managing Director: Ir Th G. Martijn
Deputy Managing Director: Ir J. Schilperoord
Principal sources of finance
Research Department is financed through funds from the water industry. Testing Department is self supporting, through fees for approval marks and payment for other testing activities.
Publications
Research papers are published in professional or scientific journals; amongst them the journal H_2O, the official journal of the Dutch water industry

Voortgangsbericht (Proceedings): results of research and special activities

Priced publications such as test requirements are issued by KIWA itself.

NZL.379 **FM**

Building Research Association of New Zealand (BRANZ)
BRANZ Private Bag, Porirua, Wellington
Telephone: Wellington 357-600
Telex: NZ 30256
Cables: Branz Wgton
Date of establishment
Incorporated 1 April 1969

The aim of BRANZ is to encourage good building in the interests of the community. To maximize effectiveness, services are specifically directed to the building industry – to architects, engineers, builders, sub-contractors, manufacturers, technical educators, and central and local government personnel. Assistance is given to people outside the industry only when industry resources cannot help.

BRANZ is an independent, non-profit-making research association governed by a board representing all groups regularly involved with building and appropriate government departments.

Main fields of work
Research undertaken to meet the building industry's information needs and to assist with product and building systems development. The range of activities includes:

- Investigating the physical performance of building components. Current emphasis is on energy conservation in such areas as insulation, solar heating, air leakage, condensation and moisture control, and weather penetration into building cavities.
- Examining the type and extent of the loads applied to buildings and the performance of individual elements and the total structure under these loads. Emphasis is given to the effect of wind and earthquake loads on houses and other low-rise buildings.
- Assessing the effects of internal and external environments on the durability of building materials with particular interest in corrision of metals, changes in properties of non-metallic materials during service, and the performance of surface coatings.
- Conducting research and sponsored investigations into the performance of building materials and assemblies when exposed to fire. Results of the Association's fire tests are accepted in New Zealand, Australia, Singapore, Hong Kong, the United States and by international maritime authorities.
- Evaluating new and innovative systems products and building. A BRANZ Appraisal Certificate is issued when a product or system has been examined and is considered to be fit for its intended purpose.
- Maintaining an information and advisory service promoting technology transfer to and within the building industry.
- Providing library services and maintaining specialist holdings on building and related topics accessible to the building industry and research workers.

Total staff: 75
Professionally qualified staff: 44
Director or Chief Executive
Director: P.K. Foster
Heads of department
Advisory and Appraisal Divisions, Publications and Publicity: Assistant Director M.J. Sansom
Components, Materials, Structures, and Fire Divisions, Computing and Statistical Services: Assistant Director (Research) R.S. Whitney
Library, Administration, Secretarial and Accounts: Assistant Director (Admin) E.T. Keene
Principal sources of finance
A statutory levy on building and construction above a minimum value, supported by government grant, membership subscriptions, fees for testing, appraisal, and product development, and sales of publications.
Principal related bodies
The Board, comprising 2 Ministerial appointees, 3 government department appointees, 1 Standards Association of NZ appointee, 2 local government appointees, 12 industry sector appointees, is responsible for policy and the research programme.

Technical Advisory Committees appointed from the Board, industry and staff audit progress of research projects
Publications
These include:

- Build, a newsletter for members giving technical information and news of the Association
- Technical Recommendations, recommended technical solutions to build design and construction problems
- Technical Papers, presenting new technical information for professionals in the building industry
- Research Reports, formal presentations of research findings.
- Building Information Bulletins, short pamphlets of practical advice and information on topics of interest to the building industry
- Appraisal Certificates, favourable opinions of innovative building products and systems.
- Annual Report
- Research Programme of Work, an annual outline of research projects being undertaken
- Selected Building References, selection of documents and journal articles indexed in the preceding month.

NZL.380 **AM**

Forest Research Institute (FRI), Forest Products Division
Private Bag, Rotorua
Telephone: (73) 475 899
Telex: NZ 21080
Cables: FRESTRA
Date of establishment
1947

The FRI is the Research Division of the New Zealand Forest Service which is a government department. Its function is to carry out research into all aspects of the production and utilization of forest products for the benefit of all state departments, industry and private individuals.

Main fields of work

- Timber Engineering
- Wood Quality and Conversion
- Timber Drying
- Wood Preservation
- Pulp and Paper
- Wood Chemistry
- Adhesives and Composites
- Biotechnology

Total staff: 64
Professionally qualified staff: 64
Director or Chief Executive
Director: Dr A.J. McQuire
Heads of department
Timber Engineering: G.B. Walford
Wood Quality and Conversion: J.M. Harris
Timber Drying: J.A. Kininmonth
Wood Preservation: J.A. Butcher
Pulp and Paper: J.M. Uprichard
Wood Chemistry: J.M. Uprichard
Adhesives and Composites: G.B. Walford
Biotechnology: J.A. Butcher
Principal sources of finance
100% government.
Principle related bodies
Advisory committees, Forest Research Institute: Production Forestry Division; Protection Forestry Division
Publications

- *Journal:* Bulletin
- *Research reports:* In Bulletin; What's New
- Publication list

NZL.381

Ministry of Works and Development Central Laboratories
PO Box 30845, Lower Hutt
Telephone: Wellington 683119

Telex: NZ 3844
Date of establishment
1957

The Laboratory performs research and development testing in support of the Ministry of Works and Development's design, construction and maintenance activities.

Main fields of work
Experimental testing in support of design, construction and maintenance including:
- aerodynamic modelling
- bitumen
- building materials
- soil mechanics, rock mechanics, foundation engineering
- hydraulic modelling
- experimental structural analysis

Total staff: 69
Professionally qualified staff: 24
Director or Chief Executive: J.H. Wood
Heads of department
Aerodynamics: J. Docherty
Bitumen: G. Ball
Building Science: P. Thorby
Concrete: J. Clelland
Geomechanics: P. Millar
Hydraulics: R. Croad
Instruments: A. Spinks
Structures: S. Thurston
Principal sources of finance
Financed through Central Government funds. Approximately 30% of costs recovered against specific works done on behalf of other Government departments. Balance provided from MWD funds.
Publications
- *Research reports:* Work is reported directly to clients. Copies of all reports are held by Central Library, MWD, and are generally available through normal library channels.
- *Bibliographies/catalogues:* Annual list of reports circulated within MWD and to other interested organizations.

NZL.382
New Zealand Concrete Research Association
13 Wall Place, Porirua
Postal address: Private Bag, Porirua
Telephone: Wellington (04) 328-379
Cables: CONSEARCH
Dates of establishment
1969; reorganized under present name 1972

To provide cement and concrete users with a technical advisory and educational service backed by a concrete and association materials research facility.

Main fields of work
- Technical information and advisory service to cement users on all aspects and applications of the material in construction
- Wide ranging programme of training courses for the concrete design and construction industries to keep abreast of developments in techniques, materials and equipment as well as underlining good fundamental concrete practice
- Research undertaken into concrete and concreting materials to provide technical information largely of a practical nature in support of the above activities.

Total staff: 15
Professionally qualified staff: 7
Director or Chief Executive Director: David P. Barnard
Principal sources of finance
Funded primarily by the cement industry and central government
Principal related bodies
New Zealand Portland Cement Association; promotion of cement and concrete
Publications
- *Journal:* New Zealand Concrete Construction, monthly, except January
- Information bulletins on a wide range of concrete usages, techniques and practices.

NZL.383
Road Research Unit, National Roads Board (RRU NRB)
PO Box 12 041, Wellington North
Telephone: 729 929
Telex: 3844
Date of establishment
1962

A committee of the National Roads Board (NRB), which is a government agency which directs expenditure of road user taxes for road system improvements.

The principal objectives of the RRU are as follows:
- To commission, organize and encourage research that will contribute to the community, and an efficient transport system by improving the quality and efficiency of the road system.
- To inform interested parties and the community of recent developments in the administration, operation, maintenance and development of the road system and particularly results of research.
- To recommend and administer an annual budget to meet these objectives, recognizing the need for continuity of work amongst established road research agencies.
- To foster academic interest in road research through grants for study, travel and university scholarships.

Main fields of work
Pavements Committee: Interested particularly in unbound granular basecourse pavements, their design, construction and performance, material properties and environmental effects on pavement behaviour. Laboratory and field testing methods; thin-coat bituminous surfacings; Asphaltic-concrete layered pavements.

Traffic Committee of Roads: Safety aspects of road environment; operational characteristics of intersection types; transportation study overviews; road user characteristics.

Administration Committee: Economic aspects of road transport, parameters for economic evaluation of road improvements, value of time, vehicle operations and cost of accidents; appraisal of administration processes for New Zealand roading authorities; road maintenance management.

Structures Committee: Particularly concrete bridges both reinforced and prestressed; dynamic aspects of piers and beam/deck systems under traffic, wind and earthquake loads; temperature and creep in prestressed elements; scour of riverbeds at piers; miscellaneous small structures, guardrails, lighting columns, retaining walls, earth and rotten rock slopes.

The Unit meets every three months to consider the results of research projects, progress and proposals submitted by the four technical committees. The Unit is serviced by the Senior Planning and Research Engineer, a permanent secretary and an advisory officer. The technical committees are each serviced by part-time secretarial and technical officers. Research projects are carried out by government agencies, local government, consulting engineers, contractors, and universities.

Total staff: 6 full-time or part-time
Professionally qualified staff: 4
Director or Chief Executive
Chairman of Road Research Unit: E.J. Burt, Chief Highways Engineer, Ministry of Works and Development, PO Box 12-041, Wellington North
Heads of department
Chairman, Pavements Committee: M. Batten
Chairman, Traffic Committee: Professor A.R. Forbes
Chairman, Structures Committee: R.L. Preston
Chairman, Administration Committee: A.F. Stacey
Principal sources of finance
Financed by grants from the National Roads Board; or by contract agreements.

Principal related bodies
Pavement Committee; Traffic Committee; Administration Committee; Structures Committee
Publications
Road Research Unit Newsletter, quarterly; *Technical recommendations:* No 5; Occasional papers as required; Project reports

NZL.384

School of Architecture Auckland University
22 Symonds Street, Auckland 1
Postal address: Private Bag, Auckland
Telephone: (09) 737-999
Telex: UNILIB NZ21480
Cables: University Auckland
Date of establishment
University established in 1883, School of Architecture in 1917.

The School is a Department of the University of Auckland, primarily engaged in teaching, but with teaching and research laboratories and associated units undertaking applied and commissioned research and testing.
Main fields of work
Main research units are the Acoustics Institute (all aspects, especially the acoustics of auditoria), and the Urban Research Unit (interdisciplinary studies). Laboratories include acoustics (3 reverberant chambers, 1 anechoic), electronics, design aids, computing including CAD, architectural psychology, structures, materials, services, also wood and metal workshops. Other interests include passive solar energy, building economics, timber technology, ethnic influences, design methodology, architectural history, land economy and valuation.
Total staff: 52
Professionally qualified staff: 26
Director or Chief Executive
Professor Allan Wild
Heads of department
Director, Acoustics Institute: Professor A. Harold Marshall
Co-ordinator, Urban Research Unit: Professor Peter J. Bartlett
Chairman, Laboratories and Workshops Management Board: Professor A.H. Marshall
Librarian, Architecture and Town Planning Library: Mrs Wendy Garvey
Principal sources of finance
Government funded through University Grants Committee, administered through the University. Individual research grants from University Research or University Grants Committee or private contracts.
Principal related bodies
University of Auckland Council and Senate and their Committees.
Publications
- *Technical standards:* Occasional papers
- *Research reports:* Some 70 study papers for sale on a variety of topics. Undergraduate and graduate theses available in the Library. Reprints of published papers are available from authors
- *Bibliographies:* The Library publishes annual accession lists.

NZL.385

Department of Civil Engineering, Canterbury University
Private Bag, Christchurch
Telephone: 482 009
Date of establishment
1881

The Department is responsible for the teaching of, and research into civil engineering at the University of Canterbury. Students for Bachelors', Masters' and Doctorate degrees are enrolled in the Department. Some research is also conducted under contract for Government and private organizations.
Main fields of work
Main areas of research are:
- *Materials:* properties of concrete and timber
- *Structural engineering:* frame and wall structures of reinforced concrete, prestressed concrete, structural steel, timber, reinforced masonry.
- *Earthquake engineering:* earthquake effects on buildings, earthquake-resistant structures.
- *Foundation engineering:* piles, retaining walls, liquefaction potential
- *Structural design:* design procedures for buildings, building code developments.

Total staff: 49
Professionally qualified staff: 19
Director or Chief Executive
Head of Department: Professor R. Park
Principal sources of finance
New Zealand Government; research contracts from Government and industry
Principal related bodies
The Department of Civil Engineering is part of the Faculty of Engineering of the University of Canterbury.
Publications
Research reports: several are issued each year, generally reprints of Masters' Project Reports and Doctoral Theses as well as staff research. Also, Research in Civil Engineering, a summary of the Department's research, is published at 5-year intervals. The University of Canterbury Calendar (issued annually) contains a list of all technical papers and other publications by members of the Department.

NZL.386

Victoria University of Wellington, School of Architecture
Private Bag, Wellington
Telephone: 721 000, ext 961
Date of establishment
1974

The School of Architecture conducts teaching and research in the field of architecture and building science. The School was established as a separate faculty by the Victoria University of Wellington.
Main fields of work
The building research work of the School relates directly to the course work of the six teaching departments. Particular emphasis has been placed on energy in architecture, on programming and post-occupancy evaluation of buildings and on building controls. Three major research groups have been established to investigate these topics.

New facilities and proposals are now being developed with respect to urban simulation, artificial lighting and architectural aerodynamics.

Members of staff are actively researching in the fields of acoustics, architectural history, building construction failures and the design of timber structures.

The Architectural Sciences Laboratory, established in 1977, now provides an extensive range of facilities for laboratory based research and field work.
Total staff: 32
Professionally qualified staff: 23
Director or Chief Executive
Chairman: Professor Gerd E. Block
Dean: Dr John A Webster
Heads of department
Co-ordinators:
Architectural Method and Design: David Kernohan
Communication and Management: Helen Tippett
Environmental Control: George Baird
Construction Technology: Tom Jarman
Structures: John Webster
History of Architecture: Russell Walden
Principal sources of finance
Annual grants from Victoria University; contract research grants
Publications
Research reports: The School publishes three series (a full list is available on request): Research Papers (RP1–RP6 available); Working Papers (WP1–WP82–14 available); Contract Research Papers (CRP1–24 available); Contract reports are also published by the sponsor organizations.

NGA.387 **FM**

IMO State Housing Corporation

Private Mail Bag 1224, Owerri, Imo State

Telephone: 230341; 230733/2

Date of establishment

By Act No.14 of 1976

The aims of the Corporation are:

- To build houses for sale to the public
- To develop estates for the erection of houses
- To provide estate management expertise for the houses built
- To provide home ownership through mortgages
- To provide rental houses for medium and low-income groups
- To provide serviced plots for industrial and commercial purposes
- To provide consultancy services for architectural design and construction
- To award contracts for the development and construction of access roads and other facilities of like nature that are necessary in modern layouts of housing estates or other buildings owned or managed by the Corporation.
- To implement housing schemes of the State Government in all the local government areas.
- To carry out maintenance services for all the estates under the charge of the Corporation.

Act No.14 of 1976 empowers the Corporation to negotiate for purchase, take on lease and payment for land required for any of its purposes. This is done within the Public Lands Acquisition Law. Some local governments have set aside sites for the development of housing estates in their local government areas. With the new development whereby the Corporation provides for itself, land acquisition will be jointly pursued by both the Corporation and Ministries of Works and Housing and Lands, Survey and Urban Development.

Main fields of work

Responsibilities are to ensure:

- That implementation of housing programmes be undertaken by official bodies at the State and Federal levels
- That housing be provided for the whole spectrum of income groups in the country's growth population
- That to meet the total housing needs, standards should not exceed available resources
- That the States will formulate, review and implement their own housing policies bearing in mind their varying experiences, needs, resources and cultural environments
- That direct control for targeting development areas, land acquisition and development management will reside at State level
- That steps are taken to encourage local production of building materials as a means of lowering housing costs and reduce dependency on imports
- An improvement in the quality of rural housing and rural environment, through integrated rural development programmes.

Total staff: 274

Professionally qualified staff: 137

Director or Chief Executive

General Manager: P.C. Chikwendu

Heads of department

Secretary/Head of Administration Division: E. Nwankwo

Chief Estate Officer, Estate Division: O.A. Kalu

Financial Controller, Finance Division: A.S.D. Nwachuku

Chief Town Planning Officer, Town Planning Department: C.C. Nkume

Senior Architect, Artectural Department: B.P. Nwaogu

Engineer I, Engineering Department: N. Okoro

Principal sources of finance

Commercial banks

Principal related bodies

Ministry of Housing and Environment: supervising Ministry; Fed. Mortgage Bank of Nigeria: partly financing

Publications

Journal: Housing News, quarterly

NOR.388

Acoustics Laboratory

Akustisk Laboratorium/ELAB
N 7034 Trondheim NTH
Telephone: (07) 59 26 36
Telex: 55552 elab
Date of establishment
Established by SINTEF (the Engineering Research Foundation at the Norwegian Institute of Technology) in 1960.

The Acoustics Laboratory is part of the Electronics Research Laboratories, ELAB, at the Norwegian Institute of Technology.

In 1976 it was officially appointed the Norwegian National Centre for Acoustic Noise Research.

Main fields of work
The Laboratory carries out basic and applied research in the fields of electro-acoustics, noise and vibration studies, psycho-physio-acoustics, and all branches of architectural acoustics.

Laboratory facilities include an anechoic room, reverberation rooms, ISO-standardized rooms for sound insulation measurements, and test-rigs for the measurement of ventilation noise and duct silencers.

Total staff: 20
Professionally qualified staff: 16
Director or Chief Executive
Director: Jens Trampe Broch
Principal sources of finance
The Norwegian Council for Scientific and Industrial Research, governmental institutions, industrial firms, including offshore oil companies.
Principal related bodies
The Norwegian Institute of Technology: the Research Council for the Electronics Research Laboratory.
Publications

- *Journal:* ELAB Annual Report
- *Research reports:* Most reports and publications are in Norwegian. However, several reports are available in English covering noise from motor vehicles, noise screens, punch press noise, noise-induced hearing damage and noise nuisance.

NOR.389

National Institute of Technology, Norway, Building and Civil Engineering Department

Statens teknologiske instituut, Bygg- og anleggsavdelingen (STI)
Akersvn 24c, Oslo dep, postboks 8116, Oslo 1
Telephone: (02) 20 45 50
Other addresses:
Main institute has eleven district offices (not specifically in the field of building)
STI, North-Trøndelag, Strandveien 7, 7800 Namsos
STI, South-Trøndelag, Postbox 55, 7079 Flatåsen
STI, Møre og Romsdal, Blindheim, 6020 Vegsund
STI, Sogn og Fjordane, Hafstadvn 21, 6800 Førde
STI, Hordaland, Lars Hillesgt 22, 5000 Bergen
STI, Rogaland, Sandviksveien 9, 4000 Stavanger
STI, Agder, Markensgt 20, 4600 Kristiansand
STI, Telemark, Vestfold, Kverndalen 5, 3700 Skien
STI, Buskerud, Kirkegt 9, 3600 Kongsberg
STI, Hedmark, Oppland, Strandgt 13A 2800 Gjøvik
STI, South-Østland, Økernveien 145, Oslo 5
Date of establishment
The main institute was established in 1917; the Building and Civil Engineering Department in 1935.

A government established institute under the Ministry of Industry, dealing with development, information, education and consulting work for industry.

Main fields of work
The main institute consists of 12 technical departments, two non-technical (organization and economy, pedagogic information) and district offices. If with technical centres each department has its own leadership, plans and budget.

The Building and Civil Engineering Department is divided into three sections, building, construction and civil engineering, water supply and sewerage (communal services) and energy economy.

Like most of the divisions, the Department for Building and Civil Engineering has laboratories and carries out testing and development work on request and at its own initiative.

Work is mainly devoted to materials, performance of new principles, and the planning and guidance of the building process directed towards the building construction firms. Blasting and land-surveying are other activities together with work on sewer systems, pipes and sewerage plants (purification). The main materials covered are timber, concrete and concrete products, brickwork and plaster. Another department deals with plastics, painting and surface treatment. Other activities are the development of education plans and teaching material for the ordinary trade schools and the instruction of teachers within these schools (vocational training). The division is trying to work in close contact with the Norwegian Building Research Institute and other research institutes, adapting and spreading the research results for the benefit of the consumers.

Total staff: Main institute approximately 300, district offices aprroximately 100, Department of Building and Civil Engineering approximately 20
Director or Chief Executive
Administrative Director: Alf Hernes
Heads of Department: Perc. Corneliussen
Publications

- Reports and papers for official use are not so numerous
- Instructional books and compendia for educational purposes are numerous and cover a variety of fields
- Some articles appear in the branch press.

NOR.390 AM

Norwegian Building Centre

A S Byggtjeneste
Haakon Vii'sgt. 2, Postboks 1575, Vika, Oslo 1
Telephone: (02) *41 72 30
Date of establishment
1956

A non-profit company owned by five major national associations within the building industry (architects, professional engineers, builders, manufacturers and building materials distributors). The Centre is self-supporting.

Main fields of work
The Centre is a central service organization for the building trade, operating on an internal financing non-profit basis. All available funds are employed for purposes aimed at improving the knowledge and insight of members of the trade and the general public. The organization maintains an information department which holds exhibitions, provides guidance and arranges courses and lectures.

In addition, the Centre issues topical information on building and construction, particularly through the building materials catalogue Byggkatalogen, the bulletin on Government building regulations, Statlige Byggebestemmelser and the bookshop Byggtjenestes Bokhandel. Besides maintaining close contact with public authorities and institutions and with the building trade's own organizations, the Centre participates in Scandinavian and other international collaboration.

The Centre is organized in three main departments: exhibitions, publications and book shop plus departments for marketing (sales) and economy/administration.

Total staff: ca. 30
Professionally qualified staff: ca. 12

Director or Chief Executive
Managing Director: Endre Hoflandsdal
Principal sources of finance
Exhibition rent; fees for editing and publishing data sheets; subscriptions; sale of technical literature
Principal related bodies
Board of directors
Publications
- Byggkatalogen (The Building Material Catalogue) consisting of edited data sheets on building materials available on the Norwegian market.
- Statlige byggebestemmelser (Governmental building regulations), a combined booklet and loose leaf publication on building regulations, codes and by-laws, amendments to these and lists of approved materials and components.

NOR.391 **FM**
Norwegian Building Research Institute
Norges byggforskningsinstitutt (NBI)
Forskningsveien 3 B, Oslo 3
Postal address: PO Box 322, 0314, Oslo 3
Telephone: (02) 46 98 80
Cables: Byggforsk
Date of establishment
1953 (a preliminary Office of Building Research existed 1949–1953)

An independent institute within the Royal Norwegian Council for Scientific and Industrial Research (NTNF), a semi-governmental organization attached to the Ministry of Industry.

NBI is a public institution at the disposal of central and local authorities, industry and individuals.
Main fields of work
- General theory and methods
- Building industry and the community: studies of employment, cost analyses, building activities, etc.
- User's requirements: building environment, special requirements, functional studies, performance
- Climatic and physical conditions: outdoor and indoor climate, light, noise and vibration
- Building and installation techniques: building materials, constructions in and above ground, installations and building systems (there are no activities on electrical installations)
- Building process: information and operations, management, production methods and equipment
- Administration of building stock: administration, operation, control, rehabilitation, etc.
- Information: publications, mass media, library, national and international collaboration, personnel, etc.

Future research and development will be aimed at and channelled into one of the following defined programmes:
- Energy economization and alternative energy resources in building
- Management, maintenance and rehabilitation of buildings
- Dwellings of the future: costs, quality and design
- Analyses of building costs and measures to curb further increases
- Ventilation of buildings
- Computer application in constructions

Total staff: 150
Director or Chief Executive
Director: Aager Hallquist
Principal sources of finance
Financed to approximately 30% by income from sponsored research, service to industry and central and local authorities, sales of publications, etc. The rest, 30%, official grant channelled through NTNF (1982 figures)
Principal related bodies
Board: Chairman and six members.
Technical advisory committees: Building Techniques; Functional Studies; Production Techniques; Heating, Ventilation, Sanitary Installations; Data sheets – Building Details.
Publications
- *Technical standards:* Building Research Design Sheets comprising two series: Architectural Data and Building Details, are issued twice yearly and distributed to 10 000 subscribers
- *Research reports:* Research results are given in traditional reports (summary in English) and in reprints of articles published in journals and periodicals
- Handbooks, describing more fundamental subjects (summary in English)
- Design Manuals, giving advisory information
- Annual Report, describing previous year's activity
- Collection of NBI's Testing Methods
- User's Manual to EDP programmes and other miscellaneous publications.
- Annual lists of publications.

NOR.392
Norwegian Geotechnical Institute (NGI)
Norges Geotekniske Institutt
Sognsveien 72, Oslo 8
Telephone: (02) 23 03 88
Telex: 19787 ngi n
Cables: Geoteknikk
Date of establishment
1 January 1953

NGI is a research and consulting organization with assignments in many different parts of the world. Basic research and development work is partly sponsored by the Royal Norwegian Council for Scientific and Industrial Research.
Main fields of work
- Soil mechanics and foundation engineering
- Rock mechanics, engineering geology and rock blasting
- Snow mechanics and avalanche investigation
- Site investigation and measurement techniques
- Earth- and rockfill dams
- Offshore structures and marine geotechnology

The Institute has specialized in field instrumentation and prototype performance observations. In recent years about 35% of NGI's activities have been related to offshore structures and fixed oil platforms.

Terzaghi Library: A few years before his death, Professor Karl Terzaghi, the 'Father of Soil Mechanics', bequeathed his professional correspondence, technical reports and papers to NGI. The collection is open to research workers and engineers from all parts of the world.
Total staff: 160
Director or Chief Executive
Director: Kaare Høeg, DSc
Heads of department
Dam and Rock Engineering: Bjørn Kjaernsli
Foundation Engineering: Kjell Karlsrud
Marine Geotechnology: Knut Schjetne
Avalanches/Slides: Karstein Lied
Special Problems: Knut H. Andersen
Field Instrumentation and Performance Observations: Elmo DiBiagio
Equipment Development: Arild A. Andresen
Site Investigation: Ivar Johannessen
Laboratory Testing: Toralv Berre
Geochemistry: Tor Løken
Administration: Roar Bøhn
Information: Finn Jørstad
Principal sources of finance
Financed 25% from the Norwegian Government; 75% from other sources in Norway and around the world.
Principal related bodies
Royal Norwegian Council for Scientific and Industrial Research
Publications
- NGI Publication Series, published approximately 5 times a year. Each

issue contains several articles related to geotechnical engineering and snow mechanics.
- Numerous papers to International Conferences on Soil Mechanics and Foundation Engineering
- Numerous papers to international symposia
- Numerous papers published in scientific journals

Principal sources of finance
Financed by annual fees from the industry, by contributions from public research funds, and by work sponsored by the industry.
Publications
Research reports: Meddelelser; Utredninger; Rapporter; Tekniske småskrifter

NOR.393

Norwegian Hydrodynamics Laboratories, Norges hydrodynamiske laboratorier, Divisional River and Harbour Laboratory (NHL VHL) Divisjon Vassdrags- og havnelaboratoriet
PO Box 4118, Valentinlyst, N 7001 Trondheim
Telephone: (07) 592300
Telex: 55435 vhl n
Cables: VASSLAB
Date of establishment
VHL 1955; NHL 1980

NHL is an independent self-financing organization.
Main fields of work
The work involves research and development comprising testing and consultative assistance within the fields of hydrodynamics, hydraulics and hydrostatics.
Total staff: 155
Professionally qualified staff: 100
Director or Chief Executive
Managing Director: Haavard Berge
Director: Jon Sletthaug
Heads of department
Divisional Ship and Ocean Laboratory:
Director: Kjell Straume
Principal sources of finance
Mainly fees from assignments.
Principal related bodies
NHL is a joint venture of the Ship Research Institute of Norway (NSFI) and the Foundation of Scientific and Industrial Research at the Norwegian Institute of Technology (SINTEF)
Publications
Journal: NHL publication, at irregular intervals

NOR.394 **AM**

Norwegian Institute of Wood Technology
Norsk Treteknisk Institutt (NTI)
Forskningsveien 3b, Blindern, Oslo 3
Telephone: (02) 46 98 80
Telex: 18171 forsk n
Date of establishment
1949

A research and information institute for the wood industry, based on voluntary membership by the Norwegian lumber manufacturers. Established on the initiative of the sawmilling industry.
Main field of work
- Sawmill and planning technology
- Wood technology and anatomy
- Wood structures and other wood products
- Timber protection
- Gluing and surface treatment
- Chipboard

Total staff: 38
Professionally qualified staff: 18
Director or Chief Executive
Managing Director: Professor Rolf Birkeland
Heads of department
Lumber department: Sverre Tronstad
Wood construction department: Erik G. Aasheim
Chemical department: Karl Mørkved

NOR.395

Norwegian Precast Concrete Federation
Norges Betongindustriforbund (NBIF)
Postboks 70, Kjelsåsveien 160, Kjelsås, Oslo 4
Telephone: (02) 15 84 50
Date of establishment
1960

A federation of individual member firms.
Main fields of work
- Information
- Service
- Research in the field of precast concrete.

Director or Chief Executive
Arne Skjelle
Principal sources of finance
Membership fees
Publications
- *Journal:* Betongprodukter, a technical magazine issued 4–5 times a year
- Technical Reports
- Brochures

NOR.396 **AM**

Royal Norwegian Council for Scientific and Industrial Research, Committee on Building Research
Norges Teknisk-Naturvitenskapelige Forskningråd (NTNF), Komite for bygg- og anleggsteknisk forskning
Sognsvn 72, 0855 Oslo 8
Postal address: Box 70 Tåsen, 0801 Oslo 8
Telephone: 02 23 76 85
Telex: 76951 ntnf n
Date of establishment
1965

The Committee is part of the NTNF system, which is a non-profit public body. The main task for this body is to distribute governmental research funds to institutions, institutes, firms and individuals, according to plans and priorities laid down within the body.
Main fields of work
The main task for the Committee is to plan the spending of research funds within the building sector, follow the various projects and evaluate the results.
Total staff: 2
Professionally qualified staff: 1
Director or Chief Executive
Secretary: Jan F. Reymert
Principal sources of finance
The Royal Norwegian Ministry of Industry
Principal related bodies
The activities are followed by the Committee, which consists of 7 members from different parts of the building sector.
Publications
Journal: Ba-forskning (Building Research) (in Norwegian), quarterly

NOR.397 **AM**

SINTEF Department of Architecture and Building Technology

SINTEF Arkitektur og byggteknikk
Alfred Getz vei 3, 7034 Trondheim-Nth
Telephone: 47- (0) 7-592620
Date of establishment
1970

The Foundation for Scientific and Industrial Research at the Norwegian Technical University (SINTEF) is a non-profit research foundation integrated with the Norwegian Technical University, carrying out contract research and development for industry and public bodies. SINTEF's Department of Architecture and Building Technology is affiliated with appropriate divisions of the Technical University's Department of Architecture and Department of Civil Engineering. SINTEF's purpose is to exploit the University's potential for contract R and D and serve as a link between the University and the outside world.

Main fields of work

The general field of work is building science and research within the architect's field of work. The topics covered are:

- The energy issues
- Solar energy and daylighting
- Rehabilitation and retrofitting
- Environmental psychology
- Functional aspects of architecture
- Computer-aided architectural design
- Product development, component design

Many research results are implemented in experimental buildings designed and evaluated by the department.

Total staff: 13
Professionally qualified staff: 11
Director or Chief Executive
Chairman of Board: Professor Øyvind Aschenhoug
Manager: Sigmund Asmervik
Principal sources of finance
The Royal Norwegian Council for Scientific and Industrial Research (NTNF), other public research funds and organizations, and industry.
Principal related bodies
The Supervisory Board is responsible for long-term policy: all three affiliated University divisions are represented. The regular manager is also responsible to the main SINTEF bodies: board, council and directors.
Publications
Research reports: All projects are reported; for publically financed projects the reports are available through SINTEF's information department. Most reports are in Norwegian.

Some research results are also described in articles for the technical press.

NOR.398 **AM**

SINTEF Norwegian Fire Research Laboratory

SINTEF Norges branntekniske laboratorium (NBL)
N 7034, Trondheim-Nth
Telephone: (07) 59 51 90
Telex: 55620 sintf n
Cables: FIRETEST
Date of establishment
1934

NBL is a research and testing department within the Foundation for Scientific and Industrial Research at the Norwegian Institute of Technology (SINTEF).

Test reports from NBL form the basis of fire technical classifications and approvals given by the Ministry of Labour and Municipal Affairs, the National Inspectorate of Fire Services, the Norwegian Maritime Directorate, Det norske Veritas, the Norwegian Petroleum Directorate, etc.

Furthermore, NBL is a national centre for fire research.

Main fields of work

Since NBL is the only laboratory of its kind in Norway, its activities include commercial fire testing, research projects and information. All these activities cover fire prevention as well as fire fighting.

The aim of NBL is:

- On a neutral basis to perform testing, control and evaluation of materials, constructions and equipment as a basis for classifications and approvals. As a consultant to perform statements and professional advisory activity within the fire technical field
- To initiate, co-ordinate and perform research work
- To contribute to the solving of educational tasks at the Norwegian Institute of Technology and to give scientific information.

Total staff: 23
Professionally qualified staff: 12
Director or Chief Executive
Director: Kjell Schmidt Pedersen, MSc
Heads of department
Research: Per Jostein Hovde, MSc, Dr Ing
Laboratory: Harald Landrø MSc, Dr Ing
Principal sources of finance
The Ministry of Labour and Municipal Affairs; the Norwegian Council for Scientific and Industrial Research; customers, e.g. oil companies, industry, etc.
Principal related bodies
SINTEF
Publications

- Annual reports
- *Research reports:* Materials reaction to fire; fire resistance of construction elements and structures; toxicity; Passive fire protection; Active fire protection; Modelling of different types of fire

PAK.399 **FM**

Building Research Station (BRS)

F 40, Sind Industrial Trading Estate, Hub River Road, Karachi 28

Telephone: 29 24 57

Date of establishment

1 July 1955

A research and demonstration centre for building science and technology established under the Ministry of Works. Performs testing upon request. It was taken over by the Council for Works and Housing Research in 1970.

Main fields of work

BRS undertakes research and testing in the following fields:

- Building materials and components
- Concrete and structures
- Soil engineering
- Building physics and fire
- Building maintenance

The Council for Works and Housing Research was created in 1964 under the Ministry of Agriculture and Works with the aim of initiating and promoting the institutions of research, studies, testing and documentation relating to housing problems in Pakistan. The housing problems in Pakistan are many and varied in nature. Conservation of natural resources and the use of agricultural wastes, meeting the challenges of war and natural calamities and providing low cost houses for labourers and farmers are worth particular mention.

Total staff: 85 of whom around two-thirds are supporting staff

Director or Chief Executive

Gul Mohammad Khan Afghan, Chairman, Council for Works and Housing Research

Dr M. Sulaiman, Director, Building Research Station

Heads of department

Concrete and Structures, Building Physics and Fire and Workshop and Stores: Dr Ataullah Maher

Soil Engineering: Miss Khalida Khan

Building Material and Components: Nadir Mansoor

Building Maintenance, Personnel and Transport: S.A.A. Hashmi

Library: Sarfaraz Hussain

Budget & Accounts: Rafiq Haider

Principal sources of finance

Ministry of Science and Technology, Government of Pakistan.

Principal related bodies

Governing Body: sanctions all budgetary and administrative policy; Council: recommends budgetary and administrative measures; Research and Development Advisory Committee: reviews past progress and recommends future research programmes

Publications

- Annual Report
- Current Papers, published frequently
- *technical recommendations:* Recommendations are made to the Pakistan Standards Institutions.
- *Research reports* published in 1983: Two experimental low cost rooms at Building Research Station; Experimental and demonstration low cost house built with rice husk ash and lime as cement; Load tests on the experimental and demonstration low cost house built with rice husk ash and lime as cement; Granulated blast-furnace slag as a substitute for cement; Low cost room built with slag at Building Research Station; concrete mix proportioning and characteristics of Karachi aggregates

PAK.400 **AM**

Building Research Station Lahore (BRS Lahore)

PO Box No 1230, Near New University Campus, Lahore

Telephone: 852685

Date of establishment

1961

A Government Testing and Building Research Establishment under the Communication and Works Department, Government of Punjab, Lahore.

Main fields of work

The day-to-day working of this station is carried out under the Director, who is assisted by the Assistant Director and Senior Research Officer. The Assistant Director is responsible for four sections through two Assistant Engineers: General Section; Accounts Section; Brick and Soils Section and Library Section.

The Senior Research Officer looks after the work of the following sections through three other officers: Concrete and Aggregate Section; Physics Section; Statistics and Instrument Section; Lime and Gypsum Section and Photographic Section.

The aims are as follows. To conduct research on indigenous materials to reduce the cost of building construction and prepare specifications for low cost housing. To advise the field engineers on their construction problems and carry out field and laboratory tests for determining the strength of materials being used in various works including the testing of the bearing capacity of soils for foundation purposes.

Total staff: 105 of whom 63 are support staff.

Director or Chief Executive

Director: Qamar-ur-rehman Mufti

Principal related bodies

Board of Advisory Committee, which guides and frames research programmes for this institution.

Principal sources of finance

Financed solely by the Government of Punjab.

Publications

Numerous papers on research and experimental work carried out by this station and on various other technical topics relating to building science are issued as publications of this station. Some of these papers have also been published in International Magazines. Useful technical information extracted from various publications is also issued in the form of Library Bulletins for guidance of field engineers.

The station also publishes progress reports covering work done during six monthly periods.

PAK.401 **UM**

Maymar Housing Services Ltd

170 A, Block 3, Sir Syed Road, PE CH Society, Karachi 3

Telephone: 430191

Telex: 25518 MYMAR PK

Cables: MAYMARCO

Date of establishment

17 July 1973

A firm of builders and contractors, mostly concerned with the construction and sale of flats and bungalows and the sale of plots developed on raw land.

Main fields of work

Planning of suburban townships including construction of roads and laying of other services on privately acquired land.

The company has also built for its clients some buildings in a hospital complex and some industrial buildings.

Total staff: 170

Professionally qualified staff: 50

Director or Chief Executive

Executive Director: M.G. Siddiqui

Heads of department

Directorate of sales and purchase: Mrs Rahat Afza, Owner Director; M.A. Pirwani, Director

Directorate of administration: Syed Mazhar Ali, Director
Executive Directorate: M.G. Siddiqui, Director

Principal sources of finance

Partly from clients' advance payment and partly from the House Building Finance Corporation arranged by the company on behalf of clients.

Principal related bodies

Karachi Development Authority; House Building Finance Corporation.

Publications

Bibliographies/catalogues

PHP.402 **AM**

Building Research Service, National Engineering Center, University of the Philippines

Rms. 101–102 National Engineering Center Building, University of the Philippines Diliman, Quezon City, Philippines

Telephone: 97 60 61 Local 211

Date of establishment

29 July 1971 Action of U.P. Board of Regents 811th meeting

Building Research Service is a unit attached to the National Engineering Center, University of the Philippines System. It undertakes research on building technology and materials, their production, component design, building use and user requirement analysis for the built environment; dissemination of information and data on the above-mentioned activities. It conducts and manages seminars, colloquia and conferences for and on behalf of organizations having similar functions to those initiated by the organization itself.

Main fields of work

- Research programs covering the fields of building theory, design, and practice conducted in co-ordination with units of the University or local and foreign agencies.
- To participate in conferences, symposia, workshops and seminars both local and international on subjects related to building, building science, architecture, construction, human settlements and building materials.
- To bring research to the notice of those who can put the findings into actual use, through publications etc.
- To initiate tie-ups with government and private research entities involved in the building industry.
- To conduct seminars and workshops.
- To provide extension and advisory services.

By the very nature of the service, the agency can call on any faculty member of the College of Engineering, the College of Architecture or Consultants in the private sector to undertake expert advice and research and to conduct investigations when the research projects need them.

Total Staff: Staff averaging 8 per research project and 2 projects annually.

Director or Chief Executive:

Director: Professor Geronimo V. Manahan

Heads of department

Building Research Service: Professor Geronimo V. Manahan

Materials Testing Laboratory: Ulpiano Ignacio

Principal sources of finance

UP Engineering Research and Development Foundation, Inc. Funds from the University of the Philippines for its administrative support.
Grants and research contracts.

Principal related bodies

Advisory Board of the National Engineering Center and the Board of Directors of the University of the Philippines Engineering Research and Development Foundation, Inc.

Publications

- Precis (at six monthly intervals)
- Site Planning Manual for Marginal Settlements, College of Architecture, U.P.
- Sariling Sikap sa Pagpapabahay (Self-Help Housing) by Carlos E. Guico
  Passive Cooling Technology for Hot-Humid Localities by G.V. Manahan
- Bibliography for Passive Cooling Technology by G.V. Manahan
- Laws and Decrees in housing planning and buildings in the Philippines 1972–1980.
- The Health Circuit, by G.V. Manahan,
- Guidelines for Energy Conservation in Buildings,
- Coping with the Philippine Economic Scenario,
- U.A.P.'s Strategic Role for the Practicing Architect, by G.V. Manahan,
- Guidelines for Reducing Flood Damages, by G.V. Manahan.

POL.403 **FM**
Instytut Techniki Budowlanej
Building Research Institute
ul Filtrowa 1, 00–950 Warszawa
Telephone: 25 13 03
Telex: 813022
Other address: The Branch of the Building Research Institute, ul Katowicka 5, 44–101 Gliwice.
Date of establishment
1945

A scientific and research institute that was created by the Ministry of Reconstruction as the Research Institute for Building. In 1949 reorganized as the Building Research Institute under the Ministry of Building and Building Materials Industry
Main fields of work
The Institute carries out a scientific and research activity: fundamental and applied research concerning the standardization and application of new types of materials and structures, approval of new building materials, problems of physics and acoustics in buildings, protection of buildings against corrosion and fire, improvement of research methods.
The Institute is authorized to approve new building materials, components and structures, as well as methods of erection, for general use in building (technical approval).
Within the Institute 17 scientific and research sections are active: the majority are located in Warsaw, with 3 in Gliwice for the building problems of mining areas.
Total staff: 555
Professionally qualified staff: 246
Director or Chief Executive
Director: Assistant Professor Marian Weglarz
Heads of department
Structures Division: Assistant Professor L. Runkiewicz
Geotechnique Division: Professor Dr A. Piaskowski
Light Partitions Division: Dr O. Korycki
Thermal Physics Division: Professor Dr W. Ptoński
Fire Research Division: Dr M. Kosiorek
Concrete Division: Assistant Professor Dr K. Stępień
Finishing Techniques Division: Z. Rydz
Building Protection Division: Assistant Professor Dr Z. Ściślewski
Acoustics Division: Professor J. Sadowski
Principal sources of finance
The Institute conducts its research on the basis of contracts with the institutions, often with the Ministry of Building and Building Materials Industry.
Principal related bodies
The Scientific Council, which estimates plans and the execution of research problems of the Institute, estimates scientific and research works of the Institute and assesses the qualifications of scientific staff for scientific promotion.
Publications
- *Journal:* Prace ITB-Kwartalnik (BRI Quarterly); Dziatalność ITB (Activity of the BRI), annually
- *Technical standards*/recommendations: Instrukcje ITB (Instructions of the BRI); Świadectwa ITB (Certificates permitting the use of new solutions in building)
- *Research reports:* Prace Naukowe ITB (Scientific Works of the BRI); Serie: Monografie (Monographs); Rozprawy (Dissertations); Konferencje (Conference papers).

POL.404
Main Centre for Building Information
Centralny Ośrodek Informacji Budownictwa (COIB)
ul Senatorska 27, 00–950 Warszawa
Telephone: 27 24 49
Telex: 813736 Wa Pl
Cables: CINBUD
Date of establishment
1960; Law No 17 of 17 September 1960 of the President of the Committee of Building and Architecture

Main centre for scientific, economic and technical information for the building sector and the building materials industry.
Main fields of work
All activities in the field of information, co-ordination of the various information departments and automation of information. Permanent building exhibition; Professional training; International co-operation. Member of CMEA (Council for Mutual Economic Assistance) and UICB (International Union of Building Centres).
Total staff: 215
Professionally qualified staff: 173
Director or Chief Executive
COIB Manager: Henryk Watcerz
Heads of department
Co-ordination and Planning, International Co-operation Section: B. Andrzejewski
Documentation: Z. Bartnik
Information: F. Malec
Branch Information System for Building SINTOB, International Co-operation with International Branch Information System for Building in CMEA countries: G. Rządkowska
Building Catalogue: J. Jabtonowski
Product File: J. Kartowicz
COIB Department in Katowice: A. Hupert
Principal sources of finance
Part proceeds from the sale of information and publications, part subsidy
Principal related bodies
Programming Committee; Editorial Committee
Publications
- Building Catalogue, available on exchange basis
- *Research reports:* Pamphlets on Building Problems available on exchange basis; Building Bulletin available on exchange basis
- Polish Building Abstracts, free of charge

POL.405
Institute for Building Mechanization
Instytut Mechanizacji Budownictwa (IMB)
Racjonalizacji 6/8, 02-673 Warszawa
Telephone: 43 02 01
Telex: 814631 IMB pl
Date of establishment
1961

An industrial self-supporting institute, whose aims are
- Training of technical staff and management of training centres for heavy machine operators
- Publishing centre for scientific and technical information including periodic and aperiodic publication
- Organization of technical and scientific meetings and symposia
- Complex mechanization of building operations: development, testing and evaluation of earth moving, construction and building materials industry machines, awarding to manufacturers the right to use of quality seals
- Abatement of noise and vibration of construction
- Machine exploitation methods (INCL, operational machinery research).

Main fields of work
Mechanization of construction works reconnoitring, researching, plan-

ning and implementing works concerning construction machinery (means of transport included), construction materials industry technology lines; building machinery management and utilization; forecasting of building machinery development; education of building mechanizers; scientific information; rationalization and co-operation with foreign countries.

Total Staff: 350

Professionally qualified staff: 245

Director or Chief Executive

Director: Eugeniusz Budny, DSc

Heads of departments

Division of the Institute in Wrocław: Chief Kazimierz Banyś lecturer

Division of the Institute in Poznań: Henryk Zaradniak

Principal sources of finance

Central funds (government and branch programmes) as well as controllers of enterprises

Publications

- *Journal:* Prace Instytutu
- Scientific and research works devoted to problems in the Institute's field of activities

POL.406

Institute for Organization, Management and Economics of the Building Industry

Instytut Organizacji, Zarządzania i Ekonomiki Przemysłu Budowlanego (ORGBUD)

ul Filtrowa 1, 00 925 Warszawa 58

Telephone: 25 52 81

Telex: 813906

Branch address: ul Cieszyńska 2, 30 015 Kraków

Date of establishment

1951; present title since 1973.

Government research institution.

Main fields of work

- Forecasting, programming and planning the development of the building industry and the construction materials industry
- The economic and financial systems of economic organizations in the building industry and the construction materials industry
- Managerial and organizational systems of the building industry and the construction materials industry
- Materials and energy management in the building industry
- Social and economic personnel-managing systems in the building industry and the construction materials industry
- Rules and conditions of investment realization (new building and modernization)
- Building standards, costs and prices.
- Organizational and economic consultancy service work; improvement of building industry and construction materials industry managerial staffs.

The Institute has 21 research departments, 2 centres for managerial staff training, and 1 centre for building standards and prices.

Total staff: 377

Professionally qualified staff: 107

Director or Chief Executive

Professor Edward Świrkowski

Principal sources of finance

Incomes based on the whole activity of the Institute, i.e. sale of scientific research works, fulfilment of orders paid for by building organizations and other institutions.

Principal related bodies

Scientific Council

Publications

Problemy Rozwoju Budownictwa (Building Development Problems), quarterly.

POL.407

Institute of Refractories

Instytut Materiałów Ogniotrwatych

ul Toszecka 99 44–101 Gliwice

Telephone: 31 92 61

Date of establishment

1 September 1953

A public institution at the disposal of all industries that are producing or using refractory materials.

Main fields of work

The aim of the Institute is to conduct research work into the technology, high-temperature properties and applications of refractories. The main activity is to improve the performance of refractories to meet the specific needs of miscellaneous industries.

Total staff: 460

Professionally qualified staff: 250

Director or Chief Executive

Docent Dr Wactaw Szymborski

Docent Dr Mieczystaw Drozdz

Heads of department

Scientific Division: Docent Dr Mieczystaw Drożdż

Technical Division: Docent Dr Wactaw Szymborski

Pilot Plant Division: Henryk Piechaczek

Principal sources of finance

Financed by sponsored research work

Publications

Przegląd Dokumentacyjny Materiatów Ogniotrwatych, monthly

Biuletyn Informacyjny Instytutu Materiatów Ogniotrwatych, quarterly

POL.408

Polish Academy of Sciences, Institute of Fundamental Technological Research

Instytut Podstawowych Problemów Techniki Polskiej Akademii Nauk (IPPT)

ul Swiętokrzyska 21,00–049 Warszawa

Telephone: 26 12 81

Telex: 815638 Ippt PL

Date of establishment

1953, affiliated to the Polish Academy of Sciences (PAS)

A research institute, devoted to the fundamental technological research in the field of mechanics and acoustics, not only in the field of building and civil engineering.

Main fields of work

The Institute conducts basic research work orientated towards solving important problems for the national economy. This research is centred on four groups of subject:

- *Fields:* this area of scientific activities is concerned with the study of the static and dynamic fields of mechanical stresses and strains, the distribution of acoustic and ultrasonic fields and selected problems in electrodynamics.
- *Materials:* this group of themes includes the theory of material microstructure, theoretical and experimental investigations into composites, the improvement of measuring methods for material properties and, in particular, non-destructive testing methods.
- *Structures:* the research is centred on calculation and design methods for machines and building methods for structures, and includes the optimization of these methods.
- *Signals:* this area comprises physical and technical aspects of atomic frequency standards, the development of measuring and control apparatus and the general analysis of the structure of acoustic and electromagnetic signals.

Total staff: approximately 500, of whom about 300 are support staff.

Director or Chief Executive

Director: Professor H. Frąckiewicz

Principal sources of finance

Financed through research projects on a customer/contractor basis

Publications

Journal: Archives of Mechanics, bi-monthly; Rozprawy Inżynierskie (Engineering Transactions), quarterly; Journal of Technical Physics, quarterly; Uspekhy Mekhaniki (Advances in Mechanics), quarterly
Polska Bibliografia Analityczna Mechaniki (Polish Analytical Bibliography of Mechanics), quarterly
Non-linear Vibration Problems, annually
Fluid Dynamics Transactions, annually
Scientific Activities of the PAS IFTR, annually
IFTR Reports (average 75 copies a year)

Books series: Applied Mechanics Series (AMS); Acoustics and Ultrasonics Series (AUS); Monographs in Theoretical Electrotechnics (MTE) Mechanics and the Computer (MC).

POL.409

Research and Design Centre for Industrial Building
Centralny Ośrodek Badawczo-Projektowy Budownictwa Przemystowego (BISTYP)
ul Parkingowa 1, 00–518 Warszawa
Telephone: 219211; 281145
Telex: 813623 bst pl
Cables: bistyp
Date of establishment
The Centre was founded in 1951 by the Ministry of Building.

Independent and autonomous state enterprise carrying out research and design works, as well as services for local and foreign organizations.

On the basis of multilateral contracts it organizes the voluntary co-operation of building design units including the co-ordination of joint export activities.

The leading centre in the field of studies and design, inclusive of standard designs, for industrial building in Poland.

Main fields of work

- Scientific research work and complete designs in the field of building
- Architectural, structural and installation designs for industrial enterprises, power plants, service and general purpose buildings constructed in Poland and abroad
- Site works and starting of production supervision
- Studies, testing, expertise and opinions
- General contractor for export activity in the field of design, consultancy and supervision services, as well as completion of deliveries necessary for realization of targets.

The Centre has three departments: Power Engineering Design and Experimentation; New Design Methods; and Industrial Building Development; and eleven divisions: 4 architectural/structural building; building partitions; refrigeration; foundations; building structures; acoustics; sanitary engineering; electrical installations.

Total staff: 600
Professionally qualified staff: 400
Director or Chief Executive
Director General: J. Bajszczak
Principal sources of finance
Income from work performed on the basis of contracts.
Principal related bodies
Advisory Council, General Director's advisory body in the scope of organizing and economic problems; Technical Board, advisory and consultative evaluation body on the merits of technical problems; Committee for Evaluation of Standard Designs and Centre for Co-operation in Building Design by BISTYP
Publications

- *Journal:* Budownictwo Przemystowe, bi-monthly review of research and design work
- Poradnik Projektanta Przemystowego, aperiodic publication containing advisory/aid items meant for design engineers
- Catalogues of construction systems and elements, and of standard solutions

POL.410

Research Institute for Environmental Development (RIED)
Instytut Ksztattowania Srodowiska (IKS)
ul Krzywickiego 9, 02–078 Warszawa
Telephone: 25 42 56
Telex: 813493
Date of establishment
February 1974

A government research establishment which reports to the Ministry of Administration and Spatial Economy. It implements projects at the order of central and local administration, industrial organizations and other bodies concerned, as well as carrying out an independent research programme within its own financial budget.

Main fields of work

The general goal for RIED is the formulation of research findings for settlement policies and strategies and environmental protection policies and strategics; the preparation of guiding principles, planning/design studies and methodology for settlement planning practice and technology for human settlements. RIED attempts to foster planning theory and practice as well as knowledge on human settlement ecosystems and natural ecosystems within the general framework of the ecosystem complex: population – resources – environment and related issues.

RIED embraces the following activities related to planning and the development of human settlements and environmental protection:

- Guiding the diagnostic identification of state needs and problems, leading to structural change in the built environment
- Focusing on the maintenance of environmental resources
- Fostering comprehensive development of the environment, aimed at the reconstruction of environmental resources
- Supporting action on the improvement of human settlements
- Concentrating on the conservation and prevention of settlement ecosystems and natural ecosystems
- Increasing the cultural and aesthetic value of the environment
- Improving the technology for human settlements
- Upgrading the organization of analytical and institutional planning and resource management
- Setting up planning legislation, environmental indices, technological standards and a procedural framework
- Strengthening the rationale of decision-making in planning, designing and management of the environment.

The activities of RIED are particularly extensive and include:

- Work programming
- Research and study
- Planning theory and practice,
- Studies and design of new and prototype technologies
- Forecasting
- Pre-planning studies, planning and design and field work
- Reports, expertise and consultancies, etc.

Total staff: about 1300
Professionally qualified staff: 199
Director or Chief Executive
Director General: Assistant Professor Pawet Btaszczyk
Deputy Director: Assistant Professor Hanka Zaniewska
Heads of department
Organizational Sector: Dr Pawet Btaszczyk
Environmental Protection Sector: Dr Pawet Btaszczyk
Scientific Secretary Sector: Dr Hanka Zaniewska
Spatial Economy Sector: Dr Elżbieta Wysocka
Housing Economy Sector: Professor Edward Kuminek
Financial Sector: Dr Leokadia Lobert
Principal sources of finance
The Institute is an economically independent unit, with its own financial budget.
Principal related bodies
The Scientific Board: shaping of R and D policy; supervising the implementation of research programmes and staff policy.

Publications
Journal: Bulletin of the Research Institute on Environmental Development, monthly; Housing Problems, quarterly; Man and the Environment, quarterly.

POL.411
Research Institute of Cementing Materials for the Building Industry
Instytut Przemystu Wiazacych Materiatów Budowlanych (IPWBM)
ul Oświęcimska 21, 45–641 Opole
Telephone: 392 01/07, 396 61
Telex: 033309
Date of establishment
1954

A government institution carrying out research, development and implementation works to meet the requirements of the cement, lime, chalk and gypsum industries.
Main fields of work
- Technology of cement, lime, chalk and gypsum manufacture
- Automation of industrial control systems
- Development and implementation of new industrial equipment and installations
- Environmental protection and industrial hygiene
- Technology of complex alumium oxide and cement production using non-bauxite raw materials
- Energy and raw material saving methods, industrial waste utilization
- Methods and conditions of raw material exploitation

The Institute has a division in Cracow and two workshops. It is organized into 10 departments.
Total staff: 438
Professionally qualified staff: 228
Director or Chief Executive
Director: Privatdocent Dr Bronistaw Weryński
Principal sources of finance
The main sources are commissions from the State and from industrial cement, lime and gypsum enterprises
Principal related bodies
Board of Directors; Academic Council
Publications
- *Journal:* Papers are published in Cement Wapno Gips (Cement, Lime Gypsum) monthly
- *Research reports:* Some research reports are published by the Institute in Prace Instytutu PWMB, issued at irregular intervals (2–4 titles annually)

POL.412
Road and Bridge Research Institute
Instytut Badawczy Dróg i Mostów (IBDM)
ul Stalingradzka 40, 03–301 Warszawa
Telephone: 11 32 31, Secretariat extn 227
Telex: 815 390 IBDM PL
Other addresses:
Road Operation and Maintenance Division,
ul Stalingradzka 32a, 03–301 Warszawa
Telephone: 11 21 32
Bridge Foundations Division,
ul Stalingradzka 32a, 03–301 Warzsawa
Bridge Division,
ul Stalingradzka 32a 03–301 Warszawa
Telephone: 11 30 97
Bridge Division, Section of Prototypes,
ul Mielczarskiego 123, Kielce
Telephone: 5 00 38
Aggregates Division,
ul Grochowska 26, Wroclaw
Telephone: 21 22 21
Date of establishment
1955

A research institute within the Ministry of Transport. Established as Instytut Budownicta Drogowego (Road Construction Institute). Then it was changed to Centralny Ośrodek Badań i Rozwoju Techniki Drogowej (Road Research and Development Centre). The present name was given in 1973.
Main fields of work
Research: physical and chemical investigations of materials used in road and bridge building; technical requirements of materials, application in road and bridge building, and the principles of utilization of substitute materials; principles of design and technical control of roads, streets, airport pavements and bridges; road and bridge construction and maintenance techniques; road geotechniques and bridge foundations; registration methods of the riding quality of roads; road traffic and adaptation of roads to the requirements of traffic safety and protection of the environment; road economics; organization of the technical base and services for road construction; prototypes of equipment necessary for research; organization and mechanization of road construction works; investigation of road construction machines.

Implementation activity and services rendered by IBDM: IBDM contracts to develop projects in road and bridge building, the results of which are then applied in all branches of the national economy. It carries out research into roads, bridges, deep foundations, foundation and road building machines, develops construction techniques, provides scholarly superintendence of the execution work, draws on expertise of road design, as well as of road and bridge construction and reconstruction methods.

IBDM co-operates with design offices and executors in introducing new structures, techniques, equipment and the principles of design and control to road and bridge building. It also participates in major road projects.
Director or Chief Executive:
Managing Director: Professor Mieczystaw Rybak
Principal sources of finance
IBDM is financed through funds paid for the development of projects commissioned by governmental bodies and other establishments.
Publications
- Bulletin of the Road and Bridge Research Institute, quarterly
- News in Road Construction Techniques, occasional
- Studies and Materials series, occasional
- Information series, occasional
- Express information, monthly
- Information Bulletin, bimonthly.

POL.413
Wood Technology Institute
Instytut Technologii Drewna
ul Winiarska 1, 60–654 Poznań
Telephone: 22 40 81
Telex: 0413505 ITEDE PL
Date of establishment
1952

A government organization within the Ministry of Forestry and Forest Industries. Established by the Ministry of Forestry and Forest Industries; research work is carried out subject to contracts with the wood processing industries.
Main fields of work
The Institute's fundamental aim is to carry out research work in the field of: structure, physical and mechanical properties of wood; preservation and hydro-thermal treatment of wood; chemistry and chemical processing of wood, wood waste gasification; mechanical processing of wood including sawmilling, furniture, packaging, strength grading of sawn timber; technology of wood-based products; improvement of wood and wood products including surface finishing, preservation against des-

tructive biological agents; gluing of wood and development of new phenolic and urea formaldehyde glues; application of wood and wood products in building construction, application of pine bark as an insulating and filling material in wall elements; economics of woodworking industry; labour safety in production plants.

The Institute has 14 scientific research departments.

Total staff: 271

Professionally qualified staff: 229

Director or Chief Executive

Director: Professor Dr R. Babicki

Deputy Director: Professor Dr M. Wnuk

Principal sources of finance

Half self-financing, half government budget

Publications

Journal: Prace Instytutu Technologii Drewna, quarterly

PRT.414
Forest Products Institute
Instituto dos Produtos Florestais (IPF)
Rua Filipe Folque, 10–J, 1000 Lisboa
Other addresses:
Av. Antónia Augusto de Aguiar, 163, 3°, Esq, 1000 Lisboa
Rua Braamcamp 14, 1°, 1200 Lisboa
Av. dos Aliados 22, 2°, 4000 Porto
Telephone: 559115/6/7, 533484, 558066
Telex: Proflor
Date of establishment
1 January 1973–DL 428/72, de 31OUT

A Government establishment within the Ministry of Agriculture, Commerce and Fisheries with activities in the field of cork, resinous products and wood and derivatives. Endowed with juridical and administrative capacities and financial autonomy.
Main fields of work
- Economic studies
- Laboratory support for enterprises
- Technological and commercial promotion for cork, resinous and wood-based products
- Studies for standards on characteristics and test procedures for products
- Legislative rules for the trade of forest products
The IPF is organized in 6 services: Economy and Statistics Service; Technical Service; Quality Control, Research and Promotion Service; Surveying Service; Administration Service; Finance Service.

Total staff: 232
Professionally qualified staff: 50
Director or Chief Executive
President: Manuel Pinheiro Ferreirinha
Directors: Francisco de Matos Chaves, Lawyer; Jaime Salazar Sampaio, Alfredo Baptista de Andrade
Principal sources of finance
Financed through service taxes paid by the industry and others.
Publications
- Boletim de Cortica (a monthly publication on cork)
- Boletim dos Resinosos (a quarterly publication on resinous products)
- Boletim de Madeira e Derivados (a quarterly publication on wood and derivatives)
- Yearly report of the IPF activities and budget
- Boletim Bibliográfico e Informativo (a monthly publication)

PRT.415 **FM**
National Laboratory of Civil Engineering
Laboratório Nacional de Engenharia Civil (LNEC)
Avenida do Brasil 101, 1799 Lisboa Codex
Telephone: 88 21 31/7
Telex: 16 760
Date of establishment
1946

A Government research institution within the Ministry of Public Works.
Main fields of work
LNEC is one of the Directorates-General of the Ministry of Public Works. It is organized into 7 main Departments, each containing several Divisions:
- *Building Materials Department:* Concrete and Binders Division; Ceramics and Plastics Division; Chemistry Division
- *Structures Department:* Structural Analysis Division; Structural Behaviour Division; Applied Dynamics Division; Timber Division; Observation of Structures Division
- *Geotechnique Department:* Special Geotechnical Studies Division; Foundations Division; Site Exploration Division
- *Hydraulics Department:* Estuaries Division; Structural Hydraulics Division; Hydrology and River Hydraulics Division; Sanitary Hydraulics Division; Harbours and Beaches Division
- *Buildings Department:* Acoustics and Lighting Division; Architecture Division; Building Behaviour Division; Building Parts Behaviour Division; Building Economy and Productivity Division; Building Methods Division; Division of Thermal Comfort and Fire Protection in Buildings
- *Transportation and Network Department:* Road Geotechnique Division; Road Pavements Division; Traffic and Road Safety Division

The LNEC is a research institution intended to support the construction industry, i.e. the conception, design and construction of buildings – especially houses, bridges, dams, roads, airfields, tunnels, harbours, etc. Specific interests are shown by the Departments and Divisions.

The LNEC has also helped to train Portuguese technical cadres and to improve overall scientific and technical knowledge in Portugal.
Total staff: almost 1000
Professionally qualified staff: 240
Director or Chief Executive
Director: Artur Ravara
Deputy Director: Ricardo de Oliveira
Deputy Director: Jao Bau
Heads of Department
Building Materials: A. Sousa Countinho
Structures: Vitor Monteiro
Dams: J. Oliveira Pedro
Geotechnique: E. Maranha das Neves
Hydraulics: J. Pores Castanho
Building: J. Teixeira Trigo
Transportation Networks: Fernando Branco
Principal sources of finance
Government grant (about 30% of total expenses) and own income (about 70% of total expenses).
Principal related bodies
Director – Administrative Council.
Publications
As a research institution engaged in solving problems in the vast domain of civil engineering, the LNEC has been constantly concerned with making the results of its studies available to all interested parties, in such a way that they can most easily and efficiently be used. To this end it has made use of the various available communication media, such as publications, meetings, exhibitions and demonstrations of equipment and techniques. However, publications are the most widely used medium for making known the results of studies, and the knowledge and experience accumulated by the LNEC. Bearing in mind the wide range of information involved, the type of utilization that is intended, and the public to whom such information is directed, various types of publication have been created.
- Proceedings of meetings
- Courses
- Technical papers
- Technical information
- Standards
- Agrément documents
- Classification documents
- Automatic computation programs
- Specifications cards
- Information on deterioration of materials
- Information on costs
- Technical information – Buildings
- Tests and equipment
- Tables
- Monthly information bulletin
- Portuguese building bibliography.

PRT.416 **AM**

South of Portugal's Building and Public Works Association

Associacão de Empresas de Construcão e Obras Públicas do Sol (AECOPS)

Rua Antónia Enes 9–5°, 1000 Lisboa

Telephone: 533193/533194/533195/533196

Telex: 12453 AECOPS P

Date of establishment

21 April 1950

AECOPS is a private association (of about 3000 members) that supports the Building and Public Works contractors, intervening in the regulation of the construction industry and mediating in wages policy.

Main fields of work

AECOPS is organized into four main sections:

- *Juridical:* gives legal advice to members and deals with all legal problems related to the construction industry
- *Economics:* prepares economic studies on which the dialogue with governmental institutions is based
- *Work Relations:* studies the yearly collective work agreement revision; advises members on legal procedures with the workers; compares work conditions in Portugal with those of other countries; studies and implements special systems for the analysis and qualification of work functions and roles; provides vocational training and professional select for the engagement of personnel
- *Technical:* provides technical support in a wide range of civil engineering applications including construction materials; supports the technical regulation of price revision, budget organization and contract execution; provides technical advice for classifying and identifying the construction companies.

Total staff: 58

Professionally qualified staff: 36

Director or Chief Executive

President: Vitor Marques dos Santos, Architect

Secretary:General: José J. Tomaz Gomes, Lawyer

Heads of department

Juridical Department: J. Ferreira da Cunha, Lawyer

Economics Department: A. Godinho Correia, Economist

Work Relations Department: J. Monteiro, Psychologist

Technical Department: F. Mauricio Garcia, Engineer

Principal sources of finance

Membership monthly fees.

Principal related bodies

General Assembly: Auditors (elected bodies that fix the policy and budget of all departments).

Publications

- *Journal:* Indústria da Construcão, bi-monthly
- *Research reports:* Economic studies on the Building and Public Works Industry
- Special Anniversary Magazine
- Portuguese Contractors Working Abroad (catalogue)
- Work Contract (book).

PRT.417

Technical Association of the Cement Industry

Associação Tecnica da Industria do Cimento (ATIC)

Av 5 de Outubro, 54–2°, Dt°, 1000 Lisboa

Telephone: 54 75 38 / 54 79 70

Cables: CIBETÃO

Date of establishment

August 1965

A non-profit-making association engaged in the study, transmission of knowledge and development of cement and mortars and their areas of application.

Main fields of work

Transmission of knowledge and development of cement and mortars and their areas of application at all levels.

Total staff: 5

Professionally qualified staff: 2

Director or Chief Executive

Manuel Lourenço Antunes, Ing

Principal sources of finance

Financed by annual contributions from members (who are the national cement manufacturers); the amount being based on their sales on the domestic market during the course of the previous year.

PRT.418 **AM**

Porto University Faculty of Engineering, Building Construction Section

Eaculdade de Engenharia da Universidade do Porto, Gabinete de Construções Civis

Rua dos Bragas, 4099 Porto Codex

Telephone: (Porto) 27505 or 380760

Date of establishment

Decree (Decreto) No. 24.966, 23 January 1935

The Building Construction Section is one of several departments concerned with a variety of fields within the University of Porto Faculty of Engineering. It was created to support teaching and research activities in the field of Building Construction. It carries out studies and testing upon request for other official services, industry and private individuals. It is part of the Civil Engineering Department of the Faculty.

Main fields of work

- Building economics and productivity
- Hygrothermic, acoustica, lightning and fire behaviour of buildings

Currently operating are a research group on Hygrothermics and an Acoustical Laboratory. An extension of work is planned in other non-structural fields of building studies and activities.

Total staff: 16

Professionally qualified staff: 13

Director or Chief Executive

Director: Professor A. Barbosa de Abreu

Heads of department

Research Group on Hygrothermics: Assistant Professor Victor Abrantes

Acoustical Laboratory: Assistant Professor L.C.P. Palermo de Faria

Principal sources of finance

Financed by Government grants, and sometimes by service rates from external work.

Principal related bodies

The Scientific Council of the Civil Engineering Department, Faculty of Engineering; the Centro de Engenharia Civil da Universidade do Porto (CECUP), a branch of the Instituto Nacional de Investigação Cientifica (INIC), Lisboa.

Publications

Boletim do Gabinete de Construções Civis, non-periodic (7 numbers have been issued recently).

ROM.419

Building Design Institute for Typified Constructions
Institutul de Proiectare pentru Construcţii Tipizate (IPCT)
70132 Bucuresti, Sectorul 2, Str Tudor Arghezi 21
Telephone: 11 38 40
Telex 10248 CERCO
Date of establishment
1956

A state institution that supplies the whole of the national economy with design types for dwellings and industrial buildings.

Main fields of work
- Industrial and agricultural constructions: halls and constructions for industrial production; storehouses and warehouses; offices; laboratories and cloakrooms; silos for grain; water towers; water and oil tanks; thermal power stations; cooling towers
- Civil buildings: single-family houses; apartment blocks; educational buildings; social and cultural buildings; sports halls; office buildings, etc.
- Use of unconventional energy sources: solar power stations; installations for heat recovery from industrial waste water; installations for use of thermal waters, etc.
- Computer aided programs: structure and construction unit calculus; establishment of seismic effects over foundations and structures; calculus of thermal installations, etc.
- Sociological studies: present and prospective study of the social and economic development of land; social and demographic analyses for special objectives; specific computer aided programs, etc.
- Technical assistance: analysis and control of design (structures, seismic protection, etc.); consultancy services for detailing typification programmes; technical assistance on the working site; advanced technical and postgraduate training, etc.

Total staff: 600
Professionally qualified staff 260
Director or Chief Executive
Director: Dr Eng Petru Vernescu
Heads of department
Industrial buildings: Eng Anatolie Cazacliu
Engineering structures: Eng Cruciţa Iacob
Socio-cultural buildings and dwellings: Eng Marin Lazăr
Principal sources of finance
Design contracts with the users of design types.
Publications
- *Journal:* Informative Journal, quarterly
- Design types catalogues
- Non-periodical publications concerning different events organized by the Institute, such as: colloquia, conferences, symposia.

ROM.420

Building Research Institute
Institutul de Cercetări în Construcţii şi Economia Construcţiilor (INCERC)
Şos Pantelimon 266, Bucuresti 2
Telephone: 27 40 85
Telex: CERCO R 10240
Date of establishment
1950

State organization aiming to promote new technologies and testing methods in the field of the building industry and to contribute to their introduction in current practice. It is also the specialized body of the state for deciding long and short term policy in the development of the building industry.

Main fields of work
- Technology of concrete, reinforced concrete and prestressed concrete
- Technology of steel structures
- Seismological engineering
- Foundations and geotechnics
- Testing of structures
- Acoustics, hygrothermics and fire
- Protection against corrosion
- Sanitary, electrical and ventilation equipment
- Mechanization of buildings
- Economy of building
- Technology for finishing

Total staff: 800
Professionally qualified staff: 320
Director or Chief Executive
General Director: Valeriu Cristescu
Heads of department
Sector for Structures and Protection of Building: Dr eng Romulus Constantinescu
Sector for Mechanization, Organization and Economy of Building: Dr eng Frimu Tomşa
Principal sources of finance
State budget and contracts with building enterprises, precasting factories and industry.
Principal related bodies
Central Institute for Research, Design and Guidance in Building (direct tutelar body); Ministry of Building Industry (highest tutelar body).
Publications
- Studii si Cercetări INCERC
- *Journal:* Construcţii (review), monthly
- *Technical standards:* Most of the existing standards, and norms in the field of building
- *Research reports:* Summaries are published annually: INCERC Research, activities, achievements (Cercetări, activităţi, realizări); INCERC Leaflet
- *Bibliographies/Catalogues:* CIB, RILEM; CEB; FIP.

ROM.421

Central Institute for Building Research, Design and Directives, Cluj–Napoca Branch
Institutul Central de Cercetare Proiectare şi Directivare în Construcţii, Filiala Cluj-Napoca
Calea Floreşti 117, 3400 Cluj–Napoca
Telephone: 951/61988; 61018
Date of establishment
1962

A technical building research and study centre at the disposal of all state, provincial and municipal services, civil services, industry and private individuals.

Main fields of work
The research and study activity develops, in close collaboration with the specialized councils of the Department of Civil Engineering Polytechnic Institute of Cluj–Napoca, in the following fields:
- Structural mechanics
- Reinforced concrete, prestressed concrete
- Lightweight concrete
- Foundations
- Civil constructions (engineering)
- Building technology
- Bridges and roads.

The Branch also performs tests on materials, components and structures, on prototypes, models and large-scale constructions. It conducts research, studies and technological development activities. It elaborates technical documents and standards. It offers expertise and technical assistance.

Total staff: 57 (there are also 122 university staff and 15 researchers from the Department of Civil Engineering of the Polytechnical Institute)
Professionally qualified staff: 32

Director or Chief Executive
Branch Director: Ioan Dobra
Heads of department
Lightweight and Special Concretes Laboratory: Doina Mircea
Structural Engineering Laboratory: Ioan Botez
Utilization of Local Materials and By-products in the Field of Constructions, Working Group: Magdalena Toma
Principal sources of finance
State budget (40%); Ministry of Construction; contracts with building companies, study bureaus, and others.
Publications
- Buletin Informative (Information Bulletin), annually
- Research results and technical articles are published in the INCERC Bulletin, Studii şi Cercetări INCERC, and in the review, Construcţii (Construction).

ROM.422
Documentary Information Office for Building, Architecture and Town Planning
Oficiul de informare documentară pentru construcţii, arhitectură şi sistematizare (ODCAS)
Boul 1848 no 10, CP 1–139, Bucureşti cod 70058
Telephone: 15 22 21
Date of establishment
Established in 1957, by continuation of the former Documentation Centre for Building, Architecture and Town Planning

A national organization within the Central Institute for Building Research, Design and Directives.
The Documentary Information Office provides technical and scientific information, bibliographic documentation on building, architecture and town planning for the construction industry users.
Main fields of work
The Office prepares papers on building, architecture and town planning with a view to providing the specialist in these branches with scientific and technical information regarding the new materials and components for building, construction technologies and industrialized systems, installations, building machines and equipment, design, typification, new methods of calculation, as well as other data needed for design and construction process.
The principal divisions are: Technical Information; and the Library.
Director or Chief Executive
Chief of the Office: Mrs Aurelia Dobrescu
Principal sources of finance
Documentary works carried out for building institutes and organizations.
Principal related bodies
The Ministry of Industrial Building.
Publications
- Documentary information papers on building and architecture themes entitled Documentare ODCAS, 6 issues a year
- Bibliographies on building, architecture, installations, building materials, etc.

ROM.423
Hydraulic Engineering Research Institute
Institutul de cercetari hidrotehnice (ICH)
Splaiul Independenţei 294
Telephone: 49 29 95
Date of establishment
1976

A Government research institution within the Central Institute for Research, Design and Guidance in Construction. The results of the studies performed are applied in the most important hydraulic structures completed in the country, solving problems in hydraulics, structural analysis, stability and building techniques.
Main fields of work
Research studies and investigations are carried out in three main divisions:
- Hydraulics: hydraulic engineering schemes and integrated river basin development works; river and canal hydraulic engineering; equipment and structures within hydro-electric schemes, reclamation works and other water uses; ship design hydrodynamics, navigation structures, harbours, shipyards and coastal engineering, off-shore structures.
- Hydraulic Engineering, Soil and Rock Mechanics: structural and mechanical characteristics of soil and rock; computation and surveying of constructions (including manufacturing and installation of measuring instrumentation); seepage studies; drainage and impervious linings, construction technologies and hydraulic structures and foundations, canals, dykes and earth dams; grouting with cement, bentonite, sodium silicate and synthetic resins; new materials (reinforced earth, geo-textiles).
- Equipment for Water Supply and Sewage Disposal Laboratory; design parameters of structures and installations for water supply and canalization; treatment of feed and waste water; field studies; chemical and biological instability of waater; treatment of silts.

Included in the staff are hydraulic, civil, electronic and mechanical engineers, physicists, mathematicians and chemists.
Total staff: 550
Professionally qualified staff: 150
Director or Chief Executive
Director: Professor Simion Hâncu
Heads of department
Hydraulics Division: Valer Jeler
Hydraulic Structures, Soil and Rock Mechanics Division: R.J. Bally
Equipment for Water Supply and Sewage Disposal Laboratory: Mihai Lesnic
Principal sources of finance
Entirely financed by design institutes and construction organizations on the basis of research contracts.
Publications
- Hydraulic Studies
- Studies on Hydraulic Engineering, Soil and Rock Mechanics
- The number of volumes of these publications issued annually varies.

ROM.424 **FM**
Research and Design Institute for Physical Planning, Housing and Municipal Engineering
Institutul de cercetare şi proiectare pentru sistematizare, locuinţe şi gospodărie comunală (ISLGC)
Str. Snagov 53–55, 70136 Bucureşti
Telephone: 12 43 42
Date of establishment
1949 (with various changes of name)

A self-management institute at the disposal of all states, depending (at government level) on the Committee for People's Councils' Problems (CPCP).
Main fields of work
Research and studies for:
- Country planning
- Town and communal planning
- Elaboration of the national programme for town and country planning
- Elaboration of the national programme for housing (prospective studies concerning the quantitative and qualitative requirements and the future technologies, requirements in building materias
- Functional studies for social cultural buildings
- Studies in economics
- Town facilities

Design for:
- Town and country planning (models for the other regional institutes)

- Important social and house buildings
- Town facilities

The Institute is organized into seven sections: 1 section for town and country planning (central planning body for the whole country); 2 complex sections for major housing and social building and for design and research; 3 sections for town facilities: water, sewage, water purification, communal transport, hygiene; 1 section for geotechnical and topographical studies. ISLGC also has management and technical services.

Total staff: 910
Professionally qualified staff: 820
Director or Chief Executive
Director: Dipl civ eng Adrian Alecsandrescu
Principal sources of finance
Contracts with State institutions for research and design; contracts with Romanian foreign trade institutions for research and design in various foreign states.
Principal related bodies
Scientific Council; General Assembly. Role: to direct the general policy of the Institute's activities, evaluate and endorse the results obtained.

ROM.425
Research and Design Institute for Water Resources Engineering
Institutul de cercetări şi proiectări pentru gopspodarirea apelor
Splaiul Independentei 294, Sector 6, 7000 Bucureşti
Telephone: 49 20 37
Date of establishment
1957

State institute under the National Water Council.
Main fields of work
Research Department:

- Research on water management, inland and marine water quality, water supply, treatment of drinking and waste water, treatment and utilization of sludge
- Studies on water quality; parameters for treatment stations
- Study of specific demands; economic effects of the use of water and of the development of water resources; optimization of water distribution
- Elaboration of frame schemes for the arrangement of water sheds; topographical, geological, hydrological and hydrogeological studies; location determination studies; techno-economic documentation of investments for dams, storage reservoirs, canals, dykes and regularization of beds and other hydraulic constructions.

Design Department:

- Completed projects for storage reservoirs with an overall capacity of more than 2 billion m³ created by concrete rock fill or earth dams of 10–130 m height
- Completed projects for river training and flood levees
- Design and implementation of water management information systems (for flood warning, water quality monitoring and water resources management)
- Mathematical models and computer programs for adequate data processing.

Total staff: 1536
Professionally qualified staff: 500
Director or Chief Executive
General Manager: Florin Iorgulescu
Heads of department
Research Department:
Water Quality Protection: R. Antoniu
Water Supply and Treatment: V. Rojanschi
Waste Water Treatment: V. Popescu
Sludge Treatment and Disposal: Veturia Ghederim
Design Department:
Field Surveys: St. Nica
River Basin Development Schemes: R. Amaftiesei
Dams: I. Tecuci
River Regulation and Embankment Works: Lăzărescu Fl
Computer Centre: Gh. Indrei
Principal sources of finance
Self-financing through contracts.
Publications

- *Journal:* Hidrotehnica Review, monthly
- Research Reports.

SGP.426 **AM**

Construction Industry Development Board

133 Cecil Street 09-01 Keck Seng Tower, Singapore 0106, Republic of Singapore

Telephone: 2256711

Telex: RS 20818 CONDEB

Date of establishment

2 March 1984 (Act of Parliament)

A statutory board within the Ministry of National Development; it was set up as an agency responsible for planning, co-ordinating and overseeing the development of the construction industry towards a more rationalised and mechanised approach.

Main fields of work

The functions of the Board, as embodied in The Construction Industry Development Board Act 1984, are

- to promote the development, improvement and expansions of the construction industry
- to facilitate and assist in the mechanisation of the construction industry
- to advise and make recommendations to the Government on matters affecting or connected with the construction industry
- to promote proper standards and efficiency in the construction industry by encouraging the standardisation and improvement of construction techniques and materials
- to provide consultancy and advisory services with respect to construction works
- to keep under review the training requirements of the construction industry and assist in providing training facilities
- to promote the advancement of the skills and expertise of persons engaged in the construction industry; and
- to promote or undertake research into any matter relating to the construction industry.

Director or Chief Executive

Chairman: Yap Neng Chew

General Manager: Ng Chee Yoong

Deputy General Manager: Chow Kok Fong

Heads of department

Technology Development Division: Lam Siew Wah

Industrial Development Division: Ong Thiam Huat

Manpower Development Division: Tan Tian Chong

Principal sources of finance

Government funded.

Principal related bodies

Housing and Development Board – the provision and maintenance of public housing

Public Works Department – the provision and maintenance of public works and infrastructure, building and development control

Jurong Town Corporation – the development of industrial and commercial premises.

Publications

- CIDB Plant and Equipment Directory – Yearly
- CIDB Review – Quarterly
- Construction Market Report – Monthly
- Prefabricated Concrete Components

SGP.427 **AM**

Singapore National University, Faculty of Architecture and Building

Kent Ridge Campus, Singapore 0511

Telephone: 7756666, ext 100

Telex: UNISPO RS33943

Cables: UNIVSPORE

Date of establishment

1 April 1969

The faculty is one of the eight faculties of the University. The main role of the faculty is to educate some 700 undergraduate and postgraduate students leading to the following degrees. First degrees: BA (Architectural Studies); BArch; BSc (Building); BSc (Estate Management). Higher degrees: MSc (Building); MSc (Building Science); MSc (Estate Management); MArch; PhD.

Main fields of work

Research is conducted in the following areas:

- School of Architecture: Urban Planning; Architectural Design; History of Architecture
- Department of Building Science: Building Structures; Acoustics; Thermal Behaviour of Buildings; Lighting; Energy Conservation and Renewal Sources of Energy
- Department of Building and Estate Management: Building Economics; Project Management; Estate Management

Faculty staff members also offer a consultancy service to the construction industry.

Total staff: 92

Professionally qualified staff: 56

Director or Chief Executive

Dean: Professor Bill BP Lim

Heads of department

School of Architecture: Associate Professor David JK Lim

Department of Building Science: Professor Bill BP Lim

Department of Building and Estate Management: Associate Professor Michael Greaves

Principal sources of finance

Government.

Publications

The following are the latest annual publications:

- School of Architecture Journal (No 3/1982)
- Reprint 1980/81, Department of Building and Estate Management
- Reprint 1981, Department of Building Science
- Note: Reprints are papers written and published by the various Departmental staff members.

ZAF.428 **FM**

National Building Research Institute (NBRI)
PO Box 395, Pretoria 0001
Telephone: 27 12 86-9211
Telex: 3-630 SA
Cables: NAVORSBOU
Other addresses: Regional Offices
601 Saambou Building, 6th Floor, cnr Burg and Castle Streets, Cape Town 8001
Telephone: 437847
PO Box 17001, Congella 4013, Natal
Telephone: 255531
PO Box 1124, Port Elizabeth, 6000
Telephone: 532131
PO Box 2192, Windhoek, 9000, South West Africa/Namibia
Telephone: 37580
Date of establishment
1946

A research institute of a statutory organization responsible to the Minister of Trade, Industry and Commerce. Established by the Council for Scientific and Industrial Research (CSIR) as one of its national laboratories.

Its management is carried out under the Chief Director and two Directors who form the Directorate and who are each responsible for several of its Divisions or Regional Offices. The Directorate and Heads of the Divisions meet regullary as a general management committee.

The Chief Director is also administratively responsible for the Technical Secretariat of the Agrément Board of South Africa, whose members are also appointed by the Ministry of Community Development.

Main fields of work
The Institute's research and related activities cover most aspects of building, construction and civil engineering (except roads). In general, the main effort is given to research on problems of national interest.

Architectural: development of principles, processes and products relevant to the briefing, planning, design and procurement of buildings to accommodate various functions, particularly in the fields of housing, education and health services.

Building Services: research into plumbing, sewer design and flow data, water reticulation, water supply and water conservation in buildings.

Energy: Research into energy consumption and conservation in buildings of all types, development of DEROB (Dynamic Energy Response of Buildings), computer simulation program and validation using comprehensive building thermal performance studies and solar water heater development and evaluation.

Environmental Engineering: research into indoor environmental conditions for comfort, health and work performance; climate and building design; thermal performance of buildings; condensation control and moisture penetration; lighting; acoustics and natural ventilation.

Evaluations: technical evaluations of innovations in the building industry involving materials, devices, components or building systems and techniques; Agrément and building regulations.

Fire: research into fire testing, fire regulations and fire in relation to housing.

Concrete Engineering: research into concrete technology, including creep, quality control, cracking of concrete and concrete-making properties of materials.

Geotechnics: research into expansive soils and collapsing sands, their effects on buildings and methods of controlling them; weathering of rock minerals; shearing phenomena in soils; soil mapping, soil physics; foundation design; stability of slopes, excavations and waste material dumps.

Inorganic materials: research into the development, uses, quality and properties of inorganic building materials, particularly cements, lime, aggregates, ceramics, precast products, natural building stones and waterproofing agents, deterioration and dimensional change of materials; chemical corrosion of cement products; use of waste products.

Organic Materials: research into the properties and durability of organic materials used in building, such as paints and other surface coatings, rubbers, synthetic resins, polymers, plasticizers, fire retardants, elastomeric sealants, waterproofing sheets and roofing, flooring and insulation materials.

Structural Engineering: field and laboratory research into the performance and safety of structures; structural model stidies; development of design criteria; investigations into basic design parameters such as loading on structures, including the structure of wind and wind loading.

Economics: research into the organization and planning of the building process and construction, particularly economic, communication, statistical and managerial aspects; mechanization in building; operational research techniques; computer applications and electronic data processing.

Research Application: activities aimed at bridging the gap between research and practice through the promotion of effective communication by means of publications, audio-visual media, conferences and symposia, technical courses, serving on technical committees, assisting in the preparation of building regulations, and by providing an information and advisory service.

Contract investigation: specific long and short term problems in building and construction are investigated on a contract basis.

Total staff: 242
Professionally qualified staff: 130
Director or Chief Executive
Chief Director: J.F. can Straaten
Directors: J.A.P. Laurie, R.J. Page-Shipp
Heads of department
Divisions of:
Architecture: K.P.J. Napier
Building Services: P.R. Crabtree
Energy: J.A. Basson
Environmental Engineering: J. van Wamelen
Evaluations: D.R. Meyer
Fire and Concrete: V. Boardman
Structures and Geotechnics: A.A.B. Williams
Inorganic Materials: Dr J.E. Krüger
Organic Materials: Dr H. Saayman
Economics: Dr T.W. Miners
Research Application: R.K. Walker
Principal sources of finance
About 70% of the funds are derived from Parliament and the remainder from earnings through contract investigations and grants-in-aid.
Principal related bodies
Building Research Advisory Committee with Regional subcommittees for Natal, Eastern Cape, Western Cape, Transvaal and Central Regions.

Research Steering Committees: Hospitals; Housing; Indoor Environments for Human Habitation; School Building; Sewer Research; Solar Energy Conservation in Buildings and the Built Environment.

Publications
Research Reports, Special Reports, Symposia Proceedings, Computer Abstracts, NBRI Information Sheets, Regional Technical Notes, an Annual Report and popular-type booklets or introductory guides on technical aspects mainly relating to domestic buildings. Many papers are published in technical or scientific journals and reprints are available.

A list of publications (over 1000 items) and a catalogue of technical films (over 70 titles) held in the film library are available.

Certain publications are issued free of charge and the others are charged for.

ZAF.429

National Institute for Transport and Road Research (NITRR) of the Council for Scientific and Industrial Research

PO Box 395, Pretoria 0001

Telephone: 27 12 86 9211

Telex: 3-630 SA

Cables: Navorspad

Date of establishment

1951 (as a unit); 1955 (became an institute); 1976 (changed name)

The NITRR is one of the research institutes of the South African Council for Scientific and Industrial Research, a statutory organization responsible to the Ministe of Industries, Commerce and Tourism. Research is done on behalf of the Department of Transport, the Provincial Roads Departments, the National Road Safety Council, municipalities and other sponsors. Contract research is also done for industry.

Main fields of work

The broad objective of the Institute's work is to provide solutions to the problems experienced by national, provincial and local authorities, suppliers, consultants and contractors, in their efforts to finance, plan, design, build, maintain and operate transport systems, including roads, to carry people and goods conveniently and safely, taking into account cost, energy, climate and environmental factors. The Institute's research falls into three main fields: Transportation, Roads and Safety.

The Transportation Branch was started in 1975 and consists of five groups: Transport Planning, Transport Analysis, Transport Operations, Traffic Engineering and Rural Transport. Research in this branch covers a wide range of subjects, from ways of attracting patronage for bus transport to methods for counting traffic volumes on rural roads.

The Roads Branch is the oldest branch of the Institute. It is concerned with research into the design, construction and maintenance of roads and, to some extent, bridges. It also develops suitable instrumentation and equipment for road construction, testing and monitoring, and offers various testing and consulting services, on contract, to the road authorities.

The main task of the Safety Branch is to investigate ways of reducing the high accident rate on South African roads. Research is done on the three main factors involved in all road accidents: the road user, the vehicle and the road itself. Accident statistics are also collected and analysed, and are particularly important for identifying trends and for evaluating the effects of any change in the road situation

Total staff: 306

Professionally qualified staff: 142

Director or Chief Executive

Director: Dr S.H. Kühn

Assistant Directors: Dr G.L. Dehlen; R.N. Walker; Dr J.R. Odendaal; Dr C.P. Marais

Heads of department

Transportation Branch: Dr M.M. Slavik, Dr P.J.D. du Toit, P.R. Mann, Dr C.J. Bester

Roads Branch: Dr C.P. Marais, Dr M.A.G. Duncan, Dr F. Netterberg, Dr C.R. Freeme

Safety Branch: R. Fieldwick, R.F. Fernie, R.K. Deppe, J.C. Hillman

Principal sources of finance

South African Department of Transport; Provincial Administrations of: the Transvaal, the Cape Province, Natal, the Orange Free State: the Treasury; the National Road Safety Council.

Principal related bodies

The Transport and Road Research Advisory Committee, the National Transport Commission, the National Road Safety Council, the Executive of the CSIR – all guide on the choice of projects and priorities assigned to them.

Publications

- Transport and Road Digest, information sheets published at irregular intervals
- *Technical standards/recommendtaions:* Two series published: Technical Recommendations for Highways (TRH); Technical Methods for Highways (TMH)
- *Research reports:* NITRR series: Bulletins (also known as CSIR Research Reports) contain end-results of large research projects; NITRR series: PAD: include NITRR's Annual Report and the annual publication Transport Statistics; NITRR series: Technical Reports; interim reports, results of surveys, experiments, etc.
- VIA, NITRR's abstracts bulletin, published three times a year, available free of charge

  A publications list is also available, free of charge on request
- NITRR series: Technical Manuals (K)
- NITRR series: Computer Program Manuals (P)
- NITRR Reprints series (RR). Reprints of papers presented at conferences or published in periodicals by staff members.

Note: About 27 000 publications are distributed annually. Information sheets published at irregular intervals Transport and Road Digest

ZAF.430

Portland Cement Institute (PCI)

Portlandsementinstituut (PSI)

PO Box 168, 1685 Halfway House

Telephone: 805-2137

Regional offices

PO Box 12, 7495 Vrijzee

PO Box 13019, 6013 Humewood

PO Box 90, 3630 Westville

Date of establishment

1938

An establishment set up by the cement industry of South Africa to provide a technical advisory service on the properties and applications of cement and concrete. Its staff of qualified engineers act as consultants and this service is supplemented by the publication of books, pamphlets and leaflets, the provision of a specialized technical library, open to all, and four laboratories.

Main fields of work

- Dissemination of technical information regarding cement and concrete and their applications
- Provision, free of charge, of a technical advisory service to cement consumers
- Routine testing of concrete materials at nominal charges
- Provision of training courses at all levels from foremen to engineers
- Ad hoc research.

Total staff: 93

Professionally qualified staff: 15

Director or Chief Executive

Executive Director: Dr D.E. Davis

Director: S.W. Norton

Heads of department

School of Concrete Technology: P. Vogel

Laboratories: G.R.H. Grieve

Publications: B.J. Addis

Library: Mrs G.R. Goldstein

Principal sources of finance

Financed by the cement producing companies.

Publications

A list of PCI publications is available on request.

ZAF.431 **AM**

Specifile (a division of the Communications Group)

Cnr Handel Rd and Northern Parkway, Ormonde Ext 5, Johannesburg 2001

Telephone: (011) 835-2221

Telex: 8-2735 SA

Postal address: PO Box 7870, Johannesburg 2000

Date of establishment

1960

A private company formed to supply a commercial information system, primarily for the construction and mining industries.

Main fields of work
Dissemination of product and service information supplied by local and overseas manufacturers and agents to specifiers such as: architects, quantity surveyors, engineers, buyers, etc. Numerous methods of dissemination are used ranging from annual compendia and hard copy library services to software supply and viewdata systems. Emphasis at present is that of supplying integrated information systems by cross-referencing from hard copy systems in use to the newer computer based systems that have been and are being developed, hence supplying complementary information systems.
Total staff: 30 full time, 50 part time
Professionally qualified staff: 4
Director or Chief Executive
Chief Executive: M. Bateman
Heads of department
Research and Information Technology: A. Michie
Production: L. Pietersen
Distribution: E. Murphy
Administration: L. Verspey
Principal sources of finance
Revenue received from manufacturers for the dissemination of their information via various Specifile services; subscriptions from professionals in building and mining industry who use these services.
Principal related bodies
Specifile is a division of the Communications Group who also publish monthly trade and technical magazines and operate a company that organizes trade exhibitions and conferences.
Publications
Journal: Compendium of Building Products, annually; Compendium of Mining and Industrial products, annually; Mediafoucs (Media in Education and Training), annually.

ESP.432

Agromàn Empresa Constructora, SA, Division de Investigacion y Metodos

Agroman Constrction Co, Ltd, Division of Research and Methods

Raimundo Fernandez Villaverde 43, Madrid 3

Telephone: 2 53 58 00/2 53 32 00

Telex: 23332 AGREC E

Cables: AGROMAN

Date of establishment

1927

The Division of Research and Methods of Agroman is dedicated to solving all problems relating to the activities of the company in the field of civil, industrial and building construction.

Main fields of work

- Investigation into construction development
- Operational research
- Fundamental research
- Materials, machine and construction methods
- Work planning
- All problems in the construction field
- Co-operation with Spanish universities and technical schools
- Collaboration with foreign universities, agencies and individuals dedicated to research (USA, West Germany, etc.).

Total staff: 11 000

Professionally qualified staff: 750

Director or Chief Executive

President: General Manager Jose Maria Aguirre Gonzalo (Agroman)

Director, Division of Research and Methods: Manuel Galan Elgart

Principal sources of finance

Agroman Empresa Constructora SA; Agroman Construction Co. Ltd.

Publications

- Obras, a review, 3 times per year
- Almanac, annual
- Internal review of articles and books, with summaries, monthly
- Internal Reports of studies and/or research for all jobs.

ESP.433

Asociacion Tecnica Española de Climatizacion y Refrigeracion (ATECYR)

Spanish Technical Association for Air-Conditioning and Refrigeration

Conde de Penaluor. 38–3°05

Telephone: 28006 MADRID

Date of establishment

Authorization of the Spanish Ministry of Domestic Policy of 16 February 1977

Professional association with non-profit objectives devoted to research, development and information about HVAC matters. The Association has 850 members.

Main fields of work

Building Services; Heating and Cooling; Air Conditioning and Ventilation; Refrigeration; Plumbing and Sanitation; Heat Transfer; Combustion and Energy; Solar Energy; Corrosion; Comfort and the Environment; Noise and Vibrations; Thermal Insulation; Regulations and Standards; Climatological Test Data for Design.

Director or Chief Executive

President: José de Benito

Principal sources of finance

Annual fees of members.

Principal related bodies

President and Board, Working Groups and Technical Committees

Publications

- Monthly Bulletin for members and international associations
- Books: Translations or lectures of seminars.

ESP.434 **AM**

Institut de Tecnologia de la Construcció de Catalunya

Bon Pastor 5, 4, Barcelona 21

Telephone: 200 93 36/209 60 99

Date of establishment

Established in 1978 as a private foundation. The official registration was published at the Boletín Oficial del Estado in July 1979

A non-profit foundation at the disposal of all state, autonomic and municipal services, civil services, industry and private individuals.

The Institute co-operates and works in co-ordination with all institutions and public bodies connected with the building and construction fields.

Main fields of work

- Promotion in the development of local technolgy and a contribution to the adaptation of technoilogy that needs to be imported
- Co-operation with the public administration and industry in order to make possible construction programmes that guarantee the relationship between technology and productive capacity
- A contribution to collective knowledge and the necessary information on construction technology to offer to the user
- Improvement of professional qualifications (courses, seminars, lectures and publications)
- Documentation service, library and data bank.
- Principal divisions are as follows:

Research areas: Construction Economics and Administrative Management; New Construction; Old Construction; Construction Quality and Regulations. Services: Administration; Permanent training; Publications; Control.

Total staff: 51

Director or Chief Executive

President: Ferran Espiau

Director: Josep Maria Valeri

General Secretary: Pere Teixidó

Principal sources of finance

The Institute is supported partly by public and private subventions and by the financing of studies developed in agreement with the public administration. Fees from teaching activities and sales of publications.

Principal related bodies

The management of the Institute is carried out by members of the foundation: representatives of autonomic administration, university centres, professional centres and the production union.

Publications

- *Journal:* ITEC news, quarterly
- *Technical standards/Recommendations:* New recommendations in Building Schools; Constructive Solutions in Building Rehabilitation; Architectural Construction: 1. Techniques.
- *Research reports:* Construction Lexicon.

ESP.435

Centro de Estudios Hidrográficos (CEH)

Centre for Hydrographical Studies

Paseo Bajo Virgen del Puerto 3, Madrid 5

Telephone: 2656800

Telex: 22029 opceh e

Date of establishment

Decree of 7 July 1960

The Centre is a government organization within the Ministry of Public Works and is one of the seven establishments attached to the CDX (Centre for Studies and Experimentation of Public Works).

Main fields of work

The Centre is particularly devoted to the identification and evaluation of the hydraulic resources of the country, in quality and quantity. It deals with sea-water desalination and quality of water resources.

Its laboratory effects practical experimental studies in relation to rivers and hydraulic structures in the country both for public authorities and for private concerns. Concurrently the laboratory undertakes the research work relevant to these experimental studies.

The Centre also collaborates in postgraduate teaching.

Total staff: 302
Professionally qualified staff: 86
Director or Chief Executive
Dr José María Martín Mendiluce
Heads of department
Experimental Hydraulics (Hydraulic Structures, Fluvial Hydraulics, Instruments and Equipment, Models and Special Operations): Dr Manuel Díaz de Rábago
Hydrological Investigation (General Hydrology, Hydrometry, Hydrogeology): Dr Rafael Heras
Studies (Techniques of Planification, Techniques of Exploitation, Techniques for Study of Qualities, Projects, Water Treatment): Dr Delfín Pulido
Principal sources of finance
Financed by State subsidies and by the commercial activities it develops.
Publications
Lectures, courses, reports, recommendations and directories.

ESP.436 AM

Consejo Superior de los Colegios de Arquitectos de España
Higher Council of Architectural Colleges of Spain
Paseo de la Castellana 12, Madrid 1
Telephone: 435 22 00
Telex: 46004 AROS E
Cables: CONSARQUI
Date of establishment
1931 (Decree-Law of 13 June 1931).

The Council is the organization representing at national and international level all Spanish Colleges of Architects, unifying their conduct and co-ordinating their regulating actions for the professional practice of all Architects of Spain, guaranteeing this practice in the same conditions for every architect.

The Council establishes a code of professional ethics, judges appeals on corrective punishments imposed by the professional Courts of Colleges, organizes national and international congresses, participates with the Administration in all legislative matters regarding the profession of architect and proposes to it the formative characteristics that concur with architectural education.

Main fields of work
Architecture: building; standard architecture; monumental architecture; interior and exterior decoration of buildings; furnishing and ornamentation of premises; scenographic, industrial and artisan design; landscape architecture; buildings and monument conservation; legalizations.

City planning: in all planning modalities; city planning works.

Expert studies relating to architecture; advice; arbitration; measurement; valuations.

The principal divisions are: Technical Secretariat; National Department; International Department; Legal Office; Administration and Personnel.

Total staff: 18
Professionally qualified staff: 4
Director or Chief Executive
Secretary: José Juan Garcia Sanchez
Principal sources of finance
Official Colleges of Architects of Spain.
Publications

- *Journal:* Arquitectos, 11 issues per year
- *Technical standards/Recommendations:* All governmental, official and Council dispositions affecting the profession of architect
- *Research reports;* All reports on professional themes (laws, etc.).

ESP.437 AM

Entrecanales y Tavora, SA
Juan de Mena 8, Madrid 14
Telephone: 232 82 00
Telex: 489 66 Eyta E
Date of establishment
1931

Main fields of work
A private construction company devoted to all types of civil, industrial and building construction with important projects in Spain and many foreign countries.
Director or Chief Executive
President, Director-General: José María Entracanales

ESP FM

Instituto Eduardo Torroja de la Construccion y del Cemento
Building and Cement Institute Eduardo Torroja
Serrano Galvache, s/ PO Box 19.002–28080 Madrid
Telephone: 202 04 40
Date of establishment
1934

The institute is devoted to research and development in the technology of materials and structures. It forms part of the High Council of Scientific Research (CSIC) on which it officially depends.

Main fields of work
There are ten research teams currently operating with the principal thrusts being on: Energy conservation in the manufacture of cement; durability of concretes; cements with active additions, industrialized techniques for social housing and schools; energy conservation in construction; concrete under ultrasound action; analysis of rectangular elements of reinforced or prestressed concrete subjected to bending and shear; behaviour of reinforced concrete structures on a linear regime; and energy conservation in communications.
Total staff: 200, of whom 60% are research workers
Director or Chief Executive
Professor Dr Juan Murcia
Principal sources of finance
Mainly financed by grants from the CSIC
Principal related bodies
The top management team is composed of the Director, Vice Director, Secretary and the Board of the Institute
Publications
Journals: Construction Reports (Informes de la Construcción); Construction Materials (Materiales de Construcción)

More than 500 monographs, handbooks, etc. have been produced.

ESP.438 AM

Instituto Nacional para la Calidad de la Edificación (INCE)
National Institute for Quality in Building
Plaza de San Juan de la Cruz, Madrid 3
Telephone: 34 1 253.39.55
Date of establishment
1932

Official institution of national level attached to the Ministry of Public Works and Town Planning. Latest legal status issued in 1972.

Main fields of work

- General assistance in all subjects concerned with building control and technology
- Creation and support of laboratories for building control and testing of materials
- Establishment of quality marks for building materials and elements, and their registration
- Accreditation and surveying of private laboratories acting in the official building sphere.

- Assistance for the issuing of building standards.
- Publication of subjects related to INCE's field of work.

The Institute has four departments: Information and Studies; Control and Laboratories; Homologation and Research; and Social Design and Exhibitions.

Total staff: about 500
Professionally qualified staff: about 100
Director or Chief Executive
Managing Director: Gonzalo Ramirez Gallardo
Principal sources of finance
Basically financed through the State's General Budget.

ESP.439 AM

Instituto Técnico de Materiales y Construcciones (INTEMAC)

Technical Institute of Materials and Construction
Monte Esquinza 30, Madrid 4
Telephone: 410 51 58-62-66
Date of establishment
1967

A private organization devoted to the quality control of design, materials and construction, with a laboratory for testing materials and structures; also research activities.

INTEMAC is entitled by an Official Agreement from the Ministry of Public Works and Town Planning in all the existing Official Classifications of Agreement.

Main fields of work
INTEMAC's work covers the control of projects both in civil engineering and building through the Department of Design Control. It also covers site control of constructional work and equipment in all aspects through the Department of Field Control. The Central Laboratory with its different Divisions ranges over a wide field including building pathology, testing of building materials and research work on the same; it also covers the field of soil mechanics and geotechnics. The Institute is developing training courses and lectures on construction through the Department of Technical Advice together with technical publications on research performed by the Central Laboratory.
Total staff: 108
Professionally qualified staff: 33
Director or Chief Executive
General Director: Professor José Calavera
Heads of department
Control Division: (Director) Professor E. Gonzalez
Control Division: (Director) L. Benito
Laboratory Division: (Director) Professor A. Delibes
Technical Advice Division: (Director) F. Blanco, Tech Eng
Principal sources of finance
Private funds.
Principal related bodies
Committee of Directors.
Publications
Research reports: non-periodical, sent free on request.

ESP.440

Laboratorio de Carreteras y Geotecnia 'José Luis Escario' (LCG)

Road and Soil Mechanics Laboratory
C/Alfonso XII 3, Madrid 7
Telephone: 4 68 24 00
Date of establishment
1954

A government research establishment within the Ministry of Public Works and Town Planning; present status: governmental autonomous body, one of the eight research establishments belonging to the CDX (Centre for Studies and Experimentation in the Field of Public Works and Town Planning).

Main fields of work
Research work covers all aspects of the following areas:

- *Geotechnics:* soil and rock mechanics, foundation engineering, soil and rock testing and test standardization, field surveys and special field tests, technical assistance and advice
- *Roads:* earthworks, drainage works, stabilization, construction procedures, in-situ experimentation and field tests
- *Pavements:* materials, bituminous mixtures and PC concretes, construction, testing of road materials and test standardization
- *Traffic:* theoretical models and practical aspects of urban and inter-urban traffic flow, safety in road engineering, safety barriers
- *Transportation:* general planning studies, economic decision-taking procedures, integrated transport planning.

Further activites of LCG are concerned with:

- International co-operative research programmes (OECD, RILEM, AIPCR, etc.)
- Technical information, library and documentation services for all engineers depending on the General Directorate for Highways (Ministry of Public Works), in co-operation with different international programmes (IRRD scheme, Latin-American Road Documentation Network, etc.).

Total staff: 164
Professionally qualified staff: 51
Director or Chief Executive
Acting Director: Dr Eng V. Escario
Heads of department
Assistant Director for Geotechnics: Dr Eng V. Escario
Assistant Director for Roads, Traffic and Transportation: Dr Eng G. Carrillo
Principal sources of finance
60% by the Ministry of Public Works and Tow Planning; 40% other private bodies.
Publications

- *Journal:* Information Bulletin, bi-monthly
- Technical Standards issued by the Laboratory.

ESP.441

Laboratorio de Ingenieros del Ejercito

Military Engineering Laboratory
Serrano Jover, 2 Madrid 8
Telephone: 31 1 2484800
Date of establishment
Founded in 1897 and reorganized in 1927.

Governmental institution, principally engaged in testing and other assignments commissioned by the Ministry of Defence.

Main fields of work
The main task of the Laboratory consists in co-oerating in performing tests and technical studies with other bodies in the Ministry of Defence.

The principal fields in which action is undertaken are:

- Equipmont and material for the Military
- Material for the Infantry
- Construction of barracks and other premises for military use.
- Editing Military Standards
- Co-operation with Inter-Ministerial Commissions for editing Standards and Regulations with a national character concerning different construction materials.

Total staff: 68
Professionally qualified staff: 20
Director or Chief Executive
Colonel Director D. Francisco Baratech Zalama
Heads of department
1st Section: Non-metallic building materials: D. Rafael Sánchez Garcia
2nd Section: Physics and Chemistry: D. Emilio Hidalgo Sánchez
3rd Section: Mechanical Testing and Metallic Materials: D. Jaime Saborido Cursach

4th Section: Electro-technics and Electronics:D. Rafael Gónzalez Gónzalez
5th Section: Soil Mechanics: D. Antonio Siminiani Ruiz
Principal sources of finance
Financed by the State, through the Ministry of Defence.
Principal related bodies
The Laboratory is subordinate to the Directorate General fgr Armaments and Material in the Ministry of Defence. It works in conjunction with the other Laboratories and Centres in the Ministry.

ESP.442
Laboratorios de Ensayos e Investigaciones Industriales 'L.J. Torróntegui'
Industrial Testing and Research Laboratories 'L.J. Torróntegui'
Av Montevideo s/n, Apartado 1234, Olabeaga, Bilbao
Telephone: 94 441 93 00
Date of establishment
Orden Ministerios Educación e Industria 31 August 1955

A public institution dependent on the Industry and Education Departments of the Basque Country Government. It has close relations with the Faculty of Engineering of the Basque Country University.
Main fields of work
Tests and research programmes in the fields of:
- *Materials and civil engineering:* Soil mechanics and conglomerates; Field control and testing materials; Metallic materials, structural behaviour and non-destructive testing
- *Environment:* Air, water, solids and equipment
- *Chemical Analysis:* Water, mineral, ferrous, non-ferrous materials
- *Electrotechnics:* Measuring apparatus, materials, domestic, industrial equipment
- *Electronics:* Microelectronics, automation, equipment testing, climatic and interference testing
- *Mechanics:* Metrology, machine tools, acoustics, vibration, hydraulics, turbine model testing
- *Nuclear:* Nuclear testing equipment, dosimetry, radiation measuring.

Total staff: 90
Professionally qualified staff: 50
Director or Chief Executive
Director: José Luis Aguiriano Araco
Heads of department
Division of Applied Research: D. José Luis Ramírez Ortiz
Division of Technical Assistance: D. Jon Zarrabeitia Miñaur
Department of Construction: D. José Manuel Bárcena Diaz
Department of Chemistry: D. José Granda Secades
Department of Electricity: D. Santiago Goyarrola Arrien
Department of Electronics: D. Juan Antonio Fernández Tellechea
Department of the Environment: D. Millán Millán y D. Lucio Alonso
Departemnt of Polytechnics: D. José M. Alcalde (Metrology), D. Juan Angel Zalbidea (Hydraulics), D. Jesús P. Cabellero (Metrology), D. Santiago Torre Enciso (Nuclear)
Principal sources of finance
Through the work on testing or by giving technical assistance and by running applied research programmes.
Principal related bodies
Board of Directors for the Laboratories: Principal responsibility for Laboratory decisions
Executive Commission: Close management delegated from Patronato
Management Committee: Assesses the Direction of Laboratories in management matters.
Scientific Committee: Assesses the Direction of Laboratories in scientific matters.
Publications
- *7Journal:* Ensayos e Investigación, aperiodic
- *Research reports:* Problemática resistente en el Hormigón Armado: origen, detección y análisis y remedios (Profs. Ramírez and Bárcena).

ESP.443
Sistemas Industriales de Construcción (SIC Internacional)
Industrialized Systems of Construction
C/Colon 82, Valencia 4
Telephone: 3 229181

A private institution
Main fields of work
All types of project in the architectural field;
- Urban planning
- Development of construction systems;
- Participation of CIB working groups Symposium on Computer Languages and Symposium on Box Units Construction

The principal divisions are: Architectural Office; Engineering Office; Research Plant ; Supervision Office (control of works in situ).
Total staff: 20
Director or Chief Executive
Director: Fernando M. Garcia Ordoñez, arch
Publications
Many publications in technical reviews and in CIB review.

ESP.444
Universidad de Santander, Escuela Tecnica Superior de Ingenieros de Caminos, Canales y Puertos
Santander University, School of Civil Engineering
Av de los Castros s/n, Santander
Telephone: (942) 275600
Date of establishment
1966

A public institution at the disposal of all state services.
Main fields of work
The School is divided into 14 departments:
- Mathematics
- Applied engineering science
- Surveying and drawing
- Soil mechanics
- Construction
- Structural engineering
- Technology of concrete and steel; bridges; buildings
- Highways
- Urban planning
- Hydraulics
- Oceanography and harbours
- Energy
- Economy and regional science
- Design.

Total staff: 115
Professionally qualified staff: 100
Director or Chief Executive
Director: Professor Miguel Angel Losada Rodriguez
Principal sources of finance
Financed by the Ministry of Education and Science.
Publications
Internal principally
Other publications are devoted to: Teaching (courses and seminars); PhD theses; Technical assistance to planners, industry and contractors.

SWE.445

Building Standards Institution
Byggstandardiseringen (BST)
Drottning Kristinas väg 73, 114 28 Stockholm
Telephone: 08 23 72 50
Telex: 17453 sis–s
Date of establishment
1942

Affiliated as a Technical Division to Sveriges Standardiserings-kommission, SIS (Swedish Standards Institution).

Main fields of work
The technical work is carried out by some 100 standards committees and working groups in which more than 500 experts from the building field take part. Participation in the committees is voluntary as is the application of Swedish Standards.

Standardization in the building field comprises:

- Fundamental standards for building and civil engineering structures, e.g. dimensional and modular co-ordination, tolerances, joints, building drawings
- Wide-ranging standards for groups of products concerning preferred dimensions, performance requirements, general test methods, etc
- Specific standards for specific building products, materials or components concerning properties, test methods, etc.

Total staff: 10 technical officers and a secretariat staff.
Director or Chief Executive
Chairman of the Council and the Executive Board: Gunnar Rosenborg
Director: Carl Eddie Lund
Principal sources of finance
About 60% through the Swedish Council for Building Research, 10% through the Government, 15% through the members and 15% through the sales of standards and other publications.
Publications

- Swedish Standards
- Handbooks and yearbooks.

SWE.446

IFM Akustikbyrån AB
Warfvinges Väg 26, 112 51 Stockholm
Telephone: (08) 13 12 20
Telex: 173 12 IFM AKUSTIKBYRAN
Cables: Instimil, Stockholm
Date of establishment
1959

A private company, owned by the employees, consulted by the State, local authorities and industry. R and D is performed.

Main fields of work
All aspects of industrial, building and vehicle acoustics including research, measurements, consulting and design of means of abatement. Measurement and analysis of shocks in connection with road works and construction and in submarines at underwater bursts. Use of finite elements method in shock and vibration problems. Product development. Environmental engineering. Development of new techniques in the field of acoustics, vibrations and shock.
Total staff: 50
Professionally qualified staff: 40
Director or Chief Executive
Managing Director: Kjell Spång, MA
Heads of department
Structural Dynamics and Environmental Engineering: Åke Samuelson
Building Acoustics and Community Noise: Göran Westerberg
Industrial Noise: Matz Öhlén
Office in Göteborg: Juha Plunt
Office in Uppsala: Olev Sarapik
Office in Lund: Leif Cederfeldt
Principal sources of finance
Income from consultancy work plus financial support from R and D funds from the Government.
Principal related bodies
Around 50% partnership in Suomen Akustiikkakeskus OY, Helsinki, Finland (consultancy and R and D); 20% partnership in Mipa (Miljö Industriplanering AB) (an innovation company in the field of industry and environment).
Publications

- Results are given in reports owned by the customers
- Articles are published in technical magazines (mainly Swedish, English and American)

Some of the reports on the results of research work can be obtained from the Swedish Board for Technical Development and from the Swedish Council for Building Research.

SWE.447

Laboratory of Construction Ergonomics
Byggergonomilaboratoriet (BEL)
KTH, S 100 44 Stockholm
Telephone: 08 7877000
Telex: 10389 kthb Stockholm
Cables: Technology
Date of establishment
August 1977

The laboratory is an independent part of the institution of industrial ergonomics at the Royal Institute of Technology, Stockholm.

The main purpose of the laboratory is to carry out research and development concerning the work environment of building sites.

Main fields of work
Within the area of ergonomics of the building site a number of research and development projects have been tackled, such as brick laying, mounting of corrugated metal roof and facade sheetings, materials handling at the building site, devibration of manual tools, physiological load in pipeline construction, ergonomics of rock drilling, etc.
Total staff: 6
Professionally qualified staff: 5
Director or Chief Executive
Director: Professor Ulf Åberg
Principal sources of finance
Swedish Council for Building Research; Swedish Work Environment Fund.
Publications
Research reports: Reports from the Swedish Council for Building Research; Reports in the TRITA-BEL-series of KTH; the reports are mostly in Swedish.

SWE.448

Lund Institute of Technology, Computer Centre of the School of Civil Engineering
Väg- och vattenbyggnads datacentral (VDC)
Box 725, S 220 07 Lund
Telephone: 046 107000 direct 107376
Date of establishment
October 1978

Provides common computer facilities at the School of Civil Engineering together with those of the Lund University computing centre.

Main fields of work

- Providing possibilities for connection of terminals, microcomputers and other peripherals
- Three terminal rooms for education and research (25 CRTs and 2 CAD work stations) and 60 terminals at the departments connected to the VDC computers and the local computer net at the university

- Advice to users in connection with purchase and installation of new equipment
- Service on some terminal types and microcomputers
- Courses in basic computer use and world processing
- Development of programs to control measuring equipment and to process and present data
- General assistance on user and system level
- (Computers NORD 10/S and Vax 11/780)

Staff consists of a part-time manager and one electronics engineer.

Total staff: 15

Director or Chief Executive

Dr Per Christiansson

Principal sources of finance

User fees for running expenses; university, research and industry funds for investments.

Publications

- Information sheets in Swedish
- User manuals in Swedish.

SWE.449

Lund Institute of Technology, Department of Structural Engineering

Avdelningen for Båtande konstruktioner

Box 725, S 220 07 Lund

Telephone: 046 107000

Telex: 33533 Luniver S

Date of establishment

July 1964

One of 13 departments of the School of Civil Engineering, which is one of 5 sections of the Lund Institute of Technology, a technical faculty of Lund University.

Main fields of work

Research and education concerning the design of structures and structural elements composed of steel, concrete, wood and other materials such as aluminium and composites. The research focus is on non-material topics such as the forces acting on structures, the optimization, reliability and quality assurance of structures, and design work, information handling and the control of computer design aids.

Total staff: 14

Director or Chief Executive

Professor Lars Östlund

Principal sources of finance

Research councils and university funds.

Principal related bodies

Department of Structural mechanics (same division).

Publications

Research reports: Series TVBK at Lund University.

SWE.450 AM

National Board of Physical planning and Building

Statens Planverk

Västerbroplan, Box 12513, S 102 29 Stockholm

Telephone: 468 737 55 00

Telex: 12 841 Havkom S

Cables: Plan-S

Date of establishment

July 1967, Instruction SFS 329/67, Law of 9 Jne 1967

Central authority for planning and building matters, attached to the Ministry of Housing.

Main fields of work

The Board supervises planning and building in Sweden. It assembles and evaluates planning and building experience and knowledge. The Board provides general recommendations and also issues building regulations. It also issues general approvals for building materials, types of construction and installations. The pace of recent technical developments in the building sector and the increased demands for management of natural resources have given the Board new tasks. For example, political decisions on energy conservation have been translated into the Board's regulations for energy saving in buildings. The Board is also interested in how local authorites and others plan for vacation cottage development and recreation facilities, and in initiatives to preserve and supplement or renew older development. The Board is responsible for co-ordinating the guidelines set up by the Government for national physical planning and for providing advice and guidance to local and county authorities on these matters.

Included on the staff are architects, civil engineers, economists, lawyers and surveyors.

Swedeplan, the overseas agency of the National Board of Physical Planning and Building, is responsible for the Board's international contacts and co-operation and for the international marketing of expert services in the fields of physical planning and building.

Total staff: 192

Director or Chief Executive

Director General: Dennart Holm

Heads of department

Building Department: Technical Director Ulf Thunberg

Planning Department: Technical Director Peter Heimbürger

Administration Division: Nils Anegren

Swedeplan: Technical Director Tomas Lindencrona

Principal related bodies

Ministry of Housing

Publications

- Information publications free of charge, such as Planning and Building in Sweden, 1983
- Plan och Bygg (Planning and Building), 6 issues per year
- *Technical standards/Recommendations: Svensk Byggnorm 80* (Code of Statues) and other statutory publications on building; Reports on physical planning and building.

SWE.451

National Board of Public Building

Byggnadsstyrelsen (KBS)

Karlavägen 100, 106 43 Stockholm

Telephone: 08 14 10 40

Telex: 104 46 build S

Other addresses:

Byggnadsförvaltningen i Lund, Box 7032, 22007 Lund

Byggnadsförvaltningen i Göteborg, Box 31072, Göteborg 24

Byggnadsförvaltningen i Stockholm, Box 27 163, 102 52 Stockholm

Byggnadsförvaltningen i Uppsala, Box 185, 751 04 Uppsala

Byggnadsförvaltningen i Umeå Box 5056, 9005 Umeå

Date of establishment:

Established in 1918 in its present form (historically dates back to 1618); reorganized in 1967.

A Government establishment that administers publicly owned real estate and provides premises for the public sector; supervised by the Ministry of Finance.

Main fields of work

The main responsibilities are to administer government owned properties and to provide premises for public civil institutions. The properties are procured through purchase, renting or new constructions. KBS is also the central expert body for the State Administration concerning matters on real estate administration, construction activities, maintenance, etc., according to Government instructions of June 1982

Director or Chief Executive

Director General: Hans Löwbeer

Total staff: About 2100 of whom about 700 are in the main body at Stockholm.

Principal sources of finance

The premises are financed through rents based on business lines; a small part of the activity of the establishment is financed by special subsidies.

Publications

Reports are published on development projects which are carried out within the establishment and are available through most bookshops.

Instructions are published for the consultants working for the National Board and are available; their content is mostly technical.

SWE.452

National Swedish Environment Protection Board

Statens naturvårdsverk (SNV)

Management, Administrative Department, Technical Department, Planning and Information Department, Smidesvägen 5, Solna

Postal address: PO Box 1302, 171 25 Solna

Telephone: 08 98 18 00; 08 98 13 20

Telex: 11131

Cables: Environ S

Other addresses:

Department of Natural Resources, Vintergatan 1, Sundbyberg

Research Department

Research Secretariat and Water Laboratory in Solna, Ekensbergsvägen 113, Solna

Water Laboratory in Uppsala, Götavägen 4 and Villavägen 3 Uppsala

Air Laboratory, Studsvik, Nyköping

Brackish Water Toxicological Laboratory, Studsvik, Nyköping

Special Analytical Laboratory, Wallenberglaboratoriet, Lilla Frescati, Stockholm

Grimsö Research Station, Riddarhyttan

Date of establishment

1 July 1967

A non-commercial authority.

Main fields of work

The SNV was set up as the central administrative authority for environmental protection. The range of responsibilities includes environmental protection. The range of responsibilities includes environmental protection, water and air conservation, public cleansing, waste management, active open-air life, fauna protection and hunting. SNV examines certain cases falling under the Environment Protection Act and is in charge of and promotes research and investigations in the field of environmental protection. SNV is also responsible for certain supervision of the handling of products control. Within the framework of these duties the aims of SNV are, among other things, to:

- Help to ensure that the cultural and scientific development proceeds with due regard to the need for recreation and open-air facilities
- Safeguard the interests of nature conservation in social planning of various kinds
- Work for the planning and conduct of other enterprises to be effected with proper respect for their environmental consequences
- Combat pollution of water and air and work for noise abatement
- Direct planning in connection with active open-air life; protect beaches and shores
- Act as an enforcement authority under the Environment Protection Act and the Act on Products Hazardous to Health and to the Environment
- See that the results of research and investigation in the field of environmental protection are made use of in the everyday environmental protection work

SNV is subordinated to the Ministry of Agriculture. It is led by a Board of Directors nominated by the Government and is headed by a Director General.

The Board of Directors has nine members: representatives of The Swedish Confederation of Trade Unions (LO), The Federation of Swedish Industries, the municipalities, the popular movements and Parliament. Two personnel representatives are on the Board of Directors.

The Administrative Department is in charge of organizational, general administrative, financial and legal matters. The Department includes the Product Control Division, responsible for the Board's assignments according to the Act on Products Hazardous to Health and to the Environment.

The Department of Natural Resources deals with matters concerning nature conservation and landscape protection, land acquisition, sports and open-air facilities, pleasure boating, fauna protection and hunting. The Department also administers the Falsterbo Canal.

The Technical Department deals with matters related to water and air protection, noise abatement, public cleansing and waste management, municipal sewage processing and to the allocation of state grants. One unit is responsible for examining cases coming under the Environment Protection Act.

The Planning and Information Department is responsible for planning and co-ordination, primarily regarding matters of general nature and the external information activities. The Department is also responsible fro investigation work as for natural resources, e.g. within the framework of national physical planning.

The Research Department is responsible for research and investigations about diffusion, occurrence and effects of pollutants and consequences of other encroachments on the natural environment. The research programme is mainly carried out at scientific institutes outside the Board whereas the survey activity for the most part takes place at the Board's own laboratories. The assignment of grants to the research programme is decided by the Research Committee.

Total staff: 640 (1982)

Director or Chief Executive

Director General: Valfrid Paulsson

Principal related bodies

The Products Control Board is an independent authority, but linked administratively to the National Environment Protection Board by its secretariat – the Products Control Division. The Board determines matters coming under the Act of Products Hazardous to Health and to the Environment.

The co-ordination of State efforts for tourism is made by the Committee for Tourism and Recreation. The Committee's Secretariat is placed at the Environment Protection Board.

The Research Committee has both advisory and decision-making functions. Most of the decisions concern the assignment of research grants.

The SNV is assisted in its work by three advisory bodies: the Nature Conservation Council, the Water Protection Council, and the Air Protection Council. These bodies include representatives of authorities, organizations, research institutes, etc., particularly concerned with environment protection. The three Councils follow the development within their subject fields and assist the Board in matters of current interest.

The Committee for Acquiring Land assists the Board in matters of land valuation and negotiation.

The Committee for Subsidizing Sports Facilities administers matters concerning State grants for sports facilities.

Publications

The advice and instruction thus compiled by the Board are published: about 20 publications are issued every year.

Regulations and bulletins of the Products Control Board are published in the Board's statute-book, the SNFS.

The SNV also sponsor a wide range of informative activities. Information material published includes booklets, leaflets, films, etc. concerning environmental protection. The Board also arranges Naturum – visitors' centres in nature reserves. A monthly journal, Miljöaktuellt, is issued.

Reports and bulletins of SNV are also published in the PM series: about 100 are issued every year.

SWE.453 **FM**

National Swedish Institute for Building Research
Statens institut för byggnadsforskning (SIB)
Office: Södra Sjötullsgatan 3, Gävle
Postal address: Box 785, S-801 29 Gävle
Telephone: 026 10 02 20
Telex: 47396 Byggfo S
Cables: Institute gaevle
Other addresses:
Laboratory for Water Supply and Drainage, c/o Atomteknik AB Studsvik
Postal address: Fack, S-611 01 Nyköping
Telephone: 0155 800 00
Branch Office of Division 5:
Skånegatan 26,
S–412 51 Göteborg
Telephone: 031 81 13 58, 031 81 13 75, 031 81 13 92
Urban and Regional Research, Lund
Tunavägen 35, S-223 63 Lund
Telephone: 046 12 25 41, 046 13 34 36, 046 14 49 56
Cables: Institute Lund
Date of establishment
1960

The Institute is subordinated to the Ministry of Housing and Physical Planning (Governmental instructions in SFS 673/1965, amendment 682/1978). It is a State-Financed research organization in the building and housing sector. Its task is to undertake research relating to building and planning problems encountered by the community and relating to established policy objectives and to the policy instruments defined for the solution of problems and the attainment of objectives. Another task is to identify problems and abuses which have escaped the notice of other research bodies. The Institute's prime concern is with problems within the purview of the Ministry of housing and physical planning, i.e. housing policy, physical planning and building activities. The aims of SIB's research activities are defined as follows: to contribute towards a well-informed discussion of desirable goals and standard; to describe and explain the courses of problems, deficiencies and abuses; to test and propose various practical ways of solving problems and eliminating deficiencies and abuses to evaluate reforms in relation to defined objectives and the instruments employed.

Main fields of work
The research activities are sub-divided under the following headings:

- Community planning (local authority planning)
- Building planning (planning, use, management)
- Housing
- Building technology
- Energy conservation

The main tasks of the Institute are to analyse shortcomings and qualities within the building sector in order to promote improvements to buildings and living conditions and provide a factual basis for a debate on better ways to plan and build. In the main the activities are orientated towards the needs of the public sector, but the results that are produced also benefit the whole building industry.

The Institute works across many disciplines – across divisions and projects – on a continuous basis. Large, broad projects can be extended over a relatively long time, allowing the researchers to arrive at reliable results.

The Institute is not an educational establishment. The researchers have previous research qualifications obtained at technical universities, etc. Testing of building materials is undertaken by the National Testing Institute, Borås, Sweden.

The Institute is represented on 11 Working Commissions/Steering Groups of CIB.

The Institute is divided into 9 research divisions. Each division is engaged in different projects, totalling more than 40. The research divisions are:

Division 1. Living conditions and housing planning: Housing design and use. Housing standards. Overcrowding and excessive standards. Housing stock, housing production and changes in these. Evaluation of housing and housing areas. Housing sociology.

Division 2. Housing market and housing policy: Scope and direction of housing construction. Tenancy, tenant-ownership and freehold tenure. Housing costs and building costs. Housing agencies and residential segregation. Tenants' rights.

Division 3. Habitat – housing and settlement studies. Patterns of housing and settlement. Housing supply. Migration. Energy conservation. Communication.

Division 4. Urban and regional research: Land use and land policies. Local planning under changing external conditions. Institutional and social relations in housing areas.

Division 5. Impact of policy instruments on building: Economic and political control. Role of the municipalities in energy supply. The building committee and the builder. Supervision and advisory activities. Controls and incentives. Public technology procurement. The environmental problem on the building sites.

Division 6. Materials and construction: Ageing of materials and building components. Environmental strains. Maintenance cycles. Testing and control. Tolerances, deformations and stability.

Division 7. Building climatology and installations: Effects of climate on people. Environmental criteria. Settlement planning with regard to climate. Energy conservation. Input documentation for standards and planning. Heating, ventilation, water supply and drainage installations. Regulations and inspection procedures.

Division 8. Energy conservation: Energy conservation in existing buildings. Effects of energy conservation measures. Evaluation in relation to national objectives. Measurements of the indoor climate. Statistical sample surveys.

Division 9. Measuring and computer services: Measurement-instrumentation-data processing. Workshops for wood and metal work.

To Division 6: Geodetic test field at Måartsbo, Gävle.

To Division 7: Heating and ventilation technology laboratory.

Water supply and drainage laboratory, Studsvik, Nyköping (address: see above).

Human criteria laboratory.

Wind tunnel laboratory.

Mobile measuring units for indoor and outdoor climate.

Other divisions at the Institute: Administration; Information, including the International Secretariat

Total staff: 170
Professionally qualified staff: 100
Director or Chief Executive
Director: Professor Nils Antoni
Heads of department
Division 1: Ola Sikiö
Division 2: Lennart Lundquist
Division 3: Lars Erik Borgegård
Division 4: Lars Kjarne
Division 5: Gosta Frederiksson
Division 6: Christer Sjöström
Division 7: Hans Allan Löfberg
Division 8: Urban Norlén
Division 9: Per Wessling
Geodetic Test Field: John van den Berg
Heating and Ventilation Technology Laboratory: Mats Sandberg
Water Supply and Drainage Laboratory: Eskil Olsson
Human Criteria Laboratory: David Wyon
Wind Tunnel Laboratory: Bengt Wirén
Mobile Measuring Units: Olle Holmgren, Eng; Mauritz Glaumann
Administration: Hans-Gunnar Johansson
Information: Christer Meyer
International Secretariat: Olof Wallmén
Principal sources of finance
Financing: Ministry of Housing and Physical Planning. Administration: Swedish Council for Building Research. To some extent by temporary contracts.
Principal related bodies
Board nominated by the Government; bi-weekly meetings with heads of

Divisions and the Director;
reference groups of external experts
Publications
- Research results are published continuously in a series Meddelanden (about 25 annually, in Swedish), some of them in English as Bulletins
- Literature lists, annually
- Åarsbok (Yearbook) in Swedish
- New projects are reported in Swedish Building Research News, published by the Swedish Council for Building Research.

SWE.454

National Swedish Road and Traffic Research Institute
Statens Väg- och Trafikinstitut (VTI)
S 581 01 Linköping
Telephone: 013 115200 (Int. + 4613115200)
Telex: 50125 VTI S
Date of establishment
1 July 1971

The Institute is an independent State Authority, under the Ministry of Transport and Communications. Its principal aim is to carry out research and development work on roads, road traffic and road safety, needed to meet the intended standards of traffic and traffic safety policies, and to communicate these results to the decision-making and executive authorities.
Main fields of work
- *Road Research:* Earthworks, foundations, geology, bearing capacity, pavement engineering, all sciences concerned with construction and maintenance problems
- *Road User and Vehicle Research:* Road user behaviour and capacity, biomechanics and crash-tests including medical aspects, vehicle engineering, ergonomics, driving simulation
- *Traffic Research:* Traffic engineering, accident analysis, safety measures, traffic economics and traffic simulation
- *Research Administration:* Measurement techniques and field studies, data processing, mechanical engineering service, national special library, Swedish Documentation Centre of the OECD IRRD (International Road Research Documentation) system.

Total staff: 195
Professionally qualified staff: about 100
Director or Chief Executive
Director General: Hans Sandebring
Heads of department
Road Research: Ingemar Bronge
Road User and Vehicle Research: Kåre Rumar
Traffic Research: Börje Thunberg
Principal sources of finance
Partly (at present approximately 55%) through commissions, partly (45%) through Government grants.
Principal related bodies
In Sweden, besides the Ministry of Transport and Communications, engagement in planning bodies concerning R and D work. Internationally, bodies such as OECD (Road Research Programme, especially the IRRD documentation system), PIARC, ECE, IRF, ISO and various expert bodies; also through the Swedish International Development Authority (SIDA); in Scandinavia bodies such as the Nordic Governmental Committee for Transport Questions and the Nordic Traffic Safety Council.
Publications
- *Journal;* VTI Aktuellt (in Swedish), 6 times per year; VTI Topics (in English), 6 times per year
- *Research reports:* VTI Rapport (Report) (normally approximately 25 titles per year); VTI Meddelande (Bulletin) (normally approximately 50 titles per year); VTI Särtryck (Reprints) (normally approximately 10 titles per year)
- *Bibliographies/Catalogues:* VTI Publications (list of approximately 1500 titles in stock); VTI Nyförvärv (Library acquisitions), approximately 20 issues per year, the periodic print-out version of the library database catalogue, which is also accessible on-line from terminal
- Annual Report (in English).

SWE.455 AM

National Testing Institute of Sweden
Statens provningsanstalt (SP)
Brinellgatan 4, Box 857, 501 15 Borås
Telephone: 033 16 50 00
Telex: 36252 Testing S
Date of establishment
1920

Central Administration authority for official testing and inspection, also including general and legal metrology; in its present form since 1972.
Main fields of work
SP organizes and co-ordinates official testing. It carries out testing of materials and constructions as well as research work within the technical and scientific field, developing testing methods. Together with other authorities and institutions in Sweden, it participates in international co-operation within testing and metrology, harmonizing methods in these fields. The aim of the Department is to provide technical knowledge concerning materials and structures especially regarding safety, serviceability and economy.

The principal departments are: Administration; Building Technology and Mechanics; Chemistry; Physics; Centre for Testing; Centre for Metrology.
Total staff: 550
Director or Chief Executive
Director General: Henric Bildt
Principal sources of finance
Financed partly through the annual Parliamentry vote of funds for the Department of Industry, partly through testing fees; research programmes are also financed by other institutions.
Principal related bodies
Board of Directors appointed by the Government, with staff representatives.
Publications
SP–RAPP: technical reports; SP–INFO: information; SP–Met: testing methods; SP–FÖR: directions and guidance.

SWE.456

Nordic Institute for Studies in Urban and Regional Planning
Nordiska institutet för samhällsplanering
Skeppsholmen, S 111 49 Stockholm
Telephone: 08/24 63 00
Date of establishment
1 July 1968

A Nordic institute with the Nordic ministry council as the supreme decisive authority.
Main fields of work
The Institute is concerned partly with providing an advanced, post-graduate course for planners, and partly with promoting and carrying out research and research education within the planning field. Stress is put on the inter-Nordic inter-disciplinary aspects of planning and its societal prerequisites.
Total staff: 33
Director or Chief Executive
Director: Eva Hedman
Principal sources of finance
Financed by the five Nordic countries of Denmark, Finland, Iceland, Norway and Sweden.
Publications
Various research reports written in Danish, Norwegian or Swedish containing a short English summary.

SWE.457 **AM**

Royal Institute of Technology, Division of Building Technology

Kungl Tekniska Högskolan, Institutionen för byggnadsteknik
Brinellvägen 34, S 100 44 Stockholm
Telephone: 4687877000
Telex: 10389 kthb Stockholm
Cables: Technology

The Division is responsible for basic and advanced education at university level as well as research and technical development in the field of building technology.

Main fields of work

Research during latter years has been primarily directed towards aspects as moisture movement, thermal insulation, sealing behaviour and other properties of the building envelope.

The great importance of energy conservation both in existing and new buildings means that a large portion of this research work is devoted to technologies, systems and products for more efficient use of energy in buildings. This includes for example testing and evaluation of the efficiency and economy of solar heating systems.

In the near future research will be directed towards the important area of rebuilding, operation and maintenance of existing buildings.

The Division is equipped with modern laboratory facilities, including a climate simulator allowing tests of building components under controlled environmental conditions.

Total staff: 14–18
Professionally qualified staff: 8–12
Director or Chief Executive
Professor Ingemar Höglund
Principal sources of finance
Education is grant-aided by the Government; research work is mainly financed by the Swedish Council for Building Research (BFR).
Principal related bodies
Monitoring Centre for Energy Research, providing an instrumentation, measurement and data acquisition service.
Publications
Research reports: Bulletins and working reports. List of publications available on request.

SWE.458

Siporex Central Laboratory

Siporex AB, Centrallaboratoriet (SCL)
Strömgatan 11, Box 21053, 200 21 Malmö
Telephone: 040/293010
Telex: 32802 (siporex s)
Date of establishment
1945

Laboratory for control, research and development in the field of steam cured concrete.

Main fields of work

- *Chemical section:* process development, pilot plant studies, raw materials, corrosion, chemical structure
- *Building section:* structural properties, reinforcement, surface treatment.

In addition to research and development, an important part of the work is the provision of a continuous service to the Siporex factories all over the world, the control of processes and products and the optimization of resources.

Total staff: 13
Professionally qualified staff: 13
Director or Chief Executive
Øystein Kalvenes, Civil Engineer
Heads of department
Chemical Section: Anders Wastesson, Chemical Engineer
Building Section: Nils Edgren, Civil Engineer
Principal sources of finance
Nearly 100% financed by Siporex AB.
Publications
Research reports: Generally the results of research work are circulated in internal reports but not published; they are available to Siporex licensees in all countries.

SWE.459

Studsvik Energiteknik AB

S 611 82 Nyköping
Telephone: 0155 800 00
Telex: 64013 studs s
Cables: studs nykoping
Date of establishment
16 January 1948

State-owned research and development company in the energy field.

Main fields of work

- *Non-nuclear:* fossil fuels (combustion, environment, characterization (analyses), gasification); district heating (piping, component testing, planning); heat storage; heat pumps; solar energy; materials research (steel, plastic); energy planning; passive fire protection systems.
- *Nuclear:* fuel research; waste and environment; safety.

Total staff: 1000
Professionally qualified staff: 200
Director or Chief Executive
Managing Director: Kjell Håkansson
Heads of department
Energy Division: Bertil H. Sjöholm
Nuclear Division: Stig O.W. Bergström
Principal sources of finance
The main income comes from contracts awarded from state agencies, communities, companies both domestic and abroad.
Publications
Research reports.

SWE.460 **FM**

Swedish Building Centre

AB Svensk Byggtjänst
Office: Västra Vägen 7A, Solna
Postal address: Box 7853, S 103 99 Stockholm
Exhibition address: Regeringsgatan 42, Stockholm
Telephone: 4687305100
Telex: 8125070
Cables: JVEBYGG
Date of establishment
Established 1934 as a permanent exhibition of building commodities.

The Centre is owned by some 40 authorities, institutions and organizations covering all aspects of the Swedish construction industry.

The Centre provides information in virtually all fields related to building. The goal is to contribute to the creation of a good built environment, through developing and offering information tools adapted to the user's needs, through collecting, processing and disseminating knowledge between different members of the building trade and the general public and formally acting for the co-ordination of the structure and content of building information.

Main fields of work

The Centre is a major building information institution in Sweden. It works in close co-operation with other institutions and authorities, e.g. the Swedish Building Standards Institution, and the National Board of Physical Planning and Building, which is the main authority for issuing building regulations in Sweden. The Centre also works in close co-operation with the regional building centres in Malmö, Göteborg and Umeå

The Permanent Exhibition of the Centre offers a systematically arranged show of building and services commodities. Other forms of information on building products are the Swedish Building Product Catalogue and the Building Commodity File. The Swedish Building Product Catalogue, consisting of six binders, contains information edited to a uniform format for ease of assimilation and comparison between products of individual manufacturers. The Building Commodity File is a computer register covering all building materials and components available on the Swedish market. The data base of the Building Commodity File can be accessed by a user terminal for those seeking direct access.

The Centre publishes AMA (General specifications of material and workmanship), which is used as a reference document in compiling job specifications. AMA is available for site work, house building, heating and plumbing, refrigeration and electrical installations. It contains tried and tested optional variants of different designs. There are also joint administrative specifications.

The Technical Department carries out technical studies as a base for the AMA, the Swedish Building Product Catalogue, etc. The Department also carries out work on co-ordination of building documents including maintenance and development of the BSAB system. The Centre is also the Swedish SfB Agency working together with the international SfB Bureau in Dublin.

The Literature Service makes available all current and frequently used building literature. The literature is recorded in a computer file together with a brief synopsis. Continuous information on recently published books is given in a special journal in Swedish.

The Centre also carries out education of building professionals through courses all over the country.

There also exist three autonomous regional Building Centres which are mainly permanent exhibitions with bookshops:

Southern Building Centre (a CIB associate member)
Skånsk Byggtjänst
Studentgatan 4, S–211 38 Malmö
Telephone: (46–40) 70955

Western Building Centre
Byggcentrum
Örgrytevägen 2, S–412 51 Göteborg
Telephone: (46–31) 810085

Northern Building Centre
Norrlands Byggtjänst
Kungsgatan 73, S–902 45 Umeå
Telephone: (46–90) 125910

The Building Centres in Stockholm, Malmö, Göteborg and Umeå together own Byggtjänstorganens Service AB (Bygginfo) which is a field service organization with regional offices at several places in Sweden.

Total staff: about 100

Professionally qualified staff: about 50

Director or Chief Executive

Managing Director: Jan Lindgren

Heads of department

Secretariat Technical: Henry Karlsson
Administration: Hans Hansson

Principal sources of finance

The share capital amounts to 1 710 000. Shareholders do not demand a dividend on their holdings. The object is to ensure that operations return a surplus to be used for developing and consolidating the company. Turnover in 1984 was SEK 78 million. The operations of the company do not receive any funds from the state, with the exception of a grant or less than SEK 2 million a year from the Council for Building Research for technical studies and for development of co-ordinated classification tables and systems for the joint trade.

Publications

- *Technical standards/Recommendations:* AMA (General Specifications of material and workmanship) which consists of books on site work, house building, heating and plumbing, refrigeration and electrical installations. There are also joint administrative specifications
- *Research reports:* Reports on the technical studies
- *Bibliographies/Catalogues:* The Swedish Building Product Catalogue which is impartially edited by the Centre's own technical editors. It contains product data sheets, technical surveys and market surveys. Parts of the information in the catalogue are published according to a special quality description system called Byggvarufakta (Building Product Facts)
- 4 Product Indexes for site work, building plant and machinery, house building, heating and plumbing and electrical installations containing concise technical information on all materials and components in those fields available on the Swedish market
- A set of recommendations regarding classification and coding systems for the building trade including recommendations for numbering of drawings as well as recommendations for trade literature
- A catalogue of important publications used in the building trade containing about 4000 titles

Publications also include manuals, handbooks, educational aids, publications aimed at laymen, etc.

SWE.461

Swedish Cement and Concrete Research Institute

Cement – och betonginstitutet (CBI)

Drottning Kristinas väg 26, 100 44 Stockholm

Telephone: 08 14 32 40

Date of establishment

Established in 1942 by a donation from the cement industry and allied enterprises

A semi-governmental institute.

Main fields of work

Since 1972 research at CBI is co-ordinated with most other concrete research projects in Sweden. Co-ordination is mainly carried out through the Programme Council for Swedish Concrete Research. The three year programme for CBI's research and for the support given by the Swedish Concrete Research Foundation to research at other institutions is drawn up through discussions between the Foundation, the Swedish Council for Building Research, the Swedish Board for Technical Development, the Programme Council and relevant institutions.

The work of CBI ranges over all aspects of concrete technology from basic research of a materials science character to production and control. The structural behaviour of concrete buildings and components is not within the scope of the Institute. In connection with the consulting activities of the Institute, several research projects of an applied character and a relatively short duration are also conducted.

Commissions are received from state, municipal and private organizations and companies, mostly in Sweden but also abroad. The Institute arranges courses in the field of concrete work. The technical staff of the Institute is engaged in national and international work on standard specifications for concrete.

Total staff: 40 of whom approximately half are support staff.

Director or Chief Executive

Director: Professor Sven G. Bergström
Deputy Director: Dr Bo G. Hellers

Principal sources of finance

Financed by the Swedish Concrete Research Foundation, the Swedish Council for Building Research, the Swedish Board for Technical Development, revenues from commissions and course fees.

Publications

- CBI forskning (research): reports documenting research results CBI rapporter (reports): summaries of research results (such as reprints of articles of a high technical level)
- CBI rekommendationer (recommendations): series meant for technicians in practical work
- Reports from the Consultant Section, for approximately 100 investigations annually
- Course publications
- Annual report
- Directory of all publications available.

SWE.462

Swedish Corrosion Institute (SCI)
Korrosionsinstitutet (KI)
Drottning Kristinas väg 45, 47 D and 48, Stockholm
Postal address: PO Box 5607, S 114 86 Stockholm
Telephone: Int + 46 8 22 25 40
Telex: 14375 STURES S
Date of establishment
1 July 1965

The Institute was founded by industry and government for the purpose of combating corrosion in Sweden. Its principals are the Government, through the Swedish Board for Technical Development, STU, and the Foundation for Technical Research on Corrosion, a group of trade associations and industrial enterprises.

Main fields of work
The objects of the Institute are, among others:
- To conduct research and development and to take the initiative to such work at other research laboratories in Sweden
- To undertake commissioned research and provide a consultancy and advisory service
- To disseminate knowledge through information and education.

A substantial amount of the Institute's work is in the building sector.

In direct or close association with the building sector the Institute's R and D Department works on such matters as anti-rust painting, galvanizing, aluminizing, anodizing and other surface coatings, corrosion conditions in soils, electrochemical protection of metallic structures against corrosion in soil and water, atmospheric corrosion on buildings, corrosion in heating, ventilation and sanitary installations and agricultural buildings.

The Consultancy Department provides a telephone advisory service free of charge and undertakes such commissions as corrosion prevention studies and inspection of drawings, consultation in connection with project engineering and selection of materials, case histories, soil corrosion studies, e.g. in connection with the design and planning of water and sewage systems, investigation of causes of damage in heating, ventilation and sanitary installations and water and sewage networks, and cathodic protection of structures in soil and water. The Department has its own mobile laboratory for work in the field.

The Information and Education Department produces manuals, teaching aids, etc. in the corrosion field and arranges courses and conferences relating to the building and other sectors.

Total staff: 43
Professionally qualified staff: 31
Director or Chief Executive
Director: Professor Einar Mattsson
Heads of department
Research and Development Department: Dr Eng Vladimir Kucera
Consultancy Department: M Met Eng Mats Linder
Information and Education Department: M Chem Eng Gösta Svendenius
Principal sources of finance
The Institute is financed through a collective grant from the Swedish Board for Technical Development and the Foundation for Technical Research on Corrosion, through project grants, income from consultancy work, field stations, sale of literature, courses, etc.
Principal related bodies
Many of Sweden's corrosion specialists are associated with the Institute via its Technical Council, seven committees and a large number of working groups. Some industrial branches have also formed co-ordination committees for co-operation with the Institute.
Publications
- *Journal:* KI-nytt & korrosionsaktuellt från Norden (SCI News and Corrosion News from Scandinavia) issued three times a year as a newsletter with notices on matters of current interest in the corrosion field in the Scandinavian countries (in Swedish, Danish, Norwegian, Finnish)
- *Technical standards:* The Institute is participating in the production of national and international standards within SIS (the Swedish Standards Institution) and ISO
- Results of projects and other investigations are published in the Institute's green series of reports. In the event of a need for higher quality of print and reproductions, and for larger issues, the reports are published in the form of bulletins. The bulletin series also includes handbooks on special subjects
- List of literature (published annually)
- Summaries in Swedish and English are drawn up for reports, bulletins, articles, etc. They are distributed in large numbers to participants in the activities of organs of the Institute, and to libraries, abstract journals and literature banks in Sweden and abroad.

SWE.463 FM

Swedish Council for Building Research
Statens Råd För Byggnadsforskning (BFR)
Sankt Göransgatan 66, S 112 33 Stockholm
Telephone: 08 54 06 40
Telex: 10398 BFR
Cables: buildcouncil
Date of establishment
1960

A governmental agency responsible for the programming, co-ordination and funding of research in the fields of urban planning, architecture, housing and building construction and energy use for buildings.

The Council does not carry out any R and D activities.

Main fields of work
The field of responsibility of the Council embraces all physical aspects of housing, building and planning, through a multi-disciplinary approach. The main task of the Council is to initiate, encourage and co-ordinate research in areas considered to be important with a view to attaining a number of societal goals. In addition to the review and granting of applications for research grants and loans for experimental building projects the Council is also responsible for the documentation and dissemination of research results.

The Council's activities are divided into eight main programmes and a number of sub-programmes. The main programmes are:
- Urban and regional planning
- Building design
- Building technology
- Renewable energy sources
- Energy conservation
- Information
- International cooperation
- Secretariat

The Council's research departments are responsible for the preparation of research policy and research programmes, processing of applications for research grants (and loans for experimental building projects), information and documentation activities, international relations and general follow-up work.

Approximately 1700 research applications and 50 loan applications are processed every year, an average of 1000 being approved. The stock of current research projects financed by the Council and monitored by the research departments amounts to 5000 including loans for experimental building projects.

The Council is active on a number of IEA Committees.

Total staff: approx. 90
Professionally qualified staff: 37
Director or Chief Executive
Director General: Rune Olsson
Heads of department
Department of Urban Planning, Housing and Architectural Design: Leif Ljung
Department of Building Technology and Energy Conservation: Bengt Olsson
Department of Energy Systems and Energy Supply: Torbjörn Boström

Department of Information and International Relations: Dr André Teikmans
Department of Planning, Administration and Publishing: Bo Peterson
Principal sources of finance
The Council's budget (for the fiscal year 1984/85 million SEK 214.8) is provided by the Ministry of Housing and Physical Planning and for the programme Renewable Energy Sources by the Ministry of Industry.
Publications

- *Journal:* Tidskriften Byggforskning, 9 issues per year in Swedish), free of charge; Swedish Building Research News, 3 issues per year in English, free of charge
- Research reports (rapporter R), in Swedish; Documents (D), research reports, in English
  Special publications or books (T), generally in Swedish
- General information material in Swedish or English (G publications), such as activity plans, research programmes, lists of publications, project lists, etc, free of charge
- Synopses, abstract publications issued 8–9 times/year, in English, free of charge
- Referat, abstract publications issued 8–9 times/year, in Swedish, free of charge.

SWE.464

Swedish Council for Precast Concrete Control
Kontrollrådet för betongvaror (KRB)
Kronobergsgatan 49, S 112 33 Stockholm
Telephone: 00946 8 531780 or 533980
Date of establishment
1950

A non-profit-making voluntary association of government and municipal authorities, producers and consumers for quality checking of factory-made concrete products.
Main fields of work
The inspection and control of buildings in Sweden are carried out by government authorities and by the municipal authorities at the place where the building is to be erected. As a basis for this inspection, there are general regulations and norms, usually laid down by a government authority.

It is not practical to supervise and inspect the materials and workmanship which all the different regulations assume that the authorities at the place of use will carry out, when the products are manufactured in several different places. To carry out the necessary checks at the Swedish concrete products factories, the authorities, the manufacturers and the users have therefore formed a joint organization, KRB, which carries out quality checks at the factories affiliated to it.
Total staff: Board: 10 members
Professionally qualified staff: 7
Director or Chief Executive
Chairman: Gunnar Essunger
Secretary and Chief Executive: Ivar Magnusson
Principal sources of finance
The costs of the Council's activities – which yield no appreciable profit – are defrayed by means of fees paid by the associated manufacturers.
Principal related bodies
Members of the Board: National Swedish Board of Physical Planning and Building, National Swedish Road Administration, Association of Swedish Municipal Building Inspectors, National Board of Agriculture, Swedish Water and Waste Water Association, Swedish Precast Concrete Federation, Cementa AB.
Publications

- *Journal:* Meddelande, annually
- List of affiliated factories, annually.

SWE.465

Swedish Environmental Research Institute
Institutet för Vatten- och Luftvårdsforskning (IVL)
Hälsingegatan 43, PO Box 21060, S 100 31 Stockholm
Telephone: 08 24 96 80
Telex: 15581 IVL S
Cables: Waterresearch
Date of establishment
1 January 1966

IVL is a independent research organization supported by the Swedish Government and Swedish industry.
Main fields of work
IVL has concentrated mainly on the problem of industrial pollution of air and water. But its activities have extended to the following fields as well:

- Development of methods for handling and conversion of solid waste
- Development of methods for sampling and analysis relating to the internal industrial environment
- Development of methods for the characterization of risk chemicals in order to be able to evaluate their effects on health and the environment
- Development of forecasting methods and of models for evaluating the effect of stress factors on ecosystems
- Development of recycling and for low-energy methods for external treatment of liquid and solid waste
- Reclamation of contaminated land areas.

IVL has broad experience of problems related to oil pollution. IVL can assess and map ecological resources, in addition to acting as a consultant in contingency planning and oil combating operations.
Total staff: 200
Professionally qualified staff: 100
Director or Chief Executive
Managing Director: Allan Gustafsson
Director of Research: Professor Arne Jernelöv
Principal sources of finance
IVL is financed jointly by the Swedish Government and industry (the Government is represented by the Ministry of Agriculture and industry by an association). IVL is also financed by special grants, from commissioned research of an applied nature and from activities such as the investigation of receiving waters and chemical analysis.
Publications

- *Journal:* IVL-Nytt, quarterly, in Swedish
- *Research reports:* IVL A-publ; IVL B-publ.

SWE.466

Swedish Geotechnical Institute (SGI)
Statens geotekniska institut (SGI)
581 01 Linköping
Telephone: 013 11 51 00
Cables: Geotekniska
Date of establishment
1944

SGI is a government agency under the jurisdiction of the Ministry of Transportation.
Main fields of work
SGI deals with geotechnical research, information and consultancy.
Fields of activity:

- Building of highways, bridges and harbours
- House building, i.e. reinforcement of foundations of old buildings
- Land-use planning
- Land slides
- Laboratory testing
- Development of equipment, site investigations, field measurements

Total staff: 75
Director or Chief Executive
Dr Jan Hartlén

Heads of department
Department of Highways and Bridges: Allan Ekström
Department of Buildings: Göran Holm (Deputy)
Department of Land-use Planning: Per Ahlberg
Department of Laboratory Testing: Carl-Eric Wiesel
Department of Site Investigations and Field Measurements: Ulf Bergdahl
Principal sources of finance
State subsidy and commissions
Publications
- Reports, 5 per year
- Accession list, twice monthly

SWE.467 FM

Swedish Institute of Building Documentation
BYGGDOK, Institutet för byggdokumentation
Hälsingegatan 49, 113 31 Stockholm
Telephone: 08 34 01 70
Telex: 12563
Cables: builddoc stockholm
Date of establishment
1966

BYGGDOK is a non-profit foundation sponsored by the National Swedish Council for Building Research and it provides user-orientated services at cost price.
Main fields of work
The Institute is the central national body providing information and documentation services on building construction, architecture, regional and urban planning, civil engineering, building services and environmental topics related to building. It serves practitioners and research workers with information searches in those and related fields. It maintains a computerized information system, the BYGGDOK database.

Begun in 1975, the BYGGDOK database is representative of Scandinavian bibliographic data on building and its related fields, covering publications from Sweden, Denmark, Finland and Norway. Some non-Scandinavian material deemed to be of interest for readers in the Nordic countries is also represented. Every document mentioned is directly available at not less than one of the Nordic partners taking part in the input to the database.

Customers who prefer to access the database from their own computer terminals instead of through BYGGDOK's staff are offered annual contracts for such external connections. This is relatively the most rapidly growing service at present.

There is a monthly SDI service tailored to individual subscribers, a photocopying service, direct and inter-library loans and purchase of special material for clients.

Training activities at different levels, in addition to being for future engineers and architectural students at technical universities, aim at making practitioners realize the value of documentation services.
Total staff: 24
Professionally qualified staff: 15
Director or Chief Executive
Managing Director: Bengt Eresund
Principal sources of finance
Financed by charges for services and by Government grants.
Principal related bodies
Board of Directors.
Publications
- *Journal:* The abstracts journal Byggreferat, 12 issues per year
- Two directories are published at intervals of a few years, Bygginstitutioner, about building authorities and institutions in Sweden, and Byggnormindex, comprehensive index of building regulations and standards in Sweden, updated by Nya byggnormer
- A series of country profiles intended for construction contractors and consultants on the export market also published.

SWE.468

Swedish Institute for Steel Construction
Stålbyggnadsinstitutet (SBI)
Drottning Kristinas väg 48, S 11428 Stockholm
Telephone: 46–8/243330
Telex: 12442 fotex s steelconstruc stockholm
Date of establishment
1 January 1968

An independent research and development institute operated by an industrial research foundation.
Main fields of work
- Research programmes cover all aspects of steel construction: material properties; elements; details; joints; design and calculation rules; steel frame design; sheet steel construction; fire protection; corrosion protection; planning and design of buildings using steel construction; steel bridges and other steel structures; code and standardization work
- Information
- Advisory work for the building industry and authorities.

Total staff: 14
Professionally qualified staff: 10
Director or Chief Executive
Dr L.E. Wallin
Principal sources of finance
Financed jointly by industry and government research funds.
Publications
- *Technical standards/Recommendations:* Handbooks; Manuals
- *Research reports*
- General publications on steel construction.

CHE.469
Battelle, Centres de Recherche de Genève
Battelle, Geneva Research Centre
7 route de Drize, 1227 Carouge/Genève
Telephone: 022/439831
Telex: 423472 CH
Cables: Battelle-Genève
Date of establishment
1952

The Centres constitute one of the five major components of the Battelle Memorial Institute, an internationally operating applied research and development organization.

Research services are conducted on a contract basis for clients in industry, for government agencies, associations and private groups.

Main fields of work
- *Industrial Technology Centre:* Materials technology, surface treatment and corrosion; Metallurgy, glass and ceramics; Industrial chemistry, electro-chemical and food technology; Mechanical engineering, energy and environmental technology; Optics, electronics and bio-analytical instrumentation; Computer sciences
- *Toxicology and Biosciences Centre:* Biotechnology and applied biochemistry; Pharmacology; Drug metabolism and pharmacokinetics; Toxicology and pathology; Genetic toxicology and carcinogenicity; Inhalation toxicology and physiology
- *Applied Economics Centre:* Economics and social sciences; Capital goods and agrofood industries; Resources, materials and environment; Business strategy and R and D planning.

Total staff: 420
Professionally qualified staff: 170
Director or Chief Executive
Director General: Dr Valentin Stingelin
Heads of department
Industrial Technology Centre: M. Vanderlind
Toxicology and Biosciences Centre: R. Haroz
Applied Economics Centre: M. de Vos
Principal sources of finance
Funds from contract research.
Publications
Information Bulletin, 3 to 4 times a year.

CHE.470
Betonstrassen Ag
Routes en Béton SA
SA Strade in Calcestruzzo
Concrete Roads Ltd
Lindenstrasse 10, CH-5103 Wildegg
Telephone: 064 53 13 74/75
Date of establishment
1929

Research and consulting institution of the Swiss Portland cement industry.

Main fields of work
- Research
- Consulting
- Expertise covering: material technology; road structural design; road construction applications
- Tuition (at training centre): special courses
- External instruction: educational and lecturing programmes
- Information

The principal divisions are: Concrete Pavements Department (for roads, airport runways and aprons, industrial yards, etc.); Soil Stabilization Department; Central Laboratory; Concrete Pavement Maintenance Team; Administration and correspondence.

Total staff: 25
Professionally qualified staff: 7
Director or Chief Executive
Director: W. Wilk, Dipl Eng ETHZ
Principal sources of finance
30% of total revenue is covered by the annual contributions of the Swiss Association of Cement Factories; the balance consists of revenue derived from professional activities.
Publications
An information bulletin published in German under the title Betonstrassen and in French under that of La Route en Béton is distributed free of charge to the most appropriate interested audience such as: engineering offices, technical tuition circles, contractors as well as student members of technical faculties, Swiss federal and cantonal road construction authorities, etc.

Special publications: Publications in technical periodicals on the subject of research activities and results achieved.

CHE.471
Dokumentation Wohnen
Living in Houses and Apartments: A Documentation Service
Winterthurerstr 52, CH-8006 Zurich
Telephone: 01-362 66 96
Date of establishment
1978

Independent private working group.

Main fields of work
- *Documentation Services:* on the use of living space in homes and apartments. Library with card-index; bibliographies
- *Information Services:* distribution of books, illustrative and working material; editors of an information bulletin entitled Dokumentation Wohnen aktuell
- *Teaching and Consulting Services:* courses for teachers and consumers; development and editing of working materials for schools; consulting services for organizations and institutions in the development and realization of courses; co-ordination of teaching activities on the use of living space; writing of commissioned articles in specialized and other magazines; public consulting and information service at the Swiss building centre; column for home advice in a weekly periodical; consulting services for private and public construction companies, as well as for owners of apartment buildings on matters of construction or renovation.
- *Working group of two interior designers:* Verena Huber, Bruno Müller-Hiestand. Outside collaborators for special tasks.

Principal sources of finance
Privately owned enterprise: sale of documents; fees for services; research grants.
Publications
- *Journal:* Dokumentation Wohnen aktuell, 3–4 issues per year
- *Bibliographies/Catalogues:* Index of literature; Prospectus for school material.

CHE.472 **UM**
Dr Ing T. Koncz, Consulting Engineer for Industrialized Building
8053 Zurich, Witikonerstrasse 297
Telephone: 55 38 22
Telex: 54009

Consulting Engineer – independent of any organization or state.

Main fields of work
Design, Research and Development of industrialized building systems and machines for the manufacturing of precast concrete constructions.

Sessions of know how and technical assistance.

Three departments: Structural, Mechanical and Research.

Principal sources of finance
Consulting engineering fees.

Publications
- *Research reports:* Reprints of articles in various journals; Books: Manual of Precast Concrete Construction in 7 languages; Industrialized Building in 3 languages
- *Bibliographies/Catalogues:* Activities report.

CHE.473

Ecole Polytechnique Fédérale de Lausanne, Département de Génie Civil, Institut de Statique et Structures (ICOM – Construction Métallique)
Swiss Federal Institute of Technology, Lausanne (ICOM – Steel Structures)
GCB (Ecublens), CH-1015 Lausanne
Telephone: 021/47 11 11, 021/47 24 25
Date of establishment
1969

ICOM is part of the Civil Engineering Department of the Swiss Federal Institute of Technology, Lausanne (EPFL). The aims are education, research and services in the field of steel structures.
Main fields of work
- Teaching of constructional steelwork
- Research in the fields of steel structures (fatigue, stability, coldformed steel) and composite steel–concrete construction (building and civil engineering)
- Studies for third parties

Total staff: 25
Professionally qualified staff: 17
Director or Chief Executive
Director: Professor John C. Badoux
Principal sources of finance
Financed by the Swiss Federal Institute of Technology, Lausanne, the Swiss National Science Foundation and other miscellaneous public and private sources.
Publications
- Theses (doctorate)
- Monographs
- Scientific papers
- A list of publications is available
- Series of courses entitled: Conception des charpentes métalliques.

CHE.474 FM

Ecole Polytechnique Fédérale de Lausanne (EPFL), Institut de Recherche sur l'Environnement Construit (IREC)
Swiss Federal Institute of Technology, Lausanne, Research Institute for the Built Environment
14 av Eglise-Anglaise, 1006 Lausanne
Telephone: 021/47 32 91
Telex: 25 507 epla ch
Date of establishment
1971

The Institute is part of the Swiss Federal Institute of Technology and provides research services to public (Federal, cantonal, municipal) bodies.
Main fields of work
IREC has two main fields of activity which are teaching and research.
Teaching concerns the following subjects: economics, sociology (introduction), urban economics and sociology, economics of energy, etc. Lectures and seminars are given in the Department of Architecture and other Departments of the EPFL.
The principal aim of the research activity is to provide interdisciplinary knowledge in the following fields:
- Interaction between human behaviour and settlements
- Urban development, in particular its economic, physical and sociological aspects
- Construction and management of the built environment.

Total staff: 20
Professionally qualified staff: 16
Director or Chief Executive
Director: Professor Dr Joseph Csillaghy
Professor Dr Michel Bassand
Principal sources of finance
Swiss Federal Institute of Technology; National Foundation for Scientific Research; Research Committee for Housing; Cantonal and Municipal Administrations.
Publications
- *Books:* Series I: Etudes urbaines et régionales, Ed. Georgi, ST. Saphorin, Suisse; Series II: Villes régions et société; Ed. Presses Polytechniques Romandes, 1015 Lausanne, Suisse
- *Bibliographies/Catalogues:* available on request.

CHE.475

Ecole Polytechnique Fédérale de Lausanne, Laboratoire des Matériaux de Construction
Swiss Federal Institute of Technology, Lausanne, Laboratory for Building Materials Science
Chemin de Bellerive 32, CH-1007 Lausanne
Telephone: 021/47 28 43
Telex: 24 478
Date of establishment
Founded in 1981 as Concrete Section of the Materials Testing Laboratory.

The Laboratory for Building Materials Science of the Swiss Federal Institute of Technology, Lausanne, is both a university research centre and an official laboratory for testing materials.
Main fields of work
Research activities: A major aim is to formulate realistic material laws; research is concentrated on creep and shrinkage of concrete; finite element analysis is applied to simulate and analyse the composite structure of concrete. The applicability of fracture mechanics to concrete is studied; special emphasis is placed on strength under impact loading conditions. Mechanisms of shrinkage and crack formation in a composite structure are studied. Polymer concrete and adobe concrete are treated in special projects. Durability of concrete structures is studied both experimentally and numerically.
Testing of materials: Routine tests such as the determination of strength on standard cubes or cylinders. Testing of frost resistance, water permeability, porosity, etc. Testing of elements such as concrete tubes, beams, etc. Analysing of damaged structures. Concrete technology in developing countries.
Teaching: Members of the laboratory are involved in teaching building materials science and technology in four different departments of the Swiss Federal Institute of Technology, i.e. Civil Engineering, Rural Engineering, Materials Science and Architecture. Teaching activities involve basic courses, specialized courses, practical exercises, projects, master's and PhD theses.
Total staff: 22
Professionally qualified staff: 11
Director or Chief Executive
Director: Professor Dr Folker H. Wittmann
Heads of department
Technology and Testing of Materials: F. Alou
Mechanics of Building Materials: P. Roelfstra
Physics of Building Materials: C. Ferraris
Chemistry and Durability: Y. Houst
Principal sources of finance
Government; Science Foundation; industry
Publications
- *Research reports:* Beiträge zur Anwendung der Aeroelastik im

Bauwesen (Reports on the application of aeroelasticity to structures). So far 17 reports have been published

- Autoclaved Aerated Concrete, Moisture and Properties, Elsevier (1983)
- Fundamental Research on Creep and Shrinkage of Concrete, Martinus Nijhoff (1982)
- Fracture Mechanics of Concrete, Elsevier (1983)
- Creep and Shrinkage in Concrete Structures, Wiley (1982).

CHE.476 AM

Eidgenössische Forschungskommission Wohnungswesen (FWW)

Commission Fedérale de Recherche pour le Logement (CRL)

Federal Housing Research Commission

Weltpoststrasse 4, CH-3000 Bern 15

Telephone: 031 61 24 44

Date of establishment

1976 (law of 14 October 1974)

Advisory body to the Federal Department of Housing. Established 1976 (previously Eidgenössische Forschungskommission Wohnungsbau (FKW)).

Main fields of work

As an advisory body to the Federal Department of Housing, the Commission is responsible for the elaboration of research programmes in the field of housing, the allocation of public research funds, the selection and monitoring of research projects as well as the dissemination of research findings.

The Commission itself is not engaged in research activities. It develops a network of scientific communications on a national and international level. It stimulates the exchange of information and sponsors conferences and workshops. It detects emerging research topics and identifies gaps in current activities.

Main research topics from the programme 1982–1984:

- Modernization and rehabilitation
- Housing cost reduction and special aspects of the rental dwelling sector
- Housing market analyses, programming of housing

The structure is as follows:

Commission: 12 members; Executive Board: 3 members; President; various ad hoc working groups; Secretariat directed by the Research Section of the Federal Department of Housing.

Total staff: $2^1/_2$ (man years)

Professionally qualified staff: 1 $1^1/_3$ (man years)

Director or Chief Executive

Head of Research Station, Federal Department of Housing: Dr oec. Peter Gurtner

Principal sources of finance

Financed by public funds; the annual outlays of ca. SF 450 000 are contained in the ordinary budget of the Federal Department of Housing.

Principal related bodies

Federal Department of Housing, to which, by law of 1974, the Research Commission is attached as an advisory body.

Publications

- Schriftenreihe Wohnungswesen (Bulletin du logement; Housing Series)
- Arbeitsberichte Wohnungswesen (Rapports de travail dur le logement; Working papers on housing).

CHE.477

Eidgenössische Materialprüfungs- und Versuchsanstalt (EMPA)

Laboratoire Fédéral d'Essai des Matériaux et Institut de Recherches

Laboratorio federale di prova dei materiali ed isituto sperimentale

Swiss Federal Laboratories for Materials Testing and Research

Ueberlandstrasse 129, Postfach, CH-8600 Duebendorf

Telephone: 01 823 55 11

Telex: 53 817

Cables: Empaprufung

Date of establishment

1880

EMPA serves as a neutral and functionally independent institute connected with the Swiss Federal Institute of Technology, outside the sphere of economic competition, for the official testing of:

- Raw materials, building and manufacturing materials, fuels and auxiliary materials with regard to their properties and practical application
- Manufactured products, processes and finished elements or entire structures and installations with regard to their suitability

The EMPA participates in the drawing up of standards, guidelines and regulations within its fields of activity.

Within its fields of activities, the EMPA conducts independent scientific research, particularly for the development and improvement of testing methods. EMPA may be assigned, either totally or partially, the enforcement of or controlling functions under federal regulations.

EMPA staff members may be authorized to take on teaching assignments in the study and testing of building and manufacturing materials at the Swiss Federal Institute of Technology, the University for Economics and Social Sciences at St Gall, technical schools and for specific courses.

Main fields of work

- Research and control on raw materials and building materials in order to define their properties and features for us in practice, on finished products or pieces in progress, on constructive works or construction elements and on installations
- The Laboratory participates in the elaboration of standards, directives or regulations in its field of activity
- It carries out independent scientific research aiming particularly at developing and perfecting testing methods
- It may be in charge of the whole or partial execution of federal regulations or of their control operations.

Total staff: 400

Professionally qualified staff: 120

Heads of department

Building Materials: U. Meier

Metals: Professor R. Steiner

Chemistry: Dr W. Jutzi

Special Services: R. Burkhardt

Principal sources of finance

Two-thirds of the budget is provided through private contracts.

Publications

- *Journal:* Publikationen (articles from scientific reviews), 3–4 issues per year
- EMPA-Bericht (EMPA Report).

CHE.478

Eidenössische Technische Hochschule Zürich, Baustatik und Stahlbau

Swiss Federal Institute of Technology, Zurich, Applied Statics and Steel Structures

ETH-Hönggerberg, 8093 Zürich

Telephone: 01 377 31 74

Telex: 54 354 ehopz ch

Date of establishment

Established around 1930.

A research unit.

Main fields of work

- Teaching in the field of applied statics, steel, wood and aluminium structures
- Consultancy for bridge and structural engineering
- Expert's reports in cases of structural failures
- Research in the following fields: steel structures (especially buckling problems of stiffened plated structures); composite structures (including effect of creep and shrinkage); wood structures (connections of high performance, glued laminated timber of hardwoods, space frame system, prestressing of timber); aluminium structures (cabins for ropeways, especially their suspension)
- Courses for practising engineers
- International connections with: International Association for Bridge and Structural Engineering (IABSE), European Convention for Construction Steelwork (ECCS), Council on Tall Buildings, Deutscher Stahibau-Ausschuss, etc.

Total staff: 10 (academic personnel: 7)

Director or Chief Executive

Director: Professor P. Dubas

Principal sources of finance

Financed partly through annual grants (budget) of the Board of the Swiss Federal Institute of Technology and partly through contributions from industry.

Publications

- Veröffentlichungen Baustatik and Stahlbau, irregular
- Numerous papers in scientific and professional journals.

CHE.479

Eidgenössische Technische Hochschule Zürich, Institut für Baustatik und Konstruktion (IBK)

Swiss Federal Institute of Technology, Zurich, Institute of Structural Engineering

ETH-Hönggerberg, Gebäude HIL, CH-8093 Zürich

Telephone: (01) 377 31 55

Telex: 54 354 ehopz ch

Date of establishment

1930

Research institute in the field of structural engineering.

Main fields of work

- Education at the Civil Engineering Department
- Further training for practitioners and scientists
- Theoretical and applied research in the following fields: application of information processing to civil engineering (girders, slabs, shells); safety of works (accident analysis, analysis of the concept of safety); shear and cross bending; slabs (prestressed with non-adherent wire, bending behaviour, punching, influence of openings near columns); concrete shear walls; masonry walls (rotational capacity, shear); fatigue (reinforced and prestressed concrete beams); earthquake loading (damping behaviour of prestressed beams, soil–structure interaction with bridges, gravity dams, material modelling); impact loading (energy absorption capacity, dynamic plasticity)
- Participation in various standard commissions (SIA, CEB, ACI)
- International collaboration with different institutes
- Expertise, reports, adjudications.

Total staff: 38

Professionally qualified staff: 32

Director or Chief Executive

Head: Professor Dr H. Bachmann

Principal sources of finance

Financed by the Swiss Federal Institute of Technology, Zurich, federal institutions, foundations, funds, etc., the cement, iron and steel industries.

Publications

- Course and advanced course manuscripts
- Technical reports
- Articles in specialized reviews.

CHE.480

Eidgenössische Technische Hochschule Zürich, Institut für Baustoffe, Werkstoffchemie und Korrosion

Swiss Federal Institute of Technology, Zurich, Institute for Building Materials, Materials Chemistry and Corrosion

ETH-Hönggerberg, CH-8093 Zurich

Telephone: 01 377 44 11

Telex: 54 354 ehopz ch

Date of establishment

1974

The Institute is part of the Department of Materials Sciences.

Main fields of work

Building Materials Section: mechanical technical performance of building materials such as metals, concrete, bituminous materials, plastics and wood. Special research fields: frost/thaw protection of concrete; creep behaviour of fibre-reinforced plastics; rheology of building materials.

Materials Chemistry and Corrosion Section: Chemistry of building materials and other construction materials; corrosion of metallic and non-metallic materials; protection against corrosion. Special research fields: localized corrosion; passivity; pitting and crevice corrosion; intercrystalline corrosion; stress corrosion cracking; corrosion in a non-aqueous environment.

Total staff: about 25

Professionally qualified staff: about 20

Director or Chief Executive

Director: Professor Dr H. Böhni

Deputy Director: Professor Dr A. Rösli

These two positions alternate every 2 years.

Heads of department

Building Materials Section: Professor Dr A. Rösli

Materials Chemistry and Corrosion Section: Professor Dr H. Böhni

Principal sources of finance

Financed partly through annual grants (budget) of the Board of the Swiss Federal Institute of Technology and partly through contributions from governmental offices.

Publications

- Several papers are published every year in professional or scientific journals
- International reports are prepared and circulated to the most appropriate audience.

CHE.481

Eidgenössische Technische Hochschule Zürich, Institut für Grundbau und Bodenmechanik (IGB)

Swiss Federal Institute of Technology Zurich, Institute of Foundation Engineering and Soil Mechanics

ETH-Hönggerberg (HIF), 8093 Zürich

Telephone: (01) 377 4411

Telex: 54 354

Date of establishment

1970

Institute attached to the Civil Engineering Department. Previously it was called the Soil Mechanics Department of the Hydraulics and Soil Mechanics Research Laboratories connected to the Federal Institute of Technology Zurich, founded in 1935.

Main fields of work

- *Education:* courses, seminars, laboratory, further education and further training for practising engineers
- *Research:* theoretical and applied research on the behaviour of soils under static and dynamic stress
- *Technical assistance:* technical advice for federal, district and communal authorities and for private parties; sub-soil studies, samples investigations; complete studies for the design and construction of earth and/or rock-fill dams; advice for the works progress; damage examination; legal expertize
- Participation in various technical commissions and in the elaboration of standards.

Total staff: 58

Professionally qualified staff: 29

Director or Chief Executive

Director: Professor H.J. Lang

Principal sources of finance

Financed through working credit from the State, research funds and expertise.

Publications

- Technical and scientific publications in various Swiss and foreign reviews
- Bulletin of the Institute, in the form of off-prints from publications or PhD theses issued when need arises (20 available)
- Conferences in Switzerland and abroad
- Courses.

CHE.482 FM

Eidgenössische Technische Hochschule Zürich,

Swiss Federal Institute of Technology, Zurich, Institute for Building Research

ETH-Hönggerberg, CH-8093 Zürich

Telephone: 01 377 2937

Date of establishment

1 January 1969 (authorized by the Swiss School Council)

Public research institute of the Swiss Federal Institute of Technology, Zurich, carrying out theoretical and practical research into all aspects of the building process with particular emphasis on housing.

The main focus for research is on subjects not yet explored by other university institutes or treated by the building industry. Account is also taken of the human sciences, education and practice in the research activities.

In addition to the research work, particular attention is paid to the education of the architect and civil engineer at the ETH through close co-operation with the teaching staff.

Main fields of work

Research is concentrated mainly on design and planning, construction and use; current projects include:

- Housing renovation
- Defects and quality in building
- Energy requirements in building
- Building cost estimation
- Structure of the building industry
- Project management

Work is carried out principally in small groups composed of members of the Institutes, supplemented by specialists from various disciplines.

Total staff: 19

Professionally qualified staff: 12 (architects) 5 attached

Director or Chief Executive

Director: Professor B. Hubert

Principal sources of finance

Financed mainly through public funds; also by expertise or services rendered for the public and private building sectors.

Principal related bodies

Swiss Federal Institute of Technology (ETHZ), Department of Architecture; Department of Civil Engineering teaching establishment.

Publications

- *Technical standards/recommendations:* according to the project
- *Research reports* (mainly in German): Current research (Red series); Working reports (internal); Teaching material (Blue series)
- *Bibliographies/Catalogues:* in conjunction with specific research work.
- Building Defects: Elimination and Prevention
- MQB publication series, Vols 1–6 (published); further issues under discussion; language: German and French.

CHE.483

Institut pour l'Aménagement National, Régional et Local (Institut ORL) de l'Epez

EPF (Federal Polytechnic) Hönggerberg, 8093 Zurich

Telephone: 01 377 29 58/377 44 11

Date of establishment

1 April 1961. The Institute took over the role of the Institute for National Development at the EPF, founded in 1943. It was necessary to develop teaching and research in the field of town and country planning.

Status of institution: University Institute

Probable future development of institution: as a permanent institution, with its development centered mainly around teaching (including advanced courses) and research.

Main fields of work

Field of activity: national, regional and local development (town and country planning) means planning with regard to the organization of the area. It focuses on the living environment and the many ways the environment relates to the State, the economy and to culture, as well as to natural living conditions. The main objectives of town and country planning are the prudent use of the soil, a rational system of occupancy of the land, and the growth of the economic, social and cultural life of each of the country's regions.

Research in this field concentrates on obtaining scientific knowledge about the interrelation and development of the principles of land planning, appropriate measures aimed at influencing planning, and the strategy and methods used in town and country planning.

Functions: the main functions of the ORL Institute are:

- research in the field of national, regional and local development (town and country planning) and in related fields;
- teaching EPF students, particularly in the context of the curricula of the departments of architecture, civil engineering, forestry, agriculture, rural engineering, surveying, and also natural sciences;
- postgraduate teaching in town and country planning for students who already have a basic university training;
- organizing advanced courses;
- setting up consultancies which allow the Institute contact with practitioners in the field – such contact is very valuable from a scientific point of view;
- advising federal, county and community authorities, and planners.

Total staff: 55

Director or Chief Executive

Professor Jakob Maurer

Directors: Professor B. Huber, Professor M. Lendi, Professor W.A. Schmid

Principal related bodies

General structure: the Institute is governed by a directorate made up of four professors, assisted by the Institute Board acting as an advisory body. Each member of the directorate is responsible for one or more research areas. The central departments provide their services internally as well as externally.

Research areas: site planning, planning or urban and suburban areas, regional economy, law and politics, methods used in town and country planning.

Central departments: TEI/Computer service, information, secretariat.

Publications

A list of the Institute's publications can be obtained from the Information Department. It consists largely of the following series of publications (as of 1983):

- DISP (Documents and information on national, regional and local development – quarterly periodical published since 1965)
- ORL Collection (30 volumes)
- Reports (45 volumes)
- Studies (53 volumes)
- Directives (1968–75)
- Handbook (every 2–3 years)
- Teaching material.

CHE.484

LIGNUM, Schweizerische Arbeitsgemeinschaft für das Holz

Union Suisse en Faveur du Bois

Swiss Study Group for Wood

Falkenstrasse 26, 8008 Zürich

Telephone: 01 47 50 57

French Switzerland office: En Budron, rte de Cugy, 1052 Le Mont-sur-Lausanne

Telephone: 021 32 62 23

A service set up by the forestry economy, the industries, the craftsmen and the wood trade.

Main fields of work

- General and directed advertisement; information
- Advice and information on technical or general problems in the use of wood
- Support to regional working communities on behalf of wood
- Assistance to education concerning the use of wood and by-products
- Assistance in scientific research on wood
- Representation of the group to public authorities and in international organizations having the same aims
- Exchange of experience with national and international organizations
- Any other useful means
- International co-operation with, among others: CEI-Bois, the European Confederation of Woodworking Industries; the West European Timber Information Councils Organization
- The wood research bodies in Austria and in the Federal Republic of Germany.

Total staff: 11

Professionally qualified staff: 4

Director or Chief Executive

Chairman: Dr Ernst P. Grieder

Director: P. Hofer

Principal sources of finance

Financed by fees paid by the members and the wood professional bodies, including the Swiss Wood Fund (a voluntary mutual aid organization).

Publications

- Documentation, bulletins and general advertisement brochures
- Technical publications: Documentations Bois; Technical books; Technical Notes; Offprints of technical and technological articles
- Press and public relations department.

CHE.485 AM

Schweizer Baudokumentation

Documentation suisse du bâtiment

Swiss Building Documentation Service

CH-4249 Blauen

Telephone: (61) 89 41 41

Telex: 62 681 docu ch

Date of establishment

1967. Effectively in operation since 1930 (editorial responsibility for the Schweizer Baukatalog)

As an information centre for the construction industry, the Service sets out to transmit information – adapted to practical needs – between

The building industry, the manufacturers of products, building research and standardization institutes, universities and scientists and

Building planners, builders and building owners

Status: Body incorporated under private law.

Main fields of work

- *Documentation:* Publication of bilingual (German/French) loose-leaf documentation which consists of a library of 39 files, divided into two principal groups: Product information, arranged according to international guidelines, classified in at present 31 red files; Theory and basics for planning, design, work planning, execution and costing, classified in 8 blue files. The information is kept up to date by means of five supplements per year, which are sent to 5000 subscribers by post and inserted in the files by the filing staff
- *Information supply:* Complementary information supply based on a data-bank on building products and services. Literature investigations by direct access to foreign data-bases (IRB)
- *Customer advice:* Competent advice on all fields of information in the construction industry. International intermediary service by liaison with European Documentation Centres
- *Book Mail Order:* Sale of all architectural literature available in the book trade
- *Production:* Editing of texts including translations by qualified building experts. Graphic design and production of publications
- *Periodical literature:* The periodical Docu-Bulletin is published ten times per year in an issue of 8000 copies. It provides information on current developments in the building industry.

Total staff: 65

Professionally qualified staff: 28

Director or Chief Executive

Managing Director: C. Weisser

Heads of department

Administration: H.R. Rupp, Director

Sales: P. Bachmann

Production: G. Lampert/M. Bornand

Subscriptions: E. Christen

Information Processing: M. Bornand

Information Service: B. Mäder

Data Processing: J. Suroweic

Principal sources of finance

Participation fees (Documentation); subscription fees; charges for other Services.

Principal related bodies

The Service maintains relations with professional associations, building centres, research establishments and related institutions on a national and international level, thus ensuring a constant mutual exchange of information and experience.

Publications

- *Journal:* Docu-Bulletin, ten issues annually, 8000 copies
- *Bibliographies/Catalogues:* Swiss Building Documentation.

CHE.486

Schweizerische Lichttechnische Gesellschaft (SLG)
Union Suisse pour la Lumière (USL)
Swiss Illumination Engineering Society
Postfach, CH 8034 Zürich
Telephone: 01 391 86 37
Date of establishment
1922 (new name and structure 1970)

A professional association.

Main fields of work
- Study of particular problems of lighting and light by the study committees
- Elaboration of Swiss standards for lighting (in German and French)
- Functionary as the Swiss National Committee of the International Commission on Illumination (CIE)
- Organizing conferences, conventions, seminars, courses on lighting
- Publishing various reports, guidelines, etc. on lighting
- Gathering information on lighting, etc. for members, for publication in newsletters.

Total staff: 2
Professionally qualified staff: 2
Director or Chief Executive
Secretary: A.O. Wuillemin (President of the Swiss National Committee of CIE)
President of Board of Administration: Dr P. Blaser
Principal sources of finance
Membership fees (individual and collective members).
Principal related bodies
Swiss National Committee of CIE.
Publications
- Newsletter: SLG-Mitteilung; Communication de l'USL, 4 to 5 issues per year
- *Technical standards:* Swiss standards on lighting
- *Research reports:* Other guidelines; Reports on conferences, conventions.

CHE.487 **AM**

Schweizerische Zentralstelle für Baurationalisierung
Centre suisse d'études pour la rationalisation du bâtiment (CRB)
Centro svizzero di studio per la razionalizzazione dell'edilizia
Swiss Research Centre for Building Rationalization
Zentralstrasse 153, CH-8003 Zurich
Telephone: 01 241 44 88
Telex: 813 668 crbch
Cables: CRB Zurich
Date of establishment
1959/61/65 registered in Zurich under Swiss law

Private association on the development and publication of documents in different fields in the building industry of the private and public sectors.

Main fields of work
Research, design and production of manuals and other publications for planning and construction.

CRB produces, edits and publishes under its own copyright the following documentation:
- BCC: Building Cost Classification
- CSBR: Catalogue of Standard Building Descriptions (standard phraseology for bills of quantities)
- Bauhandbuch: a building price book (annual publication)

Initially developed for manual processing the above documents are also available on magnetic tape or disc for computer processing. The three publications form an integrated system for project administration and cost control.
- LKP: Liegenschaftenkontenplan (Real Estate Accountancy)
- Standards on Modular co-ordination and dimensional standardization produced in collaboration with other national and international institutes
- Standard on construction for the disabled

Lectures and courses are held on the above subjects.

CRB has obtained the rights for the marketing and distribution of the Swedish NCS Natural Colour System. Frequent seminar-workshops are held on the NCS in co-operation with external experts.

Management of the International Secretariat of INTERNET, International Project Management Association

Organization of courses and seminars on project management and related subjects, with the emphasis on education, training and development.

Total staff: 17
Professionally qualified staff: 11
Director or Chief Executive
Managing Director: Heinz Joss
Principal sources of finance
Marketing of publications; membership fees; royalties; proceedings of courses and seminars
Principal related bodies
General Assembly; the Board; Steering Committee.
Publications
Journal: Bulletin CRB, 2 to 4 issues per annum; INTERNET Newsletter, 2 issues per annum.

CHE.488

Schweizerische Zentralstelle für Stahlbau (SZS)
Swiss Institute of Steel Construction
Seefeldstrasse 25, CH-8034 Zürich
Telephone: 01 47 89 80
Date of establishment
1967

The industrial association has 60 member firms which are qualified enterprises specializing in steel and light metal construction. Representing 90% of Swiss steel construction capacity, its principal objective is to promote steel construction at home and abroad.

Main fields of work
- Provision of an objective information service free of charge for builders, architects, and engineers concerning the application of steel as a building material
- Promotion of research and development
- Promotion of standardization
- Standardization of construction details
- Publication of aids for design and execution of steel constructions
- Promotion of skilled and technical manpower recruitment
- Promotion of continuous training for architects and engineers
- Introduction of contacts, on an objective basis free of charge

The advice and Information Service's first duty is to inform architects and contractors about steel applications; but it does not undertake the study of any project to avoid competing with the consulting firms; the service is free of charge and confidential. Within the Technical Commission (TK), the Economic Commission (KWK) and their subcommissions, most of the work is carried out according to a militia system by persons belonging to member companies, consulting and architectural firms, schools of further education and scientific institutions; independent persons or institutions have been entrusted with certain problems.

Total staff: 6
Professionally qualified staff: 3
Director or Chief Executive
Director: U. Wyss
Heads of department
Administration: P. Haltiner
Technical: F. Bindschädler
Principal sources of finance
Contributions by member firms.

Publications

- *Journal:* Bauen in Stahl (Construire en acier), 6–10 issues per year
- List of publications

CHE.489 AM

Schweizerischer Ingenieur- und Architekten-Verein
Société suisse des ingénieurs et des architectes
Società svizzera degli ingegneri e degli architetti
Swiss Society of Engineers and Architects
SIA Postfach, CH-8039 Zürich
Telephone: 01 201 15 70
Cables: Ingenieur
Date of establishment
1837

A private-law association, SIA is the Swiss national association of professional engineers and architects.

Main fields of work

SIA has the following objectives:

- To represent the technical and professional interests of its members
- To guard the high standards of ethical and professional conduct set for its members
- To encourage excellence and to further education in the field of engineering and architecture
- To improve communication at national and international levels
- To work out standards and recommendations concerning construction
- To publish the results of research and development arrived at by its Specialized Groups for: industrial engineering, bridge and structural engineering, forestry engineering, rural engineering and surveying, production and chemical engineering, foreign activities, architecture, industrialized construction and underground construction.

Total staff: 32
Professionally qualified staff: 10
Director or Chief Executive
President: Dr Adolf Jacob
General Secretary: Dr Ulrich Zürcher
Principal sources of finance
Membership fees; sales of Standards and other publications.
Publications

- *Journal:* Schweizer Ingenieur und Architekt, (weekly); Ingénieurs et architectes suisses, bi-weekly; Revista tecnica della Svizzera italiana, monthly
- *Technical standards/Recommendations:* National Standards on construction, issued at irregular intervals
- *Research reports:* several bulletins (Industrialized Construction, Architecture, Underground Construction, Environment, Activities Abraod); SIA Documentation Series: mainly proceedings of symposia, at irregular intervals
- List of SIA members.

CHE.490

Technische Forschungs- und Beratungsstelle der Schweizerischen Zementindustrie (TFB)
Service de recherches et conseils techniques de l'industrie suisse du ciment
Research and Advisory Services of the Swiss Cement Manufacturers
Postfach, CH 5103, Wildegg
Telephone: 064 53 17 71
Date of establishment
1933

Main fields of work

- Advisory services on: concrete technology; concrete applications; concrete handling
- Documentation services
- Education, courses.

Total staff: 24
Professionally qualified staff: 10
Director or Chief Executive
Head: W. Wilk, CEng
Heads of department
Advisory services: G. Cimma, EEng
Research Laboratories, Testing: E. Koelliker, Dr Sc Techn
Information Documentation, Education: U.A. Trub, DriScTechn
Principal related bodies Financed by the Association of Swiss Cement Manufacturers in proportion to the volume of their production.
Publications
The same monthly publications in three languages:

- Cementbulletin, in German
- Bulletin du Ciment, in French
- Bolletino del Cemento, in Italian.
- Overseas Swiss concrete constructions, exceptional works for FIP Congresses (4-yearly).

CHE.491

Verband Schweizerischer Generalunternehmer VSGU
Association Suisse des Entrepreneurs Généraux ASEG
Association of Swiss General Contractors
Auf der Mauer 11, Postfach/Case postale 8023 Zürich
Telephone: 01 47 86 14
Date of establishment
1970

Main fields of work

- To safeguard the interests of members of the Association.
- To engage in appropriate public relations and press contacts.
- To represent the members of the Association to the authorities and to other bodies.
- To provide an information service to members.
- To represent members in cases involving litigation.
- To dispense information without charge to those interested in building.

Staff: A Secretary General and an Assistant.
Heads of department
A central secretariat which encompasses all the activities.
Principal sources of finance
Financed by members' annual dues.
Publications
No publications as such are produced. Items of information are issued as and when called for.

Others: Brochures: Construire avec un entrepreneur général ASEG; Modèle de contrat; Le maître de l'ouvrage et l'entrepreneur général ASEG.

TZA.492

National Construction Council
Baraza la Taifa la Ujenzi
PO Box 40465, Dar es Salaam
Telephone: 20542
Telex: 41392
Cables: JENGA
Date of establishment
Established by Act of Parliament No. 20 of October 1979. Became operational on 17 August 1981.

The Council is a service parastatal organization under the Ministry of Works.

The Council is charged with the co-ordination of all activities related to the construction industry with the prime objective of promoting the development of the industry to sustain national growth.

Main fields of work
- Research in the field of building materials, building design and construction management
- Survey and establishment of schedule of construction rates and cost indices
- Arbitration services
- Engineering consultancy services.

Governed by a Council consisting of 22 members appointed by the Minister for Works. These members are from various ministries, parastatal organizations and private firms in the construction industry.

Total staff: 34
Professionally qualified staff: 13
Director or Chief Executive
Executive Secretary: Dr George Mlingwa
Heads of department
Contracts and Arbitration: R. Balasuperamaniam
Research and Development: Y.N. Issa
Registration and Training: J.W. Kafanabo
Documentation and Information: not yet operational
Principal sources of finance
Government subvention through the Ministry of Works.
Principal related bodies
All Ministries and parastatal organizations; all consulting firms (architectural, engineering and quantity surveying); all contractors in the construction industry; all research institutions in the construction industry.
Publications
- Seminar papers and reports
- Building Materials Seminar
- Role of Contractors in National Development Seminar
- Architects' Role in the Construction Industry
- Role of Consulting Engineers in the Construction Industry.

TZA.493 FM

National Housing and Building Research Unit
Kituo cha Utafiti wa Nyumba Bora na Vifaa vya Ujenzi
PO Box 1964, Dar es Salaam
Telephone: 48106/7
Date of establishment
Economic Committee for the Cabinet, Paper No. 7 of 1970

Building Research Unit (BRU) is a division under the Ministry of Lands, Housing and Urban Development. Its day-to-day administration is carried out under the Director.

The overall task of the unit is to research, develop and encourage the building of decent affordable houses using as much as possible the locally available materials. Rural housing is accorded first priority.

Main fields of work
To clarify the problems within the field of housing and building, and to seek optimal and relevant solutions.

The main areas covered include: research on the availability and suitability of local building materials; testing of various materials and structures; determination of housing needs and housing conditions; development of designing criteria for low-cost houses with special reference to social, cultural, economic and environmental aspects; complilation of building cost indices; and emphasizing the role of building and construction to the national economy.

The Unit also disseminates its research findings to the target group. The process involved includes the construction of demonstration (model) houses; use of news media and press (radio, newspapers and other forms of publication) and finally conducting seminars on better housing.

The principal divisions are: Technical Section; Human Requirement Section; Building Economy Section; Information Section; Administration.

Total staff: 72
Professionally qualified staff: 20
Director or Chief Executive
Director: A.L. Mtui
Principal sources of finance
Local funds from the government of Tanzania; foreign aid, mainly from Norway (NORAD).
Publications
Detailed lists of available publications are produced regularly.

TZA.494 AM

Wagala Onye Jeli and Partners, Chartered Quantity Surveyors, Building Economists and Construction Costs Consultants
Plot 354 United Nations Road East Upanga, PO Box 1611
Dar es Salaam
Telephone: 20467
Telex: 41364 WOP TZ
Date of establishment
1978

Consulting firm of quantity surveyors and building economists with offices in the following cities: Dar es Salaam Tanzania; Kampala, Uganda; Calabar, Nigeria; Enugu, Nigeria; Lagos, Nigeria.

Main fields of work
Consultancy work in quantity surveying, collection, analysis and assimilation of construction costs data.
Total staff: 13
Professionally qualified staff: 5
Director or Chief Executive
Principal Consultant: J. Coai-Ojara

TUR.495

Instanbul Technical University, Materials and Structures Laboratories

Instanbul Teknik Universitesi, Yapi Laboratuarlari
Inşaat Fakültesi Ayazağa, Maslak, Istanbul
Telephone: Istanbul 169 07 45, 46, 47
Telex: 23 706 ITU TR
Date of establishment
Established 1917 (as Building Materials Testing Laboratory of the ITU), Re-organized 1984, Law no. 2547 of 4 November 1981

The Laboratories form part of the Civil Engineering Faculty of the Instanbul Technical University and perform tests upon request and as directed by the principal related bodies noted in Section 11.

Main fields of work

The Laboratories are mainly concerned with tests and research on materials and in particular building materials and components and also structural models. In-situ or laboratory tests are performed upon request for contractors and engineers on building materials and components.

Static and dynamic testing of building components and structural models are to be carried out on the 1000 m² prestressed concrete strong floor using the high capacity hydraulic activators at the new university campus.

Research programmes are in progress on the following topics:

- Non-linear inelastic behaviour of normal and lightweight concrete under compression
- Non-destructive testing of damaged concrete
- Mechanical properties of ferrocement
- The effect of addition of blastfurnace slag and pumice on the chemical resistance of Portland clinker based cements
- Abrasion resistance of concrete
- Electro-osmotic properties of fresh concrete
- Adobe blocks stabilized with gypsum as low-cost materials
- Prediction of service life of multilayer roofing felts
- Creep and relaxation of elastomers under combined loading
- Creep of concrete.

Total staff: 27
Professionally qualified staff: 18
Director or Chief Executive
Director: Abdurrahman Güner
Heads of department
Building Materials: Mehmet Uyan
Structures: Reinforced Concrete: Ilhan Eren
Steel and Timber: T. Seno Arda
Photoelasticity: Yalçin Aköz
Building Components and Work Machines: Sedat Ersoy
Principal sources of finance
Faculty of Civil Engineering, ITU; research grants from other institutions; income from tests and expert work done for industry; subsidies.
Principal related bodies
Structures and Earthquake Research and Development Centre, ITU (Yapi ve Deprem Uygulama Araştirma Merkezi, ITÜ), Materials and Production Technologies Research and Development Centre, ITU (Malzeme ve Üretim Teknolojisi Uygulama-Araştirma Merkezi, ITÜ) (as Academic Council and Supervisory Boards of Directors).
Publications

- *Journal:* Bulletin of the Technical University of Istanbul ITU, quarterly (in English, French or German); Teknik Üniversite Dergisi, ITÜ, monthly (in Turkish, with abstracts in English, French, or German)
- Technical Reports published by the Faculty of Civil Engineering (in Turkish, with abstracts or summaries in English, French or German)
- Lists of bibliographies are published by the Faculty of Civil Engineering
- The publications lists with abstracts are stored in a computerized retrieval system at the Computer Centre, ITU.
- PhD theses, the experimental parts of which have been performed in the Laboratories are also published by the Faculty of Civil Engineering.

TUR.496 FM

Scientific and Technical Research Council of Turkey, Building Research Institute

Yapi Araştirma Enstitüsü, Türkiye Bilimsel ve Teknik Araştirma Kurumu (TÜBITAK YAE)
Bilir Sok 17, Kavaklidere, Ankara
Telephone: (41) 27 81 50; (41) 26 76 17
Telex: 43 186 BRAK TR
Date of establishment
1970 (Law No. 278 of 17 July 1963)

The Institute is a government research institution within the framework of the Scientific and Technical Research Council of Turkey. It conducts in-house research, research sponsored by public institutions and acts in an advisory capacity to various government departments.

Main fields of work

Research activities of the Institute are carried out by four research units. Names and fields of interest of the research units are summarized as follows:

- *Environmental Systems Research Unit:* Energy conservation and hygro-thermal comfort in buildings; Illumination; Acoustics and noise control; Building ergonomics; Fire protection
- *Structural Systems Research Unit:* Statics of structures; Dynamics of structures; Engineering seismology; Performance and economy of building structures
- *Building Planning Research Unit:* Planning and programming; Design; Building economics; Building information
- *Building Construction Research Unit:* Building techniques/technology; Mechanization in building; Construction management, quality control; Building materials, building components

Total staff: 56
Professionally qualified staff: 24
Director or Chief Executive
Director: Dr Mustafa Pultar
Heads of department
Environmental Systems Research Unit: Dr Cengiz Yener
Structural Systems Research Unit: Alkut Aytun
Building Planning Research Unit: Yalçin Ersoy
Building Construction Research Unit: Ender Arkun
Principal sources of finance
National budget (allocated from the budget of the Scientific and Technical Research Council of Turkey); earnings of the Institute (funded research projects, sale of publications).
Principal related bodies
Institute Council (consisting of the Director, the Deputy Directors for Research and Administration and the heads of the four research units).
Science Council (governing board of the Scientific and Technical Research Council of Turkey).
Publications
Journal: Teknik Bülten (Technical Bulletin), quarterly, in Turkish
Available publications (all are in Turkish).
Cezik, OA.: Quality and Its Control in Building Construction
Coker, B.G.: Proposals for the Solution of Problems Related to Information Flow at the Design Stage of the Industrialized Building Process in Turkey
Avsar, R.: A Method for Determining Costs of Wood-framed, Single-glazed Windows
Baytin, N.: Wet Spaces in Houses
Pişirici, E.: Determination of Housing Cost Based on Functional Building Elements
Tokman, B.: Models of Urban Layout in Terms of Infrastructure and Services in Low-Cost Housing Areas
Ersoy, Y.: Building Process for Higher Educational Facilities in Turkey
Korkut, K.: Distribution of the Earthquake Base-Shear Over the Height of Buildings with Stiffness Discontinuities and Mass Concentration
Celik, A.P.: Conservation of Energy for Heating Buildings
Özen, Y.Ö.: Building Industrialization and the Current Situation in Turkey
Sarisözen, C.: Building Process for Educational Facilities
Coker, B.G.: A Proposal for an Information System to Secure Flow of

Information Among Users of Industrialized Building Systems
Aktüre, T.: Cost Control in Building
Akçali, Ü.: A Handbook of Building Codes and Regulations
Akçali, Ü.: Unit Prices in Building Construction (1971–1980)
BRI (TÜBITAK): Contemporary Building Systems (Series conferences)
BRI (TÜBITAK): Unit Prices in Building Construction for the Year 1982
BRI (TÜBITAK): Methods of Evaluation for Industrialized Housing Systems; Proceedings of the Colloquium Organized in the Memory of the Centennary Birthday of Atatürk, 9–10 December 1981, Ankara.

SUN.497

All-Union Research Institute for Information on Construction and Architecture (VNIIIS of USSR Gosstroy)

Всесоюзный научно-исследовательский институт информации по строительству и архитектуре - ВНИИИС

VNIIIS, USSR Gosstroy, 12 Marx Avenue, 103828 Moscow
Telephone: 251 17 95

The Institute is under the State Building Committee of the USSR (USSR Gosstroy).

Main fields of work
Scientific and technical information, popularization, information systems and building catalogue.

Departments for developing and improving scientific and technical information systems in building and architecture; Mathematical and linguistic back-up; Building catalogue; Information flow organization and analysis; Handling and preparation of information publications, etc.

Total staff: 330
Professionally qualified staff: 20
Director or Chief Executive
Director: I.E. Evgenyev
Principal sources of finance
Financed from the state budget and from revenues for information publications and services rendered to other organizations.
Principal related bodies
Scientific and Technical Council; Council of the united system of scientific and technical information in building, architecture and building materials.
Publications

- Abstract journal, in the form of a journal and cards in 12 subject areas, 144 issues per year; Express information, bulletins of expanded abstracts of publications on the most important scientific and technical achievements and results of reseqrch works in 10 subject areas, 144 issues per year
- Analytical and comparative reviews on the most important scientific and technical problems in construction (in 10 subject areas, 35 issues per year)

All in all the Institute publishes 460 issues per year.

SUN.498

Central Research, Design and Experimental Institute for Industrial Buildings and Structures (TsNIIpromzdaniy of USSR Gosstroy)

Центральный научно-исследовательский и проектно-эксперимен-тальный институт промышленных зданий и сооружений -ЦНИИПромзданий Госстроя СССР

TsNIIpromzdaniy, USSR Gosstroy, 12 Marx Avenue, 103828, Moscow
Telephone: 482 45 06

The Institute is under the State Committee of the USSR on Construction.

Main fields of work
Typology, architectural treatment, planning approach and design of industrial buildings and structures; general layout of industrial units and enterprises; interbranch and branch unification of industrial buildings and structures fundamentals; fundamentals of building standardization; heating, ventilation and air-conditioning systems and facilities in industrial buildings.

Laboratories: Unification of factory and storage buildings; Inner enclosing structures; Outer enclosing structures; Coverings and roofing; Translucent enclosing structures; Scientific fundamentals of standardization; Typification of wooden structures; Architecture of enterprises of machine-building industry, light industry and agrocomplexes; Structures for multi-storey buildings; Computer-aided design of industrial buildings; Sanitary engineering; Architecture of enterprises of chemical and petrochemical industry; Floors; Protection of buildings and structures from fire; Architecture of iron and steel and building industries; Full-scale studies; Heating and ventilation; Local and central air-conditioning; Computer-aided microclimate systems; etc.

The Institute runs postgraduate courses.

Total staff: 1164
Professionally qualified staff: 127
Director or Chief Executive
Director: Yu. N. Khromets
Deputy Directors: Chief Engineer V.V. Granev, S.N. Glikin, N.N. Kim, V.I. Prohorov
Principal sources of finance
Financed from the state budget and from revenues for its research work carried out under contract.
Principal related bodies
Scientific and Technical Council.
Publications
The Institute prepares publications of an instructive and regulatory nature on construction: Rukovodstvo, published by Stroyizdat Publishers.

SUN.499

Central Research, Design and Experimental Institute for Organization, Mechanization and Technical Assistance in Construction

Центральный научно-исследовательский и проектно-экспери-ментальный институт организации, механизации и техничес-кой помощи строительству - ЦНИИОМТП

TsNIIOMTP, USSR Gosstroy, 12 Marx Avenue, 103828 Moscow
Telephone: 216 03 04

The Institute is under the State Building Committee of the USSR (USSR Gosstroy).

Main fields of work
Methods of organization, technology, mechanization and automation in construction; organization of machinery maintenance; systems of building machinery, mechanisms and power tools; metrology, quality control systems and safety engineering as regards civil engineering and erections.

Departments of: Construction organization; Construction of buildings and structures; Machinery and mechanization methods in construction; Concreting; Excavations; Finishing work and roofing; Accident prevention.

Total staff: 650
Professionally qualified staff: 126
Director or Chief Executive
Director: E.A. Dolginin
Deputy Directors: V.M. Kazarinov, V.D. Topchy, V.V. Shahparonov
Principal sources of finance
Financed from the state budget and from revenues for research work carried out under contract.

SUN.500

Research Institute for Building Physics (NIISF of USSR Gosstroy)

Научно-исследовательский институт строительной физики Госстроя СССР - НИИСФ

NIISF, USSR Gosstroy, 12 Marx Avenue, 103828 Moscow
Telephone: 482 40 76

The State Research Institute is a constituent institution in the system of the State Building Committee of the USSR (USSR Gosstroy).

Main fields of work
Building heat engineering, acoustics, illumination engineering and climatology; insulation and sun-protection of buildings and built-up areas; methods of heat, light and acoustic engineering studies of buildings, structures and components; durability problems of enclosing structures and buildings.

Laboratories: Daylighting; Electric lighting; Building climatology; Light and colour environment; Industrial and urban noise control; Aerodynamic noise control in buildings; Sound insulation in buildings and noise absorbing structures; Architectural acoustics; Computer-aided calculations; Microclimate and thermal stability of buildings; Thermal insulation of buildings; Thermo-physical tests of enclosing structures; Durability and moisture control of enclosing structures; Heat engineering of the lamellar envelope and technical economic studies

The Institute runs a postgraduate course.

Total staff: approximately 400
Professionally qualified staff: 100
Director or Chief Executive
Director: Professor V.A. Drozdov, Merited Scientist of the RSFSR
Deputy Directors: Professor F.V. Ushkov, M.I. Krasnov
Principal sources of finance
Financed from the state budget and from revenues for research work carried out under contract.
Principal related bodies
Scientific and Technical Council; Degree Committee.
Publications
The Institute's employees regularly publish monographs on individual problems relating to building physics, as well as articles in such journals as Zhilischchnoye Stroitelstvo (Housing Construction), Promyshlennoye Stroitelstvo (Industrial Construction), Vodosnabzenie i Sanitarnaya Tekhnika (Water Supply and Sanitary Equipment), etc.

SUN.501

Research Institute for Concrete and Reinforced Concrete (NIIZHB of USSR Gosstroy)

Научно-исследовательский институт бетона и железобе- тона - НИИЖБ

NIIZHB, USSR Gosstroy, 12 Marx Avenue, 103828 Moscow
Telephone: 171 26 69

NIIZHB is a constituent institution in the system of the State Building Committee of the USSR (USSR Gosstroy).

Main fields of work
Developing the theory and improving the methods of analysis of concrete and reinforced concrete structures; studies into effective reinforcement for ordinary and prestressed reinforced concrete; research into different types of concrete: heavyweight, lightweight, cellular, special, heat-resistant; developing and studying new types of prefabricated and composite, ordinary and prestressed load-carrying structures and enclosures for housing, industrial and agricultural buildings and structures; special concrete constructions; methods for protecting concrete and reinforcement from corrosion and increasing the durability of concrete; questions of adaptability of reinforced concrete structures to manufacture, general problems of reinforced concrete economics.

Laboratories: Reinforced concrete theory; Reinforced concrete structures; Prestressed constructions; Spatial constructions; Lightweight concretes and constructions; Heat-resistant concrete and structures; Heavyweight concretes; Reinforcement; Cellular concretes and constructions; Corrosion; Physical and chemical analysis of concrete; Fire resistance of concrete and reinforced concrete structures; etc.

The Institute runs postgraduate courses and operates a design and technological bureau.

Total staff: 950
Professionally qualified staff: 190
Director or Chief Executive
Director: Professor K.V. Mikhailov
Deputy Directors: Professor B.A. Krylov, N.N. Korovin
Principal sources of finance
Financed from the state budget and from revenues for research work carried out under contract.
Principal related bodies
Scientific and Technical Council; Degree Committee; Co-ordination Committee.
Publications
Journal: Beton i Zhelezobeton (Concrete and Reinforced Concrete), monthly.

SUN.502 **FM**

V.A. Kucherenko Central Research Institute for Building Structures (Kucherenko TsNIISK of USSR Gosstroy)

Центральный научно-исследовательский институт строительных конструкций им.В.А.Кучеренко - ЦНИИСК им.Кучеренко Госстроя СССР

TsNIISK, USSR Gosstroy, 12 Marx Avenue, 103828 Moscow
Telephone: 171 26 50

The Institute is under the State Building Committee of the USSR (USSR Gosstroy).

Main fields of work
Theory of construction and building mechanics; principally new forms of metal, stone, wooden, asbestos-cement structures constructed of new materials on the basis of development of construction theory; technology of their production; earthquake engineering; fire resistance of buildings and structures; computerized systems of research.

Departments: Structural analysis; Structural dynamics; Earthquake resistance of structures; Strength and new forms of metal structures; Lightweight structures; Wooden structures; Testing structures; Computing engineering; Phosphate materials. Also a pilot plant.

Total staff: 1000
Professionally qualified staff: 300
Director or Chief Executive
Director: Corresponding Member of USSR Academy of Sciences, Professor A.F. Smirnov
Deputy Directors: Professor N.N. Skladnev, Professor S.V. Polyakov, A.M. Chistyakov
Principal sources of finance
Financed from the state budget and from revenues for research work carried out under contract.
Principal related bodies
Scientific and Technical Council; Degree Committee.
Publications
Journal: Journal of USSR Gosstroy, Stroitelnaya Mekhanika i raschet sooruzhenly (Building Mechanics and Structural analysis), bi-monthly.

GBR.503 **UM**

Ashridge Management College

Little Gaddesden, Berkhamsted, Herts HP4 INS

Telephone: 044 284 3491

Telex: 826434 Ashcol G

Cables: Ashcol, Berkhamsted

Date of establishment

1954

An independent educational trust, to provide a wide range of post-experience education for managers in the UK, Europe and worldwide, financed almost entirely by course fees, with some support for research activities derived from various foundations and companies.

Main fields of work

The college provides a wide range of open and tailor-made management courses to all industries including construction, from which it has a number of significant clients. Courses are aimed at middle and top management, and can be either specialist or general in nature.

Since 1969 the college has supported a long-term programme of research into the efficiency of the construction industry. Some of this work has been carried out for government departments. This has covered such aspects as leadership styles, organization structure, problem solving skills, corporate planning, marketing and market evaluation, environmental analysis. The most recent programme of research has focused on the response of the UK construction industry to recession. Other work has involved the development of real world simulation models for management development and business planning.

Total staff: 200

Professionally qualified staff: 45

Director or Chief Executive

Principal: P.J. Sadler

Heads of department

Building Industry Group: P.R. Lansley

Principal sources of finance

Course fees and research grants.

Publications

Numerous papers are published in the academic, scientific and professional press.

GBR.504

Aslib

3 Belgrave Square, London SW1X 8PL

Telephone: 01 235 5050

Telex: 23667

Date of establishment

1924

A membership organization in the field of information management and use.

Membership includes industrial organizations, government departments, research laboratories, universities and colleges, institutions, public and national libraries and totals 2000 organizations, of which nearly 700 are outside the UK.

Main fields of work

Aslib covers all aspects of information management, handling, retrieval and use in all subject fields and is the major British focus of activity in this field. It is the British member of the International Federation for Documentation and represents both information users and information producers in discussions with government and international bodies. It undertakes a programme of research into information handling activities and provides consultancy and advisory services for any organization wishing to set up or improve its information handling systems. It trains personnel for work in information departments and runs a series of conferences, workshops, symposia and meetings on various aspects of information handling.

It also operates, through its Information Department, a referral and inquiry service for members, covering all subject fields, but with emphasis on technological, engineering and scientific fields. It has co-operated with a number of organizations in the construction industry in developing information handling and retrieval systems in this field.

Total staff: 50

Professionally qualified staff: 30

Director or Chief Executive

Director: Dr Dennis A. Lewis

Heads of department

Information and Consultancy Group: Peter Vickers

Publications Group: Richard Coleman

Professional Development Group: Daphne Clark

Marketing Manager: Douglas Wood

Principal sources of finance

Membership subscriptions 20%; Government grant for research, etc. 10%; publications, courses and conferences 70%.

Publications

- *Journals:* Aslib Information, monthly; Aslib Proceedings, monthly; Journal of Documentation, quarterly; Program, quarterly; Aslib Booklist, monthly; Forthcoming International Scientific and Technical Conferences, quarterly; Index to Theses presented for higher degrees in British Universities
- Aslib Directory of Information Sources in the UK, vol. 1: Science, Technology and Commerce, 5th edn
- Aslib Handbook of Special Librarianship Information Work, 5th edn, 1982
- Monographs, research reports, bibliographies, etc.

GBR.505

BHRA Fluid Engineering Centre

(British Hydromechanics Research Association)

Cranfield, Bedford MK43 0AJ

Telephone: 0234 750422

Telex: 825059

Date of establishment

1947

Independent, non-profit-making research establishment.

Main fields of work

Fluid engineering in a wide context, but specializing in: physical and mathematical modelling, computer software, physical modelling and specialist advice in all kinds of fluid dynamics and control; high pressure water jetting; pneumatic and hydraulic transport of solids; pumps; seals; transient flow and multiphase flow; and information services.

Contract research and development. Courses. Conferences. Publisher of conference proceedings, abstract journals, bibliographies, reviews, leaflets.

Total staff: 230

Professionally qualified staff: 80

Director or Chief Executive

Director of Research: G.F.W. Adler, OBE

Heads of department

Hydraulic Engineering: D.S. Miller

High Pressure Engineering: M.J. Fisher

Pumping and Pipeline Technology: P.J. Baker

Information Services: H.S. Stephens

Sealing Technology: B.S. Nau

Marketing Development: D.C. Bain

General Administration: R. Chadwick

Central Services: D.H. Buxton

Fluid Engineering Unit: R.C. Baker

Principal sources of finance

Consultancy; Contract research and development; licensing and sales of publications; conferences and course fees; and minimally, membership subscriptions.

Publications

- *Journal:* Nine various abstract journals (devoted to specific areas of fluid engineering)

- Bibliographies and Reviews
- Volume of Conference Papers
- Collections of papers on specific subjects and Books.

GBR.506

Brighton Polytechnic, Building Department
Mithras House, Lewes Road, Moulsecoomb, Brighton
Telephone: (0273) 693655 ext 2360
Date of establishment
Department of Building 1978

A public institution providing education for all qualified applicants and provider of services to industry, local government and private individuals.
Main fields of work
The Department of Building is concerned with the education of undergraduates in the areas of building technology and management. This includes the subjects of Building Technology, Environmental Science and Building Services, Materials Science and Materials, Quantitative Studies and Building Economics, Structures and Soil Mechanics, Surveying and Field Course, Mathematical Studies, Balancing Studies, Law, Integrated Studies and Project Control.

Students spend one year in industry which is intrinsic to the course and are required to write reports during this year. The Department also undertakes research in the areas of Construction Management, Building Technology, Third World Development Services, Fibre Reinforced Cement, Building Material Analysis, Environmental Standards Analysis, non-destructive testing thickness and building services and energy.
Total staff: 20
Professionally qualified staff: 13
Director or Chief Executive
Head of Department: P.J. Rutland
Principal sources of financw
Government funding.
Principal related bodies
Faculty of Engineering and Environmental Studies; The Chartered Institute of Building; Construction Industry Research and Information Association.
Publications
- *Occasional Papers:* Department of Building, quarterly
- *Research reports:* On completion of research projects.

GBR.507

Brighton Polytechnic, Department of Civil Engineering
Cockcroft Building, Lewes Road, Moulsecoomb, Brighton BN2 4GJ
Telephone: Brighton 693655
Date of establishment
1962

Laboratories for research and development in civil engineering. NATLAS approved.
Main fields of work
All aspects of the structural use of timber and concrete materials; Research, development and testing; Hydraulic machine and models; Coastal studies; Transportation planning; Traffic engineering. Computer applications; Geotechnical investigations and testing; Pollution monitoring and control and the structural applications of epoxide resins.
Total staff: 30
Professionally qualified staff: 20
Director or Chief Executive
Dr Barry W. Staynes
Heads of department
Structural Engineering Division: Dr Peter Rodd
Geotechnics Division: Dr Philip Drummond
Environmental Division: John Morfett
Construction Division: Malcolm Rees
Computations Division: William Matthews
Principal related bodies
Department of Mechanical and Production Engineering; Department of Electrical and Electronic Engineering; Department of Architecture; Department of Building.
Publications
- The Development of New Structural Beam Systems in Timber
- Vacuum Impregnated Epoxide Resin Concrete
- Behaviour of Rockbolts
- Traffic Modelling for Highway Appraisal
- The Redevelopment of Contaminated Land
- The Strength of Dowel Joints in Timber
- Planning for Rural Public Transport.

GBR.508

Bristol Polytechnic, Department of Construction and Environmental Health
Ashley Down Road, Bristol BS7 9BU
Telephone: Bristol 41241
Date of establishment
September 1973

The Department is an institution of higher education in the public sector that has its own articles, instruments and government.
Main fields of work
Higher Technician, undergraduate, postgraduate, professional students, research and consultancy relating to the construction industry and its related professions.

The principal divisions are Higher Technician Building Studies, undergraduate, postgraduate and professional studies in Building Technology and Management; Environmental Health; Research Consultancy.
Total staff: 50
Professionally qualified staff: 34
Director or Chief Executive
Dr M.M. Cusack
Principal sources of finance
Public Sector of Higher Education.
Principal related bodies
Council for National Academic Awards who have oversight of undergraduate and postgraduate studies in the public sector.
Publications
Polytechnic Annual Research Report.

GBR.509

Bristol Polytechnic, Department of Surveying
Unity Street, Bristol BS1 5HP
Telephone: 0272 23016
Date of establishment
January 1973

A public sector higher education establishment.
Main fields of work
Teaching, consultancy and research is undertaken in:
- Quantity Surveying/Construction Economics
- Valuation and Estate Management
- Housing
- Building Maintenance

Total staff: 45
Professionally qualified staff: 45
Director or Chief Executive
Head of Department: A.H. Spedding
Heads of department
Associate Head: Mrs A. Colborne

Quantity Surveying: A.N. Ash
Valuation and Estate Management: R.C.D. Netting
Housing: G.A. Paxton

Principal sources of finance

Public sector funds plus consultancy and research grants.

Publications

Papers concerning the Department's activities.

GBR.510

The British Board of Agrément (BBA)

PO Box 195, Bucknalls Lane, Garston, Watford, Herts WD2 7NG
Telephone: Garston (092 73) 70844

Date of establishment

Established by the Minister of Housing and Local Government in 1966.

A Government sponsored organization.

Main fields of work

The BBA reports direct to the Minister for Housing and Construction through its Chairman. A Council, representative of the construction industry and allied professions (public and private sectors), is appointed by the Minister to decide policy. It is assisted in this by Assessors representing the interests of other Government departments and the British Standards Institution.

The Agrément Board is principally concerned with Quality Assurance in the construction industry and provides independent assessments of materials, products, components and processes:

Materials Division: assessments of materials including insulation products, roofing, flooring finishes, surface coatings, timber preservatives, damp-proofing and waterproofing materials and systems

Structures Division: assessments of structural and engineering products including cladding systems, partitions, ceilings, building systems, drainage products, services, structural floors, timber connector plates, fixings, reinforcement, windows and doors

Physics Division: assessment of thermal performance, condensation risk, computer analysis of structural design, computer produced thermal profiles, nuclear shelter design.

Assessments are normally made following evaluation of submitted data, testing (usually at the Boards own Test Unit), site inspections and factory visits.

The Agrément Certificates are used by architects, engineers, contractors and authorities administering building regulations as an important source of design and performance data.

The Assessment Reports give similar data to Agrément Certificates but concern the performance of prototype products and specific buildings. The reports are confidential to the client and not published.

The Test Reports are produced at the request of manufacturers or designers to provide assistance in the development of a product which it is anticipated will achieve full Agrément Certification later. Close working links are established with the Building Research Establishment and the Transport and Road Research Laboratory of the Department of the Environment, British Standards Institution, and other construction industry bodies.

The Board is a member of the European Union of Agrément (UEAtc) and contributes to the work of establishing European Common Directives for testing and assessment procedures.

Total staff: approximately 70.

Director or Chief Executive

Chairman: W.J.W. Courtney
Director: T.P.R. Lant

Principal sources of finance

Financed by its own fee income and also by a grant in aid from the Government. Non-profitmaking.

Publications

- Agrément Certificates
- Methods of Assessment and Testing (MOATS)
- Information Sheets provide concise technical information concerning product assessments
- Newsletters bring together news and information on all Agrément activities.

GBR.511

British Calibration Service (BCS)

National Physical Laboratory, Teddington, Middlesex TW11 0LW
Telephone: 01 977 3222
Telex: 262344

Date of establishment

25 April 1966

BCS is a government-backed organisation set up to provide a service to industry for the authenticated calibration of instruments, gauges and standards, and for the certification of a wide range of measurements. BCS is a service of the National Physical Laboratory, where its headquarters are located.

Main fields of work

BCS consists of over 100 laboratories throughout the country which have been assessed and approved by the BCS headquarters organization at NPL. Calibration, measurement and compliance with specifications and standards is provided in the following fields:

- DC and 1f electrical measurements
- HF and microwave measurements
- Dimensional and other mechanical measurements including force and impact
- Measurements on fluids including flow rate, pressure and density
- Measurements of temperature and thermal conductivity
- Measurements of acoustic noise
- Other specialized measurements including hardness, chronometer rating, comparative concrete cube testing, optical transfer function and calibration of evidential breath testing machines.

Total staff: 19

Professionally qualified staff: 12

Director or Chief Executive

Head, BCS: J.L. Wilson

Heads of department

Dimensional, Mechanical and Optical Measurements including Hardness and Concrete Cube Testing: C.F. Dietrich, A.N.J. Uren
Electrical Measurements: L.J.T. Hinton, A.E. Drake, C.J. Kemp
Fluid and Temperature Measurements: H.R. Whitehead
Thermal Conductivity, Acoustic Noise, Breath Testing Machines: I.W. Goodier
Audit Officer: J. Osborne
Promotion and Publicity Officer: G.P. Gorman

Principal sources of finance

Self-financing through fees charged to member laboratories. Some initial support from Department of Industry for establishment of new fields.

Principal related bodies

National Physical Laboratory. BCS is a service of the NPL which is the national standards laboratory of the UK.

Publications

- *Technical standards/Recommendations:* Criteria documents for laboratory approval; Guidance documents relating to each field of measurement
- *Bibliographies/Catalogues:* Directory of Laboratories, updated quarterly, on subscription; About the British Calibration Service (in 17 languages); Publications; Measurements Leaflet.

GBR.512 **AM**

British Ceramic Research Association Ltd (BCRA)
Queens Road, Penkhull, Stoke-on-Trent, Staffs ST4 7LQ
Telephone: 0782-45431
Telex: 36228 BCRA G
Date of establishment
1921 (BRRA), 1937 (BPRA), merged in 1948 to form BCRA

A company limited by guarantee, not trading for a profit, carrying out scientific and technological research in ceramics, co-operatively on behalf of members, or on a contract basis for individual clients, and operating with Government support in many areas. Services are avaible to all countries and are available to non-members.

Main fields of work
Research in ceramics and other inorganic non-metals covering raw materials, manufacturing processes, product improvement, and where appropriate installation and usage of product.

Branches of the industry served include:
- Whiteware (tableware, sanitaryware, tiles, electrical porcelain)
- Refractories (high temperature ceramics, insulating ceramics, etc.)
- Industrial ceramics (silicon nitride, silicon carbide, alumina, zirconia, etc.)
- Structural clay products (bricks, clay pipes, roof tiles, floor quarries)

Support for all branches in the fields of energy usage, industrial hygiene, microprocessor applications and engineering services.

Physical and chemical testing services are available to any client. BCRA is NATLAS accredited.

A comprehensive library and information service is operated.

Consultancy services are available.

Total staff: 250
Professionally qualified staff: 125
Director or Chief Executive
Chief Executive: Dr D.W.F. James
Heads of department
Pottery: Dr R.C.P. Cubbon
Refractories and Industrial Ceramics: Dr G.C. Padgett
Structural Clay Products: H.W.H. West
Analysis and Testing Services: H. Bennett
Communications and Marketing: W.P. Howells
Principal sources of finance
Subscriptions from members; fees for contract research, testing and consultancy; Government funding for many projects.
Principal related bodies
Working closely with Department of Industry, Department of Environment, Building Research Establishment and other agencies of HM Government and all trade bodies and societies of the ceramic industries.
Publications
- *Journal:* British Ceramic Abstracts, monthly; In Brief Newsletter, quarterly, free of charge
- *Technical standards:* Various
- *Research reports:* Lists of open reports available on request
- *Bibliographies/Catalogues:* Index of Research Project titles available; catalogues of services available.

GBR.513 **AM**

British Gas Corporation Research and Development Division, Watson House Research Station
Peterborough Road, London SW6 3HN
Telephone: 01 736 1212
Telex: 919 082
Date of establishment
1926

Watson House is one of six research stations making up the Research Division. The role of Watson House is to provide the British Gas Corporation with the technical support required to pursue its objectives in the domestic and small commercial markets.

Main fields of work
The Research Station is concerned with the research, development and testing for the domestic and small commercial markets. The programme is divided along the following lines:
- *Utilization efficiency:* Projects in this area are concerned with space and water heating appliances and controls and the overall performance of the systems. Research for the longer term includes work on condensing appliances and heat pumps. Work in this area also encompasses heating, conservation and catering in small commercial premises
- *Product improvement:* The concern of work in this area is the maintenance and improvement of the reputation and marketability of gas. Projects cover work on materials, metering, cooking appliances, ignition developments and conventional flues
- *New housing:* Projects in this area cover the provision of gas supplies, metering, appliances and their installation for the smaller types of house which are being built. These houses are of different construction and have lower energy requirements than conventional dwellings
- *Engineering:* This area provides a broad-based engineering resource in support of the whole research programme, as well as taking the lead in monitoring and improving reliability in appliances and components and in implementing improved production techniques
- *Line service work:* This area is of similar size to the research function and covers component and appliance approval testing, quality assurance and codes of practice.

Total staff: 420
Professionally qualified staff: 110
Director or Chief Executive
G. Percival, OBE
Heads of department
Utilization efficiency: G.A. Pickup
Product Improvement: Dr A.J. Miles
New Housing: Dr R. South
Engineering: J. Newcombe
Approval: W.J. Bennett
Quality Assurance: J.C.M. Elliott
Standards: F. Hughes
Principal sources of finance
Financed by British Gas Corporation as shown in the Corporation's Annual Report and Accounts.
Principal related bodies
Othe research stations within the Division: Engineering Research Station (heavy engineering); Midlands Research Station (large commercial and industrial utilization); London Research Station (corporate laboratory for British Gas corporation); On-Line Inspection Centre (pipeline inspection); Westfield Development Centre (development and demonstration of coal gasification).
Publications
Journal: Watson House bulletin, quarterly, private circulation, magazine devoted to the use of gas in the domestic and commercial markets. Most of the articles tend to be the results of laboratory work. The bulletin is supplemented from time to time with reports on appliances.

Research reports: Papers are presented at meetings of the Institution of Gas Engineers, as well as at other conferences. Items are also published in scientific and technical journals.

GBR.514

British Steel Corporation (BSC)
9 Albert Embankment, London SE1 7SN
Telephone: 01 735 7654
Telex: 916061
Date of establishment
1967

Public corporation.

Main fields of work
- Development of steels to meet the various needs of the building and

civil engineering industries
- Developing the uses of steel through improved designs and studies of new applications
- Fire protection and fire engineering
- Technical services and contract research facilities are provided to external organizations in the above fields of work.

Contact: Manager, External Services, Sheffield Laboratories, Moorgate, Rotherham S60 3AR, Telephone: 0709 60166, Telex: 547279.

Total staff: Research: 1300 (R and D on building and associated topics, approximately 50)

Director or Chief Executive

Managing Director Technical: Dr F. Fitzgerald

Director, Research and Development: Dr K.J. Irvine

Publications
- *Journal:* Steelresearch, biennial review of selected R and D topics
- Papers published in the technical literature.

GBR.515

Building Advisory Service (BAS)

18 Mansfield Street, London W1M 9FG

Telephone: 01 636 2862

Telex: 265763

Other address:

Ashridge Green, Enniskerry, Co Wicklow, Eire

Date of establishment

1954

The Building Advisory Service, an integral part of the Building Employers Confederation (formerly the NFBTI), has the objective of promoting efficiency and profitability in the construction industry. Its services and courses are available to all companies and other organizations in the construction and related industries.

Main fields of work

Management consultants have carried out several thousand assignments in the construction industry, both in the United Kingdom and overseas.

Contractual consultants advise on, prepare and negotiate claims on behalf of clients – they have achieved hundreds of satisfactory settlements. They also help contractors to identify and where possible avoid claims situations and advise on procedures, records and communications.

Planning consultants provide a full project planning service including use of the BAS Computer Bureau.

Computer consultancy team provides impartial advice and guidance on hardware and software selection, and installation, including: issue of information on available systems; preparation of feasibility studies system specifications and invitations to tender; implementation advice and assistance.

Buildax, a complete costing and accounting system for the construction industry, results from a combination of consultants' experience in designing systems for the construction industry and the computer expertise of a leading software house.

The principal divisions are: Training Division; Safety and Health Specialists; Appointments Division.

Total staff: 28 consultants, 20 support staff

Director or Chief Executive

P.H.R. Few

Principal sources of finance

BAS operates commercially and derives its income from the sale of its professional and training services and publications.

Publications
- Brochures listing courses and seminars arranged or proposed
- Surveys of Salaries and Fringe Benefits in the Construction and other associated industries
- Construction Safety Manual and Construction Safety Information Service
- BAS Bulletin (monthly)
- Various pamphlets issued from time to time on subjects of current interest to the industry.

GBR.516

Building Centre

26 Store Street, London WC1E 7BT

Telephone: 01 637 1022

Telex: 261446

Cables: INBILD

Date of establishment

1931; incorporated as a Limited Company in 1963 under The Companies Act 1948.

When the Building Centre was established in 1931 it was the first such Centre in the world. Its function is to provide displays and information to specifiers and users of building materials, components and services. In addition to London there are now permanent centres in Bristol, Cambridge, Durham, Glasgow and Manchester.

The concept of the Building Centre has proved to be so successful that it has been used as a model for many similar organizations throughout the world.

The Centre is an independent body, and receives no grants from either Government or industry.

Main fields of work

Essentially, the Building Centre is a forum for communication. Firstly, by making maximum use of modern communication techniques to bring to architects and the building team a range of information and advisory services; and secondly, by extending these services to all visitors, irrespective of whether they are part of the industry or the general public as home owners. Today the Building Centre welcomes visitors at the Centre at the rate of ¼ million per annum and looks forward to serving increasing numbers in the following ways:
- *Permanent display of products:* An area of more than 40 000 sq ft on four floors is devoted to the permanent display of products of nearly 400 manufacturers. These products are representative of almost all sectors of the industry and range from structural components to decorative finishes
- *Free information and advisory service:* The Building Centre holds the technical literature of all companies exhibiting at the Centre. Professional and trade users can obtain technical data by telephone or post, free of charge
- *The Building Bookshop:* It offers the largest selection of building and architectural books in the UK
- *Consultancy services:* Those presently available are: Heating and Energy Conservation Kitchen and Bathroom Planning; Building Extensions; Building Problems Climic
- *Building Centre Computers Limited:* Supplies practical computer systems for building component manufacturers, builders merchants, builders, building services contractors, quantity surveyors, engineers and trade and professional associations
- *Information search service:* A subscriber service to architectural practices, contractors and building merchants who lack an in-house library
- Video Services
- *Overseas services department:* Gives export advice; overseas market briefings. Specialist subject briefings; also translation services through the Construction Industry Translation Unit.

Additional facilities include Seminar, Reception and Exhibition Space and the Building Centre Club.

Total staff: 50

Professionally qualified staff: 4

Director or Chief Executive

Chief Executive: John F. George

Heads of department

Finance: B.F.D. Cassidy, Group Comptroller

Administration: J.E. Tye, Company Secretary

Information Services: A. Kirk, Technical Advisor

Marketing Services: N. Newsum, Marketing Advisor

International: Ms U. Kaye, Manager Overseas Services

Principal sources of finance

Exhibition fees; fees for information services.

Principal related bodies

The Building Centre Trust (parent organization and an educational charity); the Building Bookshop Ltd (provides books and other literature pertinent to building); Building Centre Computers Ltd; Architectural Newsreel Ltd (provides newsreels of products/developments of interest to Architects for showing within their offices); the Centre for Construction Market Information Ltd (provides construction market information and research); Barbour/Builder Ltd (provision of telephone/visitors product information service. Publishes the Barbour Compendium); Construction Industry Translation Unit (translation services).

Publications

Fact sheets covering some of the more common building problems.

GBR.517 FM

Building Research Establishment (BRE)

Building Research Station, Garston, Watford WD2 7JR

Telephone: (Garston, Herts) 0923 674040

Telex: 923220

Other addresses:

Fire Research Station, Borehamwood, Hertfordshire WD6 2BL

Telephone: 01 953 6177

Telex: 8951648

Princes Risborough Laboratory, Princes Risborough, Aylesbury, Buckinghamshire HP17 9PX

Telephone 084 44 3101

Telex: 83559

BRE Scottish Laboratory, East Kilbride, Glasgow G75 0RZ

Telephone (East Kilbride) 035 52 33001

Date of establishment

BRE was established in 1972 by the amalgamation of the Building Research Station (founded 1921), the Fire Research Station (founded 1946) and the Forest Products Research Laboratory (founded 1927).

BRE is a Government research establishment within the Department of the Environment (DOE) and carries out research and development in building and construction and allied fields primarily for the needs of Government; specific research is undertaken for other customers in both the public and private sectors.

Main fields of work

BRE research ranges over most aspects of building and much of civil engineering (exceptions are roads and waterways) and stems from Government's interests in the building and construction and allied fields – its policy and statutory responsibilities for safety and building regulations, housing, energy conservation and environmental protection, its sponsorship of the construction industry and its executive role via the Property Services Agency (PSA) as a major client for building, construction and maintenance. The main research subject areas are: foundations, structural integrity and design, fire, health and safety, conservation of fuel and power, building services, building with components, maintenance and preservation, construction costs, design and building process, materials utilization (including concrete, masonry, plastics, steel, timber) and environmental protection.

Total staff: 750

Professionally qualified staff: 225

Director or Chief Executive

Director: Dr R.G.H. Watson

Heads of department

Communications and Services Department: R.E. Jeanes, Deputy Director

Environment Department: Dr S.J. Leach, Assistant Director

Fire Research Station: K.N. Palmer, Assistant Director

Geotechnics and Structures Department: Dr J.B. Menzies, Assistant Director

Materials and Construction Department: T.J. Griffiths, Assistant Director

Princes Risborough Laboratory: J.M. Baker, Assistant Director

Principal sources of finance

BRE is financed through the annual Parliamentary Vote of Funds for DOE.

Principal related bodies

Two external committees advise Government on priorities for building and construction research. They are the Building Regulations Advisory Committee and the National Economic Development Council (EDC) Research Strategy Committee, a sub-committee of the EDC's Building and Civil Engineering Committees. Official Visitors, appointed by the Secretary of State for the Environment, advise BRE on aspects of the research and the communication and application of results.

Publications

- *Journals:* BRE News of Construction Research bi-monthly; BRE News of Timber Research, three times a year; BRE News of Fire Research, three times a year
- *Research reports:* Official research reports are published by BRE or through Her Majesty's Stationery Office. Numerous papers are published in professional and scientific journals
- *Bibliographies/Catalogues:* An Information Directory is published annually, giving details of all publications available
- The BRE Research Programme is published annually
- BRE Annual Report
- BRE Digests are published monthly
- Defects Action Sheets: 24 are issued per year

For more information about publications, which are now mostly priced, contact the BRE Publications Sales Office at Garston.

GBR.518

Building Services Research and Information Association (BSRIA)

Old Bracknell Lane West, Bracknell, Berks RG12 4AH

Telephone: 0344-426511

Telex: 848288 BSRIAC G

Cables: HEATLAB, BRACKNELL

Date of establishment

Established 1955 as Heating Ventilating Research Council; became Building Services Research and Information Association in 1975

A co-operative research association maintained by the building services industry. Member organizations include consulting engineers, contractors, manufacturers, educational establishments, commercial and public bodies. Subscriptions are supplemented by Government and industrial contracts.

Main fields of work

Research: All BSRIA research stems from the information needs of the industry, to improve its effectiveness and the efficiency of building plant. Projects deal with essentially practical problems over a wider range of areas including: energy use and conservation, design and operation of plant, co-ordination of services, temperature, air movement and comfort, emergency lighting, prefabrication of systems in buildings, building communication systems, site and design office organization, computer applications, district heating networks.

Projects are conducted either at the Bracknell laboratories or on site. Both the short and long-term needs of the industry are catered for by a series of Specialist Units working in key areas: system-design and performance; pipes and ducts; statistics and forecasting; project information and technical management; building operating

Information: Providing technical information and advice is one of BSRIA's main functions. Over 8000 enquiries a year are answered and all are treated in confidence.

BSRIA enquiry officers are professional engineers with wide industrial experience. They also have the backing of the specialist expertise of all the research staff, a library which is second to none in its field, and close contact with information sources throughout the world.

The BSRIA library serves staff and members alike, and a busy loan and photocopy service demonstrates its value. International literature (books, journals, conference papers, research reports and manufacturers' catalogues) is collected and assessed. The database IBSEDEX, stored on BSRIA's Prime computer is the only English language database in the world devoted to building services. The most important references are announced in International Building Services Abstracts, which is the standard review of recent world building services documentation. The library also maintains: trade name indexes; UK agents of foreign companies; translations; bibliographies on specific topics.

Contract Services: Much of BSRIA's work is done under contract either to the Government or industry. The results are confidential to the client. Those services are provided to members at preferential rates and cover a wide range; Sponsored Research (for individual or groups of organizations, public bodies and Government Departments); Design Evaluation (scale model or full-size tests are carried out, including analysis of the test data and recommendations for design modifications); Site and Laboratory Investigations (for equipment and system performance testing, trouble shooting, fault and failure diagnosis, environmental surveys of temperature, air or fume movement, illumination and noise); Equipment Testing (extensive laboratory facilities exist for a wide range of equipment testing).

Total staff: 90
Professionally qualified staff: 60
Director or Chief Executive
D.P. Gregory
Heads of department
Administration: L. Crooks
Contracts: G.J. Baker
Membership: A.G. Foster
Technical: L.J. Stewart
Information Centre: A.R. Eaves
Principal sources of finance
Members' subscriptions; fees for contract and sponsored projects.
Publications
- Application Guides, present the results of research in a readily usable form
- Project Reports, with theoretical arguments and descriptions of experimental rigs
- Technical Notes, collate and digest information already published or stemming from other BSRIA research
- International Building Services Abstracts, bi-monthly journal of world literature on building services, compiled by Rehva (Representatives of European Heating and Ventilating Associations)
- Bibliographies on relevant subjects.
- Statistics Bulletin.

GBR.519 AM

Cement and Concrete Association (C & CA)
Wexham Springs, Slough SL3 6PL
Telephone: Fulmer (028 16) 2727
Telex: 848352
Date of establishment
Set up in 1935 by British cement industry

Research association fo the cement industry in the United Kingdom. Financed by the UK cement manufacturers but some contract work undertaken.
Main fields of work
Research, advice and training services for the whole of the building and civil engineering fields concerning the use of concrete.
Total staff: 410
Professionally qualified staff: 170
Director or Chief Executive
Director-General: Dr R.E. Rowe
Heads of department
Research and Development: Dr G. Somerville, Director
Advisory: Dr R. Sharp, Director
Information: P.J. Witt, Director
Training: Dr R.P. Andrew, Director
Administrative Services: B.W. Shacklock, Director
Principal sources of finance
UK cement industry.
Publications
- *Journal:* Magazine of Concrete Resaerch, quarterly; Concrete Quarterly
- Technical Reports, Advisory Publications

A catalogue is produced: Part 1 lists Publications, Slide Sets and Films; Part 2 lists Research, Development and Technical Reports.

GBR.520

Central Electricity Research Laboratories (CERL)
Kelvin Avenue, Leatherhead, Surrey KT22 7SE
Telephone: Leatherhead 374488
Telex: 917338
Date of establishment
1950

Central research laboratories for the electricity supply industry.
Main fields of work
Branches: Applied Physics; Biology; Control and Instrumentation; Electrical Engineering; Engineering Sciences; Materials; Laboratory Services.

Research on all problems associated with the electricity supply industry. In the field of civil engineering this includes: design of chimneys and cooling towers; cooling water systems; concrete pressure vessels; wind loading structures.
Total staff: 350 qualified scientists and engineers, 450 technical and administrative support staff.
Director or Chief Executive
Director: Dr P.F. Chester
Principal sources of finance
Financed from part of the income derived from the sale of electricity by the Generating Board.
Publications
- Proceedings of conferences and symposia
- Papers in scientific journals
- Research reports.

GBR.521 AM

Chartered Institute of Building (CIOB)
Englemere, King's Ride, Ascot, Berkshire, SL5 8BJ
Telephone: Ascot 23355
Date of establishment
Formed in 1834, incorporated under the Companies Acts in 1884 and by Royal Charter in 1980.

CIOB is registered as an educational charity.

It is the professional institution for those engaged in the construction, alteration, maintenance and repair of buildings in building firms and other building organizations and those engaged in teaching building and in building research. It has some 28 000 members.

Its objects are: The promotion for the public benefit of the science and practice of building. The advancement of education in the science and practice of building including all necessary research and the publication of the results of all such resyarch.

The establishment and maintenance of appropriate standards of competence and conduct of those engaged or about to engage in the science and practice of building.

The governing body of the Institute is its National Council. England and Wales are divided into eleven regions each with a regional council. There are branch councils for Ireland and Scotland. Each region or branch has a number of local centres, a total of 53 in the UK and the Republic of Ireland.

There are overseas centres of the Institute of Canada, Hong Kong, Jamaica, Malaysia and South Africa. Separate autonomous Institutes of

Building or Construction, with which the Institute has close relations, exist in the following countries: America, Australia, Kenya, Nigeria, Singapore, South Africa.

Main fields of work

The Institute's examinations and practical experience requirements ensure that the individual member has a sound education and training in building appropriate to the needs of the industry.

A major part of the Institute's Final Examination is devoted to management subjects thus ensuring a sound grounding of the individual member in management skills.

The Institute's rules of professional conduct require a member to discharge his duties to his employer or client with fidelity or probity and to keep himself informed of new thought and development.

The Institute exercises its influence on behalf of the industry to raise standards of building education through the standards set in its examinations and by its system of exemption for courses and qualifications of appropriate level and coverage from City and Guilds to degree.

The members provide an unequalled body of knowledge and experience of building practice, education and research, and the Institute devotes a great deal of its resource to the dissemination of this knowledge in the form of codes of practice, technical meetings, conferences and seminars, technical papers and other publications for the benefit of members and others in the industry.

The Institute provides a link between bodies concerned with education and practice and the practising member and others. It represents the building industry on a very wide range of joint bodies concerned with building education or practice.

The Institute offers advice to government departments and other agencies on many aspects of building, so as to improve the 'climate' in which building is carried on and to secure the adoption of good building practice.

Through its Queen Elizabeth II Silver Jubilee Fund the Institute grants scholarships for individual investigation and research and publishes the results of the research for the benefit of the industry.

It makes awards for outstanding performance in the management of building projects.

Through its participation with the NFBTE in the Building Industry Careers Service, and otherwise, the Institute provides advice and information on careers in the building industry to young people, parents, careers teachers and careers officers.

Total staff: 85

Professionally qualified staff: 8

Director or Chief Executive

Dennis A. Neale, OBE, MC

Heads of department

Practice: Gordon P. Cottrell
Education: Richard J. Bland
Information: Peter A. Harlow
Membership: Peter Twine
Administration: Keith Banbury
Finance: Walter V. Clark

Principal sources of finance

Members' subscriptions.

Publications

- *Journal:* Building Technology and Management, monthly; Construction Papers, biennial
- Bibliographies, etc.

A comprehensive catalogue of CIOB publications is produced.

GBR.522

Chartered Institution of Building Services

Delta House, 222 Balham High Road, Balham, London SW12 9BS

Telephone: 01 675 5211

Date of establishment

Founded as Institution of Heating and Ventilating Engineers (1897). Royal Charter 1976. Amalgamated with Illuminating Engineering Society 1978.

An independent professional engineering institution

Objectives: the promotion of the art, science and practice of building services engineering; advancement of education and research in building services engineering.

Main fields of work

- Learned Society: Publication of data and research findings; Conferences, courses and seminars
- Qualifying Body: Setting standards of entry to the profession; Accreditation of approved courses of study; Public examinations.

Total staff: 25

Director or Chief Executive

Secretary: B.A. Hodges, OBE

Heads of department

Lighting Division: A.V. Ramsay
Registrar: D. Harries
Technical Department: K. Butcher
Conferences: Mrs A. Gibbins

Principal sources of finance

Membership Subscriptions; sale of Publications.

Publications

- *Journal:* Building Services, monthly; Building Services Engineering Research and Technology, quarterly; Lighting Research and Technology, quarterly
- *Recommendations:* over 100 titles including: CIBS/IHVE Guide, CIBS Building Energy Code, CIBS/IES Code for Interior Lighting
- *Research reports:* published frequently
- *Bibliographies/Catalogues:* details and price list available on application.

GBR.523

Concrete Society

Terminal House, Grosvenor Gardens, London SW1W 0AJ

Telephone: 01 730 8252

Telex: 261700

Date of establishment

1966

A society whose membership is open to all with an interest in concrete. Its purpose is 'to bring together all with an interest in concrete and thereby to harness, share and propagate experience and knowledge of concrete in all its aspects so as to develop and advance its proper use for the public benefit'. Membership is available for individuals, students and companies.

The Society organizes technical seminars, symposia, conferences and exhibitions. It publishes reports and other documents, produced by working parties whose membership is drawn from the construction industry.

It represents members on British Standards Institution and other external bodies. It promotes excellence in concrete through various annual awards and scholarships.

Main fields of work

Technical work is carried out principally by working parties set up under seven Steering Groups (Architecture and Environment, Construction and Civil Engineering, Education Training and Publications, International and European Affairs, Materials, Membership and Events and Structural Engineering). About 30 working parties are active at present. Specialist Committees also deal with areas of special interest (prestressed concrete, structural, fire resistance, lightweight concrete, roads and pavements, publications).

Many activities are carried out in association with other UK institutions and bodies.

Technical and social activities are organized throughout the UK by ten regions and ten clubs.

Total staff: 12

Professionally qualified staff: 2

Director or Chief Executive

Director and Secretary: T.W. Kirkbride

Heads of department

Activities: G.E. Young

Technical: J.N. Clarke

Membership: Mrs B.H. Jansz

Principal sources of finance

Membership subscriptions, income from events and publications, contribution from Cement and Concrete Association.

Publications

- *Journal:* Concrete, monthly
- *Technical standards:* 20 Technical Reports, dealing with concrete materials, design, construction. 2/3 new titles each year
- Publications catalogue available on request
- Data sheets, Technical Papers, Proceedings of Conferences and other publications.

GBR.524 AM

Construction Industry Research and Information Association (CIRIA) including the Research and Information Group for Underwater and Offshore Engineering (UEG)

6 Storey's Gate, Westminster, London SW1P 3AU

Telephone: 01 222 8891

Telex: 24224 (preface messages 2063)

Date of establishment

April 1967; formerly the Civil Engineering Research Council (CERC) 1962 and the Civil Engineering Research Association (CERA) 1964

CIRIA is an independent, non-profit-making co-operative research association serving the building, civil engineering and offshore industries. It is registered as a company with liability limited by guarantee.

The Association has a membership of more than 500 firms, organizations and educational bodies concerned with design, construction, maintenance and operation of buildings, civil engineering works and offshore installations. Members also include central and local government organizations. Membership is world-wide.

Members pay annual subscriptions related to their size and activity.

The main roles of the Association are to identify the research needs of its members, to specify, finance, commission and manage the work required; and to report, disseminate and help members to exploit the results and information obtained.

The Association has no laboratory and implements the majority of its programme by placing contracts with those best able to undertake the specific work. This provides great flexibility and allows the Council and Committees a wide choice in programme planning.

A full-time staff undertakes a proportion of the Association's R and D programme but is principally responsible for administration, project management, relations with members, presentation and dissemination of information. An important feature of the Association's work is the extensive amount of active participation by the members themselves.

Main fields of work

The Association's programme is aimed at the solution of problems met by members in connection with their work, either in the UK or overseas, in the planning, design, construction, operation and maintenance of building, civil engineering and offshore works.

Types of activity undertaken include: reviews of existing information and experience; guides for good practice; theoretical research; experimental research in laboratories and on site; full-scale experiments; development of processes and equipment; studies in management and economics; studies of contractual methods; marketing surveys.

The main areas covered are indicated by the titles of the Association's Technical Committees, as follows: Structural Design, Civil Engineering Construction, Building Design and Construction, Earthworks and Foundations, Hydraulic and Public Health Engineering, Underground Construction (tunnelling, etc.), Offshore Structures, Man-under-Water (diving, tools and underwater working), Underwater Engineering, Maintenance of Offshore Structures.

Total staff: 45

Professionally qualified staff: 20 (+100 employed by others on work commissioned by CIRIA)

Director or Chief Executive

Director: Dr L.S. Blake

Heads of department

Deputy Director: P. Pullar-Strecker

Director of Research: B.M. Sadgrove

Manager, Building and Civil Engineering Division: G.M. Gray

Manager, UEG (Underwater and Offshore Engineering): R.K. Venables

Association Secretary: M.D. Hodgkinson

Principal sources of finance

Subscriptions and contributions from members, Government sponsorship, contracts, sales and royalties.

Principal related bodies

Council (Chairman H.R. Oakley), Governing Body; Finance Committee (Chairman P. Galliford), Advisory Committee of Council; Research Committee (Chairman N.H. Searle), Advisory Committee of Council; UEG Committee (Chairman R.I. Walker), Advisory Committee of Council.

Publications

- *Journal:* Annual Report (e.g. CIRIA 82); Bi-monthly Newsletter (CIRIA News)
- CIRIA Reports (R) (100 titles)
- CIRIA Technical Notes (TN) (110 titles)
- CIRIA Guides (G) (5 titles)
- Special publications (SP) (24 titles)
- UEG Reports (UR) (20 titles)
- UEG Technical Notes (UTN) (26 titles)
- *Bibliographies/Catalogues:* List of CIRIA/UEG publications issued quarterly; Miscellaneous bibliographies.
- Piling Reports/Guides (PG) (9 titles), for the Department of the Environment
- Offshore Technology Papers (OTP) (11 titles), – for the Department of Energy.

GBR.525

Constructional Steel Research and Development Organization (CONSTRADO)

NLA Tower, 12 Addiscombe Road, Croydon CR 9 3JH

Telephone: 01 686 0366

Telex: 946372

Date of establishment

1971

Main fields of work

The four sections comprise:

- *Design Development:* involves an appraisal of properties of steel construction products and assessing how these may be best exploited in design work, together with the technical administration of the research programme and the development of design. There is close co-operation with the British Standards Institution and draft codes of practice are prepared
- *Advisory Services:* responsible for giving guidance to specifiers on the competitive use of steel; a large number of general enquiries are dealt with by this department including fire protection and building regulations problems
- *Educational Services:* responsible for the furtherance of an educational programme for the professions and in conjunction with Schools of Architecture, Universities and Polytechnics

- *Planning and Liaison:* responsible for the planning and organization of all CONSTRADO programmes, events, courses, conferences, exhibitions and the Structural Steel Design Awards Scheme. CONSTRADO is represented on the various committees of the European Convention for Constructional Steelwork (in association with the British Constructional Steelwork Association) which has an important influence on European regulations and codes.

  The library in this section contains one of the most up to date and comprehensive references to steel construction in the country.

Total staff: about 20
Director or Chief Executive
Director: K.W. Groves
Principal sources of finance
British Steel Corporation.
Publications
- Design Guides
- Handbooks
- Project Studies
- CONSTRADO Monograph Series.

GBR.526
Coventry (Lanchester) Polytechnic, Department of Civil Engineering and Building
Priory Street, Coventry, West Midlands CV1 5FB
Telephone: 0203 24166
Telex: 31469 LIBCOV G
Date of establishment
1960

Government funded polytechnic concerned with higher education, research and consultancy. The Department is part of the Faculty of Engineering of Coventry Polytechnic.
Main fields of work
Civil engineering and building in the following areas:
- Soil mechanics and the rheology of soils
- Public health engineering
- Civil engineering and building structures, especially the behaviour of reinforced brickwork
- Energy and buildings
- Ventilation of buildings
- Building economics
- Housing renewal
- Industrial noise
- Industrial building performance and design
- Construction management

  Courses leading to degrees in Civil Engineering and Building.

Total staff: 46
Professionally qualified staff: 34
Director or Chief Executive
Head of Civil Engineering and Building Department: H.E. Walker
Principal sources of finance
Central Government; research and consultancy contracts; the Science and Engineering Research Council.
Publications
Research report published every three years and available on request. This contains full details of all publications.

GBR.527 UM
Duffy Eley Giffone Worthington (DEGW)
Bulstrode Place, Marylebone Lane, London W1
Telephone: 01 486 9281
Telex: 291682
Date of establishment
1974

Architects and space planners. Working in both the United Kingdom and continental Europe, and with an overseas office in Mexico City.

The practice is concerned to link design with building use, with an emphasis on user research, programming, interior planning and space management, and evaluation studies.
Total staff: 36
Professionally qualified staff: 28
Director or Chief Executive
Partners: Dr Francis Duffy, Peter Eley, John Worthington, Colin Cave, Luigi Giffone, John Francis, Christopher Byng-Maddick
Publications
Publications have been contributed to as follows:
Architects Data: Frank Duffy, David Jenkin, John Worthington
Progetto Ufficio: Frank Duffy
Office Buildings and Organisational Change: Frank Duffy, John Francis
Building conversion and rehabilitation: John Worthington, Peter Eley
Business Property Handbook: John Worthington
Macmillan Encylopaedia of Architecture and Technological change: John Worthington
Office Book: Frank Duffy
Office Planner; Planning Office Space: Frank Duffy, Colin Cave, John Worthington
Office Furniture: Peter Ullathorne
Specification '80: Patrick Manwell
Specification '82: Patrick Manwell
Swedish Council for Building Research: John Worthington
Facilities, a monthly digest for the building administration manager
Orbit.

GBR.528
Dundee College of Technology, Department of Civil Engineering
Bell Street, Dundee DD1 1HG
Telephone: 0382 27225
Telex: 76453

The College is a Scottish Central Institution providing tertiary education up to postgraduate degree level. The College also provides a testing and consultancy service for local and national industry.
Main fields of work
Teaching: BSc Degree in Civil Engineering, Higher National Diploma in Civil Engineering; Higher Certificate in Civil Engineering

Research interests: Creep and fracture in concrete; Behaviour of concrete in torsion; Behaviour of orthotropically reinforced concrete slabs with openings; Steel portal frame connections.
Total staff: 21
Professionally qualified staff: 13
Director or Chief Executive
Dr S. Sarkar
Principal related bodies
Academic Council (College internal validation and academic policy formulation); Council for National Academic Awards (CNAA) (validation of degree courses); Scottish Technical Education Council (SCOTEC) (administration of diploma and certificate courses).

GBR.529 AM
Dundee College of Technology, Department of Surveying and Building
Bell Street, Dundee DD1 1HG
Telephone: (0382) 27225
Telex: 76453 DUNCOT G

The College is a Scottish Central Institution funded by the Scottish Education Department to prepare students to degree level in vocationally orientated subjects. The degrees are validated by the Council for National Academic Awards.

Main fields of work
Main purpose of the College is to offer full-time courses at degree and diploma level in a wide variety of engineering, science, business and other processional disciplines.

The Department of Surveying and Building offers a Degree in Quantity Surveying and a Diploma in Building.

Research areas: Cost Modelling related to Building Design; Computer Applications related to Building; Maintenance; Building Regulation an Control; Building Management; Contractual Practices.

Total staff: 12
Professionally qualified staff: 8
Director or Chief Executive
Head of Department: D.H. Jex
Heads of department
Course Leader: BSc in Quantity Surveying: H.A. Gow
Course Leader: Higher National Diploma in Building: J.M. Searle
Principal sources of finance
Funded by the Scottish Education Department.
Principal related bodies
Council for National Academic Awards – validate Degrees; Scottish Technical Council – award Diplomas.

GBR.530
Electricity Council Research Centre (ECRC)
Capenhurst, Chester CH1 6ES
Telephone: 051 339 4181
Telex: 627124
Date of establishment
1965

One of four national research laboratories of the Electricity Supply Industry in the UK.
Main fields of work
Overall research objectives are twofold:

To increase the efficient use of electricity and develop new uses and new processes appropriate to electricity as a refined energy source; if proved, engineer those ways of using electricity which show a significant economic advantage in a total costed system, i.e. conservation of total resources (capital, materials, labour and energy) across the whole range of human activities at home and at work

To improve the efficiency and economics of the existing and future distribution system so that energy is available when and where the customer requires it at the minimum long term cost.

Research on the built environment covers aspects of interaction among human beings, equipment, services and building structures.

Total staff: 300
Professionally qualified staff: 100 plus 140 technically qualified support staff
Director or Chief Executive
Centre Director: M.R. Cowan
Research Director: Dr A.T. Churchman
Principal sources of finance
From the Area Boards who retail electricity and electrical goods and services in Britain. Financed by an agreed contribution from each of the twelve English and Welsh Area Boards, two Scottish Boards and two Irish Boards to the Electricity Council's central budget.
Publications
Research reports: Copies of unrestricted reports are distributed to the British Library (Lending Division), Boston Spa, Yorkshire, and to the Technology Reports Centre, Department of Industry, Orpington, Kent, BR5 3RF

Papers are published in scientific and technical journals.

GBR.531
ERA Technology Ltd
Cleeve Road, Leatherhead, Surrey KT22 7SA
Telephone: Leatherhead 374151
Telex: 264045
Date of establishment
1920

A research and development organization, private company limited by guarantee, the guarantors being its member companies representing the electrical/electronic and allied industries.
Main fields of work
An independent contract R and D organization providing confidential services across a large part of the electromagnetic spectrum.

Current activities at ERA extend from power engineering to the design of microwave and millimetre wave antennas, and include such apparently diverse topics as the fundamental aspects of dielectric materials, engineering metallurgy and the application of microprocessors. Within these fields its expertise covers market research, the evaluation of new materials, the design and development of advanced components, systems and equipment in the electrical and electronic field, and a range of test and failure analysis services.

ERA's well equipped laboratories are based at a 35 acre (14 hectare) site at Leatherhead, Surrey. Apart from specialized facilities used in particular research projects ERA also has a number of approved testing facilities in specific areas such as electromagnetic interference, hazardous environments and short circuit testing. In addition, extensive use is made by many departments of dedicated computer facilities in their research, for product development and as a design aid. ERA works in close collaboration with other UK and international research and development establishments in providing its services to clients on a worldwide basis.

Total staff: 300
Professionally qualified staff: approximately 150
Director or Chief Executive
Chairman: F.J. Lane, OBE
Managing Director: Dr A.W. Rudge
Heads of department
Materials Sciences: Dr N. Parkman
Industrial Electronics: H.G. Riddlestone
Power Engineering: R. Stokes
RF Technology Centre: Dr N.A. Adatia
Marketing: D.W.Brooks
Engineering Materials and Metallurgy: Dr M.P.E. Desvaux
Computers and Automation: J.H. Donald
Principal sources of finance
Industry and Government.
Principal related bodies
Member: Association of Independent Contract Research Organizations.
Publications
- *Journal:* ERA Technology News, published five times a year and sent out free of charge
- *Research reports:* Numerous technical reports are published throughout the year and these are listed together in the ERA List of Reports, which is periodically updated
- Translations of articles in foreign journals are published and a list of translations is also issued annually

The ERA Projects List is issued annually and this includes all on-going or on-offer projects for the coming year.

GBR.532 **AM**
Heriot-Watt University, Department of Building
Chambers Street, Edinburgh, EH1 1HY
Telephone: 031 225 8432
Date of establishment
1821, established as College; 1966, established as a University

Technological University, supported by government but having strong links with industry, many through the University's Research Park.

Considerable research, development and development work carried out for industry, commerce and government departments.

Main fields of work

- Construction Management
- Building Technology
- Building Maintenance
- Building Economics and Quantity Surveying
- Land and Property Appraisal and Management
- Acoustics, Noise and Vibration Studies
- Energy in Buildings, including Economic Studies, Insulation, Condensation and Technology
- Materials Science and Technology
- Computer Applications, including Computer Aided Design and Draughting
- Wind Loading On Buildings
- Ventilation of Buildings
- Building Services

The Department has the following Faculties: Faculty of Engineering; Faculty of Science; Faculty of Environmental Studies (including Architecture and Town and Country Planning); Faculty of Economics and Social Studies; Faculty of Arts.

Total staff: 40

Professionally qualified staff: 28

Director or Chief Executive

William Watson Professor and Head of Department: Professor Dr Victor Brownlie Torrance

Principal sources of finance

UK Universities Grants Committee; the Science and Engineering Research Council; the Department of the Environment; the Scottish Development Department; industry and commerce.

Principal related bodies

Edinburgh College of Art (affiliated college).

Publications

- *Research reports:* Provided direct to funding organizations
- Mainly papers and mongraphs to Journals and Conferences
- A Synopsis of The Law and Practice of Arbitration in Scotland by M. Weir; Heriot-Watt University
- Brickwork and Blockwork Measurement by S. Murray; George Goodwin.

GBR.533

Hydro-Air International (UK) Ltd

Midland House, New Road, Halesowen, West Midlands

Telephone: 021 550 2767

Telex: 339459

Date of establishment

1965

A private company.

Also in Creteil, Berlin, Neu Isenburg and Phaffikon

Main fields of work

Supply of machinery, nail plate fasteners and technical back-up for the timber roof truss industry. Supply of metal fasteners for buildings.

Total staff: 80

Professionally qualified staff: 7

Director or Chief Executive

Managing Director: B.J. Griffin

Principal sources of finance

Sales of products

Publications

Technical and marketing literature.

GBR.534

Imperial College of Science and Technology, Department of Civil Engineering

London SW7 2BU

Telephone: 01 589 5111

Telex: 261503

Cables: IMPCOL

Date of establishment

The Department was established as a separate department within Imperial College in 1913.

The Department is one of the major departments in Imperial College, which was established by Royal Charter in 1907. Imperial College is one of the constituent Colleges of the University of London, of which there are fifteen multi-faculty Colleges.

Imperial College is an institution mainly financed from state funds. It exists for teaching at undergraduate and postgraduate level and has research activities closely related to the needs of industry.

Main fields of work

The Department is active in four main areas: undergraduate courses, specialized one-year postgraduate courses, research work for higher degrees and research work undertaken for the civil engineering profession.

Fundamental research is undertaken in steel structures, concrete structures, timber structures, soil mechanics and engineering seismology, public health and water technology, hydraulics, transport, systems and mechanics, and surveying.

There are laboratories, workshops, computer facilities and a library.

Total staff: 136

Professionally qualified staff: 68

Director or Chief Executive

Head of Department: Professor John Munro

Principal sources of finance

Government funds, students' fees, government funded research institutions, sponsorship of industry.

Publications

Research reports: Publications internally produced on research projects.

GBR.535 AM

Laing Design and Development Centre

Page Street, Mill Hill, London NW7 2ER

Telephone: 01 959 3636

Telex: 263271

Date of establishment

Amalgamation of John Laing Research and Development, Ground Engineering and John Laing Design Associates in January 1983

The Centre is part of the Laing Group of construction companies. Its aim is to provide a complete consultancy, contract research and testing organization for the Laing Group and for the industry at large, on a commercial basis.

Main fields of work

- Land survey
- Ground investigation
- Building and civil engineering design
- Energy consultancy
- Specialist building consultancy
- Quality assurance, testing of products and materials
- Research
- Training

The Centre is organized into the following: Design Service, R and D Service, and Geotechnical Service. Staff consists of all disciplines associated with construction.

Total staff: approximately 250

Professionally qualified staff: approximately 150

Director or Chief Executive

Director: P.D. Westwood

Heads of department
Design Service: P.D. Westwood
R and D Service: A.P. Backler
Geotechnical Service: H.G. Clapham
Principal sources of finance
The Centre provides a service for the Laing Group of companies and works on a commercial basis for external clients.
Publications
Research reports: All publications are internal to the Laing Group or confidential to the client.

GBR.536

Leeds Polytechnic, School of Constructional Studies

Brunswick Building, Brunswick Terrace, Leeds LS2 8BU
Telephone: 0532 463232
Date of establishment
1 January 1970 – previously College of Technology

The School is one of eighteen within the Leeds Polytechnic and shares a unique building complex with the Schools of Architecture and Landscape and Planning and Environmental Studies.

Constructional Studies provides sandwich courses for first degrees in Building Environmental Health and Quantity Surveying together with part-time professional courses. Additionally, short and intensive courses are organized for the benefit of industry and vocational students.

The School has four main units: Building, Civil Engineering, Environmental Health, and Quantity Surveying, and has benefited from some research awards from the Science and Engineering Research Council and consultancy from industry.

Main fields of work
Education: Provision of thick sandwich degrees in areas stated; professional part-time course for industry; specialist short courses.

Research fields include: Application of micro-computers in construction; Investigations into the technical problems in the conversion and renovation of large historic buildings; Teaching methods – construction technology and practice; Thermal performance of traditional and solar houses (live scheme based on fourteen houses).

Total staff: 50
Professionally qualified staff: 50
Heads of department
Building: C. Grisenthwaite
Civil Engineering: A. Hartley
Environmental Health: J. Bickerdike
Quantity Surveying: G. Turner
Principal sources of finance
Government and local authority.
Principal related bodies
Chartered Institute of Building; Chartered Institution of Civil Engineers; Royal Institute of Chartered Surveyors; Environmental Health Officers Education Board; Council for National Academic Awards. Participation in professional activities linking industry with teaching. Liaising on developments between the industrial field and special course and consultancy demands.
Publications
- Research papers
- Text books
- Brochures
- Curricula documents.

GBR.537

Leicester Polytechnic, School of Architecture

PO Box 143, Leicester LE1 9BH
Telephone: 0533 55155
Date of establishment
Established as a Polytechnic in 1969 from the Leicester College of Art and Technology which was founded in 1930

Educational establishment.
Main fields of work
The School of Architecture is a teaching and research department located within the Faculty of Technology and Construction offering courses leading to the award of first degrees, graduate diplomas.

Research work leading to higher awards is supervised and undertaken by members of staff and full time research students into a number of fields but with particular emphasis on environmental studies and computer aided design.

Total staff: 32
Professionally qualified staff: 26
Director or Chief Executive
Professor Theo Matoff
Heads of department
BA (Hons) Course Leader: D.G. Henderson
DA Course Leader: W.L. Stuart
Dip Arch (Part time) Course Leader: R.S. Short
Dip Conservation (Part time) Course Leader: J.S. Curl
Principal sources of finance
Central funding from local authority pool.
Publications
- *Journal:* Perspective, twice per year free of charge
- *Research reports:* Various.

GBR.538

Materials and Components Development and Testing Association (MACDATA)

Paisley College of Technology, 47 High Street, Paisley PA1 2BE
Telephone: 041 887 5737
Telex: 778951
Date of establishment
1964

MACDATA is a Unit of Paisley College of Technology which exists to provide a service to clients in the research, consultancy and testing fields. This service extends to many subjects but is primarily concerned with construction matters. Up to 600 projects a year are handled.

The Unit is regularly consulted by major companies, government and local government organisations.

Main fields of work
The Unit is heavily engaged in research, consultancy and testing in relation to a wide variety of construction problems. In particular it has for many years been pre-eminent in the testing of windows and cladding systems. A great deal of research has been carried out by the unit into problems associated with energy conservation and domestic condensation. At the present time a Research Student is engaged on a three year contract under a scholarship provided by the British Gas Council to investigate the control of the environment in houses. Considerable work concerns the investigation of constructional defects and members of MACDATA staff are frequently called upon as expert witnesses in this connection. The fields of noise and vibration together with structural testing also figure largely in the work load.
Total staff: full-time 7; part-time 50
Professionally qualified staff: 53
Director or Chief Executive
E. Downey
Heads of department
Materials Group: D. Armour

Structures Group: J. Duncan
Fluids Group: J. Younger
Vibration Group: I. Buchan
Certification/Testing: M. Arthur

Principal sources of finance

The Unit is financed entirely on the basis of fees charged. It receives no contribution from any other source.

Principal related bodies

MACDATA Fire Centre – Research, Testing and Courses in fire related topics.

Publications

Research reports: Many reports published for individual clients some of which become available on request.

GBR.539 AM

National Federation of Building Trades Employers (NFBTE)

82 New Cavendish Street, London W1M 8AD
Telephone: 01 580 5588
Telex: 265763
Other addresses:
NFBTE Eastern Region, 95 Tenison Road, Cambridge CB1 2DL
Telephone: 0223 355418
Telex: 81574
NFBTE Liverpool Region, Federation House, Hope Street, Liverpool L1 9BS
Telephone: 051 709 9916
Telex: 629385
NFBTE London Region, 18–20 Duchess Mews, London W1N 3AD
Telephone: 01 636 3891/5
Telex: 265763
NFBTE Midland Region, Federation House, 2309/11 Coventry Road, Sheldon, Birmingham B26 3PL
Telephone: 021 742 5121
Telex: 337022
NFBTE Northern Counties Region, Green Lane, Durham City, County Durham
Telephone: 0385 61367/8
Telex: 538238
NFBTE North Western Region, 2, Conyngham Road, Victoria Park, Manchester M14 5SH
Telephone: 061 224 3214/7
Telex: 668015
NFBTE Southern Region, 86 Hurst Road, Horsham, West Sussex RH12 1DU
Telephone: 0403 54226
Telex: 877192
NFBTE South Wales Region, 66 Cardiff Road, Glan-Y-Llyn, Taffs Well, Cardiff CF4 7YA
Telephone: 0222 810681
Telex: 497492
NFBTE South Western Region, 22 Richmond Hill, Clifton, Bristol BS8 1BD
Telephone: 0272 741161
Telex: 449752
NFBTE Yorkshire Region, Davidson House, Hales Road, Leeds LS12 4PW
Telephone: 0532 630607
Telex: 55293
Scottish Building Employers Federation, 13 Woodside Crescent, Glasgow G3 7UP
Telephone: 041 332 7144/6
Telex: 779657

Date of establishment

1878

Trade Association.

Main fields of work

Provides service to its members in the fields of industrial relations, training, safety research and technical information, economics, public relations, contractual, taxation.

Total staff: including Regions: 400

Director or Chief Executive

Director-General: Not appointed at time of compilation
Deputy Director-General: P.J. Povey

Principal sources of finance

Members' subscriptions.

Principal related bodies

Usual administrative committees, e.g. Council, Management, Building Research and Technical Information Committee

Publications

- Booklets
- Pamphlets
- Brochures
- Reports
- Reviews
- Guides
  covering: training-management and crafts, contract forms and related contractual publications, taxation, industrial relations, small works and maintenance, housing, technical, safety, transport, accounting, insurance, incentives, membership (annual report) and miscellaneous.
- Films
- Books.

GBR.540

National House Building Council

Chiltern Avenue, Amersham, Buckinghamshire HP6 5AP
(Registered address 58 Portland Place, London W1N 4BU)
Telephone: 02403 4477

Date of establishment

1963

The National House Building Council is a non-profit distributing insurance company approved by the Department of Trade. Its chairman is appointed by the Secretary of State for the Environment, but the NHBC receives no Government finance.

Its members are nominated by those bodies most closely associated with the housebuilding industry, namely the consumers associations, building societies, builders' organizations, architects, surveyors, town planners and the Law Society.

Main fields of work

NHBC maintains a register of housebuilders and developers whose standards are vetted before their names are entered on its register. Each house built or sold by a registered member must be covered by NHBC 10 year protection scheme. This means that both the builder or developer must pay a fee to the NHBC in respect of both inspection and insurance; if the property is completed to the satisfaction of NHBC, then the 10 year insurance scheme applies. When exchanging contracts with a purchaser, the builder or developer will enter into the standard form NHBC agreement under which he gives certain warranties relating to the completion of the property and the remedying of defects which appear, broadly speaking, during the first two years.

Both these warranties are guaranteed by NHBC in the event of the builder or developer's liquidation or bankruptcy. After the end of this first two year period, NHBC directly insures the home owner and his successors in title in respect of major damage caused by structural defects. There are financial limits on the insurance relating to the original selling price, up to a maximum currently of about £105 000, inflated throughout the ten year period in line with building repair costs (but subject to a maximum of 15% per annum compound).

In addition to carrying out spot check inspections of all dwellings registered under its scheme, and insuring them, NHBC also undertakes research into consumer preferences, building standards, and land supply.

Heads of department

Director General: B. Bean

Deputy Director General and Head of Technical Services: Anthony Cooper
Director of Conciliation and Insurance: Dennis F.C. Goddard
Head of Legal Services: R. Wyles
Financial Controller: M.A. Thewlis
Registrar: H. Dufton
Director of Standards and Research: Derek Kettlewell
Computer Manager: Colin Brett
Commercial Manager: Miss L.C. Higgins
Principal sources of finance
Fees paid in by registered builders and developers both annually and in respect of each dwelling registered under the Scheme.
Principal related bodies
The following are the nominating bodies to the Council of NHBC. House Builders Federation, Building Societies Association, Federation of Master Builders, the Law Society, the Royal Institute of British Architects, Royal Institution of Chartered Surveyors, Union of Construction and Allied Trades Technicians, Consumers Association, the Society of Community Medicine, the Federation of Owner Occupiers and Owner Residents Association, the National Council of Women, the Consumer Standards Advisory Committee of the British Standards Institution, the National Council of Building Material Producers, the Chartered Institute of Building, the Royal Town Planning Institute, the Association of Metropolitan Authorities, the Greater London Council, the Association of District Councils, the Institute of Civil Engineers. We also have observers on the Council nominated by the Department of the Environment.

GBR.541
National Physical Laboratory (NPL)
Queens Road, Teddington, Middlesex TW11 0LW
Telephone: 01 977 3222
Telex: 262344
Date of establishment
1900

NPL is one of the research establishments of the Department of Industry and is the focal point of the UK's national measurement system. It develops and maintains national standards of measurement for quantities such as length, mass and time and inter-compares them with those of other countries. NPL provides access to the national standards through a diverse range of calibration services and is continually developing new standards and measurement techniques to meet the changing needs of industry. The Laboratory advises government, public bodies and industry on all aspects of measurement and participates in the work of national and international standards-writing bodies. NPL also pioneers work in new technologies of industrial importance including the application of materials and information technology.
Main fields of work
The Laboratory is organized into six scientific Divisions (see below). Four are concerned with measurement standards and techniques for quantities relating to acoustics, angle, atmospheric pollution, certified reference materials, colorimetry, density, diameter, electrical measurements (from DC to sub-optical frequencies), force, hardness, humidity, ionizing radiations and radioactivity, length, mass, micrometrology, optical metrology, photometry, pressure, radiometry, spectrophotometry, temperature, thermal conductivity, time and frequency, ultrasonics, vacuum and viscometry.

One Division is concerned with the applications of materials from special metals, alloys and steels to polymers, reinforced performance data on the properties of industrially important materials in normal, high-temperature or corrosive environments and to develop methods for their measurement.

One Division is concerned with information technology and computing and in particular with the development of data networks, data security, the man–computer interface, microprocessor applications, high level languages and numerical analysis.

NPL also offers consultancy in all corrosion matters through its National Corrosion Service and accreditation for calibration and testing laboratories through its British Calibration Service and National Testing Laboratory Accreditation Scheme.
Total staff: 900
Professionally qualified staff: 550
Director or Chief Executive
Director: P. Dean, CB
Heads of department
Division of Electrical Science: O.C. Jones, Superintendent
Division of Mechanical and Optical Metrology: P.B. Clapham, Superintendent
Division of Quantum Metrology: M.E. Peover, Superintendent
Division of Radiation Science and Acoustics: W.A. Jennings, Superintendent
Division of Materials Applications: E.D. Hondros, Superintendent
Division of Information Technology and Computing: E.L. Albasiny, Superintendent
British Calibration Service: J.L. Wilson, Head
National Testing Laboratory Accreditation Scheme: P.G. Forrest, Head
National Corrosion Service: J.A. Bernie, Head
Principal sources of finance
NPL is part of the Department of Industry and is largely funded by government. Some of its costs are recovered as fees for calibrations, consultancy and contract research.
Publications
Numerous papers are published in professional and scientific journals. Open reports are produced by the Laboratory and are available on request from the Division concerned. An Annual Report is published. Many books on specialized topics are published through Her Majesty's Stationery Office.

GBR.542 AM
New University of Ulster
Coleraine, N. Ireland BT52 1SA
Telephone: 0265 4141
Telex: 747597 NUUCOL G
Cables: University Coleraine
Date of establishment
1967

The Energy Study Group in the Physics Department is the main centre of building research in the University.
Main fields of work
Heating systems, heat pumps, instrumentation and control, thermal modelling, heat recovery, energy analysis.

Research is the main activity, but members of the group can also act as consultants through the New University Industrial Unit. A part-time MSc course in Energy Studies is taught by members of the academic staff.
Total staff: 10
Professionally qualified staff: 4
Director or Chief Executive
Dr J.T. McMullan, Dr R. Morgan (both Senior Lecturers in Physics)
Heads of department
Heating systems, heat pump applications, thermal modelling, instrumentation and control: Dr R. Morgan (CIB contact person)
Heat pump thermodynamics, heat recovery, energy analysis, instrumentation and control: Dr J.T. McMullan
Principal sources of finance
Commission of the European Communities; Science and Engineering Research Council; Department of Education for Northern Ireland; Charitable foundations.
Publications
The University does not produce regular reports or journals itself.

The Energy Study Group edits the International Journal of Energy Research for John Wiley and Sons Ltd and publishes in several other research journals.

GBR.543 **AM**

Newcastle upon Tyne Polytechnic, School of Surveying
Northumberland Building, Northumberland Road, Newcastle upon Tyne NE1 8ST
Telephone: 0632 326002 ext 3537
Date of establishment
September 1980

The School of Surveying primarily performs an educational function. Research is carried out in support of this function.
Main fields of work
- Communication of Cost/Price Information in Construction
- Use of Computers in Communications
- Planning Law
- Value Analysis
- Cost Prediction of Building Services.

Total staff: 20
Professionally qualified staff: 20
Director or Chief Executive
Head of School: W.E. Renshaw
Principal sources of finance
Central Government/Local Authority.
Publications
Publications by staff members reprinted in professional journals.

GBR.544

Ove Arup and Partners
13 Fitzroy Street, London W1P 6BQ
Telephone: 01 636 1531
Telex: 263935
Cables: OVARPART LONDON W1
Date of establishment
1946

Independent firm providing a multi-disciplinary consulting service in the fields of civil, industrial and building engineering and undertaking special studies/investigations, backed by experts in related specialized fields.
Main fields of work
- Engineering design
- Special studies/investigations in the fields of: structural analysis; geotechnics; transportation; telecommunications; offshore engineering; seismic studies; structural dynamics; acoustics; vibration; fire engineering; materials technology; energy; mechanical and electrical development; computer applications; project planning; cost-benefit studies.

The firm has 44 offices located in 21 countries.
Total staff: 2700
Professionally qualified staff: 2100
Director or Chief Executive
Chairman: Peter Dunican
Heads of department
Central Technical Services (including Research and Development Groups). Turlogh O'Brion
Industrial and Offshore Group: Michael Shears
Geotechnics Group: David Henkel
Transportation Group: Malcolm Simpson
Arup Acoustics: Richard Cowell
Design Groups: Jack Zunz
Principal sources of finance
Fee income from consultancy.
Publications
Journal: Arup Journal, quarterly.

GBR.545

Oxford Polytechnic, Department of Construction
Gypsy Lane, Headington, Oxford OX3 0BP
Telephone: 0865 64777

The Laboratories are part of the Construction Department of Oxford Polytechnic, and perform testing and contract research upon request.
Main fields of work
- Concrete cube testing service
- Testing of structural components
- Testing of complete structures
- Mechanical tests of materials
- Analysis of hardening concrete
- Testing of mortars
- Testing of gypsum plasters
- Diagnosis of rising damp
- Testing of electroplated coatings
- Soil classification tests
- Soil compaction tests
- Soil strength tests
- Performance tests for rainwater gullies.

Total staff: 45
Professionally qualified staff: 16
Director or Chief Executive
P.D. Goddard
Principal sources of finance
Financed by the income from research contracts from various sources.
Publications
About 70 publications in various journals
Research reports are normally confidential.

GBR.546

Plymouth Polytechnic, Department of Civil Engineering
Palace Court, Palace Street, Plymouth PL1 2DE
Telephone: Plymouth (0752) 264643
Telex: 45423
Date of establishment
1969

The Department is part of the Faculty of Technology of Plymouth Polytechnic. It performs testing in its extensive laboratory facilities upon request across the whole range of civil engineering work. It undertakes also development and research work for state, county and municipal organizations, Civil Service, private industry and private individuals, either on a commercial basis or in certain circumstances upon request.

It is involved in the training of diploma students, graduates and postgraduates across the range of civil engineering studies and is actively involved in providing refresher courses for those in industry as required.
Main fields of work
Areas of interest cover all aspects of construction engineering. Major areas of research and development work include:
- Coastal protection and in particular quantification of wave forces on sea walls and the design of sea walls
- The behaviour, both quasi static and dynamic, of reinforced concrete and reinforced brickwork structures
- The engineering properties of soils, soft rocks and hard rocks, including site investigation studies
- Open channel hydraulics
- Hydrology and in particular ground water movements
- Highway maintenance including the design of structural maintenance, the development of analytical design procedures and the management and control of maintenance funds
- The testing of civil engineering materials.

Total staff: 24
Professionally qualified staff: 24
Director or Chief Executive
Head of Department of Civil Engineering: C.K. Kennedy

Heads of department
Coastal Protection: P.J. Hewson
Structural Behaviour: C. Southcombe
Hydraulics: F. Rendell
Hydrology: L. Hamill
Geotechnics: S.D.G. Johnston
Short Courses: D. Cox
Laboratory Testing: J.M. Clatworthy
Principal sources of finance
Central Government finance for the Department. Individual development and research projects supported by central and local government, private industry and the Research Councils.
Publications
- *Journal:* occasional
- *Research reports:* Reports are prepared for all externally financed research contracts. Reports are available on micro-computer programmes developed for highway maintenance and also on the behaviour under snow and wind loading of reinforced concrete portal frames. Reporting is normally achieved through international conferences and international journals.

GBR.547 **UM**
Polytechnic of the South Bank, Faculty of the Built Environment
Wandsworth Road, London SW8
Telephone: 01 928 8989
Date of establishment
1970

The Polytechnic of the South Bank is one of the largest institutions of science, engineering and technology in the United Kingdom. It is one of the five constituent Polytechnics supported by the Inner London Education Authority and is a company limited by guarantee.

The Faculty of the Built Environment is unique in covering virtually all aspects of the construction industry from planning at regional and local levels through the design and construction processes to property management and maintenance. The full sweep of disciplines represented within the Faculty ensures that cross- and multi-disciplinary initiatives in such fields as urban renewal, conservation, project management, housing, urban design and urban management are pursued on a comprehensive basis.
Main fields of work
Apart from providing technician, undergraduate and postgraduate courses across the professions of the built environment, the various disciplines are involved in specialist fields of work as follows:
- *Architecture:* History and theory of architecture; Relationship of technology and architecture; Relationship between architectural education and practice
- *Building Administration:* Project management, forecasting and control; Capital/Maintenance Energy ratios in housing; Reduction of vandalism
- *Building Economics:* Technological and economic aspects of the rehabilitation, maintenance and repair of buildings and of the provision of housing suitable for handicapped people; Implications of building legislation
- *Civil and Structural Engineering:* Properties of unreinforced masonry; Timber testing; Structural performance of timber frame/brick veneer construction; Load share in piled raft foundations
- *Institute of Environmental Science and Technology:* The transient and steady state behaviour of air conditioning plant; Acoustics; Occupational hygiene and safety at work
- *Estate Management:* Valuation methods; High rise housing; Land policy; Investment and development finance
- *Town Planning:* Information technology, employment and location; Regional uneven development and state policy; Process of housing development in the UK and USA; Residential environment in the inner-city.

Total staff: 250
Professionally qualified staff: 170
Director or Chief Executive
Dean of Faculty: John Ratcliffe
Heads of department
Architecture: Hans Haenlien
Building Administration: Leslie Wilder
Building Economics: Vacant
Civil and Structural Engineering: John Slaney
Institute of Environmental Science and Technology: David Scott
Estate Management: Barry Redding
Town Planning: Shean McConnell
Principal sources of finance
Inner London Education Authority; variety of external sources including grants and contracts from research councils, trusts, commercial organizations and government departments.
Publications
- *Technical standards:* various
- *Research reports:* summarizing all research in the Polytechnic published at 3 yearly intervals
- *Bibliographies/Catalogues:* Available from Faculty Library.

GBR.548
Polytechnic of Wales (Politechnig Cymru), Department of Civil Engineering and Building
Llantwit Road, Pontypridd, Mid Glamorgan CF37 1DL
Telephone: 0443 405133
Date of establishment
1913: Institution established as South Wales and Monmouthshire School of Mines. Subsequently it became a Regional College of Technology and in 1970 it was designated a Polytechnic by the Department of Education and Science.

The Polytechnic offers a comprehensive range of degree and diploma courses. Most of these in the technological areas are organized on the sandwich system where the students obtain supervised work experience in industry. All degrees awarded are those of the Council for National Academic Awards (CNAA). These include PhD and MPhil degrees as well as first degrees in Building and in Civil Engineering. The diplomas awarded are either Higher Technician Diplomas or Polytechnic Associateships.
Main fields of work
The Department provides the following courses:
- *Sandwich Courses:* BSc Degree and BSc Degree with Honours in Civil Engineering; BSc Degree in Building; Higher Diploma in Civil Engineering Studies; Higher Diploma in Building
- *Full-time (One year) Course:* Polytechnic Associateship in Building
- *Part-time Courses:* MSc Degree in Civil Engineering; Higher Certificate in Civil Engineering Studies; Examinations of the Institution of Highway Engineers; City and Guilds Concrete Practice; City and Guilds Concrete Technology and Construction
- *Special Mid Career-Training Short Courses:* Civil Engineering Law, Contract Procedure; Water Supply and Distribution, Open Channel Hydraulics, Water Treatment, Sewerage Systems; Microcomputing in Hydraulic and Structural Engineering; Earthworks – Construction and Placement, Bearing Capacity of Foundations; General Site Surveying, Setting Out; Design and Analysis of Prestressed Concrete Structures; The Health and Safety at Work Act in the Construction Industry; Membership Examination of the Chartered Institute of Building
- *Research and Consultancy:* The research activities in the Department include fracture mechanics, materials science, geotechnics, highway engineering, hydrology, photogrammetry and theory of structures.

A close link has been established with industry and the Department has recently been involved in projects in Coastal Engineering, Geotechnics, Structures, Surveying and Photogrammetry. It is envisaged that this will include more emphasis on information technology in the future

as the Department has extensive computing facilities ranging from microcomputers to a full CAD CAM system.
Total staff: 64
Professionally qualified staff: 35
Director or Chief Executive
Dean of Faculty of Environmental Studies: R.D. McMurray
Deputy Head of Department of Civil Engineering and Building: B.W. Preece
Principal sources of finance
The Polytechnic is financed through an annual budget provided by the Government and administered by the Public Sector.
Principal related bodies
The Polytechnic is governed by a Governing Body representing local authorities, public bodies and industry. The Polytechnic is administered by the Academic Board comprising a Director and Senior Academic Staff.
Publications
Numerous papers are published by members of staff in professional and technical journals.

GBR.549

Portsmouth Polytechnic, Department of Civil Engineering
Burnaby Building, Burnaby Road, Portsmouth, PO1 2EG
Telephone: Portsmouth 827681
Date of establishment
1958/59

Educational institution. The Department is one of 21 departments within the Polytechnic. The principal role is that of education in civil engineering but this is complemented by research and consultancy activity which can be conducted in any of its sixteen laboratories, computer/design studios or its four outstations. Currently the most active research areas are those in geotechnics, public health engineering and materials engineering but obviously in a lively department these areas can and do change.
Main fields of work
To provide the following courses: Higher Diploma in Civil Engineering; BSc(Eng) in Civil Engineering; Future Certificated BEng(Hons)
It should be noted that the research and consultancy referred to above can and often does lead to postgraduate students achieving degrees at MPhil or PhD level
Co-operative research with twelve industries in the development of new techniques and system (> £1m).
Total staff: 36
Professionally qualified staff: 33
Director or Chief Executive
P.A. Stead
Heads of department
Mathematics and Computing: M. Devane
Materials Engineering: P. Walden
Structural Engineering: G. Coleman
Geotechnics: R. Ritson
Construction: B. Curtis
Public Health Engineering: A.H. Pe
Transportation: G.A. Selvester
Research Manager: J.E. Butler
Design and Computers: R.J.C. Wren
Hydraulics: C.M. Archer
Principal sources of finance
Department of Education and Science via Hampshire County Council, for courses. Science and Engineering Research Council, County and Regional Water Authorities, private industry, Polytechnic contribution, Institute of Public Health Engineering, Department of the Environment, CIRIA, largely for research and consultancy.
Principal related bodies
As above but: SERC, research finance; County and Regional Water Authorities, small grants but large support in facilities; private industry (many firms), large research contracts (> £1m), mainly staff and equipment and materials provision; Polytechnic contribution, large voluntary staff time, currently minor fiscal support; Department of Environment adn CIRIA, small contracts.
Publications
Research reports: Over 1000 publications written by staff since 1960 30/40 books written by staff since 1960.

GBR.550 UM

Portsmouth Polytechnic, Department of Surveying
Cambridge Building, Cambridge Road, Portsmouth PO1 2LF
Telephone: Portsmouth 827681
Date of establishment
Department of Surveying established 1969.

Educational institution. Recognized by the Royal Institution of Chartered Surveyors as 'Centre of Excellence' for Surveying Education: General Practice Surveying and Quantity Surveying.
Undertakes research work on behalf of outside firms and inaugurates research related to the surveying profession.
Main fields of work
- Education: three year full time courses: BSc and BSc Honours Degree in Urban Land Administration; BSc and BSc Honours Degree in Quantity Surveying; both being the Council for National Academic Awards.
- Short courses, continuing professional development, in: General Practice Surveying; Quantity Surveying
- Research: computer application to: General Practice Surveying; Quantity Surveying; and allied subject areas.

Total staff: 37
Professionally qualified staff: 36
Director or Chief Executive
Head of Department: P.S. Brandon
Heads of department
General Practice: M.W. Warwick, Course Director
Quantity Surveying: R.G. Belcher, Course Director
Continuing Professional Development: R.J. Grover, Course Director
Research and Computer Applications: G. Wassell
Principal sources of finance
Polytechnic; Science and Engineering Research Council; private companies and firms; public institutions.
Principal related bodies
The Royal Institution of Chartered Surveyors; the Institute of Quantity Surveying; the Institute of Building.
Publications
Occasional papers published.

GBR.551

Queen's University of Belfast, Architecture and Planning Information Service (APIS)
Science Library, Chlorine Gardens, The Queen's University of Belfast, Belfast BT9 5EQ, Co Antrim, Northern Ireland
Telephone: 661111 ext.4304
Telex: 747691

The APIS is part of the Library of Queen's University of Belfast and serves the staff and students of the University as well as providing information to architects and planners in the community.
Main fields of work
Provision of an information service covering all aspects of architecture and planning. The following collections are available for use:
- Technical and trade literature
- Irish file (items dealing with all aspects of architecture and planning in Ireland)

- Slide collection
- Bibliography collection
- Publication of a weekly current awareness journal (the APIS Bulletin).

Total staff: 4
Professionally qualified staff: 1
Director or Chief Executive
Sub-Librarian: Mrs Karen Latimer
Principal sources of finance
University Grants Committee.
Publications

- *Journal:* APIS Bulletin, weekly
- *Bibliographies:* A large collection covering many topics.

GBR.552

Redland Technology Limited
Graylands, Langhurstwood Road, Horsham, West Sussex RH12 4QG
Telephone: Horsham (0403) 50222
Telex: 87442
Date of establishment
Established 1964 as Redland Development Unit; present name adopted 1974

Industrial research and development establishment within the Redland Group of Companies. Financed largely by Redland Companies to carry out projects involving new technology and product development; in addition, a small amount of contract research is undertaken.
Main fields of work
Performance of building materials and road surfacing materials in use; development of new materials and products (materials include ceramics, concrete, GRP).

Facilities for wind tunnel testing, environmental testing and a number of general duty material processing machines. These facilities can be made available to universities, test houses and other companies for undertaking processing trials.

The Centre houses the Redland Library and Information service.
Total staff: approximately 75
Professionally qualified staff: approximately 50
Director or Chief Executive
Chairman: J.E. Bailey
Managing Director: S.E. Horsley
Heads of department
Project Engineering and Technical Economics: A.W. Rawe
Product Design and Testing: R.M. Baker
Process Engineering and Energy Studies: K.W. Garrett
Cement Technology: H.A. Barker
Ceramic and Metallic Materials: G.W. Aston
Polymer Composites: J. Maxfield
New Product Evaluation: H.G. Bowden
Principal sources of finance
Financed by Redland Companies.

GBR.553 UM

Robert Gordon's Institute of Technology (RGIT)
St Andrew Street, Aberdeen AB1 1HG
Telephone: 0224 22338
Other address:
Scott Sutherland School of Architecture, Garthdee, Aberdeen
Date of establishment
1910

A higher education establishment under the administration of the Scottish Education Department.
Main fields of work
Higher education to CNAA degree and postgraduate level. Research in relation to variety of work covered by the three faculties

RGIT is the largest of 13 Scottish Central Institutions engaged in higher education. RGIT has 17 separate schools covering the arts, sciences and technology. It has separate schools of architecture and of surveying. Degree courses are offered in Building Economics, Quantity Surveying, Architecture and Urban Design. Facilities are available for research.

There are three faculties under the charge of the Deans of Faculty: Faculty of Arts; Faculty of Science; Faculty of Technology.
Total staff: approximately 277 academic staff
Director or Chief Executive
Dr Peter Clarke, CBE
Heads of department
Scott Sutherland School of Architecture: D. Kinghorn, Acting Head
School of Surveying: S. Baxter, Head
Principal sources of finance
Mainly financed through the Scottish Education Department.
Principal related bodies
Institute Committee; Academic Council; Various committees including: Research and Higher Degrees Committee, Architectural and Surveying Course Committees.
Publications
Various academic staff publications: Annual Reviews.

GBR.554 AM

Royal Institute of British Architects (RIBA)
66 Portland Place, London W1N 4AD
Telephone: 01 580 5533
Cables: Ribazo London W1N 4AD
Date of establishment
Established in 1834, incorporated by Royal Charters of William IV, Victoria, Edward VII, George V and Elizabeth II.

A professional organization. The purposes of the RIBA were expressed in the first Royal Charter granted in 1837 as the general advancement of Civil Architecture and for promoting and facilitating the acquirement of Architecture and for promoting and facilitating the acquirement of the knowledge of the various Arts and Sciences connected therewith.

The Institute now has more than 27 600 members of which some 5700 are overseas members. There are, in addition, over 3700 student members including approximately 330 student members overseas.
Main fields of work
As a learned society the RIBA seeks through its journal, sessional programme, educational policy, library and drawings collection to extend the body of knowledge upon which the practice of architecture is based.

The Research Section conducts ad hoc research survey work resulting in papers and sometimes prompting conferences and seminars.

The Statistics Section conducts a quarterly survey of workload and other statistics and an Earnings Survey of architects' earnings statistics.
Director or Chief Executive
Secretary: Patrick K. Harrison CBE
Principal related bodies
Council of 60 members, Executive Committee for Education and Practice; Executive Committee for Membership and Public Affairs; Finance Committee; Policy Committee
Publications

- RIBA Journal, monthly
- Library Bulletin, quarterly
- Journal of Architectural Research, published jointly by the RIBA and AIA (American Institute of Architects) in three issues annually
- Architectural Periodicals Index
- Individual books and management aids.

GBR.555 AM

Royal Institution of Chartered Surveyors (RICS)

12 Great George Street, Parliament Square, London SW1P 3AD
Telephone: 01-222 7000
Telex: 915443 RICS G
Date of establishment
1868. Granted Royal Charter in 1881.

A professional body governed by an elected Council with supporting Committees including research.
Main fields of work
The RICS is the leading professional body concerned with land and property and the training of men and women in land economy and the planning, use and development of land.
Total staff: 180
Director or Chief Executive
Secretary-General: Michael Pattison
Heads of department
Quantity Surveyors Division: Alex Trimmer
Principal sources of finance
Financed by membership subscriptions.
Publications
The Institution publishes a weekly journal giving an account of recent developments and items of interest to all branches of the surveying profession. In addition, a monthly journal is published for members of the Quantity Surveyors Division.

Some 250 publications are issued by Surveyors Publications at Basingstoke.

GBR.556

Scottish Farm Building Investigation Unit

Craibstone, Bucksburn, Aberdeen AB2 9TR
Telephone: Aberdeen (0224) 713622
Date of establishment
1967

Research, development and information unit for agricultural buildings for the Scottish Agricultural Colleges.
Main fields of work
- Livestock housing: environmental requirements; environmental control; space requirements; flooring
- Crop storage
- Human and animal welfare
- Publications.

Total staff: 19
Professionally qualified staff: 9
Director or Chief Executive
Head: Seaton H. Baxter
Principal sources of finance
Government (Scottish Office: Department of Agriculture and Fisheries for Scotland).
Principal related bodies
North of Scotland College of Agriculture: agricultural education, research and advice; East of Scotland College of Agriculture: agricultural education, research and advice; West of Scotland Agricultural College: agricultural education, research and advice.
Publications
- *Journal:* Farm Building Progress, quarterly
- Farm Building Cost Guide, annual
- Selected Library Additions, monthly.

GBR.557 AM

Timber Research and Development Association (TRADA)

Stocking Lane, Hughenden Valley, High Wycombe, Buckinghamshire HP14 4ND
Telephone: Naphill (0240 24) 3091
Telex: 83292
Date of establishment
Timber Development Association incorporated in 1934 as a limited company. Became TRADA in 1962.

TRADA is an independent research association with its membership drawn from professions which specify timber, those who use timber in industry, in construction and in packaging and those who supply it.

Advisory, testing and consultancy services are available to all, with reduced fees for TRADA members.
Main fields of work
TRADA carries out research and development work with the object of ensuring that timber is used correctly and economically. Since the construction and packaging industries are the largest users of timber, TRADA's work is mainly concentrated in these areas. TRADA employs all the means of communication available to disseminate information.

TRADA offers testing and consultancy services in many fields including timber frame construction, timber engineering, timber drying, timber yard operations. Overseas consultancy projects undertaken have included feasibility studies of forest and sawmill operations and design consultancy. Testing services include fire resistance tests on building components plus tests for surface spread of flame, ignitability and smoke control to ISO standards. Structural tests on building components can be undertaken and TRADA has developed an in situ test service for timber load bearing members in buildings.

Quality assurance schemes are operated for the manufacture of trussed rafters, fire resistant doors and doorsets, finger jointed timber and visual stress grading of timber.

Advisory services are available from headquarters or from ten regional officers in the UK and Republic of Ireland.
Total staff: 120 man years
Professionally qualified staff: 52
Director or Chief Executive
Director: J.G. Sunley
Heads of department
Construction and Engineering Division: K. Thomas
Industrial and Commercial Division: Dr O.P. Hansom
Principal sources of finance
Funds are obtained from members' subscriptions; from a voluntary levy on firms in the imported timber trade through the Timber Trade Federation; from fees received for commercial and consultancy work and from support received from the Government through the Department of the Environment.
Principal related bodies
TRADA Governing Council plus standing sub-committees, Construction and Engineering, Industrial and Commercial, Information and Education and Training.
Publications
Over 300 publications in print, including books, technical leaflets, design aids (span tables, etc.), Wood Information Sheets and library bibliographies. List available on request.

GBR.558

Transport and Road Research Laboratory (TRRL)
Crowthorne, Berkshire RG11 6AU
Telephone: Crowthorne 3131
Telex: 848272
TRRL Scottish Branch:
Craigshill West, Livingston, West Lothian, Scotland
Telephone: Livingston 31921
Date of establishment
1933

A Government research establishment within the Department of the Environment.

Main fields of work
The Laboratory provides technical and scientific advice and information to help in formulating, developing and implementing government policies relating to roads and transport, including their interaction with urban and regional planning.

These aims are achieved mainly by carrying out research and related activities in highway engineering, traffic engineering and safety and in more general transport subjects.

The work deals with the planning, design, construction and maintenance of roads and highway structures, particularly bridges and tunnels. Traffic and safety is concerned with the safe, efficient and convenient movement of people and goods, including the layout of roads and transport networks, the control of traffic flow, the study of road accidents and methods of reducing their frequency and severity.

The Laboratory works closely with local authorities, consultants, contractors and industry. There is also an Overseas Unit engaged on work sponsored by the Ministry of Overseas Development.

Total staff: approximately 650
Director or Chief Executive
Director: R. Bridle
Principal sources of finance
Financed very largely through the annual Parliamentary Vote of funds for the Department of the Environment; about 6% of the total funds represent work paid for by the Ministry of Overseas Development.
Principal related bodies
TRRL Advisory Committee; TRRL Advisory Committee on Highway Engineering; TRRL Advisory Committee on Traffic and Safety; TRRL Advisory Committee on Transport.
Publications
The Laboratory publishes a series of research reports and papers are frequently published in professional and scientific journals.

Priced publications are issued through Her Majesty's Stationery Office, London.

GBR.559

University College London, Bartlett School of Architecture and Planning, Building Economics Research Unit
Wates House, 22 Gordon Street, London WC1H 0QB
Telephone: 01 387 7050 ext 646 and 601
Date of establishment
1967

The Unit is part of the Building Department of the Bartlett School, and undertakes externally funded research work, both in the form of contract university research and government or construction industry bodies, and in the form of grant-funded studies supported by research-funding councils or institutions.

Main fields of work
The Unit undertakes a wide range of work within the broad areas of the economics of the construction industry and of the construction process. The Unit has special interests in questions concerning labour and the labour process in construction, questions of the capacity and mechanisms of response to demand of the industry, and questions of technological and organizational change in the industry.

Current projects comprise: a study of the determinants of the level of training provision in the construction industry; and an investigation of the costs of construction site accidents.

Immediate future plans include: work on the mechanization of the on-site construction process, and the market for and conditions of production of contractors' plant and equipment; an international study of recent developments in the organization, division and control of the construction labour process; and a study of the economic role of project appraisal and project management.

Total staff: 5
Professionally qualified staff: 1
Director or Chief Executive
Lecturer in the Economics of Construction: G.J. Ive
Principal sources of finance
Science Research Council; Building Research Establishment, Department of the Environment; Health and Safety Executive.
Publications
Research reports: Design Philosophy and Speculative Housebuilding, 1981; Construction in Economic Models, 1977; The Mechanism of Response to Effective Demand, 1975, 4 vols; The Construction Industry: its Economic Significance and its Role in Development, 1969, Capacity and Response to Demand of the Housebuilding Industry, 1981

A list of papers is available from the Unit.

GBR.560

University College London, Bartlett School of Architecture and Planning, Construction Industry Development Unit
Wates House, 22 Gordon Street, London WC1H 0QB
Telephone: 01 387 7050
Telex: 28722 UCPHYS G
Date of establishment
1 October 1982

The Unit provides an academic industrial focus within the Bartlett School for the study of improving construction industry performance in countries at varying stages of social and economic development. The Unit undertakes research, teaching and consultancy; and especially promotes education and training on an inter-disciplinary basis for the administration and management of construction in developing countries.

Main fields of work
About half the capital investment of developing countries is in construction: economic and social infrastructure and housing. Generally not enough attention is paid in these countries to developing the construction industry itself so that it can contribute fully to the achievement of national development plans. The process of satisfying client need for construction has been studied in the School for many years. The Unit has been established with industrial support to enable the work to grow in scope and influence, especially to make this knowledge and experience of the administration and management of construction available to developing countries.

The intended scope of the Unit may be inferred from some examples of recent and current activities:

- MSc in building economics and management
- Academic industrial attachments and exchanges
- Training courses, seminars and workshops
- Curriculum development
- Staff development
- Evaluative studies
- Advice to government and other institutions

These and other activities are being carried out at national, regional and local levels in the UK and countries in Africa, the Middle and Far East, often under the aegis of major international agencies. Their common purpose is to aid the construction industry to aid general development.

The Unit calls upon appropriately qualified and experienced people from the School, other academic institutions and industry, many with overseas experience, to assist in all its activities.

Total staff: 2
Professionally qualified staff: 2
Director or Chief Executive
Director: John Andrews, also Senior Adviser, Construction Management Programmes, ILO, Geneva
Secretary/Librarian: Margaret Beringer
Principal sources of finance
Industrial benefaction, student and consultancy fees.
Principal related bodies
Building Economics Research Unit, Development Planning Unit and other Departments in the Bartlett School of Architecture and Planning; and the Academic Services Unit and other Departments, University College London. The Unit has an invited international advisory panel.
Publications

- *Research reports:* The construction industries of developing countries: the applicability of existing theories and strategies for their improvement and lessons for the future: the case of Ghana. PhD Thesis; MSc Reports 1971–82: list of almost 100 titles available
- *Bibliographies:* A first critical bibliography is available

The first 3 Papers of the CIDU Series are: Education and Training for a construction profession, October 1982; Construction Industry Development: a comprehensive programme for strengthening the construction industry in developing countries, November 1982; Education and Training for the management of construction in developing countries: first teach the teachers, November 1982.

GBR.561

University College London, Development Planning Unit (DPU), Bartlett School of Architecture and Planning.

9 Endsleigh Gardens, London WC1H 0ED
Telephone: 01 388 7581
Telex: 896559 Gecoms
Date of establishment
1972

The DPU is an academic centre within the Bartlett School of Architecture and Planning of University College London. It specializes in academic teaching, practical training and research in the field of urban and regional development in developing countries. Its courses are supported by all the relevant international agencies as well as by national governments.
Main fields of work
The DPU is directly concerned with the understanding of the process of rapid urbanization, and the policy and planning response to economic development. Its basic work is to contribute to the increase of skills in the developing countries, to improve capacities so that governmental responses to the problems and potential of urbanization can be better informed and more effective. It seeks to achieve these purposes by offering long and short specialized courses in London and in the developing countries themselves. The DPU's internationally experienced staff of planners, economists and social scientists, as well as its associates within University College and elsewhere, run these programmes and undertake advisory and research projects in the field of urban and regional development. There are three major areas of work:

The London Training Programme which operates: a one-year Master's Degree Course which studies the practice, methods, and theory of urban development planning; a nine-month Postgraduate Diploma Course which provides practical professional training in urban development planning; a series of 12 week intensive courses for senior and mid-career professionals in selected aspects of urban development planning; a programme of higher degree studies.

London Centre Functions: primarily concerned with the assembly and dissemination of information on planning urban areas in the third world through seminars, conferences and publications. Visiting scholars and fellows participate in this work.

Overseas Training Advisory Services: provide advice to governments, planning and implementing institutions, international agencies and foundations on the training of professionals working in the field of urban development planning.
Total staff: 16
Professionally qualified staff: 5
Director or Chief Executive
DPU Director: Professor Colin Rosser (seconded to overseas work for two years)
Acting DPU Director: Dr Nigel Harris
Heads of department
London Centre Functions: Dr Caroline Moser, Director
Overseas Services: Desmond McNeill, Director
MSc Degree Course: Ronaldo Ramirez, Director
Postgraduate Diploma Course: Babar Mumtaz and Patrick Wakely
Short Courses: Dr Nigel Harris, Acting DPU Director
Principal sources of finance
Student fees and consultancy fees.
Principal related bodies
Bartlett School of Architecture and Planning and University College London.
Publications
H.T. Dimitriou: Urban Traffic and Transport Planning in Developing Countries, 1980
Working Papers are produced by the DPU. Recent titles include: No. 13. D. McNeill and S. Meikle (eds): British Planners in the Suez Canal Zone: the Lessons of Experience, 1983
No. 14. C.O.N. Moser (ed): Evaluating Community Participation in Urban Development Projects, 1983
N. Harris: Economic Growth and Spatial Change, 1983
B.K. Mumtaz: Readings in Action Planning, Vol. 1, 1982.

GBR.562

University College London, Department of Civil Engineering

Gower Street, London WC1E 6BT
Telephone: 01 387 7050
Date of establishment
University College London 1828

University College London is the largest and oldest school of the University of London.

The Department is engaged in teaching to first degree and postgraduate degree level and in research in the field of civil engineering.
Main fields of work
Research in the following fields:

- Concrete and Concrete Structures: Plastic methods of analysis and design for flexural and shear strength; Membrane action in slabs; Non-destructive testing; Properties of grouts
- Fluid Mechanics: Turbulence in waves and currents; Forces on cylinders and pipes in wave flow
- Geotechnical Engineering: Strength and deformation of granular soils; Use of X-ray techniques, Soil sampling techniques; Pre-stressed earth
- Marine Technology: Design of offshore structures; Wave and current loading; Foundation behaviour; Flow induced vibrations; Cementitious and resinous grouts; Reliability studies
- Public Health Engineering: Water filtration studies; Flocculation; Viruses in water; Particle size measurement
- Structural Mechanics: Buckling and post-buckling behaviour of thin shells; Non-linear dynamics and stability; Shell theory formulation
- Transport Studies: Transport planning; Analysis of traffic flow; Transport operations and control; Public transport policy; Safety and accident analysis.

Total staff: 57
Professionally qualified staff: 34
Director or Chief Executive
Head of Department: Professor K.O. Kemp

Heads of department
Structural Engineering: Professor K.O. Kemp
Fluid Mechanics: A.J. Grass
Geotechnical Engineering: J.R.F. Arthur
Public Health Engineering: Professor K.J. Ives
Transport Studies: Professor R.E. Allsop
Principal sources of finance
University Grants Committee; Science and Engineering Research Council; Government Departments and industry.
Principal related bodies
London Centre for Marine Technology co-ordinates research programmes in Marine Technology at University College London and Imperial College.
Publications
- Bi-annual Research Report
- Individual project reports.

GBR.563
Bath University, School of Architecture and Building Engineering
Claverton Down, Bath BA2 7AY
Telephone: 0225 61244
Telex: 449097
Date of establishment
1964

University department engaged in teaching and research.
Main fields of work
- Architecture and Urban Design
- Environmental Engineering
- Acoustics
- Lighting
- Soil Mechanics
- Structures
- Civil Engineering

Total staff: 22
Professionally qualified staff: 22
Director or Chief Executive
Professor Michael Brawne
Heads of department
Architecture: Professor Michael Brawne
Civil/Structural Engineering: Professor Edmund Happold
Environmental Engineering: Dr Derek Croome
Principal sources of finance
University Grants Committee; Science and Engineering Research Council.
Publications
Research reports: Members of staff publish in technical journals.

GBR.564
Birmingham University, Civil Engineering Department
PO Box 363, Birmingham, B15 2TT
Telephone: 021 472 1301
Telex: 338938 SPA PHYG
Date of establishment
University Status 1900

The Civil Engineering Department of the University performs testing in connection with building and construction and civil engineering generally upon request.
Main fields of work
Theory and design of structures; Strength of Materials; Concrete Engineering; Hydraulics and Fluid Mechanics; Water Resources Technology; Public Health Engineering; Soil Mechanics and Foundation Engineering; Surveying; Computing.
Total staff: 51
Professionally qualified staff: 25
Director or Chief Executive
Professor of Water Engineering and Head of Department: M.J. Hamlin
Heads of department
Professor of Civil Engineering (Concrete and Structures): B.P. Hughes
Principal sources of finance
University Grants Committee, research councils, authorities and industry.
Publications
Research reports: University of Birmingham Research and Publications, annually; University of Birmingham, Department of Civil Engineering, Research Report, bi-annually.

GBR.565
Bradford University, School of Civil and Structural Engineering
Bradford, West Yorkshire BD7 1DP
Telephone: Bradford 733466
Telex: 51309 UNIBFD G
Date of establishment
1966

The University of Bradford was established in 1966. The School of Civil and Structural Engineering is one of several schools within the Faculty of Engineering of this technological university. Formerly the University had the status of College of Advanced Technology.
Main fields of work
The School provides an undergraduate Civil and Structural Engineering course with a strongly vocational bias. There is particular emphasis on structural analysis and design, industrial experience, major project, complementary studies, geology and surveying field courses.
Internally and externally supported research is carried out into:
- Reinforced concrete structures, prestressed concrete, design of slender columns, shell concrete structures
- Hydraulics, flow through granular media
- Transportation
- Environmental noise and vibration
- Stability of brick tunnels

Postgraduate courses in Structural Engineering, Transportation and Civil Engineering.
Total staff: 34
Professionally qualified staff: 11
Director or Chief Executive
Chairman of the School: Professor C.B. Wilby
Heads of department
Structures: Professor C.B. Wilby
Transportation: R.J. Salter
Hydraulics: A. Waterhouse
Environmental Noise and Vibration: D.C. Hothersall
Principal sources of finance
Science and Engineering Research Council; industrial support.
Publications
- *Research reports:* Reports to sponsors
- Papers in professional or scientific journals
- Technical books intended for use by practising civil and structural engineers and as textbooks for students.

GBR.566

Edinburgh University, Department of Civil Engineering and Building Science

The King's Buildings, Mayfield Road, Edinburgh EH9 3JL

Telephone: 031 667 1081 ext 3387

Telex: 727442 UNIVED G

Date of establishment

1 January 1964 by division from Department of Engineering first established in 1869

A university department engaged in teaching and research.

Main fields of work

Main current research interests include: all aspects of structural masonry; non-destructive testing of concrete and masonry elements; timber structures; computer aided design in geotechnics; combined bending and torsion of reinforced concrete members; dynamic response of buildings to wind loads; properties of no-fines concrete.

Total staff: 38

Professionally qualified staff: 12

Director or Chief Executive

Professor A.W. Hendry

Heads of department

Industrial Liaison through Construction Engineering Research and Testing for Industry (CERTI): J. Morgan

Principal sources of finance

University Grants Committee; Science and Engineering Research Council; British Ceramic Research Association.

Publications

Research reports. No publication by Department – research reports published in books, professional and technical journals, conference proceedings, etc.

GBR.567

Glasgow University, Building Services Research Unit (BSRU)

3 Lilybank Gardens, Glasgow G12 8RZ

Telephone: 041 334 2269

Telex: 778421

Date of establishment

1960

A sub-department of the Department of Mechanical Engineering, the University of Glasgow. Established in 1959 as the Hospital Engineering Research Unit.

Main fields of work

Building Services Engineering.

Specialist experience has been built up in the following subjects: air conditioning and ventilation for hospitals, manufacturing pharmacies and industrial clean rooms; air flow through doorways; air curtains; control of airborne bacteria; laminar flow ventilation; natural ventilation; site measurements of demands for services (electricity, water and gases); thermal comfort of hospital patients; water consumption and flow rates in buildings; medical gas consumption and flow rates in hospitals; energy requirements and economy in buildings.

Total staff: 7

Professionally qualified staff: 4

Director or Chief Executive

Leader of Unit: William Carson

Principal sources of finance

Research contracts and grants supplemented by the University.

Publications

Papers are published in professional or scientific journals and as reports.

A list of publications is available on request.

A list of current and recent research projects is available on request.

GBR.568 **AM**

Liverpool University, Department of Building Engineering

PO Box 147, Liverpool L69 3BX

Telephone: 051 709 6022 ext. 2201

Telex: 627095

Date of establishment

1959

University department established on the initiative of the University Grants Committee. Teaching at undergraduate level in Building Services Engineering and Building Construction Engineering. Postgraduate research.

Main fields of work

Teaching and research. Current interests include: sound generation and transmissions through constructions and building services, room acoustics, thermal behaviour and energy consumption in buildings, control of building services, efficiency of defrost techniques, lighting design techniques, building materials, concrete and cement rheology, building operations and site productivity.

Total staff: 18

Professionally qualified staff: 6

Director or Chief Executive

D.W. Cheetham

Principal sources of finance

Government through University Grants Committee and Science and Engineering Research Council.

Principal related bodies

Other universities.

Publications

Papers are published in professional and scientific journals.

GBR.569 **AM**

Manchester University, Institute of Science and Technology (UMIST), Department of Building

PO Box 88, Manchester M60 1QD

Telephone: 061 236 3311

Telex: 666094

Date of establishment

UMIST 1824 – University Charter 1956

A university department in an Institute of Science and Technology which has awarded degrees in Building Technology since 1934. UMIST is the Faculty of Technology of the University of Manchester and has been so since 1903. Because of its independent origin, however, the Institute differs from the other university faculties in that it had its own governing body and is financed independently under its charter of 1956.

Main fields of work

The department has both postgraduate and undergraduate schools. The undergraduate degree courses are in Building Technology, Construction Management and Building Services Engineering. At the postgraduate level there is a course in Building Services Engineering and additionally at postgraduate level there is an active research school. Some of the research activities of the department are:

- The management and planning problems associated with the production and erection of buildings; the study and relationship between design and production; operational analysis and simulation; computer techniques for the modelling and analysis of the cyclic processes involved in the production of buildings
- Industrial relations; incentive payments and the wage structure of the building industry
- Financial administration of construction contracts; financial planning, econometrics and forecasting
- Performance in use of various housing projects
- Energy conservation; thermal performance of buildings; performance of heating and ventilating systems; computer aided lighting design
- Materials technology related to buildings; dynamic mechanical properties of fibre reinforced cementitious composites; non-destructive

testing; water movement in building materials; fire and materials.
Total staff: 19
Professionally qualified staff: 10
Director or Chief Executive
Professor R. Pilcher
Professor P.J. Burberry
Principal sources of finance
The Government through the University Grants Committee. Additionally research projects are also funded from other external sources such as the Science and Engineering Research Council and private and public bodies.
Publications
Research reports are published from time to time.
Bibliographies are published from time to time.
Members of academic staff regularly publish results of their research work in journals prominent within the field within which their studies are undertaken.

GBR.570

Newcastle upon Tyne University, Department of Civil Engineering
Claremont Road, Newcastle upon Tyne, NE1 7RU
Telephone: 0632 328511
Date of establishment
1963

A department of the University of Newcastle upon Tyne, with major teaching and research responsibilities.
Main fields of work
The Department of Civil Engineering offers courses leading to the degrees of BSc, BEng, MSc and PhD and to various postgraduate Certificates and Diplomas. The Department is heavily engaged in research work sponsored by industry, central and local government departments, and the Science and Engineering Research Council and the Natural Environment Research Council.
Total staff: 24
Professionally qualified staff: 24
Director or Chief Executive
Head of Department: Professor P. Novak
Heads of department
Hydraulic Engineering: Professor P. Novak
Public Health and Environmental Engineering: Professor M.B. Pescod
Structural Engineering: Professor F.K. Kong
Transport Engineering: Professor P.J. Hills
Principal sources of finance
Allocation from the University and funds arising out of research contracts. The University is mainly financed by central Government through the University Grants Committee.
Principal related bodies
The Department of Civil Engineering, together with the Department of Geotechnical Engineering and the Department of Mining Engineering, form the School of Civil and Mining Engineering.
Publications
- Numerous research papers have been published by the Department of Civil Engineering and these are listed in the Annual Reports of the Vice-Chancellor of the University of Newcastle upon Tyne
- Advanced Studies in Hydraulic Engineering
- Advanced Studies in Public Health and Environmental Engineering
- Advanced Studies in Structural Engineering
- Advanced Studies in Transport Engineering.

GBR.571 AM

Newcastle upon Tyne University, School of Architecture, Building Science Section
Newcastle upon Tyne NE1 7RU
Telephone: 0632 328511
Telex: 53654
Date of establishment
1962

A postgraduate section of the School of Architecture.
Main fields of work
Research work in the fields of heat, light and sound in relation to building environmental performance and human acceptability; also research into building energy utilization and conservation. A limited amount of design consultancy, to investigate building failures and materials testing.
Total staff: 6
Professionally qualified staff: 3
Director or Chief Executive
Professor A.C. Hardy
Heads of department
Building Environmental Performance: Professor A.C. Hardy
Lighting and Acoustics: Dr T.J. Wiltshire
Heating, Ventilating and Air Conditioning: B. Warren
Principal sources of finance
Science and Engineering Research Council; Electricity Council.
Principal related bodies
School of Architecture, Housing for Developing Countries; School of Architecture, Project Office; Department of Mechanical Engineering, Design Unit; Department of Civil Engineering Structural Laboratory.
Publications
Research reports on a variety of building performance subjects are available on request.

GBR.572

Nottingham University, Department of Architecture
University Park, Nottingham NG7 2RD
Telephone: 0602 56101

Main fields of work
P.R. Tregenza, Senior Lecturer: Daylight measurement and prediction; Computer-aided architectural design
J.K. Whittle, Lecturer: Energy and Thermal Performance Prediction by computer programmes; Computer programme validation
D.J. Jones, Senior Lecturer: Auditorium design; Noise control.
Director or Chief Executive
Professor Christopher Riley

GBR.573 AM

Reading University, Department of Construction Management
Whiteknights, Reading RG6 2BU
Telephone: 0734 875123
Telex: 847813
Date of establishment
1972

The Department is a part of the Faculty of Urban and Regional Studies of the University. Other Departments include Land Management and Development and a School of Planning Studies.
Main fields of work
Management of Construction: This includes the management aspects of all the construction professional's roles. The Department is particularly interested in project management of new construction including the client's role, management of the design process, the design/construction interface, and management of the construction process. The Department has a well-established interest in building management

and maintenance. A new but growing area of interest is in the management of organizations including professional practices and construction firms.

Total staff: 19

Professionally qualified staff: 16

Director or Chief Executive

Professor J. Bennett

Heads of department

Undergraduate Teaching:

BSc Quantity Surveying: Professor J. Bennett, Dr R. Flanagan

BSc Building Surveying: J. Sanderson, Dr N.G. Marsh

BSc Building Construction and Management: L. Holt, Professor W.D. Biggs

Postgraduate Teaching:

MSc Construction Management: Professor J. Paterson, P.E. Goodacre

MSc Construction Management (Maintenance): R. Lee

Principal sources of finance

Normal University funding from UK Department of Education and Science augmented by contract research.

Publications

Research reports: Periodic Departmental Reports; Cost Factors of Dimentional Co-ordination; US and UK Construction Industries, a comparison of design and contract procedures; Measurement of Complexity on Construction Projects; In preparation: Design/Cost theory for measuring buildings; Life Cycle cost planning; Systems for managing Hospital Estates.

GBR.574 UM

Salford University, Department of Civil Engineering

Salford, M5 4WT

Telephone: 061 736 5843

Telex: 668680 (Univ Salford)

Date of establishment

1967

University department with the following divisions: Civil Engineering; Building and Surveying; Environmental Science; offering undergraduate degree courses in: Civil Engineering; Building Surveying; Building Management; Quantity Surveying; Environmental Science; together with postgraduate research and industrial testing and consultancy.

Main fields of work

- Light gauge structural steelwork in all aspects of building construction, including light-weight framing and purlins, cladding and sandwich construction
- Building materials including the development of new products and the influence of deleterious materials
- Thermal insulation and condensation services
- Economics of building construction
- Computer aided analysis and design.

Total staff: 75

Professionally qualified staff: 55

Director or Chief Executive

Chairman of Department: E.M. Wilson

Heads of department

Civil Engineering: E.M. Wilson

Building and Surveying: J.M. Davies

Environmental Science: P. Garside

Principal sources of finance

University Grants Committee; industrial contracts.

GBR.575

Sheffield University, Department of Civil and Structural Engineering

Mappin Street, Sheffield S1 3JD

Telephone: 0742 78555

Telex: UL SHEF G 54348

Date of establishment

1892

A degree-awarding University Department in Civil and Structural Engineering. It awards three main types of degree, Bachelor of Engineering, Master of Engineering, and Doctor of Philosophy. The Bachelor of Engineering Degree is recognized for accreditation purposes by the main professional institutions in the UK as exempting its holders from the professional examinations. The Master and PhD degrees are by postgraduate study and research. The Department also carries out research in civil engineering and participates in professional training activities for the construction industry.

Main fields of work

Civil engineering education and research. Research falls into several main areas: structural steel, structural concrete, geotechnical engineering, hydraulic engineering, transport and traffic engineering. Some 80 postgraduate students are currently enrolled in the Department reading for Master and PhD degrees.

Two General Master's degrees are run, each of one year duration, one in Civil and Structural Engineering and the other in Transport and Traffic Engineering.

Total staff: 45

Professionally qualified staff: 16

Director or Chief Executive

Professor T.H. Hanna

Principal sources of finance

Government and University.

Principal related bodies

Various national and international organizations.

Publications

- Contributions to learned society journals.
- An Annual Report on the activities of the department is produced.

GBR.576 AM

Strathclyde University, Department of Architecture and Building Science

131 Rottenrow, Glasgow G4 0NG

Telephone: 041 552 4400

Telex: UNSLIB - G

Date of establishment

1876; reorganized 1964

The Department is a constituent part of the University. It is a public establishment with research facilities available, through contracts or consultancy arrangements, to public and private organizations and individuals.

Main fields of work

Architectural and building science research with particular emphasis on:

- Building performance studies
- Computer-aided building design studies
- Building energy studies
- Construction information studies
- Housing studies
- Historical studies
- Building materials and fabric studies.

Total staff: 20 teaching staff, 11 administration and support staff, 12 research staff

Professionally qualified staff: 31

Director or Chief Executive

Professor F.N. Morcos-Asaad

Professor T.A. Markus
Professor T.W. Maver
(Professoriat with Chairman in charge in rotation)

Heads of department

Building Performance Unit: Professor T.A. Markus and K. Alexander
Computer-aided Building Design (ABACUS): Professor T.W. Maver
Building Energy: Dr P.P. Yaneske
Construction Information Education Project (CIE): M. Munday
Housing (ASSIST): J.H. Johnson
Historical Studies: Dr F.A. Walker

Principal sources of finance

UK University Grants Committee; UK research councils; UK central and local government agencies; UK industry.

Publications

- *Journal:* Research Bulletin, bi-annual
- *Research reports:* Occasional Papers, irregular series.

GBR.577

University of Wales Institute of Science and Technology, Welsh School of Architecture, Research and Development Section WSA (R and D)

20–22 North Road, Cardiff CF1 3DY
Telephone: Cardiff 24733
Date of establishment
1970

A departmental research section within the University of Wales Institute of Science and Technology. Established 1970 by the appointment of Chair of Architcectural Science and Reader in Architecture, in the Welsh School of Architecture. As part of the University of Wales Institute of Science and Technology, the permanent staff and central facilities of the Research and Development Section are financed through University funds. Most of the research projects undertaken are financed by the Science Research Council and the Social Science Research Council.

Main fields of work

The Research and Development Section is a group of architects, mathematicians, physical scientists, social scientists and engineers studying buildings.

The aims of the Section are to:

- Provide field data on the physical and social nature of the environment in and around buildings
- Provide information on user attitudes to and use of buildings
- Determine whether current prediction design techniques work in practice

Main areas of current study are:

- The use of computer design models in energy conservation in buildings
- The transmission of pulsed sound through buildings
- Relationships between illuminance level, visibility, quality and energy use in the visual environment
- Patterns of space use in primary schools
- Patterns of space use in commercial office buildings
- Noise generated by traffic roundabouts.

Total staff: 35 of whom approximately half are support staff

Director or Chief Executive

Professor P.E. O'Sullivan; Professor A. Lipman

Heads of department

Professor J.E. Eynon is Head of the Welsh School of Architecture. Professor P.E. O'Sullivan has overall responsibility for the Research and Development Section and supervises work in the general area of the Physical Environment. Professor A.R. Lipman supervises work in the general area of the Social Environment. Dr P.T. Lewis carries out day-to-day management. Each research project has one or two supervisors and is staffed by Post-Doctoral Research Associates, Graduate Research Assistants, technicians and clerical staff.

Principal sources of finance

University of Wales; British Gas; Department of Education and Science; SERC.

Publications

Numerous papers are published in a wide range of journals, including pure and applied scientific, pure and applied social scientific and architectural publications. In addition, internal papers are produced on each of the current research projects which may be obtained directly from the R and D Section.

GBR.578

Water Research Centre (WRC)

Registered Office: Henley Road, Medmenham, PO Box 16 Marlow, Bucks SL7 2HD
Telephone: 049 166 531
Telex: 848632
Date of establishment
1974

The Centre is incorporated under the Companies Act as a Private Company, with liability limited by guarantee. It is financed mainly by subscriptions from its members. It also receives substantial financial and other support from the Department of the Environment, and now obtains a minor, but growing, income from commercial contract work.

The objective of WRC is to provide technical leadership primarily in support of the policies of the UK water undertakings.

Main fields of work

- WRC Engineering: Provides an intense effort on water engineering research and development. It is concerned with sewers, water mains survey techniques, leak detection, pipe renovation and replacement techniques and instrumentation, computing and automation
- WRC Environment: Research on environmental protection and water quality and health
- WRC Processes: Evaluation of water and wastewater treatment processes.

Total staff: 500

Director or Chief Executive

Chief Executive: J.L. van der Post

Heads of department

WRC Engineering (Swindon): M.J. Rouse
WRC Environment (Medmenham): Dr S.C. Warren
WRC Processes (Stevenage): R.C. Clayton
Scottish Office (Stirling): R. Fellows
International Office (London): Dr P.J. Newman
Director of Planning: V.K. Collinge
Head of Administration: C.H. Reavley
Marketing Manager: D.B. Field

Principal sources of finance

The United Kingdom Water Authorities and subscriptions from other members, support from the Department of the Environment and income from commercial contract work.

Principal related bodies

Council

Publications

- *Journal:* WRC Information, weekly
- *Research reports:* Technical Report Series, occasional; Annual Report and Accounts (1982/83); Research Programme (1983/84)
- *Bibliographies/Catalogues:* Various subjects
- Book series in conjunction with Ellis Horwood Ltd.

GBR.579

Welding Institute

Abington Hall, Abington, Cambridge CB1 6AL
Telephone: 0223 891162
Telex: 81183
Other address:
54 Princes Gate, Exhibition Road, London SW7 2PG
Date of establishment
Established 1968, by the amalgamation of the British Welding Research Association and the Institute of Welding.

An independent, specialist research organization with industrial members in 23 countries supporting a common research programme by annual subscriptions and contracts.
Main fields of work
Welding technology, materials and metallurgical problems in welding, process applications, economics, design of welded construction, consultant services to members.
Total staff: approximately 520
Director or Chief Executive
Director General: Dr A.A. Wells, OBE
Principal sources of finance
Self-financing.
Publications
- Metal Construction, a monthly periodical available on subscription and free to members
- Bulletin, a confidential publication available to research member companies only and not available on subscription
- Surfacing Journal.

GBR.580

Wimpey Laboratories Limited

Beaconsfield Road, Hayes, Middlesex UB4 0LS
Telephone: 01 573 7744
Telex: 935797
Other addresses:
Gelderd Road, Leeds LS12 6DE
Telephone: 0532 632653/4
Uphall Depot, Uphall, Broxburn, West Lothian, Scotland
Telephone: Livingston 38883
PO Box 588, Bahrain, Arabian Gulf
Telephone: Bahrain 8761/2
Telex: 8427GJ
PO Box 1811, Dubai, United Arab Emirates
Telephone: 31475
Telex: 5708 DB
Wimpey Laboratories Technical Services, 8720 30th Street, PO Box 5768, Station L, Edmonton, Alberta, T6C 46Z, Canada
c/o Salwa Contracting Engineers, PO Box B2133, Doha, Qatar
Unit 1516, International Plaza, 10 Anson Road, Singapore 0207
Wimpey Offshore Engineering Services and Structures Division, George Wimpey Inc., 3700 Buffalo Speedway, Houston, Texas 77078, USA
Wimpol Inc., 1434 West Belt Drive North, Houston, Texas 77043, USA
PO Box 358, Dhahran Airport, Saudi Arabia
c/o Hetherington Kingsbury Pty Ltd, 11th Floor, Underwood House, 37 Pitt Street, Sydney NSW 2000, Australia
Date of establishment
Established 40 years ago to service the main Company, George Wimpey and Co Limited.

A contracting civil engineering laboratory, limited company.
Main fields of work
Wimpey Laboratories Ltd is essentially a commercial laboratory. Originally founded 40 years ago, it has developed into a testing, research and contracting laboratory which will take work from any fee-paying client. It undertakes geoconsultancy, irrigation, air and water quality, waste handling management, land and marine site investigation, geophysical surveys, geochemical surveys, hydrographic surveys, acoustic navigation and location control, cement and chemical injection, rock and earth anchors, pile anchor dowelling, rock earth stabilization, building and structural repairs, mine and cavity infilling, hydraulic model studies, environmental studies including noise and vibration surveys, pollution control studies, material resources surveys, waste management; civil engineering and building materials (research and testing), concrete, asphalt, grouts, metals, chemicals, etc. Research and testing for structures.
Total staff: 830
Professionally qualified staff: 231
Director or Chief Executive
Managing Director: N.H. Searle.

GBR.581 FM

Property Services Agency, Department of the Environment (PSA)

Whitgift Centre, Wellesley Road, Croydon, CR9 3LY
Telephone: 01 686 8710
Telex: 917237
Date of establishment
1 September 1972

The Property Services Agency (PSA) is part of the Department of the Environment and its policies are determined by the Secretary of State for the Environment.

The role of PSA is to provide, manage, maintain and furnish the property required by the Government, including defence establishments, offices, law courts, Royal palaces, research laboratories and training centres. PSA is responsible for the management of the Government civil and defence estate. It also performs some of these functions for a number of Government-related bodies. PSA traces its origins back to the members of the Royal Household who looked after the King's Works in the Middle Ages. In 1851 the Office of Works became a department of central government; in 1940 it became the Ministry of Works (MOW) and in 1962 it was renamed the Ministry of Public Building and Works (MPBW), which from 1963 included the responsibility for defence works, previously carried out by separate armed services works branches. The Ministry became part of the Department of the Environment when this was set up in 1970 and the PSA was created within the Department in 1972.
Main fields of work
PSA work falls into the following main categories: Design of new buildings and engineering projects, using either PSA professional staff or private consultants, and the supervision of construction. All construction work is done by private contractors. Maintenance of buildings and installations, mainly using private contractors, and the operating of heating and other plant. Estate management – the purchase, disposal, leasing, management and allocation of land and property to meet the changing needs of Government departments. The Crown Suppliers, a distinct trading unit within PSA, manage the provision of fuel, furniture, domestic equipment, vehicles, engineering plant and interior design services.

PSA does not carry out research itself. However, because of its size and importance, PSA's specialist services and methods are of interest to the construction industry generally. Specialist branches under the Director-General of Design Services (DGDS) provide advice to PSA works offices in London and in the regions on a wide range of architectural, building, civil engineering, mechanical and electrical engineering, specification and costing problems. There are continuing developments in the areas of planned engineering maintenance, energy management and conservation, feedback systems for building defects, product quality assurance, and the application of computers to building design, maintenance and management. The methods of provision of technical information within the complex PSA organization are being studied.

PSA maintains regular contact with professional institutions and gives wide support to the British Standards Institution, being represented on some 330 BSI committees. PSA staff produce professional and technical guidance for internal use and for publication.

PSA is directly responsible for some 2000 historic buildings in use by Government departments (a greater number and variety than any other body in the UK) and is expected to set an example of the highest order in their care.

Total staff: 27 000

Professionally qualified staff: 8600

Director or Chief Executive

Chief Executive: A.G. Manzie CB

Deputy Chief Executive I: G.H. Chipperfield CB

Deputy Chief Executive 2: H.P. Johnston CB

Director-General of Design Services: J.B. Jefferson CBE

Heads of department

The executive management of PSA is the responsibility of the Chief Executive and the Executive Board. The Deputy Chief Executives divide between them the operational tasks for serving PSA's clients. The Director-General of Design Services heads the overall professional advisory services and is charged with enhancing standards of design generally throughout PSA. PSA operates through a series of Headquarters Directorates, the United Kingdom Territorial Organization (eight English regions, Scotland and Wales) and five overseas regions. In 1983–1984 PSA's expenditure on construction, maintenance, rents and purchases was £1813m.

Principal sources of finance

Government funds voted by Parliament.

Principal related bodies

PSA Advisory Board, established in December 1980, tenders advice to the Secretary of State and the Chief Executive on the work and procedures of PSA. The Board consists of thirteen members drawn from the private sector.

Publications

Construction (quarterly) published by B & M Publications (London) Ltd, PO Box 13, Croydon CR9 4NL

PSA publications available for purchase or free of charge are listed in PSA in print, revised every two years, obtainable free from PSA Library Sales office.

USA.582 **FM**

Advisory Board on the Built Environment (ABBE), National Research Council, National Academy of Sciences
2101 Constitution Avenue, NW, Washington, DC 20418
Telephone: (202) 334-2842
Telex: 248664 NASW UR
Cables: NARECO WSH
Date of establishment
The National Research Council (NRC) was established in 1916 by President Wilson as an agency of the National Academy of Sciences.

The purpose of the National Research Council is to broaden the base of experts providing advice to the government by inviting scientists and engineers outside the Academy's limited membership to serve on advisory boards and committees. The Council has become the principal operating and business arm of the Academies of Science and Engineering. As an operational unit of the National Research Council, ABBE seeks to serve the national interests by fostering the sound development of science and engineering as applied to the built environment, and by promoting their effective application for the benefit of society.

The objectives of ABBE are:

- To provide to all levels of government requested advice which is within its competence
- To facilitate the dissemination of information, the diffusion of innovation and the transfer of technology for all concerned with the design, production and management of the built environment
- To provide a continuing inquiry which anticipates, identifies and projects emerging impacts of science and technology on the professional community, the built environment and the nation and
- To encourage technical exchange and professional co-operation with other countries.

Main fields of work
ABBE offers advice on issues of science and technology as they affect the built environment. ABBE collaborates with professional societies and trade associations, but does not take positions on legislation or represent industry's viewpoint before the government.

The range of work includes the following:

- Those that advance the understanding of the relationships between the natural and the built environment
- Those that advance the understanding of human requirements as they relate to the built environment
- Those that advance the technological capabilities of those responsible for the design, production and management of the built environment and
- Those intended to develop policies for improving community and urban development.

ABBE undertakes a continuous and intensive effort to stimulate major public and private issues concerning the impacts of science and technology on the design, production and use of the built environment as it sustains and supports the quality of life.

The Federal Construction Council is supported by the fourteen federal agencies that design and build facilities or that manage research programmes affecting the built environment. This council defines and develops an agenda of technical questions based on the shared problems of designing, producing and managing federal facilities including the technological constraints associated with the use of new technologies.

The agenda is implemented in three different ways:

- Through Consulting Committees of the Federal Construction Council which are made up of technically qualified members of the sponsoring agencies
- Through Ad Hoc Advisory Committees which are made up of experts from outside the government appointed through National Research Council procedures who address scientific and technological issues that are broad in scope and require the advice of the larger community of expertise from which the academy can draw
- Through participation in specific ABBE generated issue studies.

State and Local Government Public Facilities Council consists of administrators from state and local building programmes. The Council receives advisory reports and technical studies conducted for the Federal Construction Council and identifies key concerns of their own which differ from those at the federal level.

International Technology Council represents the National Research Council's membership to CIB and includes establishing working relationships with Washington, DC based Embassy Office of Science and Technology to provide contacts between ABBE and the academies, and research organizations in other countries beyond those that are members.

Advanced Building Technology Council addresses the private sector research and development needs. This council consists of administrators, executives and research managers from industry, universities, non-profit and profit-making consulting firms. Its emphasis is to provide its members with an intelligence system on technological advances being made anywhere in the world.

Total staff: (ABBE) 10 (principal staff back-up provided by the National Research Council)
Professionally qualified staff: 4
Director or Chief Executive
Executive Director, ABBE: John P. Eberhard
Principal sources of finance
Through contracts and grants from the federal government, foundations, and private organizations.
Principal related bodies
A Board of recognized experts in building-related segments of industry, government, the design profession, academia, and research institutions.
Publications

- ABBE Research Briefs, bi-monthly
- Technical reports on separate Federal Construction Council Studies, 12 per year
- Federal Guide Specifications, 24 per year
- ITC Periodic and Annual Reports
- Proceedings of conferences/meetings
- Advisory Reports on Individual ABBE Studies
- Issue Study Reports.

USA.583

Aluminum Company of America, Alcoa Laboratories
Alcoa Centre, PA 15069
Telephone: 412/339-6551
Date of establishment
Established 1973 by merger of Alcoa Research Laboratories (founded 1919) and three development laboratories

Largest aluminium R and D organization in the world.
Main fields of work
R and D activities of Alcoa Laboratories encompass all aspects of aluminium technology: extraction, production, fabrication, properties, metallurgy, forming, joining, finishing, design, applications and end uses of aluminium and its alloys.
Total staff: 925
Professionally qualified staff: approximately 370
Director or Chief Executive
Vice Chairman: M.E. Gantz, Jr
Principal sources of finance
R and D programmes funded by the Corporation
Publications
Many papers are published in scientific and technical journals.

Alcoa's Catalogue of Literature

Most publications generated at Alcoa Laboratories are abstracted and indexed in the journal World Aluminium Abstracts, published by the American Society for Metals.

USA.584

American Concrete Institute (ACI)

22400 West Seven Mile Road, Detroit, Michigan 48219
Telephone: 313 532 2600
Telex: 810 2211454
Date of establishment
January 1905

A scientific and educational association.

Main fields of work
Through technical committees (96) and education committees (8) to further engineering and technical education, scientific investigation and research, and development of standards for the design and construction of concrete structures. The Institute promotes improved technology, technical competence and good design and construction practices. The Institute organizes the efforts of its members for a non-profit. Public service in gathering, correlating, and disseminating information for the improvement of the design, construction, manufacture, use and maintenance of concrete products and structures.
Total staff: 45
Professionally qualified staff: 16
Director or Chief Executive
Executive Director: George F. Leyh
Heads of department
Deputy Executive Director and Publisher Concrete International: Robert E. Wilde
Construction Development Department: Charles M. Ayers
Editorial Department: Robert G. Wiedyke
Education Department: Harold W. Gilley
Engineering Department: Samuel J. Henry
Administrative Services Department: William R. Tolley
Principal sources of finance
Membership dues and sale of publications.
Publications

- *Journal:* Concrete International: Design and Construction, monthly; Journal of the American Concrete Institute, bi-monthly; Concrete Abstracts, bi-monthly
- *Technical standards:* A wide variety of Standards on concrete design and construction
- *Research reports:* A wide variety of publications: Manual of Concrete Practice, books, handbooks, symposia volumes, monographs, educational bulletins, bibliographies, committee reports and miscellaneous bulletins. Catalogue available.

USA.585

American Iron and Steel Institute (AISA)

1000–16th St, NW Washington, DC 20036
Telephone: (202) 452 7100
Cables: STEEL
Date of establishment
1908

AISI is a non-profit association of the iron and steel industry. Its purpose is to engage in activities of common interest to its members. These pursuits include co-operative ventures that would not ordinarily be undertaken by individual members, and which are of general concern to all.

The stated purposes of the Institute are: to promote the interests of the iron and steel industry or any branch or part thereof; to collect statistics and other information about the industry; to engage in investigation and research; to provide a forum for the exchange of information and discussion of problems relating to the industry; and to promote the use of iron and steel.

Institute activities embrace research and technology, engineering, collection and dissemination of statistics, public distribution of information about the industry and its products, public affairs, and discussion of industrial relations including health, safety, and hygiene.

Main fields of work
The main strength of the Institute lies in its committee system, which enables AISI to draw on the special talents of its member companies' most skilled and dedicated people. Working through standing and special committees, sub-committees and task forces, they address themselves to an extraordinary range of meaningful projects. Institute staff professionals supplement and activities of the committees.

Those committees actively involved in building and construction research and technology are: Committee of Sheet Steel Producers; Committee of Steel Plate Producers; Committee of Steel Pipe Producers; Committee of Structural Steel Producers; Committee of Stainless Steel Producers; Committee of Wire Rope Producers.

The Committee on Construction Codes and Standards is responsible for industry activities relative to construction codes and regulatory standards, and for research and studies on fire protection of steel in buildings, seismic resistant design, etc. These activities benefit architects, construction engineers, and the public, and have helped introduce new standards of steel construction and the revision of many obsolete code requirements.

The Committee's studies of fire hazards and fire protection have contributed significantly to new regulatory concepts relative to the use of non-combustibles, matching protection requirements to the severity of hazards, and avoiding imbalance in fire risks. The result is greater fire safety, often at lower building costs.
Total staff: 92
Professionally qualified staff: 55
Director or Chief Executive
President: Robert B. Peabody
Heads of department
Engineering and Promotion: R.T. Willson, Senior Vice President
Manufacturing and Research: W.E. Dennis, Vice President
Construction Engineering Product Applicaton Research: J.C. Spence, Assistant Vice President
Principal sources of finance
Company member dues.
Publications
AISI Publications on Steel in Construction Catalogue is available.
A list of Current Research projects is available.

USA.586

American Planning Association (APA)

1313 East 60th Street, Chicago, Illinois 60637
Telephone: 202 872 0611
Other address:
1776 Massachusetts Avenue, NW, Washington, DC 20036
Telephone 312 955 9100
Date of establishment
APA is a new organization that has been around for many decades. It was formed when the members of th American Institute of Planners and the American Society of Planning Officials voted to unite the two organizations in 1978. Its predecessor organizations date back over 65 years.

APA is the national organization of people who are active in working for better-planned communities: APA brings together:

- The professional planner and the citizen member of a board or commission
- The transportation planner, economic development planner, housing planner, and other specialized people who see their responsibilities as part of comprehensive planning
- The elected official and the member of the community group
- Planners for big cities and small, for states, regions, and rural areas
- The planning practitioner, the student, and the teacher.

Through its chapters and divisions, APA gives members systematic ways to work on problems in common and to affect national planning policies. The American Institute of Certified Planners is a part of APA that concentrates on the development of the planning profession, its standards and continuing education.

With publications, research, and conferences, APA provides professional aid to the planner and current information to all.

Main fields of work

The APA provides a full range of services to its members, including monthly publications, research, conferences, lobbying, professional certification and job services. The major activities include:

- National lobbying, which represents the interests of the urban planning profession in national debates on federal legislation. APA's staff monitors changes in policy, regulations and emerging legislation identifying issues with important implications for the planning profession
- National conferences intended to bring members of the profession together and share experiences. APA sponsors both a national conference and regional workshops and institutes.

Total staff: 52

Professionally qualified staff: 20

Director or Chief Executive

Executive Director: Israel Stollman

Heads of department

Membership Services/Conferences: Linda Corby, Director
Publications Department: Sylvia Lewis, Director
Research Department: David Mosena, Director
Lobbying and Public Information: George Marcou, Deputy Executive Director
Book Publishing: Frank S. So, Deputy Executive Director
Finance: Charlotte McCaskill, Finance Director
Certification: Rosemary Jones, Deputy Executive Secretary for AICP

Principal sources of finance

Membership dues, research grants.

Principal related bodies

American Institute of Certified Planners: tests and certifies professional competence in the field of planning. Meeting the membership requirements of AICP stand in lieu of the licensing and registration used in other fields as the mark of a qualified professional.

Publications

- Land Use Law and Zoning Digest, 12 issues a year
- Planning magazine, covering the major stories in the planning profession; the magazine is distributed nationally to over 21 000 APA members 11 times a year; APA Journal, published quarterly, covering the research in planning practice and the planning literature
- Planning Advisory Service Reports, which are research reports published eight to ten times a year. These reports focus on a single topic related to planning practice or urban development issues
- *Bibliographies/Catalogues:* Planners Bookstore; Index.

USA.587

American Plywood Association (APA)

PO Box 11700, Tacoma, Washington 98411
Telephone: 206 565 6600
Telex: 32 7430

Date of establishment

1936

An industry trade association.

Main fields of work

APA engages in quality supervision and testing, product and application research, and promotion of structural panel products on behalf of producers, users, specifiers, distributors, designers, regulatory agencies and other key groups.

The Technical Services Division conducts research on basic physical and mechanical properties of plywood and other structural panel products manufactured in compliance with US Product Standard PS-1 and APA performance standards. Methods for improving serviceability, performance and appearance are studied. Performance standards for panel products are developed and evaluated as required by the industry. Proprietary products developed by Association members are tested. Standard test methods are developed. New and improved applications and systems are explored, particularly in construction and industrial applications. Design and fabrication methods are worked out for building components and systems. Maintenance of industry production and performance standards to accommodate changing needs is included in activities. Acceptance of APA building recommendations is pursued with regulatory agencies.

Total staff: 165

Professionally qualified staff: 22 are in Technical Services Division, of which 13 are professionals

Director or Chief Executive

Executive Vice President: Bronson J. Lewis
Vice President and General Manager: William T. Robison

Heads of department

Technical Services Division: Thomas R. Flint, Director
Quality Services Division: Bruce E. Lyons, Director
Field Services Division: Herman W. Glover, Director
Information Services Division: David L. Rogoway, Director
International Operations: M.T. Fast, Director

Principal sources of finance

Financed through dues paid by member mills, based on volume of panel production. Membership is voluntary.

Principal related bodies

Board of Directors, composed of member principals.

Publications

- *Technical standards* (major titles): Performance Standards and Policies for Structural-Use Panels; Plywood Design Specification; Design/Construction Guide – Residential and Commercial
- *Research reports:* Various titles
- *Bibliographies/Catalogues:* Publication Index; Research Literature (supplement to Publications Index)

Results of technical investigations are given in Research Reports or Technical Notes, and are listed in Research Literature. Publications of more general technical or promotional interest are listed in Publications Index. Single copies of either type are usually available on request without charge.

USA.588

American Society for Testing and Materials (ASTM)

1916 Race Street, Philadelphia, PA 19103
Telephone: (215) 299 5400
Telex: 7106701037

Date of establishment

Originated in 1889 as the American Section of the International Association for Testing Materials; became the American Society for Testing Materials in 1902; incorporated in 1904; under present name since 1961.

Main fields of work

ASTM has established the world's largest source of voluntary consensus standards for materials, products, systems and services.

Engineers, scientists and skilled technicians hold membership as individuals or as representatives of business firms, government agencies, educational institutions, laboratories.

The Society has 137 technical committees, each having 5 to 50 sub-committees. New committees are organized each year to keep pace with technological advances.

It sponsors more than 125 research projects, issues various awards for research and signifant contributions in this area and has developed more than 8500 standard test methods, specifications and recommended practices now in use.

ASTM has 30 000 members.

Total staff: 200

Director or Chief Executive

President: William T. Cavanaugh

Publications

- *Journal:* Standardization News, monthly; Journal of Testing and Evaluation, 6 times a year; Journal of Forensic Sciences, quarterly;

Cement, Concrete, Aggregates, twice a year; Geotechnical Testing Journal, quarterly; Composites Technology Review, quarterly
- Book of ASTM Standards, annual
- Numerous technical papers and reports.

USA.589

American Society of Civil Engineers (ASCE)
345 East 47th Street, New York, New York 10017
Telephone: 212 705 7496
Other address: Suite 607–608, 1625 Eye Street, NW, Washington, DC 20006
Telephone: 202 785 4454
Date of establishment
1852

A professional society, voluntary association (individual membership).
Main fields of work
All aspects of civil engineering and construction. For this report, councils and projects include structures, foundations, materials of construction, computer practices.
Total staff: 130
Director or Chief Executive
Executive Director: Dr Edward O. Pfrang
Assistant Executive Director: Don A. Buzzell
Principal sources of finance
Financed by budget and grants for projects.
Principal related bodies
Council of Committee personnel, 14 Research Councils, 11 Committees on Research.
Publications
- Civil Engineering, the Society's official monthly magazine
- Professional Issues – Journal of Professional Activities, quarterly
- Proceedings, journals of the technical divisions
- Convention and conference papers
- Transactions, annual
- ASCE Publications Abstracts, bi-monthly
- Information Retrieval
- Directory, membership list
- Official Register, printed at the end of each year
- Annual Report of the Board of Direction
- ASCE News, monthly newspaper
- Manuals and Reports on Engineering Practice.

USA.590

American Society of Heating, Refrigerating and Air-Conditioning Engineers, Inc. (ASHRAE)
1791 Tullie Circle, NE, Atlanta, GA 30329
Telephone: 404 636 8400
Telex: 705343 ASHRAE
Date of establishment
1894

To advance the arts and sciences of heating, refrigeration, air conditioning and ventilation, the allied arts and sciences, and related human factors for the benefit of the public.

Membership is open to practising professionals and students the world over.
Main fields of work
ASHRAE generates technical data through sponsorship of a research programme and disseminates information from its research programme and from the practice of members through publications, meetings and seminars. ASHRAE develops standards of assist industry and the public by offering uniform methods of testing, terminology and recommended practice. ASHRAE's fields of interest include all technologies related to heating, refrigeration, air conditioning and ventilation.
Total staff: 75
Director or Chief Executive
Executive Vice President; Andrew T. Boggs
Heads of department
Administration: Frank M. Coda
Communications: W. Stephen Comstock
Government Affairs: J.E. Cox
Publishing: Carl W. MacPhee
Technology: J. Richard Wright
Principal sources of finance
Membership dues, publications sales, research contributions.
Principal related bodies
Associated with 27 other societies with similar fields of interest and operates 140 chapters. Liaison is maintained with numerous allied organizations.
Publications
- *Journal:* ASHRAE Journal, monthly
- *Technical standards/Recommendations/Research reports:* Catalogues are published periodically
- ASHRAE Handbook, annually

Publications catalogue is available free upon request.

USA.591

Architectural and Building Research, Southwest Research Institute (SwRI)
6220 Culebra Road, San Antonio, Texas 78284
Telephone: 512 684 5111
Telex: 767357
Date of establishment
1946

A chartered non-profit corporation.
Main fields of work
- *Architectural and building research:* building systems, design concepts, construction techniques, resources and energy evaluation, modular co-ordination shelter and settlement (international) concepts, environology, urban and regional planning, building economics
- *Fire research:* fire protection and evaluation, testing hazard and fire products analysis, systems development, standards and development and evaluation
- *Environmental sciences:* environmental impact studies, oceanographic studies, resources planning and cost analysis
- *Engineering sciences:* materials development, substitutions, cost evaluation and testing, geological and meteorological studies
- *Structural research:* structural design, evaluation and testing, systems development, standards analysis
- *Electronic systems research:* non-destructive testing and evaluation, electronic systems, computer techniques and application.

Total staff: approximately 1300 (professional 450, technical 600, support 250)
Director or Chief Executive
Norman I. Turner
Principal sources of finance
Contract research and development on an individual project basis by industry, government, business or individual sponsors.
Publications
- Numerous papers by SwRI individuals are published in professional or scientific journals and in conference or seminar transactions. Some reports resulting from various research programmes are published by the sponsor
- Monographs and tests are published for general sale on occasions
- Annual Report is published by the Institute

A listing of technical publications is published every 3 to 4 years.

USA.592

Austin Company
3650 Mayfield Road, Cleveland, Ohio 44121
Telephone: (216) 382 6600
Telex: (810) 421 8540
Cables: AUSTIN
Date of establishment
1878

A closely held, private, for-profit international corporation, the minority of whose stock is owned by directors, officers and key employees.

The Austin Company was the originator of the concept of complete service in the construction industry.

The Austin Method, introduced at the turn of the century, for the first time integrated architecture, engineering and construction services in one organization. By means of a single contract, Austin assumed complete and undivided responsibility for all the professional and technical services required to develop a building program.

The original basic services have been augmented over the years by many specialized services. These services are organized in three classifications: engineering/economic; management consulting; technical planning.

Austin is active in all the principal branches of industry and commerce. The company also serves the US government, including the military services.

Austin is recognized as a world leader in the design and construction of facilities for: Aerospace/aviation; Automotive manufacturing; Broadcasting (television, radio); Electronics (semiconductors, microelectronics); finance (banking, insurance); Food processing; Health care (hospitals); Chemical processing; Pharmaceuticals (drugs, devices, diagnostics); Publishing (newspapers).

Main fields of work
The Austin Company's basic business is the design, engineering and construction of industrial, commercial, governmental and institutional facilities. Services for the foundry and cement industries are provided by the Mining and Metals Division in Cleveland, Ohio. The chemical industry is served by the Process Division in Chicago, Illinois. The Technical Services Division, located in Cleveland, develops new fields of service and originates new concepts, methods and techniques. In addition to architecture, engineering and construction services, the Company also offers its clients the following: land planning and development studies; facility location surveys; site evaluation analyses; feasibility analyses; plant layout studies, industrial engineering and systems engineering; and specification, procurement and installation of equipment.

The Austin Company operates through 8 US regional operating units, 2 special-industry divisions, 15 sales offices, 2 support divisions, 2 domestic subsidiaries and 8 international subsidiaries, a technical services group and a management consulting division.

Among the special services that Austin provides are: Building retrofit; Environmental engineering; Facilities location surveys; Interior architectural design; Manufacturing engineering; Master planning/land use studies; Materials handling engineering, Physical distribution engineering; Regulatory compliance engineering (pharmaceutical industry); Systems engineering.

Total staff: 2000.
Professionally qualified staff: 1500
Director or Chief Executive
President and Chief Executive Officer: Charles A. Shirk
Heads of department
Technical Services Division: Robert A. Will, Manager
Management Consulting Division: Milton J. Cooke, Vice President and General Manager
Process Division: Charles E. Vogel, Vice President and Manager
Mining and Metals Division: Richard Moran, Division Manager
Principal sources of finance
Internal and banks. Note: Effective January 1984, the Austin Company becomes a wholly owned subsidiary of National Gypsum Company.
Principal related bodies
Ragnar Benson Inc., Chicago, Illinois/Pittsburgh, Pennsylvania: construction subsidiary; Scandia Inc., Atlanta, Georgia/Miami, Florida: construction subsidiary.
Publications
- *Austin Update*, monthly news publication of the parent company
- *Effective Management*, monthly publication of the Management Consulting Division
- *The Exchange*, quarterly publication of the Technical Services Division.

USA.593

Battelle Memorial Institute
505 King Avenue, Columbus, Ohio 43201
Telephone: 614 424 6424
Telex: 25 5454
Other addresses:
Battelle: Pacific Northwest Laboratories, Battelle Boulevard, Richland, Washington 99352
Battelle: Seminar and Study Programs, 4000 NE 41st Street, PO Box 5395, Seattle, Washington 98105
Battelle: Geneva Research Centre, 7 route de Drize, 1227 Carouge-Geneva, Switzerland
Battelle-Institut eV: Am Römerhof 35,6000 Frankfurt/Main 90 Federal Republic of Germany
Date of establishment
1929

A scientific organization devoted to research, technology development and education. R and D projects are conducted for industry and government organizations on a contract basis; Battelle provides the sponsor with the physical facilities, equipment and personnel necessary to perform the study and, in return, the sponsor reimburses Battelle.

Main fields of work
Battelle conducts multi-disciplinary research and development studies in virtually all areas of the physical, life, social and behavioural sciences.

In the area of construction, Battelle conducts both technical and economic research programmes for the residential, non-residential and non-building sectors of the construction industry. Most of the technical programmes have been materials-orientated, such as developing materials and products for new and existing construction applications. Other areas of technical activities include pipeline, structural and seismic research, new construction methods or techniques, evaluating and developing industrialized building systems, evaluating land use and financing techniques, developing design criteria and performance specifications for new products and components, conducting business strategy research programmes, developing forecasts of construction activity.

Total staff: 7100
Director or Chief Executive
Chairman and CEO. Dr Sherwood L. Fawcett
Publications
- President's Report and Annual Review
- Battelle Today, quarterly, presents the broad interests and activities of Battelle worldwide in scientific research, education and technology development
- Published Papers and Articles, a yearly compilation of published writings by the worldwide staff of Battelle.

USA.594

California University College of Engineering, Department of Civil Engineering, Division of Structural Engineering and Structural Mechanics (CE SESM)

Room 721 Davis Hall, University of California, Berkeley, CA 94720

Telephone: 415 642 6463

Other addresses:

Earthquake Engineering Research Center, University of California, 1301 South 47th Street, Richmond, CA 94804

Fire Research Group, Room 543 Davis Hall, University of California, Berkeley, CA 94720

National Information Service-Earthquake Engineering – Computer Applications, Davis Hall, University of California, Berkeley, CA 94720

Date of establishment

About 1930

The original laboratory has undergone various organizational modifications within the College of Engineering.

Main fields of work

The Division is one of three divisions within the Department of Civil Engineering, College of Engineering. It administers the Structural Engineering Materials Laboratory, which includes a wide range of teaching and research facilities, among the leading structural research facilities in the world. Associated with the Division is the Earthquake Engineering Research Center with its main research facilities in the Earthquake Simulator Laboratory; as a part of the Center's activities they maintain the National Information Service – Earthquake Engineering – Computer Applications.

The principal research areas are in structural engineering, structural mechanics, engineering materials and earthquake engineering.

The University enters into contracts on research projects with outside agencies when it is convinced that participation will lead to the extension of knowledge or to increased effectiveness in teaching. University facilities are not generally used for investigations of a purely commercial character except when it is shown that satisfactory facilities for such services do not exist elsewhere.

Total staff: approximately 200, of whom about 150 are student research assistants

Director or Chief Executive

Chairman of the Division: G. Powell

Principal sources of finance

Approximately 95% of research funds come from contracts and grants made to the appropriate group.

Publications

Research investigations are reported either under the Structures and Materials Research series or under the Earthquake Engineering Research Center series reports. Copies of these reports are now forwarded to the National Information Service, National Bureau of Standards, US Department of Commerce, Springfield, Virginia 22151 and copies can be obtained through that office.

USA.595

Clemson University, Department of Civil Engineering

Clemson, South Carolina 29631

Telephone: (803) 656 3000

Date of establishment

1893

Clemson University is a co-educational, state supported, land grant university fully accredited by the Southern Association of Colleges and Schools. It offers 59 undergraduate and 80 graduate degree programmes.

Main fields of work

The primary mission of the Department of Civil Engineering, as well as of Clemson University, is to provide a well-balanced and fundamental educational programme. The Department also participates in an active research programme aimed at advancing the state of knowledge in civil engineering related fields.

Total staff: 21

Professionally qualified staff: 21

Director or Chief Executive

Head, Department of Civil Engineering: Russell H. Brown

Principal sources of finance

General appropriations from the State of South Carolina; student fees; research contracts.

Publications

Research reports: Internal reports are not available for general distributions. Some may be obtained by special request.

USA.596

Cornell University, Department of Structural Engineering

Hollister Hall, Cornell University, Ithaca, New York 14853

Telephone: 607 256 3760

Date of establishment

The University was founded in 1866 and included Civil Engineering at that time.

Teaching and research in structural engineering, civil engineering materials and geotechnical engineering.

Main fields of work

The Department is one of two departments in the School of Civil and Environmental Engineering. It is devoted to education and research.

Within Structural Engineering, there are four laboratories involved in research testing:

- The George Winter Laboratory is used for both large scale tests and smaller component tests. Much of the work that led to the US specification for the Design of Cold-Formed Steel Structural Members was carried out here
- The Structural Models Laboratory specializes in ultimate strength models of reinforced concrete structures. Special techniques for simulating complex behaviour up to failure have been developed under NSF sponsorship and have been applied to a variety of structural forms
- The Concrete Materials Laboratory is used as an undergraduate teaching laboratory, as a graduate teaching laboratory, as a general testing laboratory and as a research laboratory, mostly for concrete but also for construction in general
- The Geotechnical Laboratory.

Total staff: 20

Professionally qualified staff: 13

Director or Chief Executive

Professor Arthur H. Nilson

Heads of department

Structural Analysis and Design: Arthur H. Nilson

Geotechnical Engineering: Fred H. Kulhawy

Civil Engineering Materials: Floyd O. Slate

Principal sources of finance

Cornell University about 60%; outside research sponsors, mainly US government, 40%.

Publications

Most of the research projects are incorporated into thesis writing and many of the theses are also issued as Department of Structural Engineering Reports, which are available from Cornell.

Professional journal papers result from most of the laboratory research, with the bulk of the publications appearing in ASCE Journals and the Journal of the American Concrete Institute.

USA.597

Earthquake Hazard Mitigation Program, National Science Foundation (NSF)

1800 G St, NW, Washington, DC 20550

Telephone: 202 357 9545

Date of establishment

1950 by Act of US Congress

NSF is a government research agency.

Main fields of work

NSF Earthquake Hazard Mitigation Program is a part of the Division of Civil and Environmental Engineering.

It supports a broad, basic and applied research activity in all aspects of engineering, geotechnical geology, social and behavioural sciences which are aimed at developing methods and procedures for the mitigation of earthquake disasters.

It is directed toward the following objectives:

- To develop economically feasible design and construction methods for building earthquake resistant structures of all types
- To develop procedures for integrating information on seismic risk with ongoing land use processes
- To develop an improved understanding of social and economic consequences of individual and community decisions on earthquake related decisions
- To present programme results in forms usable by the affected interest communities to control the vulnerability to earthquakes.

Director or Chief Executive

William W. Hakala

Principal sources of finance

US Congress budget.

Publications

- Summary of Awards, yearly, presents brief description of projects funded during fiscal year
- Research reports published by research organizations.

USA.598

Factory Mutual Research Corporation (FMRC)

1151 Boston-Providence Turnpike, Norwood, Massachusetts 02062

Telephone: 617 762 4300

Telex: 92 4415

Other address:

Factory Mutual Test Center, West Glocester, Rhode Island 02814

Date of establishment

Incorporated in 1941 as a formal function; other Factory Mutual standards and product testing functions whose origins date to the formation of the Factory Mutual System in 1835 were consolidated in FMRC in 1969.

A private research and testing organization.

Main fields of work

FMRC conducts research, testing and standards development activities principally relating to fire and explosion and other perils, insured by Factory Mutual System insurance companies. In addition, it undertakes, under contract, similar research for external clients.

The Research Group conducts basic and applied research on fire and explosion, including the behaviour of fires and explosions, the optimization of protection and the development of cost-effective risk management procedures. This group also operates the world's largest Test Center designed and dedicated to the study of large-scale fire events. The Approvals Group conducts, on a fee basis, evaluations of the fire performance/fire resistance of materials, the design adequacy of fire protection systems and nature and other equipment designed to control hazards.

The Standards Group provides loss prevention guidelines for use by FM System engineers and the management of insured industry in the large, highly protected risk category, and in the operation and maintenance of major equipment such as turbines, generators, boilers.

Total staff: 180

Director or Chief Executive

President: J.A. Love

Chief Operating Officer: P.M. Fitzgerald

Principal sources of finance

Financed internally by the Factory Mutual System, externally by contracts with industry, industry associations and Government agencies.

Publications

- Factory Mutual Record, published quarterly, contains information of general interest in loss prevention
- A wide variety of data sheets (for FM insured members only) and instructional material is available from the Factory Mutual Resource Center/Order Processing Department
- Reports of product approval investigations must be obtained from the manufacturer

The Research Group currently publishes nearly one hundred reports annually and about half of these are available to the public.

USA.599

General Services Administration, Public Buildings Service (PBS), Office of Design and Construction

18th and F Streets, NW Washington, DC 20405

Telephone: 202 566 0574

Date of establishment

1959: Public Buildings Act of 1959

An organization within the Public Buildings Service of the General Services Administration responsible for planning, design, construction and rehabilitation of buildings to house the civilian activities of the US Government.

Main fields of work

The Office of Design and Construction (Central Office) has the responsibility for planning directing and co-ordinating the nationwide PBS design, construction and rehabilitation of buildings programme; including historic structures.

In accomplishing its mission the Office undertakes research studies and evaluates new technologies and approaches that have potential for improving quality, efficiency, and productivity in the Federal workplace. The office has developed a quality control system for the review and monitoring of design and construction from programme inception through post-occupancy evaluation.

A major focus of the Office is the development of procedures, criteria and standards that are followed in the day-to-day regional operations for the design, construction and rehabilitation of Federal buildings.

Total staff: approximately 87 in the Central Office and 539 in the 11 regional offices

Director or Chief Executive

Acting Assistant Commissioner for Design and Construction: Richard H. Field

Heads of department

Deputy Assistant Commissioner: Clifford A. Thomas (Acting)

Directors: Fine Arts: Karel Yasko

Historic Preservation: J. Walter Roth

Art-in-Architecture: Donald W. Thalacker

Divisions and Directors:

Design Management: Edward A. Feiner

Construction Management: Alexander Papademetriou (Acting)

Cost Management: Harry L. Kay (Acting)

Programme Planning and Analysis: Donald E. Parker (Acting)

Principal sources of finance

Annual appropriations are approved by Congress; the Office also accepts transfer projects from other Federal agencies on a reimbursable basis.

Publications
Research Reports funded by PBS, undertaken by the National Bureau of Standards: NBSIR–82–2605, Measurement Methods for Evaluations of Thermal Integrity of Building Envelopes; NBSIR–83–2662, Documentation and Assessment of the GSA/PBS Background and Research Plan.

USA.600 UM

Clayford T. Grimm, PE, Inc., Consulting Architectural Engineer

1904 Wooten Drive, Austin, Texas 78748
Telephone: 512/471 7259
Date of establishment
1971

Practice confined to the design, construction, maintenance and restoration of brick, stone, terra cotta, and concrete masonry.
Main fields of work
Masonry engineering.
Principal sources of finance
Clients.
Publications
More than 80.

USA.601

Honeywell, Inc. Technology Strategy Center (Honeywell, TSC)

1700 West Highway 36, Roseville, Minnesota 55113
Telephone: 612 378 4178
Telex: 290631
Date of establishment
1976

TSC is part of Honeywell, a multi-national, advanced technology company. The Center undertakes research, development and demonstrations of new technologies and product concepts to support the strategic thrusts of Honeywell business units.
Main fields of work
Ergonomics, Communications, Sensors and Controls and Manufacturing Automation are the main areas of endeavour.

Ergonomics or human factors technology is being applied to identify future control room improvements. The Center also applies new input/output techniques such as voice recognition to improve future human–machine interfaces.

The communication needs of future offices and factories are addressed through programmes in local area networks, power line carrier and circuit switched networks.

Engineers and scientists at the Centre apply advanced sensing and control technologies to develop new concepts for process monitoring and building and industrial systems.

TSC also uses image processing and optical recognition techniques in assembly and inspection problems.
Professionally qualified staff: 130
Director or Chief Executive
Director, Paul A. Anderson
Heads of department
Department of Technology: Melvin F. Mittelmark, Director
Principal sources of finance
Financed by government-sponsored, privately and Honeywell funded research and development programmes.

USA.602

Illinois Department of Transportation, Division of Highways and Bureau of Materials and Physical Research, Physical Research Group

126 East Ash Street, Springfield, Illinois 62706
Telephone: (217) 782 6732
Date of establishment
Established 1973 as a combination of former Bureau of Materials and Bureau of Research and Development when the Illinois Department of Transportation was formed.

A government institution which conducts and co-ordinates research on highways and bridges. The activities of the Physical Research Group are under the direction of the Engineer of Physical Research who reports to the Engineer of Materials and Physical Research. He in turn reports to the Deputy Director of Highways for Operations. The Research and Development Councils serve as advisors to the Director of Highways on the research and development activities of the Group.
Main fields of work
The Group is responsible for planning, conducting and co-ordinating a programme of operational research and development to improve the economy and quality of design, construction, operation and maintenance of highway and transportation systems.

The activities of the Group are carried out under seven organizational units: traffic research; bridge research; pavement research; instrumentation development; products evaluation; research evaluation; research co-ordination.
Total staff: 32
Professionally qualified staff: 16
Director or Chief Executive
Engineer of Materials and Physical Research: J.G. Gehler
Engineer of Physical Research: Donald R. Schwartz
Principal sources of finance
All activities are financed through funds appropriated by the State General Assembly. About half of the research activities are partially reimbursed by the Federal Government.
Publications
Research reports: Interim and final reports on individual studies conducted by the Group are published in a PR report series. Reports on studies conducted for the Group by other agencies are published by those agencies

A Products Evaluation Circular is published quarterly on new products tested by the Group.

USA.603

Illinois University at Urbana-Champaign, Department of Civil Engineering

1114 Newmark Civil Engineering Laboratory, 208 North Romine Street, Urbana, IL 61801
Telephone: 217 333 3814
Date of establishment
1867

A public state university.
Main fields of work
Undergraduate and graduate education in all fields of civil engineering: research in all fields of civil engineering. Continuing major research efforts include earthquake engineering, tunnelling, reinforced concrete, risk and reliability, soil behaviour, water resource systems and water resource planning, storm sewer design, diffusion and mass transport in rivers, microbiology of wastes, and highway materials.

The Department is one of ten departments in the College of Engineering. It is not organized into formal subdivisions, but has seven recognizable specialty areas with which faculty and students associate: Construction Engineering, Environmental Engineering, Geodesy and Photogrammetry, Geotechnical Engineering, Hydrosystems Engineering,

Structural Engineering (including structural materials) and Transportation Engineering. The Department conducts research in all of these areas.
Total staff: 270
Professionally qualified staff: 60
Director or Chief Executive
Head of Department: Jon C. Liebman
Principal sources of finance
Approximately 50% State of Illinois. Other major funding from National Science Foundation, Department of Defense, Department of Transportation, Environmental Protection Agency.
Publications
The results of most departmental research projects are published in professional and scientific journals. In addition, doctoral dissertations completed in the Department serve as a means of reporting research findings.

Reports of progress and completed research are issued in a continuing series of Civil Engineering Studies. Some research reports are also issued as Engineering Experiment Station Bulletins.

The Annual Summary of Engineering Research, published by the College of Engineering, lists all research projects and publications for the year.

USA.604 **FM**
International Technology Council (ITC)
Building Research, National Academy of Sciences, 2101 Constitution Avenue, NW, Washington, DC 20418
Telephone: (202) 334 3375
Telex: 248664 NASW UR
Cables: NARECO WSH
Date of establishment
1962

The International Technology Council (formerly the US National Committee for the International Council for Building Research, Studies, and Documentation) was organized in 1962 within the US National Academies National Research Council (NRC) to represent the United States in and to encourage, facilitate, and develop US participation in the technical activities of the CIB. The ITC serves as the mechanism through which the US and the international building communities exchange research findings to help meet building technology needs, to advance improved building criteria, and to make building practices responsive to worldwide needs. The ITC's secretariat for NRC membership in CIB:

- Serves as the vehicle through which US building researchers participate in CIB working commissions and exchange building research findings developed by US organizations and other CIB members
- Stimulates, through ITC counterparts to CIB activities, the generation of research studies, and the collection and dissemination of information on the state of the art of international building research and technology
- Motivates organizations to take positive action in furthering building research, documentation and communication

ITC works with representatives from the science and technology offices of foreign embassies in Washington DC and with organizations and individuals who share an interest in international affairs.

ITC performs studies as appropriate for sponsors requiring advice, technology support, and answers to building issues of public and private concern. All of ITC findings are published in the open literature and are disseminated for those concerned with the built environment.
Main fields of work
The ITC seeks the active participation of US building related organizations and offers these organizations an opportunity to become acquainted with and recognized by experts in building research throughout the world. The ITC offers a way to contribute to and benefit from progress in the world-wide research community.

ITC participants:

- Become involved in and influence ongoing programmes, directions of research, and trends within the international research community to the benefit of the US community
- Join in and personally contribute to the work of existing CIB commissions and ITC counterpart commissions and task groups whose efforts influence international programmes, practices and standards
- Stimulate exchanges of knowledge on international building research
- Become active in US and international gatherings of researchers and research-users and have access to the products of such programmes
- Have access to ITC and CIB publications as well as information about selected other international building-related documents.

ITC participates in various CIB activities and are co-ordinators to the ITC counterpart working commissions (CWC) in CIB working commissions (WC) and co-ordinators of CIB WCs:

- CWC-55 Building Economics
- CWC-62 Water Supply and Drainage
- WC-65 Organization and Management of Construction
- CWC-67 Energy Conservation in the Built Environment
- WC-73 Earthquake Hazards Reduction
- CWC-79 Controls of Building Services
- WC-79 Controls of Building Services
- WC-80 Services-life Prediction
- WC-83 Single-layer Roofing

We expect to create other CWCs and WCs during the next 12 months. Co-ordinators are selected from the US building community for their competence and experience in the field.

Participation in the work of the ITC is by invitation, and the membership comprises representatives of the US building community:

- American Concrete Institute
- American Institute of Architects Foundation
- American Iron and Steel Institute
- American Society of Civil Engineers
- American Society of Heating, Refrigerating and Air Conditioning Engineers
- American Society for Testing and Materials
- US Construction Engineering Research Laboratory
- US Department of Agriculture
- US Department of Housing and Urban Development
- Environmental Design Research Association
- US Federal Emergency Management Agency
- International Masonry Institute
- Joint Center for Urban Studies
- National Association of Home Builders
- US National Bureau of Standards
- US National Science Foundation
- US Public Buildings Service
- Architectural Research Centers Consortium, Inc.
- Southwest Research Institute
- Urban Land Institute.

Total staff: 2
Professionally qualified staff: 2
Director or Chief Executive
Executive Secretary, ITC: Noel J. Raufaste Jr
Heads of department
Full Committee of ITC: John Joyce, Chairman
Program Committee of ITC: John Joyce, Chairman
Principal sources of finance
Contracts or grants from member organizations and from various public and private organizations provide the financial support to ITC.
Principal related bodies
ITC National Research Council, National Academy of Sciences Representative to CIB, Dr R.N. Wright. Building Research Board, George Jenkins, Chairman. Building Research Board, John P. Eberhard, Executive Director.
Publications
Annual Report
Periodic Information Report, bi-monthly
Periodic proceedings of workshops.

USA.605

Kansas State University, Institute for Environmental Research

Seaton Hall, Kansas State University, Manhattan, Kansas 66506

Telephone: 913 532 5620

Date of establishment

1963

Established in 1963 following a gift of an environmental chamber from the American Society of Heating, Refrigerating and Air-Conditioning Engineers to Kansas State University and a Health Facilities Grant from the National Institute of Health for a building in which to house the chamber and provide offices and laboratory space for the staff. An electrically heated copper manikin together with a guarded hot plate are available for clothing research in addition to four smaller environmental chambers.

Main fields of work

Research and teaching in the following areas:

- Human thermal comfort
- Environmental ergonomics
- Heat stress
- Temperature related product design and testing
- Protective clothing systems and outerwear
- Consulting services to industry and government
- Innovative R and D
- Human engineering for the elderly
- Energy conservation strategies
- Subjective scaling development
- Provision of the setting, guidance and facilities for MS and PhD research.

Total staff: draws upon entire faculty

Director or Chief Executive

Director: Professor Frederick H. Rohles, Jr

Principal sources of finance

Financed by the State of Kansas and research and grants from public and private institutions.

Principal related bodies

Advisory counsel.

Publications

Research reports: Reprint series, from scientific and professional journals; Special publications.

USA.606

Manville Service Corporation, Research and Development Center

PO Box 5108, Denver, Colorado 80217

Telephone: (303) 978 2000

Telex: 45705 JOHNMAN DVR

Cables: JOHMANVL

Date of establishment

1916

A wholly owned subsidiary of the Manville Corporation; also engaged in development and testing for government and private entities.

Main fields of work

- *Building Products:* Development and testing of thermal insulation systems for buildings
- *Industrial Products:* Development and testing of industrial insulations, refractories and lighting systems
- *General:* Acoustical, fire, physical and thermal testing. Computer modelling of building systems.

Total staff: 270

Professionally qualified staff: 150

Director of Chief Executive

VP of Research and Development: J.H. Swensen

Heads of department

Research Contracts and Applied Technology: C.R. Vander Linden, Director

Fiber Glass Products and Roofing: F.A. Winterhalter, Director

Industrial Products, Filtration and Minerals: J.W. Helmick, Director

Fiber Process Development: F.Q. Van Kirk, Director

Holophane (Lighting Systems): A.B. Gough, Director

Principal sources of finance

Private and some government financing.

Publications

No regular external publications. Papers by staff members are published in professional, scientific and trade journals.

USA.607 AM

Massachusetts Institute of Technology, Department of Civil Engineering, Center for Construction Research and Education (CCRE)

Room 1–175, 77 Massachusetts Avenue, Cambridge, Massachusetts 02030

Telephone: (617) 253 7273

Telex: 921473 MIT CAM

Cables: MIT CAM

Date of establishment

February 1981

The Center is part of the Department of Civil Engineering at the Massachusetts Institute of Technology. The Center, in partnership and collaboration with the engineering and construction industry, conducts research and offers education programmes in technological, managerial, and resources aspects of construction.

Main fields of work

Research: The Center conducts fundamental research in a range of areas including: Technology: innovation, productivity, adaptation and transfer; Management: management issues and methods at the project, firm, and industry levels; Resource Development and Utilization: Manpower, construction materials and products, construction equipment, and finances.

In addition, multi-sponsor research programmes are currently being developed in the following areas: Technological Innovation and Productivity Improvement; Construction and Socio-Economic Development; Trends and Conditions in International Construction; Computers and Construction; Management of Superprojects; Construction and Energy.

Education: Graduate Degree programmes in Construction Engineering and Management; Short courses, seminars, and symposia on current issues and challenges facing construction and on recent research results; Mid-Career Masters programme for industry personnel with 10–15 years experience (under development); Senior Executive programmes (under development).

Industry affiliation: Construction Industry Affiliates programme opened to engineering and construction firms and other firms with an interest in the future of construction; Adjunct Faculty and Research Fellowships programme.

The Center draws its professional staff from the faculty and research staff of MIT's 23 academic departments with primary reliance on the 50 person staff of the Department of Civil Engineering.

Director or Chief Executive:

Director: Fred Moavenzadeh, Professor of Civil Engineering

Deputy Director: Charles H. Helliwell, Research Associate

Principal sources of finance

Sponsored research, unrestricted grants, academic funds, affiliates programme membership fees, and fees from continuing education programme.

Principal related bodies

Industry Advisory Board, provides programme guidance and evaluation by representatives from various segments of the engineering and construction industry; Policy committee, provides policy directions for the Center and its programmes by academic representatives of the Civil Engineering Department with inputs from industry representatives.

Publications

- Brochures
- Annual Report
- Announcements of Short Courses, Conferences and Seminars
- Listing of Research in Progress
- Listing of Available Research Reports.

USA.608

Michigan Department of Transportation, Research Laboratory Section of the Testing and Research Division

PO Box 33049, Lansing, Michigan 48909

Telephone: (517) 322 1632

Date of establishment:

Highway Department established circa 1914: Department of Highways and Transportation established 1973. Department of Transportation established 1978.

The Research Laboratory is a part of the Testing and Research Division and does research on a broad area of planning, design, construction and maintenance of highways.

Main fields of work

The function of the Department is the administration, direction, supervision and co-ordination of all administrative, engineering and operational functions to effect a safe, efficient and co-ordinated state transportation system.

The Research Laboratory Section of the Testing and Research Division conducts research and developmental testing for the evaluation and improvement of materials and methods used in highway construction and maintenance. It performs assigned research in soil compaction, mechanics, interpretation of pavement performance and conducts studies of engineering properties and uses of soils, aggregates and related materials. It conducts research in the broad areas of design, construction, maintenance and performance of pavements, structures and foundations. It also conducts environmental studies on air and water quality and on noise pollution.

Total staff: 62

Professionally qualified staff: 28

Director or Chief Executive

L.T. Oehler

Heads of department

Director Department of Transportation: John P. Woodford

Testing and Research Engineer: K.A. Allemeier

Engineer of Research: L.T. Oehler

Principal sources of finance

Financed by transportation user taxes.

Principal related bodies

Michigan Transportation Commission; Michigan Aeronautics Commission; Public Transportation Council, Mackinac Bridge Authority; International Bridge Authority.

Publications

- Numerous Departmental engineering and operational reports are published each year
- Twenty-seven Research Laboratory Reports for 1982

USA.609 FM

National Bureau of Standards, Center for Building Technology (NBS/CBT)

Gaithersbourg MD 20899

Telephone: (301) 921 3377

Date of establishment

NBS Organic Act created the National Bureau of Standards in 1901; the Center for Building Technology was created in 1972, formerly the Building Research Division (founded in 1947); prior NBS building research was carried out under different NBS organizational units dating back to 1901.

The Mission of the Center for Building Technology of the National Bureau of Standards, the US Government Research Laboratory, is to increase the usefulness, safety, and economy of buildings through the advancement of building technology and its application to improving building practices.

Main fields of work

The NBS/CBT produces technical bases for building practices, and measurement technologies for their application, to give owners, designers, manufacturers, builders, and regulators maximum opportunities for innovation while assuring usefulness, safety, and economy. CBT does not promulgate building standards or regulations but does provide technical support to governmental and private organizations that have standards and codes responsibilities. The Center conducts co-operative programmes with other research organizations, professional societies, standards-writing groups, testing laboratories, and educational institutions.

CBT performs work in:

- Geotechnical Engineering
- Earthquake Engineering
- Construction Engineering
- Structural Engineering
- Service-Life Prediction
- Quality Assurance
- Cement Hydration
- Thermal Analysis
- Heat and Mass Transfer
- Building Acoustics
- Lighting
- Building Equipment
- Building Controls
- Computer-Integrated Construction

Total staff: 150

Professionally qualified staff: 100

Director or Chief Executive

Director, Center for Building Technology: Richard N. Wright

Heads of department

Structures Division: Charles G. Culven

Building Physics Division: Tamami Kusuda

Building Equipment Division: James Hill

Building Materials Division: Geoffrey J. Frohnsdorff

Principal sources of finance

US Federal Government

Principal related bodies

The Center for Fire Research, NBS, performs and supports research in all aspects of fire, with the aim of providing scientific and technical knowledge applicable to preventing and controlling fires including the physics and chemistry of combustion processes and products; fire investigations; and biological, physiological, and psychological factors affecting fire victims.

Publications

CBT's results are published by the National Bureau of Standards under several publication series:

- Building Science Series
- Technical Note
- Special Publications
- Handbook
- NBS Interagency Report

Also NBS/CBT publishes in various technical, professional, and trade journals. The Center's annual listing of publications and summary of ongoing building research projects are published in:

Building Technology Publications, NBS SP 457, an annual supplement that documents CBT's reports with keywords and abstracts. Can be ordered from the US Government Printing Office, SN 003 003 02205

Building Technology Project Summaries, NBS SP 446, an annual supplement that summarizes each CBT research project. Can be ordered from the US Government Printing Office, SN 003 003 02236. US Government Printing Office, Washington, DC 20402.

USA.610

Notre Dame University, Department of Civil Engineering
Notre Dame, Indiana 46556
Telephone: 219 293 5380
Date of establishment
Established 1842 as an institute of Catholic higher learning.

Institute of higher learning offering undergraduate and graduate degrees.
Main fields of work
The Department is one of five departments in the College of Engineering. Advanced study in civil engineering includes research and professional development in the following areas of specialization: applied mechanics and engineering materials, soil mechanics and foundations, transportation, structural mechanics, structural engineering and water resources. The Environmental Health Engineering programme includes waste and wastewater treatment, water pollution control, solid waste management and environmental planning and management.
Total staff: 16
Director or Chief Executive
William G. Gray
Principal sources of finance
Tuition fees, grants, endowments
Institute of higher learning offering undergraduate and graduate degrees from students, contract research funds, i.e. National Science Foundation, Federal Highway Administration, Environmental Protection Agency, Department of Energy.
Principal related bodies
College of Engineering Departments:
Aerospace/Mechanical;
Electrical;
Chemical;
Material Science and Metallurgy.
Publications
Numerous technical papers published in scientific, professional and educational journals

USA.611 UM

O.M. Orshan, Inc. and Building Resources & Technology
8509 Aragon Lane, Chevy Chase, Maryland 20815
Telephone: (301) 587 7787
Telex:
904059 WSH USA
Date of establishment
Incorporated (Maryland, USA) 24 September 1976

A consulting firm providing services in the fields of building technology and construction economics to the construction industry in the USA and in Europe.
Main fields of work
These include:

- Consulting in construction economics and in building technologies
- Planning and scheduling of projects (time, cost, resources)
- Value engineering (analysis) and life cycle costing of buildings and building systems
- Pathology of construction management (litigation and arbitration of construction contract).

Total staff: varies
Professionally qualified staff: varies, as needed
Director or Chief Executive
Ossy M. Orshan
Principal sources of finance
A profit corporation

USA.612

Owens-Corning Fiberglas (OCF) Technical Center
Box 415, Granville, Ohio 43023
Telephone: 614 587 0610
Telex: 24 6657
Cables: FIBERGLAS GRVL
Date of establishment
Established in the late 1950s at its present site

A private research body and part of the technical establishment of Owens-Corning Fiberglas.
Main fields of work
Of the ten laboratories of the Center, three are almost exclusively concerned with the development and support of products, processes, methods, and techniques aimed at the building and construction industry:

Building and Insulation Product Technology: new product development and product application; reinforced concrete; roofing systems, ceilings, interior acoustical systems, insulation materials, insulation systems for air handling duct, piping, and pre-engineered buildings

Building and Insulation Process and Materials Technology: its emphasis is to insure the products lines are using optimal material configurations, both technically and economically

Building Research: development of building systems; energy conservation; moisture control; heat loss and vapour migration in structures; support research on the consumer safety aspects of Owens-Corning Fiberglas building products as related to flammability, smoke, and toxicity.

The Technical Center operates one of the most complete thermal insulation testing facilities in the world and a well equipped sound/acoustical laboratory.

R and D of OCF building products is characterized by an integrated approach.
Total staff: 1050
Professionally qualified staff: 600
Director or Chief Executive
Vice President and Manager, Technical Center: Roger W. Roth
Principal sources of finance
Primarily through the sponsorship of the Operating Divisions of Owens-Corning Fiberglas

USA.613

Pittsburgh Testing Laboratory (PTL)
850 Poplar Street, Pittsburgh, PA 15220
Telephone: 412/922 4000
Telex: 866735
Date of establishment
1881

An independent commercial inspection and testing laboratory performing services for both public and private organizations.
Main fields of work
PTL offers testing and inspection services to clients, generally using established standards and test methods. In the building construction

area, included are geotechnical engineering, cement and concrete materials testing, and quality control of construction practices; fabrication and erection of structural steel; non-destructive testing of welds in buildings, bridges, tanks and pipelines.

Other services are chemical analysis of materials, metallurgical and failure analysis, physical testing of materials, materials engineering and special client specified testing. Quality assurance services include the development and maintenance of programmes for manufacturers and vendor surveillance. Inspection and expediting services encompass both raw materials and complex machines and instrumentation.

PTL clients include architects, engineers, contractors, commodity buyers and sellers for industry, governments and individuals.

The Corporate Offices and Main Laboratory are located in Pittsburgh, PA with 44 District Laboratories under 4 Regional Offices.

Total staff: 1200

Professionally qualified staff: 120

Director or Chief Executive

President and CEO: M.Y. Ruyan

Heads of department

Northern Region: W.H. Levelius, Vice President
Southern Region: B.L. Couch, Vice President
Gulf Central Region: R.L. Jefferies, Vice President
Western Region: J.L. Munnerlyn, Vice President

Principal sources of finance

Financing is entirely through fees paid by clients for services rendered.

Principal related bodies

PTL is an independently owned corporation and is not related to a parent organization. PTL does have working agreements with similar organizations in various countries around the world.

Publications

PTL does not publish journals or reports to the general public; reports of examinations and findings are issued to clients who, in some cases, may use data generated in their own publications.

Many of the professional and technical staff contribute articles to technical societies or scientific journals for publication.

USA.614

Portland Cement Association, Research and Development Construction Technology Laboratories (PCA–R&D/CTL)

5420 Old Orchard Road, Skokie, Illinois 60077

Telephone: (312) 966 6200

Date of establishment

Organized in 1916 and incorporated in 1970

A non-profit corporation.

Main fields of work

PCA–R&D/CTL is the research unit of the Portland Cement Association of the United States and Canada. Through contract research the PCA offers its facilities and services to the construction industry, public agencies, private industry, cement and concrete producers, manufacturers and users of other construction materials.

Operating Departments:

- Chemical/Physical Research: Basic and applied research in the areas of cement, clinker, admixtures, aggregates and other construction materials
- Concrete Materials Research
- Engineering Development: Structural testing and research, fire research and testing, transportation development and testing, construction and product development, structural evaluation
- Administrative and Technical Services
- Occupational Safety and Health Services

Research personnel conduct laboratory and field studies and provide consultation and technical services related to factors influencing performance of concrete and other construction materials.

Total staff: approximately 130, equally divided between professional employees and support staff.

Director or Chief Executive

Group Vice President, R&D/CTL: W.E. Kunze

Principal sources of finance

Financed by dues paid by cement-producing member companies of the PCA and by revenues from research and development and engineering services work under contract with public and private agencies.

Publications

Research reports are published in professional and scientific journals, bulletins and reports of the PCA–R&D/CTL and other information papers available from the Portland Cement Association

A list of publications is included in the Cement and Concrete Reference Catalog issued annually.

USA.615

Prestressed Concrete Institute (PCI)

201 N Wells Street, Chicago, Illinois 60606

Telephone: 312 346 4071

Date of establishment

1954

A non-profit corporation.

Main fields of work

The purposes of the Institute are to stimulate and advance the common interests and general welfare of the prestressed concrete industry and the precast concrete industry; to perform all lawful and desirable activities to promote the efficient, constructive and beneficial operation of the prestressed concrete industry and the precast concrete industry.

Covered areas: research, engineering development, production, safety, marketing materials, technical publications, training courses, statistical information in industry, development of recommended practices, building and bridge code work, preparation of specifications and guidelines, product standardization, awards programmes.

Total staff: 18

Director or Chief Executive

Executive Director: Fromy Rosenberg

Principal sources of finance

Financed by dues and sales of literature plus meeting registration fees.

Publications

- PCI Journal, bi-monthly technical journal
- Ideas, quarterly
- Handbooks (architectural precast, design handbook, drafting handbook)
  Catalogue of publications
- Newsletters
- Quality Control Manuals
- Many special purpose publications.

USA.616

Rensselaer Polytechnic Institute, School of Architecture, Center for Architectural Research (CAR)

Troy, New York 12181

Telephone: 518 270 6461

Telex: 646542 rpi tro

Date of establishment

1965 (as a department within the School of Architecture)

Rensselaer Polytechnic Institute is a private university and the Center for Architectural Research is an entity within its School of Architecture.

Main fields of work

Organizationally, the Center functions as an administrative entity within the School of Architecture (one of the five undergraduate schools at

Rensselaer). Its staff members hold academic staff appointments and are integrally involved in the educational activities of school and university. Additionally, most of the staff participate in practice or consulting.

Sponsored research projects may be supported by a public agency, or by a private foundation, organization, or firm which faces a building-related problem. They may require the traditional skills in architecture and building; more commonly, they will require that the Center establish a project team of staff, consultants and advisers in several disciplines.

Over the past ten years, the Center had carried out projects in these general areas: facilities planning and design; development and delivery of the built environment; protection of property and occupants; building systems development, utilization and evaluation; energy conscious design and resources conservation.

Total staff: 4–5 (staff fluctuates, based on amount of research; a core staff of 4–5 people; additional staff obtained from University Departments as research requires)

Director or Chief Executive

Director: Walter M. Kroner

Principal sources of finance

Financed through contract research.

Publications

Reports, guides and books are available from the Center.

USA.617 UM

E. George Stem, Registered Architect (RA), Earle B. Norris Research Professor Emeritus of Wood Construction, Virginia Polytechnic Institute and State University

PO Box 361, Blacksburg, Virginia 24060

Telephone: 703 961 6639

Date of establishment

As Industrial Consultant: 1936

Consultant to University, trade associations, industries. Chairman of technical committees of American Society for Testing and Materials (ASTM), American Society of Civil Engineers (ASCE), American Society of Mechanical Engineers (ASME), National Ornamental and Miscellaneous Metals Association (NOMMA).

Main fields of work

- Wood Construction and Assembly
- Fasteners in Wood Construction
- Wooden Pallets
- Railing Systems and Rails for Buildings.

Principal sources of finance

Consulting and grants.

Publications

- Author of more than 800 publications in field of wood construction and assembly
- *Journal:* VPI&SU Wood Research and Wood Construction Laboratory Bulletins, irregular
- *Technical standards:* ASTM E06 and F16 Standards; ANSI MH1 Standards; ISO TC51 Standards
- *Research reports:* VPI&SU Wood Research and Wood Construction Laboratory Bulletins; Unpublished Reports of VPI&SU Wood Research and Wood Construction Laboratory; VPI&SU Pallet and Container Laboratory Bulletins; Unpublished Reports of VPI&SU William H. Sardo Jr Pallet and Container Research Laboratory.

USA.618

Stevens Institute of Technology, Davidson Laboratory

711 Hudson Street, Hoboken, New Jersey 07030

Telephone: 201 420 5300

Date of establishment

1935

A non-profit research laboratory which is part of a private education institution, engaged in studies related to ship hydrodynamics, transportation, and building technology.

Main fields of work

- *Applied Mechanics:* fluid dynamics; stability and control; offshore dynamics; building technology (especially innovative plumbing systems for houses)
- *Marine Craft Development:* marine craft dynamics; ship and yacht; computing
- *Transportation Research:* transportation analysis; vehicle mechanics and development

Facilities and Services.

Total staff: 40

Professionally qualified staff: 25

Director or Chief Executive

Acting Director: Daniel Savitsky

Heads of department

Marine Dynamics: P. Ward Brown

Transportation Research: I. Kamn

Building Technology: T. Konen

Principal sources of finance

Federal government; state government; industry

Publications

Numerous papers are published in professional and scientific journals.

A complete list of published reports is available from the Laboratory.

USA.619

Transportation Research Board (TRB)

2101 Constitution Avenue, NW, Washington, DC 20418

Telephone: 202 334 2934

Telex: 710 822 9589

Cables: NARECO

Date of establishment

1920

A non-profit-making unit of the National Research Council, which is the principal operating agency of the National Academy of Sciences and the National Academy of Engineering to serve government and other organizations.

Main fields of work

The purpose of TRB is to advance knowledge concerning the nature and performance of transportation systems through the stimulation of research and dissemination of information derived therefrom. Work is carried out in the following sections:

- Regular Technical Activities: economics, design, materials and construction, traffic and safety, maintenance, soils, urban transportation, air, rail, inland waterways, legal resources
- Special Technical Activities: short-term studies of transportation subjects with regard to their potential as regular activities of TRB
- Co-operative Research Programs: National Co-operative Highway Research Program (NCHRP) and the National Co-operative Transit Research Program (NCTRP): the professional staff provides a secretariat to assist the American Association of State Highway and Transportation Officials in the formulation of annual research programmes and the Urban Mass Transportation Administration respectively
- Administration

Total staff: 95

Professionally qualified staff: 47

Director or Chief Executive
Executive Director: Thomas B. Deen
Heads of department
Regular Technical Activities: K.B. Johns
Special Technical Activities: Damian Kulash
Administrative Activities: Edward C. Squires
Co-operative Research Programs: K.W. Kenderson, Jr
Principal sources of finance
State Highway and Transportation Departments; Modal administrations of US Department of Transportation and other government agencies; Association of American Railroads; supporting associates, publication sales, NAS, etc.
Publications

- Research Records, irregular
- Transportation Research News, six issues each year
- Special Reports, irregular
- TRB Directory, one each year
- NCHRP and NCTRP Reports
- TRB Circulars, irregular
- Selected Studies in Highway Law
- HRIS Abstracts and Urban Transportation Abstracts.

USA.620

Underwriters Laboratories Inc. (UL)

333 Pfingsten Rd, Northbrook, Illinois 60062
Telephone: (312) 272 8800
Telex: 4330173 (internat.)
Cables: ULINC NORTHBROOK, IL
Date of establishment
Founded 1894

Independent, non-profit organization, testing for public safety. Its objectives are to establish, maintain and operate laboratories for the investigation of materials, devices, products, equipment, constructions, methods and systems with respect to hazards affecting life and property. Then periodic factory follow-up inspections are done to audit and countercheck the manufacturers' production controls to ascertain that the applicable UL label is being affixed only to those items that continue to meet the safety standard.
Main fields of work
Scientific investigation of products for the purpose of safety certification. Investigations include electrical, fire, chemical, and other analysis of products to determine that they meet applicable safety standards.
Total staff: 2734
Professionally qualified staff: Engineering Services 1065
Director or Chief Executive
President: Jack Bono
Heads of department
Burglary Protection and Signaling: George Saunders
Casualty and Chemical Hazards: Robert Harris
Electrical: Milton Cox
Fire Protection: James Beyreis
Heating, Air Conditioning and Refrigeration: Charles Schram
Marine: Al Bernardi
Principal sources of finance
Entirely self-supporting; the costs of services are borne by the individual clients who submit products for investigation. Research work is performed on a cost reimbursement basis.
Principal related bodies
Included in the more than 300 groups in which the staff has membership are code panels of the National Electrical Code, National Fire Protection Association Committees, American Society for Testing and Materials (ASTM), Committees, American National Standards Institute (ANSI) Committees and International Electrotechnical Commission (IEC) Groups.

Publications

- Product Directories, lists of manufacturers whose products have met UL requirements
- Catalog of Standards, list of published standards
- LAB DATA, quarterly, technical and informational magazine
- UL TRENDS, quarterly newsletter on new developments in safety testing, testing equipment, safety standards and new products.

USA.621 AM

US Army Construction Engineering Research Laboratory (CERL)

PO Box 4005, Champaign, Illinois 61820
Telephone: 217 352 6511
Date of establishment
1968

CERL's role is to conduct research and development investigations to support Army programmes in military construction, military engineering, civil works, and operations and maintenance; and improve construction quality and energy efficiency with attendant environmental safeguards. In so doing, CERL supports the US Army Corps of Engineers, the largest, most active building procurement industry in the world, to plan, design, build, manage, operate, maintain and repair 150 Army installations worldwide.
Main fields of work
CERL's Energy Systems Division is responsible for research in energy systems with emphasis on development of tools, techniques and criteria for energy conservation, energy management and the use of alternative energy sources.

The Engineering and Materials Division's research studies military technology in material sciences for the improvement of design construction, maintenance and repair of military and civil works facilities. Emphasis is placed on better construction quality control, protective coatings, corrosion mitigation and innovative construction material.

The Environmental Division develops tools and techniques to decrease air, water, solid and noise pollution. Managing environmental quality during the life-cycle of military facilities and environmental assessment procedures are important research areas.

The Facilities Systems Division is responsible for management engineering for the efficient planning, design, construction and operations of military facilities, as well as methods of enhancing their livability. Programmes and products include project design using automated systems, research related to system building and the development of systems to support the managing of all aspects of military installation operations.
Total staff: 215 (9 military)
Professionally qualified staff: 135
Director or Chief Executive
Commander and Director: Col. Paul Theuer
Heads of department
Engineering and Materials Division: Robert F. Quattrone
Environmental Division: Ravinder K. Jain
Energy Systems Division: Richard G. Donaghy
Facility Systems Division: Edward A. Lotz
Principal sources of finance
US Federal Government.
Publications

- *Journal:* CERL Reports, quarterly; CERL Abstracts, quarterly, free on request
- *Research reports:* Available for purchase from: National Technical Information Service, 5285 Port Royal Road, Springfield, VA 22161 USA.

USA.622

US Army Engineer Waterways Experiment Station, Structures Laboratory (WES–SL)

PO Box 631, Vicksburg, Mississippi 39180

Telephone: 601 542 3264

Date of establishment

As Concrete Laboratory 1946; as Structures Laboratory 1978

The Structures Laboratory is responsible for conceiving, planning, and execution of scientific and engineering investigations and research and development projects in support of the civil and military missions of the Chief of Engineers and other Federal agencies, in the broad fields of structures, weapons effects, earth dynamics, and construction materials by establishing and maintaining staff and facilities for research, development, testing, and evaluation as related to design and analysis of structures to resist static and dynamic loading; defining free-field effects produced by the detonation of explosives; development of useful applications of explosives; evaluation of materials properties, applications, and behaviour in service; and defining the state of stress in soil and rock masses especially as associated with transient loadings. It investigates properties and behaviour of a wide variety of construction materials, both singly and in combinations in composite systems such as hydraulic-cement concrete and in structural elements and structures. The constituents of hydraulic-cement concretes, mortars, grouts; plastic and adhesives; polymers and reinforcing materials, bonding, and jointing materials; coatings and curing materials; stone and rock are included. It operates the US Department of Defense Concrete Technology Information Analysis Center.

Main fields of work

The Laboratory performs research and investigations that concern the physical, chemical and mineralogical properties of concrete and concrete materials, the processes used in concrete construction and the structural capabilities of concrete when incorporated in a structure. Prominent among its activities is the Civil Works R and D, with the general objective of achieving major improvements in quality and structural performance of concrete, resulting in more efficient and economical construction.

The Laboratory also serves as the Materials and Concrete Laboratory for the Lower Mississippi Valley Division and conducts tests and research on concrete and construction materials on request for other Government agencies. It is assigned as the sole Federal Laboratory responsible for sampling, inspecting, and acceptance testing of Portland cement and pozzolan used on Corps of Engineers and other Federal construction. It provides technical surveillance and maintains extensive expertise in the field of cement and cement replacement materials.

It provides capabilities in chemical research and petrographic and special concrete production facilities for all other WES laboratories.

The staff provides consulting services on almost the full spectrum of laboratory and field construction applications.

Total staff: 160

Professionally qualified staff: 90

Director or Chief Executive

Bryant Mather

Heads of department

Concrete Technology Division: John M. Scanlon

Explosives Effects Division: Jerry W. Brown

Geomechanics Division: John G. Jackson

Structural Mechanics Division: James T. Ballard

Principal sources of finance

US Department of Defense.

Principal related bodies

Next higher echelon: Office, Chief of Engineers, US Army, Washington, DC

Publications

- *Technical standards:* Handbook for Concrete and Cement (supplements issued quarterly since August 1949)
- *Research reports:* WES Technical Reports (TR); WES Miscellaneous Papers (MP); CTIAC Reports (as TRs or MPs)
- *Bibliographies/Catalogues:* Included as TRs or MPs.

USA.623

US Department of Housing and Urban Development (HUD)

451–7th Street, SW, Washington, DC 20410

Date of establishment

1966 by amalgamation of five separate organizations comprising former Housing and Home Finance Agency (HHFA)

US Federal Government agency in the executive branch.

The Department sponsors research activities in all phases of housing and urban matters.

The primary research and development activity is carried out by the Assistant Secretary for Policy Development and Research. Building technology research is carried out in the Office of Housing Studies, Building Technology Division.

Main fields of work

The Department is the US agency responsible for providing support for housing and community activities related to housing and urban development activities. It provides grants to communities, supports local authorities who provide public housing at reduced rentals to local citizens who cannot afford market housing, insures mortgages issued as security for the purchase of single family and other housing, and is responsible for special housing issues such as housing for American Indians and Alaskan Natives.

Building technology research in support of HUD programmes is carried out by the Building Technology Division, Office of Housing Studies, reporting to the Assistant Secretary for Policy Development and Research. Current activities of the Building Technology Division include management of a nation-wide series of demonstrations of 'affordable housing', showing how housing costs can be lowered through regulatory reform; studies of energy and water management issues in community activities; and research supporting the development and maintenance of standards for projects with HUD support. In general, laboratory studies and demonstrations are undertaken under contract or grant support; analytical studies may either utilize Division staff or be carried out by contractors or grantees.

There are approximately 14 000 departmental employees throughout the US only a portion of whom are technical.

The department is headed by a Presidential Cabinet level Officer called the Secretary of Housing and Urban Development.

Total Division staff: 16

Professionally qualified staff: 12

Director or Chief Executive

Division Director: Joseph Sherman

Deputy Director: David Moore

Principal sources of finance

United States Congress

Principal related bodies

Center for Building Technology, National Bureau of Standards, Department of Commerce: responsibility for basic building research; National Institute for Building Science: co-ordinating role in developing standards and promoting research; National Association of Home Builders; leading private sector group

Publications

- All research projects sponsored by HUD culminate in the development of a final report. These reports are made available to the HUD Information Office, the HUD Library, the Superintendent of Documents, US Government Printing Office, Washington, DC 20404, and the National Technical Information Service, 5284 Port Royal Road, Springfield, Virginia, 22151
- Research projects carried out under HUD sponsorship by other federal agencies such as the National Bureau of Standards, the Department of Agriculture, the National Aeronautics and Space Administration, the Oak Ridge National Laboratories, the National Academy of Sciences, etc. are available through their normal distribution channels

- HUD user: a computer-based research information service created by the Office of Policy Development and Research.

USA.624 **AM**

United States Gypsum Company Research Center
700 North Highway 45/Libertyville, Illinois 60048
Other laboratories:
United States Gypsum Company, PO Box 460, Round Lake, Illinois 60073
A.P. Green Refractories Company, Green Blvd, Mexico, Missouri 65265
Telephone: (312) 362 9797
Telex: 253 035
Cables: GYPSUM
Date of establishment
1922

Research activity is divided between the Corporate laboratories in Libertyville, the sound test facility at Round Lake, and the A.P. Green laboratories in Mexico, Missouri. Research efforts are primarily for the Corporation, its subsidiaries and divisions.
Main fields of work
The Corporate laboratories' main activities relate to product and process research in gypsum, lime, cement, paper, wood and mineral fiber, acoustical products, textures, joint compounds, adhesives, coatings, plastics, inorganic chemicals and fillers, and fabricated steel and aluminium. Also, engineering and design functions for construction systems and fire, sound, structural and environmental testing.
A.P. Green laboratories primarily are involved in product and process research relating to ceramic fiber, high temperature insulation and refractories including mortars, bricks and castables.
Total staff: 268
Professionally qualified staff: 148
Director or Chief Executive
Sr Vice President, Science and Technology: R.D. Carnahan
Vice President R and D: J.N. Walker
Vice President, Research: A.P. Green, J.A. Crookston
Heads of department
Administration: R.P. Entz, Associate Director
Major Technical Programs: J.M. Summerfield, Associate Director
Formulated Products: E. Maynard, Divisional Manager
Gypsum and Lime: T.E. Nevins, Divisional
Fiber and Acoustical Products: K.D. White
Systems and Analytical: W.C. Veschuroff, Divisional Manager
Principal sources of finance
Funded primarily from the parent Corporation.
Publications
The majority of publications are intra-company and reach the public in the form of specifications and technical publications. Papers are also published in various technical publications and through technical societies.

USA.625

US Steel Corporation Technical Center
One Tech Center Drive, Monroeville, PA 15146
Telephone: 412/825 2000

Main fields of work
Activities are devoted to developing new or improved processes for the mining and beneficiation of raw materials such as coal and iron ore and for the production of steel and chemicals; to developing new and improved products in steel, chemicals, and other product lines; to developing technology for conservation of energy, for meeting environmental regulations and for achieving higher productivity in these areas; to serving customers in the selection and use of the Corporation's products; and to generating basic scientific information for future developments.
Total staff: Professional – 300 in chemistry, metallurgy, physics, chemical, civil, mechanical and electrical engineering, mathematics; 250 technicians and auxiliaries.
Director or Chief Executive
Director, Technical Systems and Services: W.J. Murphy
Director, Chemicals Research: E.J. Nemeth
Director, Research, Steel and Related Resources: A.M. Rathbone

USA.626

United States Testing Company, Inc. (USTCO)
1415 Park Avenue, Hoboken, New Jersey 07030
Telephone: (201) 792 2400
Telex: 12 6493
Cables: USTCO
Date of establishment
Established in 1880 as a laboratory of a silk importing association; incorporated in 1907 as an independent laboratory.

A private, for profit, independent testing laboratory. A member of Société Générale De Surveillance, SA, Geneva, Switzerland.
Main fields of work
Testing and inspection in all areas of construction; these include physical testing of soils, concrete, steel, and asphalt. Examination of metals and welder qualifications. Non-destructive testing. Special flammability studies on materials of construction. Complete Quality Assurance/Quality Control Service is provided to the Nuclear/Fossil Power Plant Industry. There are many other support services provided by USTCO.
Total staff: 750
Professionally qualified staff: 150
Director or Chief Executive
Chief Executive Officer: Gordon Cantley
Heads of department
Quality Assurance Services: Gene Basile
Construction Inspection Services: Dave Edley
Engineering Services: Rudy Giglio
Principal sources of finance
Fees for services rendered
Principal related bodies
SGS, SA: owner; Nationwide Consumer Testing Institute: listing and labeling
Publications
All reports are the property of the client.

USA.627

University of Pennsylvania School of Engineering and Applied Science, Department of Civil Engineering
Philadelphia, Pennsylvania 19104
Telephone: (215) 898 7044
Telex: 710 670 0328
Cables: PNSYL Phila, PA
Date of establishment
1742 by Benjamin Franklin

Private university, training young people in the humanities, arts and sciences. At the disposal of all qualified individuals throughout the world.
Main fields of work
Education and research in:
- Structural Analysis and Design
- Resource and Environmental Engineering
- Transportation Engineering and Systems
- Housing Systems

Total staff: 10 full-time faculty members
Professionally qualified staff: 10

Director or Chief Executive
Chairman: John A. Lepore
Heads of department
John A. Lepore: Structures/housing
Edward K. Morlock: Transportation
Iraj Zandi: Resource/Environment
Principal sources of finance
Financed through tuition income, research income and endowments
Publications
Numerous papers are published in professional and scientific journals.

USA.628

University of Pittsburgh, Department of Civil Engineering
949 Benedum Hall, University of Pittsburgh, Pittsburgh, PA 15261
Telephone: 412 624 5354
Date of establishment
University founded in 1787 as Pittsburgh Academy of Engineering. Chartered in 1856.

Main fields of work
Faculty involved in teaching, research and consulting.

All research conducted by the Program is of a structural nature, not devoted exclusively to buildings. Projects consist of theoretical, computer or experimental studies.

Large structural test floor available for static or dynamic examination of large scale specimens.
Total Staff: 7
Director or Chief Executive:
Chairman: Joel I. Abrams
Dean: Max L. Williams,
Principal sources of finance
Financed through State funds, endowment, tuition.
Publications
All these available through University Microfilms, Ann Arbor, Michigan

Faculty publications in professional journals appear also as School of Engineering Reports; list available on request.

USA.629

Virginia Highway and Transportation Research Council
Box 3817, University Station, Charlottesville, Virginia 22903
Telephone: 804 293 1900
Date of establishment
1949

A government research establishment within the Virginia Department of Highways and Transportation.
Main fields of work
Economics: Environmental; Planning; Soils; Computer Services; Concrete; Industrialized Construction; Structures; Bituminous; Maintenance; Pavement; Safety; Traffic.
Total staff: Approximately 120, of whom around one half are support staff.
Director or Chief Executive
Head: Howard H. Newlon, Jr.
Principal sources of finance
Financed approximately 70% by State of Virginia research allocation, 30% by federal planning and research funds.
Publications
98 papers in a reprint series of those published in professional or scientific journals.

A summary report issued every other year lists all publications issued during the period covered.

VEN.630 **AM**

Instituto Nacional de la Vivienda (INAVI)

National Institute of Housing

Edificio Cruz Verde, Esquina Cruz Verde, Carcas (1010)

Telephone: 507 40 99/507 40 90/507 35 56

Telex: 22 560/21 407

Date of establishment

13 May 1975 (formerly Banco Obrero, 1928)

INAVI is an executive and administrative entity of the National Government for housing projects of social interest.

Main fields of work

INAVI mainly covers the field of low-cost housing by means of the construction of housing projects through different programmes, which it carries out in those regions and cities considered to be of strategic importance for the economic development of the nation.

It also controls the availability of adequate and necessary areas of land for the realization of the Institute's programmes, through advance acquisition of land. It carries out programmes of urban renewal in the country's larger cities, in those sectors which present deteriorating conditions, through Special Renewal Agencies in order to improve the living conditions and the urban environment.

It attends particularly to the squatter settlements through the provision of basic services and facilities required by the population which inhabits such areas. Its field of action does not end with housing and urban improvement, but also covers the integral promotion of the development of man and family by means of the realization of programmes aimed at community development and to obtain an improved standard of life. It also covers research areas with special emphasis on the improvement of housing quality, production processes and incorporation of new building systems. The Institute also has a Credit Programme directed at low-income classes for the acquisition of housing.

The Directorate, integrated by the President and 4 Directors, is in charge of the direction and administration of the Institute. The Institute is organized on the basis of three specialized departments:

- Management of Projects and Construction, which takes care of the entire phase of production of housing for the Institute: projects, construction contracts, execution of the works, inspection, supervision and upkeep of housing and services
- Management of Housing Administration: carries out all the activities related to the selection of grantees and adjudication of houses, minor upkeep and conservation, payments, and handling the grantees system
- Management of Land: has the task of developing the programmes of land acquisition and rehabilitation, which will enable them to be assigned to the diverse programmes of the Institute. It also takes charge of registration and control of the properties of the Institute.

For the realization of their programmes these three departments are assisted by decentralized units which operate locally throughout the country.

Total staff: approximately 4000 employees working mostly at the main offices in Caracas – the rest at the local offices throughout the country. It also has 5000 support workers.

Director or Chief Executive

President: Tito Fernando Herrera Armas

Directors: Esperanza Matos de Saad
Mario Caires Siegert
Jesus Abreu Anselmi
Victor Herrera Suarez

Principal sources of finance

The Institute is financed through contributions from the following. Public sector: The Institution's own funds, funds of the National Government, extraordinary funds, funds derived from surrender of property, capital transference from other public organizations. Private sector: National savings and loans bank, Mixed Development Programmes.

Publications

Publications of a technical and informative nature related to research work and to programmes which the Institute develops to meet the housing demand are issued regularly. Among them are:

- Memoria y Cuenta Anual de Obras Realizadas y Programmas a Realizarse
- Informative Papers, containing a synthesis of the Housing Programmes of the Institute and its Housing Policy

Other Specialized Technical Publications, for example:

- Coordinación Modular y Dimensional
- Fundamentos para la Formulación de una Politica Habitacional y de Investigación del INAVI
- Ley del INAVI y su Reglamento
- Others

The publications are distributed among interested Institutions by means of an exchange service offered by the Research Centre, through the Documentation and Information Unit of the INAVI.

VEN.631

Universidad Central de Venezuela Instituto de Materiales y Modelos Estructurales (IMME)

Central University of Venezuela, Institute of Materials and Structural Models

Apdo 50.361, Caracas 1050A

Telephone: 662 69 03; 662 69 66

Cables: IMMECU

Date of establishment

1962

University Research and Service Institute within the College of Engineering, devoted to the national construction industry.

Main fields of work

IMME research areas are mainly related to all aspects of building and civil engineering works (except waterways). Through the different Divisions it deals with the performance of materials and structures. Research is conducted in the following areas:

- New materials for low cost housing (sulphur, organic fibres)
- Concrete corrosion
- Foundations for low cost houses on expansive clays
- Seismic risk of the Caribbean area and other zones in Venezuela
- Non-linear behaviour of reinforced concrete members under axial force and bending moments (normal and lightweight concrete)
- Behaviour of precast concrete joints under alternating loads
- Liquefaction of soils under seismic loads
- TV recorded technical services
- Off-shore structures
- Information Centre, library and videocassette library on Structural Engineering and Construction
- Postgraduate courses.

Total staff: 75

Professionally qualified staff: 30

Director or Chief Executive

Director: Maria L. Diaz de Smitter

Heads of department

Materials and Soils: Dr Joaquin Porrero

Special Studies: Dr Oscar A. López

Structures: Dr José Grases

Principal sources of finance

University and Technical Services.

Principal related bodies

Technical Council; Senate of the College of Engineering.

Publications

- *Journal:* Boletín Técnico del IMME bi-annual; 1963
- *Research reports:* Technical Services: not published
- *Bibliographies/Catalogues:* Published in the Boletìn Técnico del IMME; Educational Materials, booklets and books.

YUG.632

Laboratorij za Mehaniko Tekocin
Fluid Mechanics Laboratory
Hajdrihova 28, 61111 Ljubljana
Telephone: 061 268 261
Date of establishment
1937

After separation from the Faculty of Civil Engineering, Archtecture and Geodesy of the Hydraulic Laboratory in 1954, the Fluid Mechanics Laboratory was renovated for scientific research studies of the teaching staff of that Faculty.

Main fields of work
- Experimental demonstrations and exercise training for students on the Faculty's hydrotechnical courses
- Research practice for grade III, and postgraduate students: doctoral dissertations of hydrotechnical orientation
- Research work by the teaching staff
- Hydraulic research and studies for external clients.

Total staff: 8
Professionally qualified staff: 7
Director or Chief Executive
Chief of the Laboratory: Professor Milan Pšeničnik
Principal sources of finance
Funds from the Special Education Association and Union of Research Organizations of the Republic of Slovenia as well as fees for research studies.
Principal related bodies
The Faculty of Civil Engineering, Architecture and Geodesy of Ljubljana University as principal administrative body.

YUG.633

Highway Institute
Institut za Puteve
Kumodraska 257, 11000 Beograd
Telephone: 011 466 133
Telex: 11036 YU INSPUT
Cables: INSTITUT ZA PUTEVE BEOGRAD
Date of establishment
1950

Scientific researching activity and consultancy services for highway and traffic engineering.

A working organization with funds in social property.

Main fields of work
- Traffic and Economy Department carries out plans, programmes and designs for road and street network development traffic and transport systems, engineering and technological designs for public and internal transportation, traffic arrangements and facilities, managing systems for traffic regulation signalling and safety for roads and city streets. Economic and financial studies for transportation development programmes. Project evaluation and management for transport infrastructure
- Design Department carries out preliminary and final designs for highways, roads, urban streets, intersections, tunnels, bridges, railways, aerodromes, etc.
- Research and Testing Department carries out studies, experiments and analyses in the field of geotechnique, concrete, asphalt, stone and aggregates and engineering structures. Scientific researching work in the field of geotechnique, asphalt, concrete, stone and aggregates and engineering constructions. Design and evaluation of geotechnical projects, landslide corrections, foundations, underground structures, pavement structures, hydroisolation, concrete and asphalt mixtures design, etc.
- Supervision and Technical Control Department carries out laboratory quality control under construction, maintenance and correction of highways, roads, bridges, tunnels and other objects. Analysis of behaviour of roads and objects in exploitation
- Education of senior officials from developing countries. The Institute provides this service in the country and abroad
- The Institute, singly and together with other similar organizations, organizes national and international meetings (congresses, symposia, consultations, etc.)

Total staff: 615
Professionally qualified staff: 414
Heads of department
General Manager of the Highway Institute: Radovan Radoman, BSc Eng
Departments in the Highway Institute:
Traffic and Economy Department: Katić Žarko, BSc Econ
Trasa Design Department: Krstajić Mŏmčilo, BSc Eng
Technology and Operation Department: Milovanović Ljubiša, BSc Eng
Research and Testing Department: Mitrović Petar, MSc
Administration and Services Department: Levajac Milojko, BSc Econ
Department for Business Abroad: Aleksandar Virag, PhD Econ
Principal sources of finance
Financing is based on contracts for the execution of services effected with institutions in the country and abroad.
Principal related bodies
Professional Board of the General Manager: organization and realization of the development programme of the Highway Institute; Scientific Council (of experts from the Institute): programme and realization of scientific research work.
Publications
- *Journal:* Professional journal, Institut za Puteve, twice yearly
- Work on standardization derived from the Institute's activity

A number of research reports derived from the Institute's activity are published in national and international professional publications.

YUG.634

Institute for Metal Structures
Inštitut za Metalne Konstrukcije (IMK)
Mencingerjeva 7, 61000 Ljubljana
Telephone: 061 262 661
Cables: IMEKO Ljubljana
Date of establishment
1955

An independent organization set up by the University of Technology for scientific and industrial research.

Main fields of work
- Research on metals and metal structures; theory of structures; welding of steel and aluminium; welding technology
- Technical assistance to contractors, advice and testing
- Design of steel constructions: bridges, structural engineering, sports halls
- Non-destructive testing.

IMK is organized in three departments: Research and Development; Calculation, Design and Control; Testing of Materials and Structures by Non-destructive Methods; and two divisions: Administrative Division; Publication and Documentation.

Total staff: 65
Professionally qualified staff: 21
Director or Chief Executive
Director: Ciril Šivic
Principal sources of finance
Fees for industrial research; grants for research; miscellaneous.
Publications
Information bulletin, annually.

YUG.635

Zavod za raziskavo materiala in konstrukcij Ljubljana (ZRMK) Ljubljana

Institute for Research and Testing in Materials and Structures, Ljubljana

Dimičeva 12, 61000 Ljubljana

Telex: 31449 YU ZRMK

Cables: RAZMAT

Date of establishment

1949

Official research centre and public institution.

Main fields of work

- Research into building materials, structures and structure elements: mechanical properties, corrosion resistance and protection, thermal and acoustic protection, fire resistance, chemical and structural analysis, pathology of structures; geotechnics, geomechanics, earthquake resistance, dynamic loading tests of transport engineering components, wear and wear resistance.
- Research and utilization of industrial and mine waste materials, development of special cements, concretes, refractories, deposition of radioactive wastes, development of special machinery and laboratory equipment, research into the causes of damage and carrying out of repair works
- Additional information: Observation of dams, bridges, road constructions, ropeways, etc. Pilot plant production of special materials. Technological projects, engineering and consulting. Testing.

Total staff: 480

Professionally qualified staff: 130

Director or Chief Executive

Director: Dr Jože Vižintin

Heads of department

Institute for Roads: Vinko Vodopivec

Geotechnics: Borut Gostič,

Institute for Building Physics and Repair Works: Janez Kržan

Institute for Structures and Earthquake Engineering; Ludvik Bonač

Institute for Materials: Edvard Mali

Mechanical Engineering: Borut Tomažič

Principal sources of finance

Research and development works are financed by industry and building enterprises, while the fundamental investigations are financed from republic science funds.

Publications

- Informacije ZRMK: information paper, monthly
- Annual report with bibliography of publications and research reports with summaries
- *Bibliographies/catalogues:* Bibliografija znanstvenih in strokovnih objav 1950–1979. Ljubljana: ZRMK, 1980 – 121 str. Details of 820 publications. Catalogue in print.

YUG.636

Institute of SR Serbia for Testing Materials

Institut Za Ispitivanje Materijala SR Srbije (IMS)

Bulevar vojvode Misića br. 43,11000 Beograd

Telephone: 650 322

Telex: 12403 YU IMS

Date of establishment

1948

An independent scientific research institution in the field of building and allied industries

Main fields of work

IMS deals with theoretical, experimental and practical research, testing and design in the field of building materials and construction in the country and abroad; develops research of systems and new methods of construction, studies and design concerning construction and industry, control of quality of materials, constructions and equipment, production of the special equipment and tools for the internal use of customers. The Institute, as the inventor, has at its disposal the system of prefabrication in housing and other buildings, as well as the system of precast construction of halls, bridges and many other types of structures.

Total staff: approximately 400

Professionally qualified staff: One half are engineers and technicians

Director or Chief Executive

Milõs Banić

Principal sources of finance

The Institute is financed largely through the Yugoslav Building Industry Organizations, Factories, Building Enterprises, and to a lesser extent through Federal and Republican organizations.

Publications

- Informativni Bilten, periodical
- Bilten IMS, periodical.

YUG.637

Jaroslav Cerni Institute for Development of Water Resources

Institut za vodoprivredu Jaroslav Černi (Hidroinstitut)

Bulevar vojvode Mišića 43, 11000 Beograd; PO Box 530

Telephone: (011) 649 265

Cables: Hidroinstitut, Beograd

Date of establishment

1947

A scientific research organization.

Main fields of work

- Study of water regime, its management and control
- Water resources studies and solutions
- Testing and treatment of engineering and economical problems, studies and designs in hydraulics, hydrology, irrigation and drainage, hydros, river training, flood control, navigation, water supply, sewerage engineering and water pollution control, erosion problems, rock mechanics, construction
- Investment and technical documentation
- Carrying out various measurements and testings on hydraulic structures and power plants.

The Institute's activity is carried out both in the country and abroad.

The Institute comprises eight departments: Department of Hydraulics; Department of Structures; Department of Engineering Pedology; Department for Water Supply and sewerage engineering and water pollution control; Department for Watercourse Management; Groundwater Hydraulics and Reclamation; Bureau for Graphical Activity; Department for Common Affairs.

Total staff: 300

Professionally qualified staff: 100

Director or Chief Executive

Director General: Professor Dr M. Milozador

Deputy Director: Dr S. Bruk

Assistant Director: Professor Mihailo Vojinović

Principal sources of finance

Financed through contracts for executing various assignments

Publications

- Saopštenja (Transactions), 4 vols per year
- Hidrotehnička bibliografija, 6 vols per year
- Monographs, 1 or 2 per year.

YUG.638

Water Management Institute

Vodnogospodarski Inštitut

Hajdrihova 28, 61111 Ljubljana

Telephone: 061-268 261

Date of establishment

1937 as the Hydraulic Laboratory of the Faculty of Civil Engineering, Architecture and Geodesy of Ljubljana University

Research organization formed in 1978 by the annexing of two design bureaux of the Union of Water Management Communities to the original Hydraulic Laboratory.

Main fields of work

- Model studies and research into the problems of hydrotechnical structures, river engineering, soil reclamation and irrigation, hydrotechnical equipment, maritime structures and wave action, water supply, sewerage and waste water disposal and treatment
- Hydrological studies, with particular reference to flood regimes and sediment transport in water courses
- Assistance in development and measurements for manufacturers of hydrotechnical equipment
- Assistance in special hydraulic field measurements
- Consultation in problems of maritime harbour construction: preliminary designs and measurements, feasibility studies, evaluation and revision of projects, design and static analyses of structures, specific development studies
- Industrial hydraulics and special hydraulic problems
- Assistance and facilities for research training of postgraduate students and external experts.

Total staff: 60

Professionally qualified staff: 43

Director or Chief Executive

Executive Committee President: Dušan Legiša
Management Board Chairman: Bogdan Šetina
Both bodies to be re-elected by mid-1983

Heads of Department

Research Department, Hydraulic Laboratory: Dušan Legiša
Study Department: Miloš Gnus
Maritime Structures Department, Saša Lipovšek

Principal sources of finance

Mainly self-supporting by fees for research projects; participating in funds released by the Union of Water Management Communities and the Republic budget for scientific research

Principal related bodies

Faculty of Civil Engineering, Architecture and Geodesy of Ljubljana University; Republic Union of Water Management Communities as founders

Publications

- Annual Digest of Research Work, 22 issues published
- Special Publications, at irregular intervals, 10 issues published.

YUG.639 **FM**

Yugoslav Building Centre

Jugoslovenski gradjevinski centar (JGC)

84 Bulevar revolucije, 11000 Beograd

Telephone: 436 122 G.Man. 431 422

Telex: 12007-YuJGC

Date of establishment

3 October 1962 by the decision of the Section for building construction industry of the Federal economic Chamber of Yugoslavia

The Centre is a self-supporting organization for improvement and development in the building construction and materials industry, with importance on the federal level.

Main fields of work

- Studies and researches from science, techniques, technology, economics and working organizations in the field of building industry and appropriate industries
- Symposia and seminars
- Rationalization in building construction: applied results of studies
- Technological design and technical assistance in the building materials industries
- Documentation: compiling and processing, Library
- Preparation and distribution of various kind of information; (business, science – technical and regulation)
- Permanent exhibition
- Temporary exhibition in the country and abroad.

Total staff: 168

Professionally qualified staff: 60

Director or Chief Executive

General Manager: Dragomir Maričić

Heads of department

Department of common services: M. Milosavljević, Deputy General Manager
Department for improvement and development of construction and building materials industry: V. Pecija, Deputy General Manager
Department for research, information and documentation: M. Lancoš, Deputy General Manager

Principal sources of finance

Organizations of the building construction industry; General Association of the building construction industry and building materials industry; sale of publications

Principal related bodies

Assembly of the members of the Working Collective; Workers' Council; Board of Experts

Publications

- Documentation for the building construction industry and architecture (DGA – Dokumentacija za gradjevinarstvo i arhitekturu), 12 issues a year
- *Technical standards/recommendations:* General and technical regulations in the building construction industry (collection, comments and recommendations)
- Regulation for construction works abroad (12 countries), issues every 2–3 years
- Research reports from many studies in the field of housing; rationalization of construction; technical solutions in construction etc. Reports are kept in the library of JGC.

Organizations Index

Organizations are listed after index entries alphabetically by their country code. Thus, an index reference to Spain (ESP) will be listed before France (FRA). In the Directory itself, countries are arranged alphabetically by their name.

Organizations Index

Organizations Index

Organizations Index

Organizations Index

Organizations Index

Personnel Index

Organizations are listed after index entries alphabetically by their country code. Thus, an index reference to Spain (ESP) will be listed before France (FRA). In the Directory itself, countries are arranged alphabetically by their name.

Personnel Index

Personnel Index

Subject Index

Organizations are listed after index entries alphabetically by their country code. Thus, an index reference to Spain (ESP) will be listed before France (FRA). In the Directory itself, countries are arranged alphabetically by their name.

Subject Index

Subject Index

Subject Index

For Product Safety Concerns and Information please contact our EU representative GPSR@taylorandfrancis.com Taylor & Francis Verlag GmbH, Kaufingerstraße 24, 80331 München, Germany

Batch number: 10732722

Printed by Printforce, the Netherlands